Advanced Computing Concepts and Techniques in Control Engineering

NATO ASI Series

Advanced Science Institutes Series

A series presenting the results of activities sponsored by the NATO Science Committee, which aims at the dissemination of advanced scientific and technological knowledge, with a view to strengthening links between scientific communities.

The Series is published by an international board of publishers in conjunction with the NATO Scientific Affairs Division

A Life Sciences B Physics	Plenum Publishing Corporation London and New York
C Mathematical and Physical Sciences D Behavioural and Social Sciences E Applied Sciences	Kluwer Academic Publishers Dordrecht, Boston and London
F Computer and Systems Sciences G Ecological Sciences H Cell Biology	Springer-Verlag Berlin Heidelberg New York London Paris Tokyo

Series F: Computer and Systems Sciences Vol. 47

Advanced Computing Concepts and Techniques in Control Engineering

Edited by

Michael J. Denham

Department of Computing
Plymouth Polytechnic, Drake Circus
Plymouth, Devon, PL4 8AA, UK

Alan J. Laub

Department of Electrical and Computer Engineering
University of California
Santa Barbara, CA 93106, USA

Springer-Verlag
Berlin Heidelberg New York London Paris Tokyo
Published in cooperation with NATO Scientific Affairs Division

Proceedings of the NATO Advanced Study Institute on The Application of Advanced Computing Concepts and Techniques in Control Engineering, held in Il Ciocco, Italy, September 14–25, 1987.

ISBN 3-540-50037-5 Springer-Verlag Berlin Heidelberg New York
ISBN 0-387-50037-5 Springer-Verlag New York Berlin Heidelberg

Printed in Germany

Printing: Druckhaus Beltz, Hemsbach; Binding: J. Schäffer GmbH & Co. KG, Grünstadt
2145/3140-543210 – Printed on acid-free paper

Preface

Computational concepts and techniques have always played a major role in control engineering since the first computer-based control systems were put into operation over twenty years ago. This role has in fact been accelerating over the intervening years as the sophistication of the computing methods and tools available, as well as the complexity of the control problems they have been used to solve, have also increased. In particular, the introduction of the microprocessor and its use as a low-cost computing element in a distributed computer control system has had a profound effect on the way in which the design and implementation of a control system is carried out and, to some extent, on the theory which underlies the basic design strategies. The development of interactive computing has encouraged a substantial growth in the use of computer-aided design methods and robust and efficient numerical algorithms have been produced to support these methods. Major advances have also taken place in the languages used for control system implementation, notably the recent introduction of Ada™, a language whose design is based on some very fundamental computer science concepts derived and developed over the past decade.

With the extremely high rate of change in the field of computer science, the more recent developments have outpaced their incorporation into new control system design and implementation techniques. This is particularly true in those areas which have been the subject of intensive study in the last few years under various strategic governmental programmes, many of which were started in response to the Japanese "Fifth Generation" initiative. These areas include:

* artificial intelligence
* software engineering methods and tools for distributed systems
* parallel computing algorithms and architectures

The interface between the disciplines of computer science and control engineering is growing rapidly, not just in these areas but also in the fields of data communication networks and human-computer interaction. Unless a substantial effort is devoted to the study of the potential application and benefit of these powerful new computing techniques in control engineering, as is happening now in other fields such as office automation, it is possible that control engineers will be severely limited in their access to new hardware and software tools and technology with the result that both their own productivity and the efficiency of the systems they design will be seriously degraded.

The aim of this NATO Advanced Study Institute, held 14th-25th September, 1987 in Il Ciocco, Italy, was to gather together a multidisciplinary group of experts who are at the forefront of the application of advanced computing concepts and techniques in control engineering to present a set of lectures on underlying concepts and methods and their application. This volume contains a set of papers based on these invited lectures, together with some papers contributed by the participants in the Institute.

The volume is organized in four main sections, the first three of which contain the invited papers corresponding to the three principal themes of the Institute:

I. Artificial Intelligence and Expert Systems
II. Discrete-event and Distributed Systems
III. Algorithms for Advanced Architectures
IV. Contributed Papers

Artificial intelligence techniques, in particular expert systems techniques, are finding increasing use in the design and implementation of control systems. Section I begins with a tutorial overview of some of the major expert system concepts and is followed by three additional papers which cover specific additional topics in the artificial intelligence and expert systems area together with applications of special interest to control engineers.

Section II is concerned with the fields of discrete-event and distributed systems. These fields have been studied intensely for some time now by computer scientists but a rapidly growing theory for the control of such systems is beginning to emerge. An overview paper on the control of discrete-event systems is followed by five additional papers on various aspects of the control of discrete-event and distributed systems.

Section III is devoted to the area of parallel algorithms and architectures. Computing environments for the implementation of various classes of parallel algorithms are now a reality and this technology is certain to have a dramatic impact on control engineering. An overview paper on various advanced architectures, with particular emphasis on shared-memory machines, is followed by three papers describing particular analysis and design algorithms for control and signal processing for a variety of advanced architectures including distributed memory and systolic processors.

Finally, six papers based on talks contributed by participants in the Institute are included as Section IV. These papers are arranged by topic in roughly the order of the first three sections. Although the papers in the first three sections are based on the invited talks and thus have a strong didactic emphasis, the contributed papers also have considerable tutorial value for both the research-oriented as well as practicing control engineer.

What we are unable to record in this book are the large number of informal discussions which took place both inside and outside the lecture room in the course of what we believe, and we hope all the participants will agree, proved to be a very interesting and productive two weeks. We do hope, however, that this set of papers will serve to stimulate many future discussions and new ideas for the further application of innovative computing concepts and techniques in control engineering.

M.J. Denham
A.J. Laub

March 1988

Acknowledgements

We gratefully acknowledge the support of the following organisations, in respect of the Advanced Study Institute of which this volume constitutes the proceedings:

North Atlantic Treaty Organization, Scientific Affairs Division

National Science Foundation, USA (for travel grants)

Olivetti, Italy (for the loan of computing equipment)

Table of Contents

I. Artificial Intelligence and Expert Systems

II. Discrete-Event and Distributed Systems

III. Algorithms for Advanced Architectures

IV. Contributed Papers

List of Participants

A. Aras	Ege University, Turkey
A. Barbe	Katholik Universiteit Leuven, Belgium
U. Baser	Marmara University, Turkey
G.L. Blankenship	University of Maryland, USA
J-L. Calvet	LAAS du CNRS, Toulouse, France
L. Chisci	Università di Firenze, Italy
J.R.B. Cockett	University of Tennessee, USA
P. Colaneri	Politecnico di Milano, Italy
G. Cybenko	Tufts University, USA
M.J. Denham*	Kingston Polytechnic, UK (Director)
G. De Nicolao	Politecnico di Milano, Italy
R. Eising	Hollandse Signaalapparaten B.V., Netherlands
K. Erciyes	Ege University, Turkey
J.D. Gardiner	University of California, USA
C. Golaszewski	Princeton University, USA
M. Gori	Universita di Firenze, Italy
A.C.W. Grace	University College of N. Wales, UK
S. Hem	Reed Paper and Board (UK) Ltd., UK
G. Hoffman	ETH, Zurich, Switzerland
Y. Istefanopulos	Bogazici University, Turkey
J. Kramer	Imperial College, University of London, UK
P. M. Larsen*	Technical University of Denmark
A.J. Laub*	University of California, USA (Co-Director)
L.G. Lebow	University of Maryland, USA
R. Leitch	Heriot-Watt University, UK
J. Lemos	Instituto Superior Tecnico, Lisbon, Portugal
P. Lourtie	Instituto Superior Tecnico, Lisbon, Portugal
T. Mendonca	University of Oporto, Portugal
D. Meyer	University of Virginia, USA
G. Miminis	Memorial University of Newfoundland, Canada
T. Moran	Cambridge Control Ltd., UK
E. Mosca	Universita di Firenze, Italy
J. Ostroff	York University, Canada
J.P. Quadrat	INRIA, France
P.J. Ramadge	Princeton University, USA
A. Ruano	University of Aveiro, Portugal
A.H. Sameh	University of Illinois, USA
P. Silvestrin	European Space Research Centre, Netherlands
P. Valckenaers	Katholik Universiteit Leuven, Belgium
P. Van Dooren*	Philips Research Laboratories, Belgium
W.M. Wonham*	University of Toronto, Canada

*member of the Organizing Committee

I. ARTIFICAL INTELLIGENCE AND EXPERT SYSTEMS

Aspects of Expert Systems

J. Robin B. Cockett*

Abstract

Expert system technology has grown out of the desire to produce and support systems which are expert. The technology itself, however, is more widely applicable. It is supported by a growing number of commercial expert system shells. These shells may be viewed as providing specialized and high level software development environments. Seen in this light, the lessons of expert system development may have considerable impact when applied to general software development.

A central idea in expert systems has been to try to separate the knowledge from its use. In traditional software development terms the knowledge plays the role of a specification. The purpose of an inference engine is to produce a procedural behavior directly from this specification. For such an approach to work effectively, however, there must be a smooth transition between the logic of the specification and the procedural form. A logic which allows such a transition is introduced.

The transition from logic to procedural form involves choosing an appropriate *control* regime to determine what should be done next at every stage. It is argued that the current approach, which is to employ a standard inference engine, does not address this problem effectively. The step of *knowledge shaping* is proposed as a possible resolution.

Keywords: categorical expert systems / software development / software specification / control of knowledge / knowledge representation / decision theory / decision expressions.

1 Introduction

Expert Systems (ES) represent a powerful approach to disseminating and delivering problem solving ability in a particular domain. Obviously, the fundamental ingredient required to produce such a system is the *expertise* itself. However, this is by no means the only ingredient. The raw knowledge must be engineered into a deliverable system which not only uses that knowledge effectively, but also facilitates its maintenance and refinement.

In this view, the production system is only the tip of an iceberg. To provide that whole package, from production system through to the support environment, must be the primary aim of companies and researchers in the knowledge engineering business.

There are many different sorts of expert systems; in this paper I shall be concerned almost exclusively with *categorical* expert systems and the techniques and tools required to build them. A categorical expert system is one which uses precise reasoning. This distinguishes it from systems which use approximate or heuristic reasoning, such as MYCIN [4] and PROSPECTOR [13]. While it has been argued that an expert system which does not handle imprecision should not be called an expert system, this view seems a little hard on those areas of expertise which

*Department of Computer Science, University of Tennessee, Knoxville, TN37916

NATO ASI Series, Vol. F47
Advanced Computing Concepts and Techniques in Control Engineering
Edited by M.J. Denham and A.J. Laub

do not require imprecision but obviously require considerable skill. Most people recognize the mathematical skill embodied in MACSYMA [26], and would count it as an expert system. Yet imprecision has little to do with algebraic manipulations.

This paper starts with a general discussion of (categorical) expert systems and the technology which has grown up to support them. This leads into a discussion of the tools which are required in a support environment and some examples of why they are needed. Finally the paper ends by suggesting a formalism for knowledge representation which facilitates the transition between logical and procedural forms of knowledge.

2 Expert System Technology

2.1 What is an Expert System?

In common usage the term "expert system" covers such a wide range of computer based systems that it seems unrealistic to try to pin down the notion. In fact, perhaps to the dismay of the founders of the field, almost any piece of software seems now to qualify for consideration.

If we are literal then a system should earn the title *expert* in the same manner that a human does; that is, a system should be labeled *expert* by acclaim from its knowledgeable users in recognition of a rare *skill* in a particular domain. Thus an *expert system* becomes under this definition simply a technological achievement in the computer based management and delivery of expertise.

In the same way that building submarines or rockets requires a high level of industrial sophistication, building a computer system which will be considered expert requires a high level of software technology. This software technology is commonly referred to as *expert system technology*. Many systems now use some aspects of this technology. However, it is not clear that simply because they use the technology that they should be called *expert systems*. A system should earn its "Badge of Expert Merit" rather than simply inherit the title.

Common usage, however, confounds these distinctions; the term "expert system" is a popular one which catches the mood and aim of a project. Efforts to use more sober terms like "knowledge-based system" or "rule-based system" simply do not catch the imagination. Thus, in practice an expert system is often simply a system which uses some aspect of expert system technology. In no way does this detract from the purpose of the technology itself, which is certainly to facilitate the development of *expert systems*, in the first sense mentioned above. It is clear that this purpose has many implications for the software tools required to support such developments.

For example, as a technological achievement an *expert system* is limited by the technology and the understanding of the problem area at the time that it is created. This means that if the system is to remain current, it usually must undergo constant revision and expansion by the experts; until eventually, like an old ship, it is scrapped and replaced completely. This makes it important that high level support tools be provided so that the domain experts can maintain and refine the system.

Expertise is actually quite fickle in nature. It often happens that something which seemed very complex and requiring of rare skill becomes almost routine. Part of the motivation behind creating an expert system must be an ambition to bring this to pass in the system's particular domain. However, to directly undermine its *expert* value in this way, the knowledge it represents must *actually* become widely disseminated and routine. The sheer volume of knowledge being accumulated by modern society militates against this ever happening.

This literal definition of *expert system* is rather subjective. Human judgment on matters of intelligence and expertise has a poor track record.[1] Just as the Turing test failed so miserably

[1] In Pamela McCorduck [28] there is an interesting historical discussion of these issues.

to be the touchstone of intelligence, we gullible humans will happily label mere idiot savants as experts. Furthermore, our judgment of expertise will always be prejudiced. Even when the cheapest chess playing program routinely massacres the master chess player (and can explain exactly why it won), we will still regard the master as an expert.

It is fortunate that we do not have so many qualms about testing the performance of computer systems as we do about testing human performance. This may help to remove some of the subjective element. However, comparing the performance of two complex systems is no easy task, even when they provide the same answers and do the same things, which two expert systems covering the same domain are unlikely to do. Subjective judgment is forced by complex systems.

2.2 Expert Interface

It is clear that if a system is to be considered *expert* then it should also have an expert interface. That is, it should be a robust and usable interface and it should be able to explain its actions. Furthermore, a system which is expert should be able to apply its knowledge in a manner which responds to the requirements of its individual users.

Considerable emphasis is often put on the ability of expert systems to explain their actions.[2] Explanation capabilities are important for two reasons. First, they provide a high level debugging tool; in a very high level software development environment their value is that they almost allow one to discuss the purpose of the program with the computer. This level of interaction should be the ultimate goal and if this is to be achieved then certainly computers must learn natural language. Second, explanations provide a very effective means of communicating knowledge between intelligent agents (whether machine or human).

Just as help menus are becoming commonplace in software systems it is likely that explanation capabilities will also become more common. Indeed, it is probable that we will come to view them as an integral part of any good software product.

2.3 The Bottleneck

While there is considerable publicity about expert systems, the number of full blown expert systems which are doing significant work in the real world can probably still be counted on one's fingers and toes [43]. There are many experimental, research, and prototype systems, and these prove that it is quite possible to automate problem solving for a variety of complex tasks.[3] However, it is reasonable to wonder whether there might not still be some ingredient missing otherwise there would surely be more systems out there.

Feigenbaum [15], pointed out that there is a "bottleneck problem of applied artificial intelligence", which lies in getting human experts to **say how** they solve problems. In [31], Michie takes this further by pointing out that humans often, in some sense, have compiled (or subconscious) knowledge, which they cannot easily explain. He therefore strongly advocates that experts should predominately **show how** problems are solved and let the machine *infer*[4] how the problem was solved. In fact there is considerable evidence that for some domains this method (the **say how**) of communicating knowledge is actually a more reliable and efficient method of

[2] Indeed Donald Michie, [31], believes this to be an essential defining feature of expert systems. He has referred to these capabilities as the "human window into the thinking machine".

[3] For further discussion of these points and an extensive catalogue of expert systems see [41].

[4] Michie is a strong advocate of Quinlan's decision tree inference techniques [35,34]. He points out that these techniques have not been generally accepted because they derive from the "wrong subculture", namely the work of Earl Hunt [21], a psychologist. He makes the following comment about the rule-based machine learning techniques pioneered by Michalski [29], which are (in the U.S.A.) in the right subculture: "Whether or not Michalski's algorithms, which are academically motivated, could be made into cost-effective software tools is an open question." The comment is indicative of a continuing representation and methodological debate.

transferring problem solving ability [36,30]. It is also, of course, the predominant method used by humans in transferring skills.

However, there is probably a more fundamental problem which faces expert system technology and that concerns becoming accepted as a mainstream approach to problem solving and software development.

2.4 Getting in the Mainstream

The first problem which faces this technology is simply that of finding appropriate applications for the new ideas. The existing application mechanisms must be such that improving the speed and quality of problem solving ability will in fact make a difference. If this is not the case then it is hard to justify the effort involved in applying the new ideas. Thus, the target environment must have the right technological base to provide an application which has a payoff.[5] Suppose this is the case and the management takes the plunge. If they buy a LISP machine, an expert system shell, hire consultants, and train personnel, can they be reassured that in the longer term this will be a mainstream approach? Or will they find that in a few years they have to start again?

This leads into the second problem: a software company, in providing a customer with a large computer system and its extensive support environment, will inevitably effect the terminology and the methods used for managing the expertise which the system embodies. Such a system, therefore, can have not only a performance impact but also a major social impact. It then becomes important that the methods and terminology used by the system are seen to come from the mainstream of computer science, mathematics, and the domain of the system itself. If this is not the case there is a risk that the system may simply isolate the customer. Ultimately this will almost certainly undermine his confidence in the system.

One problem is that many of the Artificial Intelligence (AI) techniques commonly associated with ES technology are regarded with a good deal of skepticism. They often seem to use concepts which are not in the mainstream of computer science or mathematics: frames, scripts, demons, semantic nets, approximate reasoning, fuzzy logic, meta-rules, etc. While it is reasonable for an area to develop its own terminology, it is also reasonable to be suspicious of ideas which too often seem to be accompanied by implicit or even explicit philosophical claims.

There is no doubt that there is still more shaking out to be done in the field before it can be considered an established or even mature technology. However, this *is* happening; a core of ideas is becoming established and the relationship between them increasingly understood. The understanding of this software technology is thus catching up with the initial dash to achieve expert performance. Last, but by no means least, the pool of people with expertise in this area is growing.

2.5 The Technology

The packaging of expertise into an effective, usable system is itself an art and requires techniques which often are not very particular to the domain itself. The attempt to abstract these techniques into generally applicable tools is a major preoccupation of computer scientists and others who work in this field. Obviously the availability of good tools for building and maintaining such systems and the understanding of how and when they should be applied is fundamental to the problem of making expertise widely available in an effective and well-supported computer based way.

---

[5] It is no coincidence that one of the most successful expert system families, R1/XCON, was employed by a computer manufacturing company, DEC.

There is now extensive literature on Expert Systems. Some of the books are simply catalogues of different techniques [18], descriptions of particular expert systems and shells [10], or extensive catalogues of expert systems [41]. Of particular note is Jackson's book [22], since it provides a very level headed overview of the technology. However, if one wishes to obtain a more detailed description of the basic techniques then it is probably worth looking in the AI literature.

There are two mainstream directions for books on artificial intelligence: those that survey techniques without explicit reference to implementation, [32,33], and those which concentrate on developing programming skills in artificial intelligence techniques using a chosen high level language, [42,1].

AI relies for many of its fundamental techniques on discrete mathematics and logic. Discrete mathematics has only recently become a major direction in mathematics. This is partly because the computer has made feasible the evaluation of combinatoric problems which had in the past simply been considered unfeasible.[6] This has meant that often the theoretical base for understanding AI problems has not been in place, let alone the algorithms for their solution.

Expert system technology inevitably also relies heavily on methods from discrete mathematics and logic. Perhaps the main reason for this is that expert systems are concerned with the *management* of knowledge and many of the skills required for this are discrete in nature. In particular, techniques from theorem proving and searching (unification, matching, backward chaining, and forward chaining [32]) traditionally play a major role. These techniques have been the topic of extensive research and are now supported at the language level. For example, Prolog supports backward chaining [5,40] and OPS5 supports forward chaining [3,17,16].

2.6 Shell or Software Development Environment?

There are now a variety of systems specifically designed to support the building of expert systems. They are often called *expert system shells*. Strictly speaking a shell need consist of only those components, such as the interface and inference engine, which are required in order to use a knowledge base. However, these systems typically include facilities for designing the interface, managing and debugging the knowledge base, and for accessing other external facilities[7].

The less expensive commercially available expert system shells often suffer from being rather limited. At the other end of the scale some systems are more like toolboxes to enhance an underlying language (such as LISP or Prolog), but lack any real coherence of their own. Almost all introduce no small amount of new terminology.

There is no doubt that an expert system shell allows one to quickly prototype a system. Furthermore, using such a shell is also a way of becoming familiar with some of the terminology of this area. However, although shells provide many pretty facilities they often do not represent a fundamental advance over the support one might expect from a high level language which implements some of the basic AI techniques. Indeed, while a greater variety of reasoning mechanisms may be present, they will almost certainly not be as efficiently or cleanly implemented as they would be at the language level.

The development of a high level language based on a particular computing mechanism will inevitably make shells, whose value is based largely on having implemented those mechanisms, less exciting. Prolog and OPS5 made backward and forward chaining, respectively, widely available; expert system shells have therefore to offer more. A feature, which shells frequently include, is the ability to rapidly design sophisticated graphics interfaces; these graphics abilities

[6] The obvious example of this is the Enigma decoding effort in the second world war. The life of Turing [20], reflected the lack of regard in which this branch of mathematics was held in the post war period.

[7] This may be rather important if the system is required to do large numerical computations. Almost certainly the language of the shell will not allow the efficient implementation of these calculations. In fact it is quite possible that it will not even support floating point arithmetic!

are frequently offered in an object oriented way. This sometimes makes shells a mixture of object oriented, logic and even functional programming styles.

Inevitably the perspective with which we view a shell will be colored by developments in programming languages and certainly the environments provided by shells should be compared to the cleanness of a programming language environment. However, the purpose of an expert system shell is definitely more extensive than that of a language. The ultimate aim of an expert system shell is to provide a very high level development and support environment for producing and maintaining expert systems. This aim, it is interesting to note, differs little from that of a support environment for large scale software development. This is hardly surprising: after all an expert system is a large software system.

However, there are two respects in which an expert system development environment differs from a general software development environment. First, to provide the level of support given in an expert system building tool involves irrevocably making many implementation choices for the programmer. Many of these choices are, at present, not regarded to be in the purview of the software environment. Second, it is not clear that, in an expert system development environment, all Turing computable functions should be treated equally. Expert systems are generally heavy on *control* and light on *calculation* and thus the environment should reflect this bias.

These points, however, only indicate that an expert system building environment is a rather specialized environment. It is therefore interesting to pursue this analogy for it is conceivable that the expert system experience can be played straight back into that commonest of tasks: software development.

3 On the Support Environment

One of the main attractions of the rule-based approach to expert systems is the claim that knowledge can be encoded into rules and that from this abstract form a procedural behavior which applies the knowledge can be inferred. The computational machine which infers the procedural behavior is appropriately enough called an *inference engine.* The most popular inference mechanism for this engine is backward chaining with forward chaining coming in a close second.

In software development terms the rules play the role of a system specification. The inference engine performs the amazing feat of producing a procedural behavior directly out of the specification. Seen in this light it is small wonder that expert system technology seems so attractive.

3.1 System Specification

This rule-base idea is very attractive for by declaring the purpose of the program carefully in such a specification one can actually obtain the program itself. The term *declarative* programming became popular to distinguish this approach from more traditional *imperative* programming techniques.

Of course, this is altogether too good to be true. Kowalski[8] and others were quick to point out that there was an important ingredient missing, namely *control.* Control, in this context, means the appropriate ordering of the events which constitute a computation. Although an inappropriate ordering of computational events have an effect which is quite apparent, it has proved to be very difficult to determine what it is that *must* be specified over and above the logic to secure control.

[8] Kowalski [24], coined the equation “Algorithm = Logic + Control”.

In forward chaining the control is, in theory, relegated to the *conflict resolution* strategy which determines which of the active rules is actually to be fired next. In practice, however, for good reason programmers abuse this mechanism in two ways.

First, the conflict resolution strategy is often used to determine the logical meaning of the program by the way it chooses the rule to be fired. For example, a common conflict resolution strategy is to always prefer more specific rules (that is those with more antecedents) over less specific rules and otherwise to generally prefer rules which have most recently been enabled. This allows the programmer to write branching behavior more succinctly: the default in the conditional branching becomes the most general case with the more specific cases caught by more specific rules. Second, because it is difficult to write conflict resolution strategies, a programmer will often write the rules to explicitly include an antecedent which determines the control context in which the rule is to be applied (the antecedent need have no *logical* effect).

In backward chaining (as implemented by Prolog for example) an even worse situation persists: the control is hidden in rule and antecedent ordering. As in forward chaining this implicit control can be abused. The logical meaning of a program can be affected by the order in which the rules are expressed[9]. Alternatively the programmer can write the rules so as to include a control context.

Expert system builders quickly realized this was a muddle and added meta-rules [11] in an attempt to keep the control information separate from their logical specification. It is not clear, in retrospect, that this approach really worked. The obvious methodological problem concerned the control of the meta-rules themselves and the threat of an infinite regression of control specifications. However, the idea of separating control information from the logical specification was clearly correct.

Developments in discrete decision theory [7], which we are discussed below, give a much clearer idea of what this control specification might look like. They indicate that it is possible (and even possibly *practical*) to separate the control specification from the logical specification. The logical specification and the control specification, thus, form two of the main ingredients of a specification of an expert system.

The full specification has three major components:

- The logical specification,
- The control specification,
- The interface specification.

The last ingredient in the specification is the description of how the system interfaces to the outside world. This may include the specification of measurements to be obtained, procedures to be run, questions which must be presented to the user, menus to be used, etc.

3.2 Knowledge Acquisition

The problem of knowledge acquisition is to fill out each of the logical, control, and interface specifications. This could always be done by the expert explicitly writing these specifications. However, this assumes, not only that he can articulate it, but also that it is a cost effective and reliable way to acquire these specifications. Evidence suggests that this is not the case.

The traditional view of knowledge acquisition concerns primarily the acquisition of the logical specification. Concerning this there will indeed be some things that an expert can articulate and this will probably be the starting point of the system. There may also be some things which he will prefer to show by example. However, there will almost certainly be some things that only become apparent in the course of building the system.

[9]This programming technique is explicitly supported in Prolog by the "cut."

The ease with which graphics interfaces can be specified has been much improved by interactive techniques (using icons and pop-up menus with a mouse for example). These allow the relatively painless construction of very respectable graphics interfaces. However, underlying this pretty interface, the system in order to be able to explain its actions, must also acquire the terminology of the area. This means that natural language capabilities are necessary.

Interactions with the user may not be the only sort of interactions the system has. In addition it may have to interact with measurement devices and other programs. These interfaces must also be specified. Current shells are often weak in this area. However, this is something which is an increasing requirement in expert systems and thus deserves high level support.

The control specification will often be the sugar on the cake. Only once the logical knowledge is mostly acquired will the problem of how it should be controlled become very critical. Some of this knowledge will only become explicit through experimenting with control strategies. Almost certainly the eventual control specification will be incomplete and allow many alternate procedural forms. However, there may well be some obvious features, such as the relative expense of making decisions and the relative likelihood of certain events, which an expert knows and can readily impart to the system; these can guide the control.

The whole support environment should in effect be an acquisition tool. Debugging and restructuring are both aspects of specification acquisition. High level explanation generation is a powerful logic specification debugging tool. The ability to play with the control of the system, for example by delaying the evaluation of certain decisions, is a control specification debugging tool.

A difficult problem is to ensure that a modification at the procedural level is translated into a correct modification at the specification level and that these two levels are generally kept in synchronization. This appears to be a particularly subtle problem for the control specification. Techniques for extracting control specifications from procedural specifications are generally in their infancy. The problems and promise of the techniques which are available will be discussed further below.

3.3 Knowledge Shaping

Traditionally once the logical specification is formed (and possibly a control specification using meta-rules) an *inference engine* is used to directly interpret this "knowledge base" into a procedure. This introduces a number of very practical performance problems, which severely restrict the complexity of the control mechanisms which can be reasonably considered.

However, if another intermediate step, called *knowledge shaping*, is introduced, this difficulty can be resolved. The idea of this step is to find interactively a procedural form which simultaneously satisfies both the control and logical specification. Knowledge shaping allows the designer to interactively explore alternative control regimes without altering the logical specification.

The result of the knowledge shaping step is a procedural form of the knowledge, which is distinct from the logical form. In this form the knowledge can be interpreted relatively efficiently. Furthermore, it can still be manipulated and this allows explanations and interactive control modification to be performed. In this form the knowledge is well-suited for use in the debugging environment.

3.4 Interpreting, Soft-Compiling, and Hard-Compiling

The main development environment is the *interpreter*. The interpreter has to be a somewhat special as it has to know how to manipulate the procedural form. This level of interpreting is sometimes known as meta-level interpreting. From this environment the engineer should have access to all the other tools of the environment so that he can interactively use, develop, and modify both the code and its specification.

When he is finally ready to produce a production system he must be able to create an executable module. This module will not need all the debugging features available in the interpreter, although he may wish to retain the explanation capabilities in the compiled form. For this purpose two forms of compiling are required: *soft-compiling*, which retains certain meta-level features, and *hard-compiling* which creates the procedure as a highly efficient black box.

The interpreter should permit incremental compiling. Thus, before the engineer reaches the fully compiled stage, he may well have compiled some portions of the system while continuing to debug other portions.

4 Problems in Knowledge Representation

It may not be immediately obvious why a logical specification does not provide all the information required to give an acceptable procedural form. In order to illustrate these problems two (albeit artificial) examples are now discussed. These examples shall also be use to informally introduce the logic which is more formally presented in section 5.

4.1 The Three Daughters Problem

Consider the following problem:

A man has three daughters: Sue, Pam and Joy.

- Sue and Joy have brown hair.
- Pam's hair is red.
- Sue always wears a skirt, while Joy never does.
- Pam may or may not wear a skirt.

One day, the owner of a candy store found out that one of the man's daughters had stolen some candy from his store. The owner of the store knew the girl's father but not the girl's name. So, he went to the man and asked for his money. The man paid the owner, and asked him the girl's name, but the owner did not know.

How can the man find out which of his daughters should be punished?

The problem, after some logical manipulation, can be described by the following set of *prime* rules, which constitute the logical specification the problem.

Rule 1: If the girl wears a skirt and has brown hair then her name is Sue.

Rule 2: If the girl does not wear a skirt and has brown hair then her name is Joy.

Rule 3: If the girl has red hair then her name is Pam.

Using the logic which is introduced in section 5, these may be more formally presented as follows:

Rule 1: wears-skirt(x)$\Diamond$yes $\wedge$ hair-color(x)$\Diamond$brown $\vdash$ name(x)$\Diamond$Sue,

Rule 2: wears-skirt(x)$\Diamond$no $\wedge$ hair-color(x)$\Diamond$brown $\vdash$ name(x)$\Diamond$Joy,

Rule 3: hair-color(x)$\Diamond$red $\vdash$ name(x)$\Diamond$Pam.

The symbol $\Diamond$ associates an attribute value or a *codomain symbol* to an attribute or a *decision*. The formula "hair-color(x)$\Diamond$red" is called a *branch instruction*. In order to fully specify the logic it is necessary to associate with each decision a set of codomain symbols:

- *codsym*(wears-skirt) = {yes,no},
- *codsym*(hair-color) = {red,brown},
- *codsym*(name) = {Pam,Sue,Joy}.

If we are using a backward chaining interpreter and we processed the rules in the order given, then we would obtain the following equivalent tree:

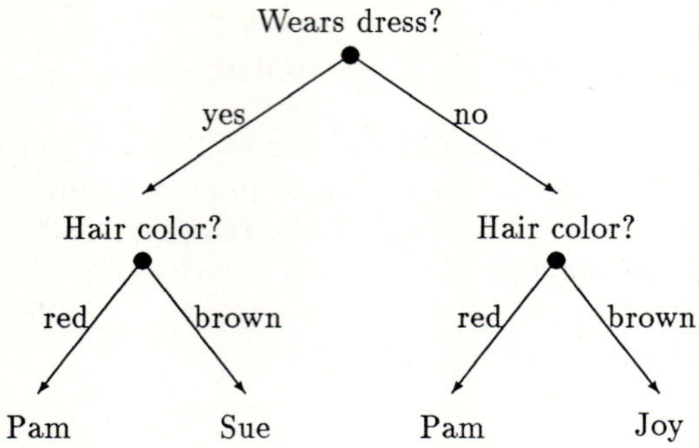

This tree can be more formally represented as a *decision expression*:

wears-skirt(x):
[yes $\gg$ hair-color(x):
    [red $\gg$ $<: y_{\mathrm{Pam}}$,
     brown $\gg$ $<: y_{\mathrm{Sue}}$
    ],
 no $\gg$ hair-color(x):
    [red $\gg$ $<: y_{\mathrm{Pam}}$,
     brown $\gg$ $<: y_{\mathrm{Joy}}$
    ]
]

The terms of the form $<: y_{\mathrm{Pam}}$ give the outputs. The whole problem is specified by equating this decision expression to:

$$\mathrm{name}(x) : [\mathrm{Pam} \gg\ <: y_{\mathrm{Pam}}, \mathrm{Sue} \gg\ <: y_{\mathrm{Sue}}, \mathrm{Joy} \gg\ <: y_{\mathrm{Joy}}].$$

Rather than writing out this identity we shall use the names directly instead of the output variables and assume that the identity is understood.

An inspection of these rules immediately tells us that if we had reordered the rules so that **Rule 3** was first we would have obtained the following equivalent tree:

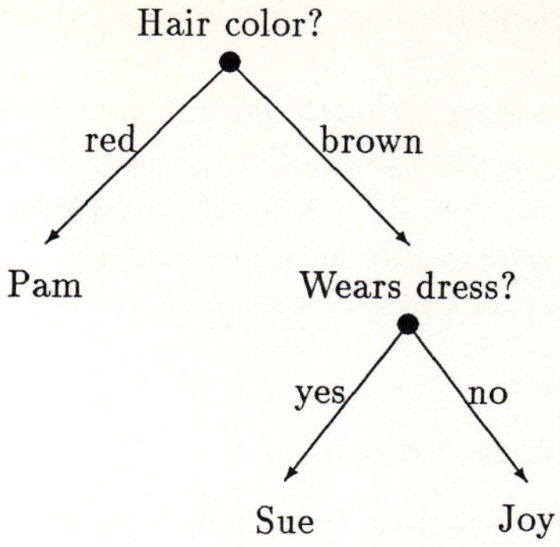

whose decision expression form is:

hair-color(x):
[red ≫ <:Pam,
brown ≫ wears-skirt(x):
[yes ≫<: Sue,
no ≫<: Joy
]
]

It is clear that this second tree is always going to be a better way of solving the problem. It may be thought that a control specification should have been added to the logical specification in order to avoid the first form. In fact, the second tree is *simple*, in the terminology of [6], which means that for these decisions this is the *only reasonable* way of solving the problem. Thus, in this example there is no need to give a control specification as there is no freedom for control after the logical specification has been met. In practice such circumstances are rare. The example serves to illustrate that control and logic are definitely not independent.

The computational behavior of straightforward backward chaining is clearly very sensitive to the rule and antecedent ordering. This illustrates how the separation of the knowledge from the performance of the system is not achieved by using a simple-minded backward chaining inference engine.

4.2 Achieving the Desired Control

From the above example it might be thought that obtaining a reasonable procedural form is simply a matter of sensibly ordering the rules and their antecedents. However, in general this is not the case. Suppose that one wanted to achieve a particular optimal behavior using prime rules and a backward chaining engine; it turns out that it may simply *not* be possible [9]. Thus, there is a nontrivial problem concerning how one turns a minimal logical specification into a procedure.

A pragmatic solution to these problems is simply not to insist on the rules being prime. However, this introduces a small problem. If one had intended to use the rules as explanations, there may now be some irrelevant antecedent information in them which was introduced to obtain a particular control regime. This problem can in turn be corrected by indicating the irrelevant antecedents in rules, by some marking device, and simply omitting them from the explanations. However, these modifications make the knowledge base contain much more than the abstract logical knowledge.

In fact, ultimately, when one inserts control information into the rules in this way, one can no longer guarantee that they can be manipulated according to the rules of logic. There is always a danger that a simple logical manipulation will actually completely change the meaning of the procedure generated under the chosen inference mechanism. This is rather unsatisfactory as the clean separation of knowledge from procedure seemed to be one of the principle advantages of the ES approach.

4.3 Control via Knowledge Shaping

It is reasonable, therefore, to ask whether there is not some other approach to these problems. Of course, there is and it concerns the stage I have referred to as knowledge shaping.

The idea is this: backward chaining is used as before to generate a procedural form of the knowledge. However, rather than simply using this procedural form it is now examined to see whether it makes good procedural sense. Obviously the control specification provided to the system is used in this determination. Contrary to popular opinion, usually the initial procedural form will not make procedural sense [9]. If it does not, then the procedure is *shaped* into one that does. This can be done either interactively or automatically using the algorithms based on the developments described in [7,6,19].

The advantage of this approach is that it recognizes cleanly that the abstract logical specification may not be sufficient to completely determine the procedural form. Therefore, it employs a separate step which uses any control information specified to obtain a procedural version of the knowledge. This allows the logical specification to be no more than a logical specification, and control information to be developed quite separately. The disadvantage is that it is another expensive computation between the knowledge and the final system.

As the knowledge shaping process is an expensive computation, it is necessary for the knowledge engineer to split the problem into small and manageable chunks. This amortizes the cost of the computation and makes it feasible to consider the use of the tool in an interactive mode. This is not an unreasonable thing to expect a knowledge engineer to do as these chunks also give the handle for explanation facilities.

The process of development then works on the precept that first the knowledge engineer will try to capture the logical knowledge of his chunks using no control constraints. This makes the procedure shaping step relatively trivial. Once he has captured the logic of the knowledge he can return to the problem of specifying the control.

Normally, at this stage the knowledge engineer would "tinker" with the system, by changing the form of his rules, in order to optimize its performance. In the unsupported process of tinkering there is a strong possibility that he will unwittingly introduce a deviation from the logical specification. Using a knowledge shaping tool he cannot do this; he can only change the control regime. If, as he changes the control regime, the system can infer what might have caused the change, then it becomes possible for the system to acquire control knowledge. This knowledge can be used in later iterations to ensure that only sensible control regimes are ever explored.

In order to illustrate these ideas a slightly more complex version of the three daughters problem can be considered.

4.4 The Three Daughters Problem Continued

The father asked the shopkeeper what color the girl's hair was and he replied that it was brown. So he asked whether she was wearing a skirt. The shopkeeper replied that he was not actually there when the incident happened: his two assistants Margaret and Nathan were tending shop. Furthermore the counter arrangement in the shop would almost certainly mean that neither Nathan nor Margaret would have been able to see whether the girl was wearing a skirt. Nathan had told the shopkeeper's wife whether or not the girl had been wearing glasses, though the shopkeeper could not remember which.

Now the father knows that both Sue and Joy occasionally wear glasses. Furthermore he remembered that Sue had had her hair in braids and was wearing a red sweater on the day in question, although he could not recall whether she was wearing her glasses. He also knew that Joy had only one red sweater with which she always wore her hair down, and usually this was the only outfit with which she wore glasses. So, if Joy was not wearing her glasses, she probably would not have been wearing her red sweater. Conversely, if she was wearing her glasses she probably would be wearing her red sweater. Also when Joy braided her hair she hardly ever wore her glasses as she said it made her look goofy, but when she was not wearing her glasses she usually braided her hair.

The father asked the shopkeeper whether the girl was wearing braids. He replied that Nathan would have noticed; although Margaret never noticed such things. He then asked the shopkeeper whether she was wearing a red sweater and he said that Margaret would probably have noticed, but that Nathan was color blind.

As they were very near to the shopkeeper's home, and the father wished to apologize to the shopkeeper's wife, the question of whether the girl had been wearing glasses or not could easily be resolved. However, Margaret and Nathan lived at opposite ends of the village. In order not to waste more of his time than was absolutely necessary, the father wanted to try and visit only one of them. How should he approach this problem?

The new rules of the logical specification contained in this description are:

Rule 4: If the girl has brown hair, wears a red sweater, and has her hair in braids then her name is Sue,

Rule 5: If the girl has brown hair and is not wearing a red sweater then her name is Joy,

Rule 6: If the girl has brown hair and does not have her hair in braids then her name is Joy.

The formal version of these rules are:

Rule 4: hair-color(x)◇brown ∧ wears-red-sweater(x)◇yes ∧ braids-hair(x)◇yes
⊢ name(x)◇Sue,

Rule 5: hair-color(x)◇brown ∧ wears-red-sweater(x)◇no ⊢ name(x)◇Joy,

Rule 6: hair-color(x)◇brown ∧ braids-hair(x)◇no ⊢ name(x)◇Joy.

The codomain symbols for the new decisions are {yes,no}.

Notice that there is no mention of "wearing glasses" as whether the girl wears glasses or not does not determine her identity. Thus, clearly there is something significant missing from this description of the problem. It is, of course, the control specification. This may be expressed in two stages. First, a description of the relative costs of each decision:

Cost 1: There is no cost associated with discovering whether the girl was wearing glasses and what her hair color was,

Cost 2: Whether she is wearing a skirt or jeans is unknown,

Cost 3: There is (significant) cost associated with discovering whether she had her hair in braids and whether she was wearing a red sweater.

Formally we may express these as follows:

Cost 1: $cost(\text{wears-glasses}) = 0$, $cost(\text{hair-color}) = 0$,

Cost 2: $cost(\text{wears-skirt}) = \infty$,

Cost 3: $cost(\text{braids-hair}) > 0$, $cost(\text{wears-red-sweater}) > 0$.

Secondly a description of the relative strengths of certain conditional probabilities:

Control 1: If the girl was Joy and she was wearing glasses it is more likely that she was wearing a red sweater than not.

Control 2: If the girl was Joy and she was wearing glasses then it is more likely that she did not have her hair in braids than she did.

Control 3: If the girl was Joy and she was not wearing glasses then it is more likely that she was not wearing a red sweater than that she was.

Control 4: If the girl was Joy and she was not wearing glasses then it is more likely that she had her hair in braids than not.

These may formally be written as inequalities between conditional probabilities as follows:

Control 1: $$prob(\text{wears-red-sweater}(x)\Diamond\text{yes}|\text{name}(x)\Diamond\text{Joy} \wedge \text{wears-glasses}(x)\Diamond\text{yes}) > prob(\text{wears-red-sweater}(x)\Diamond\text{no}|\text{name}(x)\Diamond\text{Joy} \wedge \text{wears-glasses}(x)\Diamond\text{yes}),$$

Control 2: $$prob(\text{braids-hair}(x)\Diamond\text{no}|\text{name}(x)\Diamond\text{Joy} \wedge \text{wears-glasses}(x)\Diamond\text{yes}) > prob(\text{braids-hair}(x)\Diamond\text{yes}|\text{name}(x)\Diamond\text{Joy} \wedge \text{wears-glasses}(x)\Diamond\text{yes}),$$

Control 3: $$prob(\text{wears-red-sweater}(x)\Diamond\text{no}|\text{name}(x)\Diamond\text{Joy} \wedge \text{wears-glasses}(x)\Diamond\text{no}) > prob(\text{wears-red-sweater}(x)\Diamond\text{yes}|\text{name}(x)\Diamond\text{Joy} \wedge \text{wears-glasses}(x)\Diamond\text{no}),$$

Control 4: $$prob(\text{braids-hair}(x)\Diamond\text{yes}|\text{name}(x)\Diamond\text{Joy} \wedge \text{wears-glasses}(x)\Diamond\text{no}) > prob(\text{braids-hair}(x)\Diamond\text{no}|\text{name}(x)\Diamond\text{Joy} \wedge \text{wears-glasses}(x)\Diamond\text{no}),$$

The control specification is a set of assignments and inequalities on the costs of decisions and the conditional probabilities of events. It is the relative costs of the decisions and the relative frequencies of occurrences which make one procedural form more attractive than another.

In this case, once the father has discovered the girl has brown hair, he must try to detect the fact that it is *not* Sue as soon as possible. Thus, if the girl had been wearing glasses and it had been Joy then Joy would probably be wearing her red sweater. So this question would probably not help. However, as Joy usually does not braid her hair when she wears glasses, this would be a better question. This makes the following approach to the solution optimal as, if it is Joy, then one visit will probably be sufficient.

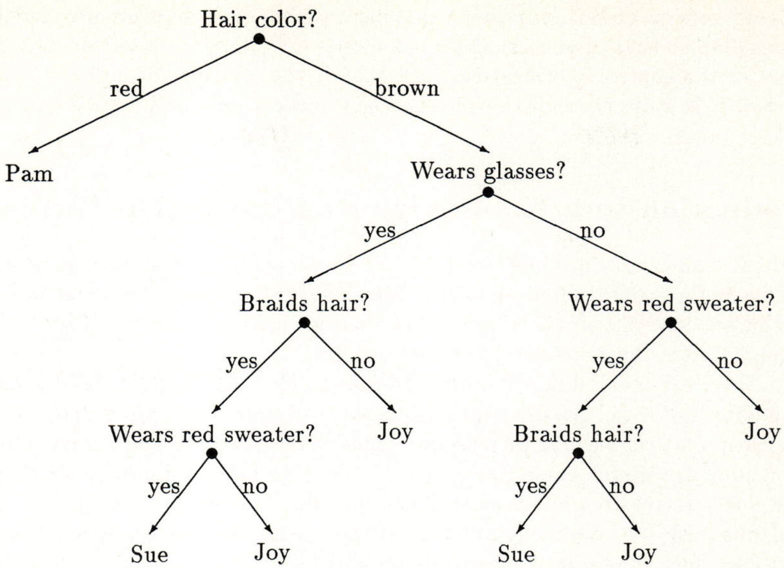

As a decision expression this may be written:

```
hair-color(x):
[red≫ <:Pam,
 brown≫ wears-glasses(x):
        [yes≫ braids-hair(x):
                [yes≫ wears-red-sweater(x):
                        [yes≫ <:Sue,
                         no≫ <:Joy
                        ],
                 no≫ <:Joy
                ],
         no≫ wears-red-sweater(x):
                [yes≫ braids-hair(x):
                        [yes≫ <:Sue,
                         no≫ <:Joy
                        ],
                 no≫ <:Joy
                ]
        ]
]
```

4.5 Is Knowledge Shaping Feasible?

It is proposed that all this should be *automatically* inferred. Clearly it is vital to this proposed approach to the control problem that this can be done with reasonable efficiency. However, this is still an open question which hinges on, among other things, the ability to solve efficiently the symbolic inequalities between probabilities and costs which arise from the procedural forms.

A restricted version of this knowledge shaping tool has been implemented in the Decision Tree Designer [38], so that in general these techniques definitely are feasible. This system uses the first part of the control specification only, that is the relative costs of the decisions, in a restricted way. It is still very much open whether a full control specification can be used in a cost-effective manner.

5 Introduction to a Knowledge Representation Formalism

The theoretical framework in which we have set the control problem would be considerably enhanced if we had a precise definition of the logical and procedural representations. This is the purpose of this section. My aim is to give a feel for some of the theoretical ideas which underlie the transition between the procedural and logical form.

Although there are many different ways of representing "knowledge" a fundamental method is to use some form of logic. As the relationship between procedural representations and logics has become clearer it has become increasingly apparent that only certain fragments of the full predicate calculus are useful computationally.[10] Isolating and exploring these fragments has been an extremely active area of research, which includes logic programming.

It would obviously be extremely valuable if there were already an agreed form in which to express knowledge. There is, for example, a fairly strong consensus in the data-base area concerning the representation of information. There would clearly be significant benefits to knowledge engineering if common standards arose [12]. However, this has yet to happen. One could argue that the "if-then" rule approach gives such a common language. However, there are serious problems with this methodology. In particular, it is very difficult to represent relationships between attributes and, of course, the semantics and treatment of negation is problematic.

The logic which I present below does overcome some of these problems. However, this advantage does not come cheaply; there is a considerable amount of formal machinery to master in the development. In this treatment I shall try to give the key definitions, however, the ideas themselves will not be developed in any depth.

For those not familiar with formal logic some words of caution are in order: setting up a logic and developing the properties of that logic is at best a mighty task; at worst, a seemingly pointless exercise in definitions. The logic I am about to set up is not particularly pretty. In fact, it will probably be rather unclear where the motivation for some of the definitions comes from. For those familiar with category theory and its connection to logic, the motivation and the method can be simply explained by saying that the logic is that of finitely complete categories with stable disjoint coproducts. This means that this logic has a natural procedural meaning given by its syntactic category.[11]

It is well-known that the methods pioneered by Gentzen [37], using the sequents, are often decidable. This makes them an attractive computational vehicle. Furthermore they actually fit the "if-then" rule paradigm remarkably well. We shall therefore use a version of these techniques in the development below.

In setting up a logic, the first job is to define the *frame*. A frame specifies the sorts (or types), the function symbols, and the predicate symbols which are available in the language. The logic I shall present is peculiar in that it has no predicate symbols; it relies on function symbols and two polymorphic *branch symbols*.

Definition 5.1 *A many-sorted decision theory,* $\mathbf{D} = (\mathbf{Fr}, \mathbf{Sq})$*, consists of a decision theory* frame, **Fr**, *and a set of valid sequents,* **Sq**.

[10] In particular the full predicate calculus is undecidable [2]. Computational semantics for predicate calculus are discussed in [14].

[11] Here I am using the terminology of Peter Johnstone [23], the connection between category theory and logic is also discussed in [27,25].

A decision theory frame, $\mathbf{Fr} = (\mathbf{S}, \mathbf{F}, \mathbf{Q})$, *consists of the following data:*

(i) *a set of* primitive sorts, $\mathbf{S}$, *from which are built recursively the following* composite sorts*:*

- *1 is a sort (the constant sort),*
- *if $S \in \mathbf{S}$ is a primitive sort then S is a sort,*
- *if S_1 and S_2 are sorts then $S_1 \times S_2$ is a* product *sort,*
- *if S_1 and S_2 are sorts then $S_1 + S_2$ is a* coproduct *sort,*

each primitive and coproduct sort, S, comes with a stock of variables

$$x_0^S, x_1^S, x_2^S, \ldots$$

The constant sort, 1, *which is also the empty product sort, only has one variable denoted* $x^{\top}$.
If $S = S_1 \times \ldots \times S_n$ is a product sort then a variable of that sort will always be expanded into its components $\vec{x} = \langle x_1^{S_1}, \ldots, x_n^{S_n} \rangle$, where these variables are distinct.
Associated with each coproduct sort, $S_1 + S_2$, are two special partial function symbols $b_0^{S_0+S_1}$ and $b_1^{S_0+S_1}$ which are called branch symbols *which have signatures respectively $b_0^{S_0+S_1} : S_0 + S_1 \rightharpoonup S_0$ and $b_1^{S_0+S_1} : S_0 + S_1 \rightharpoonup S_1$.*

(ii) *a set of function symbols,* $\mathbf{F}$, *each of which has a signature*

$$\mathrm{sig}(f) = S_1 \times \ldots \times S_n \longrightarrow S_0$$

where the domain *of f is* $\mathrm{dom}(f) = S_1 \times \ldots \times S_n$ *and the* codomain *of f is* $\mathrm{codom}(f) = S_0$ *and each of $S_0, \ldots, S_n$ is a primitive sort. The constant function symbols are symbols with domain* 1.

(iii) *a set of* decision symbols, $\mathbf{Q}$, *each of which has a signature*

$$\mathrm{sig}(q) = S_1 \times .. \times S_n \longrightarrow \vec{T}_1 + \ldots + \vec{T}_m$$

where $m \geq 2$ and the domain is a product of primitive sorts and the codomain is a coproduct of products of primitive sorts. Thus each $\vec{T}_j$ is $T_{j,1} \times \ldots \times T_{j,n_j}$ a product of primitive sorts $T_{j,i}$.

For convenience if the typing (or sort) of a variable or branch symbol is clear from the context (or does not matter) then it shall be omitted.

As usual in developing a logical language we must first set up the terms and then define the formulae. The formulae are built from assignment statements of two varieties: a functional assignment and a partial assignment, called a *branch instruction*. These primitive formulae can be combined using conjunction, existential quantification over *freely assigned* variables, and *disjoint* disjunction. Sequents are ordered pairs of formulae, $F_1 \vdash F_2$, read as F_1 *entails* F_2. The valid sequents , $\mathbf{Sq}$, give the special axioms satisfied by the theory.

The purpose of the variable restriction in the quantification is to make it a *unique* existential quantification. As an assigned variable (this is meant in the normal programming sense, see below) is always uniquely determined by its assigning expression obviously one step to ensure this is to restrict the quantification to assigned variables. However this is not quite enough as it is clearly possible to quantify assigned variables in a cycle and thus make the quantification non-unique on the remaining free variables. To avoid this we introduce the stronger concept of a *freely assigned* variable.

The restricted use of the disjunction ("or") also needs some explanation. An underlying purpose here is to mimic the effect of the "if-then-else" programming construct in the logic itself. Arbitrary disjunctions lead to a class of formulae with which it is difficult to associate a computational meaning.[12] The value of adding this construct is that it allows the direct representation of decisions which are a fundamental part of an expert system.

The formulae are then used to form a sequent calculus. The "turnstile" operator, $\vdash$, plays the role of implication in the calculus. Thus a sequent can be regarded as a prototypical "if-then" rule.

Definition 5.2

(i) A term *of a decision theory frame,* $\mathbf{Fr} = (\mathbf{S}, \mathbf{F}, \mathbf{Q})$, *is defined recursively by:*

- x_i^S *is a term of sort* $S \in \mathbf{S}$,
- *if* t_i *is a term of sort* S_i *then* $\vec{t} = \langle t_1, .., t_n \rangle$ *is a term of sort* $\vec{S} = S_1 \times ... \times S_n$.
- *if* $f : S_1 \times ... \times S_n \longrightarrow S_0$ *is a function symbol of* $\mathbf{F}$ *and* $\vec{t}$ *is a term of sort* $\vec{S}$ *then* $f(\vec{t})$ *is a term of sort* S_0.
- *if* $q : S_1 \times ... \times S_n \longrightarrow T_1 + ... + T_m$ *is a decision symbol in* $\mathbf{Q}$ *and* $\vec{t}$ *is a term of sort* $\vec{S}$ *then* $q(\vec{t})$ *is a term of sort* $T_1 + ... + T_m$. *This term is called a* decision head.

(ii) If t *is a term of sort* S *and* x_0^S *is a variable of sort* S *not occurring in* t *then* $t \rhd x_0^S$ *is an* assignment instruction *and the variable* x_0^S *is an* assigned variable.

(iii) If $b_j^{S_0+S_1}$, *where* $j = 0, 1$, *is a branch symbol and if* $z^{S_0+S_1}$ *and* y^{S_j} *are variables then* $b_j(z^{S_0+S_1}) \gg y^{S_j}$ *is a* branch instruction *and* y^{S_j} *is an* assigned variable.

(iv) A formula *of the language is now defined recursively by:*

- $\top$ *and* $\bot$ *are formulae,*
- *An assignment instruction is a formula,*
- *if* F_1 *and* F_2 *are formulae then* $F_1 \wedge F_2$ *is a formula.*
- *if* F_1 *and* F_2 *are* disjoint *formulae, as explained below then* $F_1 \overline{\vee} F_2$ *(the disjoint union) is a formula.*
- *if* x_i^S *occurs as a* freely assigned variable *in* F *in the sense described below, then* $(\exists x_i^S)[F]$ *is a formula.*

(v) Two formulae F_1 *and* F_2 *are* disjoint *if there is a proof using no more than the axioms and the inference rules described below which makes*

$$F_1 \wedge F_2 \vdash \bot$$

a valid sequent.

(vi) A variable, y, *is* freely assigned *in a formula if it is assigned from a term* t, *by an assignment,* $t \rhd y$ *or a branch instruction,* $t \gg y$, *such that all the variables of* t *are independent of* y. *A variable is* independent *of* y *if it is free and different from* y *or bound and assigned from some term* t' *all of whose variables are independent of* y.

(vii) If F_1 *and* F_2 *are formulae then* $F_1 \vdash F_2$ *is a* sequent.

A variable, y, is *dependent* on a set of free variables, X, if there is an assignment chain starting only with the variables in X and ending with y. Thus, another way of expressing the concept of *free assignment* is to to insist that there are a set of free variables different from y on which y is dependent. This ensures there is some way of "calculating" y from the free variables.

[12]For example, this is why Prolog restricts itself to Horn clauses - which are equivalent to the logic we are setting up if we remove the disjoint disjunction and existential quantification (and allow arbitrary predicates).

The test for disjunction threatens to be impossibly difficult. Fortunately it turns out to be quite reasonable. The test can be reduced to a test of formulae involving only conjunctions and disjunctions (no quantification). The test, however, involves some preparation. In particular the quantification must be removed. The formula must then be rewritten using $\triangleright$- substitution, see below, and all branch instructions with equal first argument must have their assigned variables identified. The test then is determined by:

- $b_0(y) \gg x_1$ and $b_1(y) \gg x_2$ are disjoint,
- $F_1 \wedge F_2$ is disjoint from F if either F_1 is disjoint from F or F_2 is disjoint from F,
- $F_1 \overline{\vee} F_2$ is disjoint from F if F_1 is disjoint from F and F_2 is disjoint from F.

The final ingredients of a logic are the axioms and the rules of inference:

Definition 5.3 *A decision theory* **D** *always satisfies the following axioms:*

(i) $\bot \vdash F$,

(ii) $F \vdash \top$,

(iii) $F \vdash F$,

(iv) $F_1 \wedge F_2 \vdash F_1$ *and* $F_1 \wedge F_2 \vdash F_2$,

(v) $F_1 \vdash F_1 \overline{\vee} F_2$ *and* $F_2 \vdash F_1 \overline{\vee} F_2$,

(vi) $F_1 \wedge (F_2 \overline{\vee} F_3) \vdash (F_1 \wedge F_2) \overline{\vee} (F_1 \wedge F_3)$,

(vii) $b_0(x^{S_0+S_1}) \gg y^{S_0} \wedge b_1(x^{S_0+S_1}) \gg y^{S_1} \vdash \bot$,

(viii) $\top \vdash b_0(x^{S_0+S_1}) \gg y^{S_0} \overline{\vee} b_1(x^{S_0+S_1}) \gg y^{S_1}$,

(ix) $F \vdash (\exists x)[F]$,

(x) $F_0 \wedge (\exists x)[F_1] \vdash (\exists x)[F_0 \wedge F_1]$ *when* x *is not free in* F_0,

(xi) $t \triangleright x_1 \wedge t \triangleright x_2 \vdash x_1 \triangleright x_2$,

(xii) $b_j(x^{S_0+S_1}) \gg y_1^{S_j} \wedge b_j(x^{S_0+S_1}) \gg y_2^{S_j} \vdash y_1^{S_j} \triangleright y_2^{S_j}$, *for* $j = 0, 1$,

(xiii) $b_j(x^{S_0+S_1}) \gg y_1^{S_j} \wedge y_1^{S_j} \triangleright y_2^{S_j} \vdash b_j(x^{S_0+S_1}) \gg y_2^{S_j}$, *for* $j = 0, 1$,

(xiv) $\top \vdash (\exists y)[t \triangleright y]$ *where* t *is any term.*

The inference rules for a decision theory **D** *are as follows:*

(i) $\wedge$*-elimination:*

$$\frac{F \vdash F_1 \wedge F_2}{F \vdash F_1}$$

$$\frac{F \vdash F_1 \wedge F_2}{F \vdash F_2}$$

(ii) $\wedge$*-introduction:*

$$\frac{F \vdash F_1, F \vdash F_2}{F \vdash F_1 \wedge F_2}$$

(iii) $\overline{\vee}$*-elimination:*

$$\frac{F_1 \overline{\vee} F_2 \vdash F}{F_1 \vdash F}$$

$$\frac{F_1 \overline{\vee} F_2 \vdash F}{F_2 \vdash F}$$

(iv) $\overline{\vee}$*-introduction:*

$$\frac{F_1 \vdash F, F_2 \vdash F}{F_1 \overline{\vee} F_2 \vdash F}$$

(v) $\triangleright$*-substitution:*

$$\frac{t \triangleright x \wedge F_1 \vdash F_2}{F_1[x := t] \vdash F_2[x := t]}$$

where x must not be an assigned variable in F_1 or F_2, must occur free in F_1 and where $F[x := t]$ is F with x substituted by t.

(vi) $\triangleright$*-demodulation:*

$$\frac{t \triangleright x \wedge F_1 \vdash F_2}{t \triangleright x \wedge F_1 \vdash F_2[x := t]}.$$

(vii) $\exists$*-introduction:*

$$\frac{F_1 \vdash F_2}{(\exists x)[F_1] \vdash F_2}$$

where x is not free in F_2.

(viii) cut rule:

$$\frac{F_1 \vdash F_2, F_2 \vdash F_3}{F_1 \vdash F_3}.$$

In order to get a feel for this logic it is worth discussing some of the central results. The logic has many similarities to a simple programming language: this, of course, is part of the motivation for introducing it. In particular assignment statements behave like assignment statements in an ordinary imperative programming language. In fact, the only constructs in this simple language are assignments and branches and all the programs must be acyclic.

The connecting symbols, $\vdash\dashv$, will mean that the sequent holds in both directions. If a formula is written $F(y)$ it means that the variable y occurs freely somewhere in the formula.

Lemma 5.4

$$(\exists y)[t \triangleright y \wedge F(y)] \vdash\dashv F(y)[y := t]$$

Proof. The following sequent is clearly valid:

$$t \triangleright y \wedge F(y) \vdash F(y).$$

Using $\triangleright$-demodulation we obtain

$$t \triangleright y \wedge F(y) \vdash F(y)[y := t],$$

as y is not free in $F(y)[y := t]$ we can employ $\exists$-introduction to obtain:

$$(\exists y)[t \triangleright y \wedge F(y)] \vdash F(y)[y := t].$$

For the converse using the existential axiom *(ix)* we have

$$t \triangleright y \wedge F(y) \vdash (\exists y)[t \triangleright y \wedge F(y)]$$

and now using $\triangleright$-substitution we obtain:

$$F(y)[y := t] \vdash (\exists y)[t \triangleright y \wedge F(y)].$$

□

The importance of this observation is that the assignment acts just like an assignment in a programming language and the quantification simply gives the scope of the temporary variables. This means that we can break any formula down into a *primitive* assignment form, in which each assignment only involves one primitive function symbol. If this is done common subexpression elimination can be performed using the following:

Lemma 5.5

(i)

$$t \rhd y_1 \wedge t \rhd y_2 \dashv\vdash y_1 \rhd y_2 \wedge t \rhd y_1$$

(ii)

$$(\exists y_1, y_2)[t \rhd y_1 \wedge t \rhd y_2 \wedge F(y_1, y_2)]$$

$$\dashv\vdash$$

$$(\exists y)[t \rhd y \wedge F(y_1, y_2)[y_1 := y, y_2 := y]]$$

Proof.

(i) The direction $\vdash$ is a direct consequence of the axioms *(xi)* and *(iv)* and of using the rule of inference $\wedge$-introduction. For $\dashv$ we have:

$$y_1 \rhd y_2 \wedge t \rhd y_1 \vdash y_1 \rhd y_2$$

and so $\rhd$-demodulation with $t \rhd y_1$ gives,

$$y_1 \rhd y_2 \wedge t \rhd y_1 \vdash t \rhd y_2$$

from which the result follows easily.

(ii) For, $\vdash$, we use $\rhd$-substitution to obtain:

$$(\exists y_1, y_2)[t \rhd y_1 \wedge t \rhd y_2 \wedge F(y_1, y_2)$$

$$\vdash$$

$$F(y_1, y_2)[y_1 := t, y_2 := t]$$

but from the latter form we can easily obtain the desired consequence using 5.4.
Conversely, for $\dashv$ we have:

$$t \rhd y_1 \wedge y_1 \rhd y_2 \wedge F(y_1, y_2) \vdash t \rhd y_1 \wedge t \rhd y_2 \wedge F(y_1, y_2)$$

existentially quantifying both sides by y_2 allows by the above lemma all mention of y_2 to be removed from the LHS. Quantifying both sides by y_1 then gives:

$$(\exists y_1)[t \rhd y_1 \wedge F(y_1, y_2)[y_2 := y_1]] \vdash (\exists y_1, y_2)[t \rhd y_1 \wedge t \rhd y_2 \wedge F(y_1, y_2)]$$

which with a change of bound variable gives the result.

□

Thus, for example, the common subexpression x_1+x_2 can be eliminated from $f(x_1+x_2,(x_1+x_2)-x_3)$:

$$f(x_1 + x_2, (x_1 + x_2) - x_3) \rhd y$$

$$\vdash\dashv$$

$$(\exists y_1, y_2, y_3)[x_1 + x_2 \rhd y_1 \wedge x_1 + x_2 \rhd y_2 \wedge y_2 - x_3 \rhd y_3 \wedge f(y_1, y_3) \rhd y]$$

$$\vdash\dashv$$

$$(\exists y_1, y_3)[x_1 + x_2 \rhd y_1 \wedge y_1 - x_3 \rhd y_3 \wedge f(y_1, y_3) \rhd y].$$

In a manner similar to predicate logic it is possible to reduce an arbitrary set of sequents to an equivalent set in "clausal" form. A *clause* in this logic consists of a sequent whose left-hand side is a conjunction of assignments and branch instructions (with no existential quantification) and whose right-hand side is a disjunction of existentially bound conjuncts. This can be seen from $\wedge$-introduction and elimination, $\overline{\vee}$-introduction and elimination, the distributive axiom and $\exists$-introduction.

Any assignment whose assigned variable is bound can be substituted and thus eliminated from the expression by substitution. Thus, if the right-hand side only had assignment statements, we could completely eliminate the existential quantification reducing the formula to a Horn clause (that has only one implied "literal" or assignment on the RHS). However, this situation is complicated by the existence of branch instructions. It is the branch instructions which make the logic of special interest in this context: for it is these which allow the expression of the logic of control.

Our first step is to obtain the analogue of 5.5 for branch instructions:

Lemma 5.6

(i) $b_0(z) \gg y \vdash\dashv (\exists y_1)[b_0(z) \gg y_1 \wedge y_1 \rhd y]$,

(ii) $(\exists y_1, y_2)[b_0(z) \gg y_1 \wedge b_0(z) \gg y_2 \wedge F(y_1, y_2)]$

$$\vdash\dashv$$

$$(\exists y)[b_0(z) \gg y \wedge F(y_1, y_2)[y_1 := y, y_2 := y]].$$

Proof.

(i) For $\vdash$, as $b_0(z) \gg y \vdash (\exists y)[b_0(z) \gg y]$ we have:

$$b_0(z) \gg y \vdash\dashv b_0(z) \gg y \wedge (\exists y)[b_0(z) \gg y].$$

Now

$$b_0(z) \gg y \wedge b_0(z) \gg y_1 \vdash b_0(z) \gg y \wedge y_1 \rhd y,$$

so that existentially quantifying each side by y_1 gives:

$$b_0(z) \gg y \wedge (\exists y_1)[b_0(z) \gg y_1] \vdash (\exists y_1)[b_0(z) \gg y_1 \wedge y_1 \rhd y]$$

which, using the identity proven above and the fact that the names of quantified variables can be changed, gives the desired sequent.

The converse follows directly from the axioms and $\exists$-introduction.

(ii) For $\vdash$,

$$b_0(z) \gg y_1 \wedge b_0(z) \gg y_2 \wedge F(y_1, y_2)$$

$$\vdash$$

$$b_0(z) \gg y_1 \wedge b_0(z) \gg y_2 \wedge F(y_1, y_2) \wedge y_2 \rhd y_1 \wedge y_1 \rhd y_2$$

Now $y_2 \rhd y_1 \wedge F(y_1, y_2) \vdash F(y_1, y_2)[y_1 := y_2]$ so that the above entails:

$$(\exists y_1)[b_0(z) \gg y_1 \wedge y_1 \rhd y_2] \wedge b_0(z) \gg y_2 \wedge F(y_1, y_2)[y_1 := y_2]$$

$$\vdash$$

$$b_0(z) \gg y_2 \wedge F(y_1, y_2)[y_1 := y_2]$$

which gives the desired result.
For $\dashv$,

$$b_0(z) \gg y \wedge F(y_1, y_2)[y_1 := y, y_2 .= y]$$

$$\vdash\dashv$$

$$b_0(z) \gg y \wedge (\exists y_1, y_2)[y \rhd y_1 \wedge y \rhd y_2 \wedge F(y_1, y_2)]$$

Rearranging this gives:

$$(\exists(y_1, y_2, y)[F(y_1, y_2) \wedge (b_0(z) \gg y \wedge y \rhd y_1) \wedge (b_0(z) \gg y \wedge y \rhd y_2)]$$

which easily gives the desired form.

□

To facilitate the discussion of the logic of control we shall introduce some syntactic sugar, which I have already used informally above, to allow the branch behavior to be readily expressed. Consider a decision symbol, q,

$$sig(q) = S_1 \times ... \times S_n \longrightarrow T_1 + ... + T_m,$$

then to obtain the behavior of the decision q on its jth branch, T_j, it is necessary to write down the following formula:

$$(\exists z_1, ..., z_j)[q(\langle x_1, ..., x_n \rangle) \rhd z_1 \wedge b_1(z_0) \gg z_1 \wedge ... \wedge b_1(z_{j-1}) \gg z_j$$
$$\wedge b_0(z_j) \gg \langle y_1, ..., y_{m_j} \rangle],$$

which is quite a mouthful.[13]

Definition 5.7 *If* $q : S_1 \times ... \times S_n \longrightarrow T_1 + ... + T_m$ *is a decision symbol then we may associate with* q *a set of* codomain symbols $\mathrm{codsym}(q) = \{a_1, ..., a_m\}$ *to indicate each product sort in the codomain and let*

$$q(\langle x_1, ..., x_n \rangle) \Diamond a_j(\langle y_1, ..., y_{m_j} \rangle)$$

indicate the above formula.

It is now easy to prove the following facts:

Lemma 5.8 *If* $\mathrm{codsym}(q) = \{a_1, ..., a_n\}$ *then:*

[13] I have assumed the association $T_1 + (T_2 + (... + (T_{m-1} + T_m)..)$.

(i) $q(\vec{x})\Diamond a_i(\vec{y}_i) \wedge q(\vec{x})\Diamond a_j(\vec{y}_j) \vdash \bot$, *whenever* $i \neq j$,

(ii) $\top \vdash (\exists\vec{y}_1)[q(\vec{x})\Diamond a_1(\vec{y}_1)]\underline{\vee}....\underline{\vee}(\exists\vec{y}_n)[q(\vec{x})\Diamond a_n(\vec{y}_n)]$.

There is still something too general about the logic we are developing. For example the formula

$$(\exists y)[f(x) \rhd y \wedge g(y) \rhd x]$$

cannot easily be interpreted to be a procedure. What are the inputs and outputs? To rectify this we have the following definition.

Definition 5.9

(i) *A variable x, in a formula F, is an* input variable *if x is free and does not occur as an assigned variable. A variable y is an* output variable *if y is a freely assigned variable, freely assigned from the input variables. That is dependent for its value only on the input variables.*

(ii) *A formula is an* (X,Y)-production *if X is the set of input variables and Y is the set of output variables and every other variable is bound and only assigned once in the scope of that binding.*

(iii) *A sequent $F_1 \vdash F_2$ is an* (X,Y)-production *sequent if F_1 and F_2 are (X,Y)-production formulae.*

Now it is obvious that as a production formula has well-defined inputs and outputs and it expresses a partial functional relationship between the inputs and outputs. It is partial because it is quite possible that a branch instruction may simply fail causing the whole expression not to evaluate. Thus, production sequents, which are a (generalized) formal counterpart of production rules, encode only fragments of procedural behavior. Their value of course is in the specification stage where one wants to specify fragments of behavior independently. Thus, production sequents form the basis of the specification language.

It remains to discuss the procedural form of this language. An (X,Y)-production formula, F, is *procedural* if $\top \vdash (\exists Y)[F]$. Essentially, this is the requirement that for every input there is guaranteed to be a defined output. A procedural formula can be obtained by the disjoint joining of a number of fragments of behavior. If we had a parallel computer we could give each fragment to a processor (or to a group of processors) to evaluate and we would be guaranteed that for any input only one processor would successfully finish. For a sequential process, such as obtaining answers from a user or a program on a conventional computer, we must further restrict the form to resemble that of a decision tree so that only one thing happens at a time.

Definition 5.10 *Given a decision theory frame, $\mathbf{Fr} = (\mathbf{S},\mathbf{F},\mathbf{Q})$, in which the decision symbols have been given codomain symbols, a* decision expression *is defined recursively by:*

- $<: \vec{y}$ *is a decision expression,*
- *if $f \in \mathbf{F}$ with $f : \vec{S} \longrightarrow S_0$ and d is a decision expression then $f(\vec{t}) : [y^{S_0} \gg d]$ is a decision expression, where $\vec{t}$ is a term of type $\vec{S}$ in which the variable y^{S_0} does not occur.*
- *if $q \in \mathbf{Q}$ is a decision symbol with $q : \vec{S} \longrightarrow \vec{T}_1 + ... + \vec{T}_m$,* codsym$(q) = \{a_1,...,a_m\}$, *and $d_1,...,d_m$ are decision expressions then*

$$q(\vec{t}) : [a_1(\vec{y}_1) \gg d_1, ..., a_m(\vec{y}_m) \gg d_m]$$

is a decision expression where $\vec{t}$ is a term of type $\vec{S}$ and $\vec{y}_j$ is a variable of type $\vec{T}_j$ and $\vec{t}$ and $\vec{y}_j$ have no variables in common.

We may translate a decision expression, d, into a formula, $\tilde{d}$ of our logic as follows:

- $f(\vec{t}) : [y^{S_0} \gg d] \mapsto (\exists y)[f(\vec{t}) \rhd y \wedge \tilde{d}]$,
-

$$q(\vec{t}) : [a_1(\vec{y}_1) \gg d_1, ..., a_m(\vec{y}_m) \gg d_m]$$

$$\mapsto$$

$$(\exists \vec{y}_1)[q(\vec{t}) \Diamond a_1(\vec{y}_1) \wedge \tilde{d}_1] \bar{\vee} ... \bar{\vee} (\exists \vec{y}_m)[q(\vec{t}) \Diamond a_m(\vec{y}_m) \wedge \tilde{d}_m],$$

- $<: \vec{y} \mapsto \vec{y} \rhd \vec{y}_{out}$ where this last is a conjunct of assignments for each variable in $\vec{y}$ to an output variable (which does not occur elsewhere in the formula).

A decision expression translates into a production formula which is procedural. It is a production formula because the binding of the branch variables ensures that the branch instruction gives the only assignment of that variable within the scope of the binding. Furthermore only the input and output variables are free. Identities which decision expressions satisfy can now be obtained directly from the logic. The proposition below outlines the three basic identities which decision expressions satisfy (as decision expressions are algebraic in nature $=$ is used instead of $\vdash\dashv$):

Proposition 5.11

(i) Idempotence:

$$q(\vec{x}) : [a_1(\vec{y}_1) \gg d, ..., a_m(\vec{y}_m) \gg d] = d$$

whenever no variable in $\vec{y}_i$ occurs freely in d.

(ii) Repetition:

$$q(\vec{x}) : [a_1(\vec{y}_1) \gg d_1, ..., a_r(\vec{y}_r) \gg q(\vec{x}) : [a_1(\vec{z}_1) \gg d_{1,1}, ..., a_m(\vec{z}_m) \gg d_{1,m}], ..., a_m(\vec{y}_m) \gg d_m]$$

$$=$$

$$q(\vec{x}) : [a_1(\vec{y}_1) \gg d_1, ..., a_r(\vec{y}_r) \gg d_{1,r}[\vec{z}_r := \vec{y}_r], ..., a_m(\vec{y}_m) \gg d_m]$$

(iii) Transposition:

$$\begin{aligned} & q_1(\vec{x}_1) : \\ & \quad [a_1^1(\vec{y}_1) \gg q_2(\vec{x}_2) : [a_1^2(\vec{z}_1) \gg d_{1,1}, ..., a_m^2(\vec{z}_m) \gg d_{1,m}], \\ & \quad ..., \\ & \quad a_n^1(\vec{y}_n) \gg q_2(\vec{x}_2) : [a_1^2(\vec{z}_1) \gg d_{n,1}, ..., a_m^2(\vec{z}_n) \gg d_{n,m}]] \end{aligned}$$

$$=$$

$$\begin{aligned} & q_2(\vec{x}_2) : \\ & \quad [a_1^2(\vec{z}_1) \gg q_1(\vec{x}_1) : [a_1^1(\vec{y}_1) \gg d_{1,1}, ..., a_n^2(\vec{y}_n) \gg d_{n,1}], \\ & \quad ..., \\ & \quad a_m^2(\vec{y}_m) \gg q_1(\vec{x}_1) : [a_1^1(\vec{y}_1) \gg d_{1,m}, ..., a_n^2(\vec{y}_n) \gg d_{n,m}]] \end{aligned}$$

where $\vec{x}_1$ has no variables in common with any $\vec{z}_j$ and $\vec{x}_2$ has no variables in common with any $\vec{y}_j$.

These identities show that decision expressions can be manipulated in the same way as decision trees [7]. In fact the only difference is that the variable restrictions must be respected. This allows decision tree shaping techniques such as *reduction* [6,8], and alternative form generation [19], to be applied.

6 Conclusion

The concept of an *expert system* has been a major driving force behind the development of software tools. The separation of knowledge from the procedural implementation is reminiscent of the separation of the specification from the program in the more traditional software development cycle. This led to the view of an expert system building environment as a specialized software development environment in which the rules actually form the logical specification.

This view sets some classical software problems in a rather different light. The more free-wheeling business of building an expert system in which the rule-base is continually changed suggests that developing the specification should be regarded as part and parcel of the development cycle rather than an earlier and separate activity.

A major problem in expert systems is the exercise of *control*. In this area there remain many issues to be resolved. I have suggested a framework in which the control problem can be viewed. This framework suggests that there should be a significant additional ability of the development environment, called *knowledge shaping*, which is currently missing. The computational feasibility of using a control specification in the form I have suggested to do knowledge shaping has not been resolved.

The key to closely linking specification and procedure is to have available the means to make the transition. This means that there must be a procedural and logical form with a well-defined relationship. In the last section a description of a logic and procedural form has been given together with a description of the transition between the forms. This transition exploits the fact that with many logics one can associate a syntactic category which gives the procedural form.

The procedural form described, *decision expressions*, is remarkably similar to that of decision trees. This allows the knowledge shaping abilities of that area to be transferred.

References

[1] I. Bratko (1986) *Prolog Programming for Artificial Intelligence.* Reading, Mass.: Addison-Wesley.

[2] G.S. Boolos and R.C. Jeffrey (1985) *Computability and Logic.* Cambridge: Cambridge University Press.

[3] L. Brownston, R. Farell, E. Kant, and N. Martin (1985) *Programming expert systems in OPS5: an introduction to rule-based programming.* Reading, Mass.: Addison-Wesley.

[4] B. Buchanan and E. Shortcliffe (1984) *Rule-based expert systems - the MYCIN experiments of the Stanford heuristic programming project.* Reading, Mass.: Addison-Wesley.

[5] W.F. Clocksin, C.S. Mellish (1981) *Programming in Prolog.* New York: Springer Verlag.

[6] J.R.B.Cockett (1986) *Decision expression optimization.* Fundamenta Informaticae (X).

[7] J.R.B.Cockett (1987) *Discrete decision theory: manipulations.* Theoretical Computer Science (to appear).

[8] J.R.B.Cockett (1987) *Decision tree reduction.* University of Tennessee, Department of Computer Science, Technical Report CS-87-68.

[9] J.R.B.Cockett and J.A.Herrera (1986) *Prime rule-based systems give inadequate control.* In Z.W. Ras and M. Zemankova (Eds.) Proceedings of the ACM SIGART international symposium on methodologies for intelligent systems, 441-449.

[10] R. Davis and D.B. Lenat (1982) *Knowledge-based systems in artificial intelligence.* New York: McGraw-Hill.

[11] R. Davis (1980) *Meta-rules: reasoning about control.* Artificial Intelligence, 15, 179-222.

[12] J. Debenham (1987) *Expert Systems: an information processing perspective.* in *Applications of Expert Systems* J.R. Quinlan (Ed.), Reading, Mass.: Addison-Wesley, 200-217.

[13] R.O. Duda, J.G. Gashnig, and P.E. Hart (1979) *Model design in the PROSPECTOR consultation system for mineral exploration.* In D. Michie (Ed.) *Expert systems in the microelectronic age* Edinburgh: Edinburg University Press, 153-167.

[14] M.H. Emden, R.A. Kowalski (1976) *The semantics of predicate logic as a programming language.* Journal of the Association for Computing Machinery, Vol 23, No. 4, 733-742.

[15] E.A. Fiegenbaum (1977) *The art of artificial intelligence 1: themes and case studies of knowledge engineering.* STAN-CS-77-621, Dept. of Computer Science, Stanford University.

[16] C.L. Forgy (1982) *RETE: A fast algorithm for the many patterm/many object pattern match problem.* Artificial Intelligence, Vol. 19, No. 1,

[17] C.L.Forgy (1981) *OPS5 User's Manual.* Technical Report CMU-CS-81-135, Carnegie-Mellon University, Department of Computer Science.

[18] F. Hayes-Roth, D.A. Waterman, and D.B. Lenat (Eds.) (1983) *Building Expert Systems* Reading, Mass.: Addison-Wesley.

[19] J.A. Herrera and J.R.B. Cockett (1986) *Finding all the optimal forms of a decision tree.* University of Tennessee, Department of Computer Science, Technical Report CS-86-67.

[20] A. Hodges (1983) *Alan Turing: the enigma of intelligence.* London: Urwin Paperbacks.

[21] E.B. Hunt, J. Martin, and P.J. Stone (1966) *Experiments in Induction.* New York: Academic Press.

[22] P. Jackson *Introduction to Expert Systems.* Reading, Mass.: Addison-Wesley.

[23] P.T. Johnstone (1977) *Topos Theory.* New York: Academic Press.

[24] R.A. Kowalski (1979) *Logic for Problem Solving.* New york: North Holland.

[25] J. Lambek and P. Scott (1986) *Introduction to higher order categorical logic.* Cambridge: Cambridge University Press.

[26] W.A. Martin and R.J. Fateman (1971) *The MACSYMA system.* In proceedings of the Second Symposium on Symbolic and Algebraic Manipulation, Los Angeles, 59-75.

[27] M. Makkai and G.E. Reyes (1976) *First order categorical logic.* Springer Lecture Notes in Mathematics 611.

[28] P. Mccorduck (1979) *Machines who think.* San Fransisco: W.H. Freeman and company.

[29] R.S. Michalski (1983) *A theory and methodology of inductive learning.* Artificial Intelligence, 20.

[30] R.S. Michalski and R.L. Chilausky (1980) *Knowledge acquisition by encoding expert rules versus computer induction from examples: a case study involving soybean pathology.* International Journal of Man-Machine Studies, 12, 63-87.

[31] D. Michie (1987) *Current Developments in Expert Systems.* in *Applications of Expert Systems* J.R. Quinlan (Ed.), Reading, Mass.: Addison-Wesley, 137-156.

[32] N. Nilsson (1980) *Principles of artificial intelligence.* Palo Alto, California: Tioga Publishing Co.

[33] E. Rich (1983) *Artificial Intelligence.* New York: McGraw-Hill.

[34] J.R. Quinlan (1987) *Inductive Knowledge Acquisition.* In *Applications of Expert Systems* J.R. Quinlan (Ed.), Reading, Mass.: Addison-Wesley, 157-173.

[35] J.R. Quinlan (1986) *Induction of Decision Trees.* Machine Learning, 1, 81-106.

[36] J.R. Quinlan (1986) *The effect of noise on concept learning.* In *Machine learning 2* R.S. Michalski (Ed.), Los Altos: Morgan Kaufmann.

[37] M.E. Szabo (1969) *The collected papers of Gerhart Gentzen.* London: North-Holland.

[38] J.A. Vrba and J.A. Herrera (1987) *The decision tree designer: producing efficient expert systems in general computers.* A Perceptics Corporation application note, Knoxville.

[39] A. Walker (Ed.), M. McCord, J.F. Sowa, and W.G. Wilson (1987) *Knowledge Systems and Prolog* Reading, Mass.: Addison-Wesley.

[40] D.H.D. Warren (1977) *Implementing Prolog: compiling predicate logic programs.* Research Reports 39 and 40, Department of Artificial Intelligence, University of Edinburgh.

[41] D.A. Waterman (1986) *A guide to Expert Systems.* Reading, Mass.: Addison-Wesley.

[42] P.H. Winston and B.K.P. Horn (1984) *LISP.* Reading, Mass.: Addison-Wesley.

[43] P.H. Winston (1987) *The Commercial Debut of Artificial Intelligence.* In *Applications of Expert Systems* J.R. Quinlan (Ed.), Reading, Mass.: Addison-Wesley, 3-20.

QUALITATIVE MODELLING OF PHYSICAL SYSTEMS FOR KNOWLEDGE BASED CONTROL

ROY LEITCH
INTELLIGENT AUTOMATION LABORATORY
DEPARTMENT OF ELECTRICAL AND ELECTRONIC ENGINEERING
HERIOT-WATT UNIVERSITY
EDINBURGH EHI 2HT
SCOTLAND

INTRODUCTION

There is currently much discussion, and some soul searching, on the impact of Artificial Intelligence (AI) techniques on Control Engineering. In particular, the success of expert systems technology in other fields has led to the consideration of how these techniques could be used to develop more robust control systems for complex dynamic processes. The problem is - **where** do we put the expertise ? And **what** expertise do we use ? Two distinct approaches are emerging. The first, claiming the term **Expert Control** [1], puts the expertise on top of existing control system designs, and massages the system when it goes wrong. This expertise is essentially the knowledge and skill of the control engineer, on-line. The second approach uses knowledge about the process itself to derive appropriate control action. This knowledge is usually expressed as a set of rules, consequently this approach has been called rule-based control (including fuzzy logic controllers). This approach has been generalised [2] to allow other forms of knowledge representation and is termed **Knowledge Based Control Systems** or **Qualitative Control**. In this case the knowledge used is the skill and intuitions of the process operator, with the expertise placed directly in the feedback path

NATO ASI Series, Vol. F47
Advanced Computing Concepts and Techniques in Control Engineering
Edited by M. J. Denham and A. J. Laub

The analysis of physical systems involves two steps:

1) Developing models for physical systems which accurately represent their physical behaviour.

2) Predicting the behaviour of systems based on these models.

Analytic control theory focuses on the second step and usually assumes that the models provided are sound or **wellposed**. All control methods are based, either explicitly or implicitly, on a model of the process to be controlled. Problems with control arise when either the system to be controlled is in some way ill-defined and/or it is not possible to gain access to the internal states of the system. Both of these represent a lack of information about the underlying physical system. Usually deficient information about a model can be compensated for by having access to most of the states and by designing a 'tight' feedback system. This approach is most successful when applied to the regulation problem; for servo-mechanisms the lack of information is much more crucial. In this case, an explicit model can be used to generate estimates of the behaviour that can be used to modify the dynamic behaviour to satisfy the performance specification. It is the purpose of this paper to discuss the impact of AI technology on the modelling of complex dynamic physical systems, and therefore to determine the most appropriate use of knowledge based systems techniques in the control of complex dynamic systems.

FEEDBACK

The central idea in Cybernetics [3] was systems that could observe their present state and from knowledge of the desired state generate the difference between the two states as feedback to move the system towards the desired state. In the analytic theory, it is assumed that the system state could be represented as a point in a vector space and that the desired state is also specified as a unique and distinct point in that state space. It is further assumed that there is a nice relationship between the difference and the appropriate control action; and, in fact, it is usually restricted to be linear. The vector space description led to the representation of dynamic systems by continuous, real-valued state variables which closely resemble the actual state variables observable in the physical system. It was here that Cybernetics bifurcated into what is now the essential difference between the analytic approach of established control theory and the empirical approach of knowledge based control systems. The essential difference is the definition of the fundamental unit or **atom** of knowledge. In the knowledge based system approach the basic unit is the symbol whereas analytic control theory took the continuous real-valued function as its primitive.

The adoption of the real-valued function has led to a sophisticated and comprehensive mathematical theory of feedback control. Differential or difference equations are used to model continuous-time and discrete-time dynamic systems respectively, and powerful synthesis methods have been derived [4]. Given the desired system, in terms of relationships between real-valued functions, either as differential/difference equations, or their complex frequency counterparts, a range of good methods exist for designing an appropriate feedback controller. However, in practice, the transformation of the formal specification of the desired performance or behaviour into the required mathematical description is, apart from relatively simple systems and specifications, problematic, and remains outside of the general analytic theory. In addition, the close correspondence between the physical variables and the continuous real-valued variables has led to a confusion between the actual physical system and its mathematical model. Too often the modelling assumptions necessary to the analytic method, once made, are neglected during the design and subsequent verification phase. This can result in the actual behaviour being significantly different from the desired, or designed behaviour, and failure to satisfy the prescribed specifications. Ironically, empirical methods are then used to restore the performance to satisfy the original specifications. For difficult processes the range of satisfactory operation can be extended through the use of adaptive controllers. These attempt to identify (modify) the model in real-time and to continuously update the control law. However, these adaptive algorithms are complex and require a priori assumptions and a significant amount of heuristic logic to ensure proper functioning [1]. This heuristic logic tends to negate the advantages of analytic methods.

The approach adopted within AI has emphasised, by the use of symbols, the modelling aspects of systems design. In particular, the General Problem Solver (GPS) of Newell, Shaw and Simon [6] is recognised as one of the first programs to separate its general problem solving method from the domain specific representation (model) of the problem. GPS developed the technique of means-ends analysis ,and can be viewed as a symbolic servo-mechanism. Feedback was used in that a desired or goal state was defined and the difference between the current state and the goal state used to determine the next state, such that the difference was reduced. This procedure would then be repeated iteratively until the goal was reached. Unfortunately, classifying differences and finding appropriate operators resulted in the notorious combinatorial explosion. It took until the early 1970's for AI workers to discover that the next possible states had to be constrained by domain specific knowledge. Essentially, this knowledge determines the relationships between the symbolic states in the same way as an analytic description based on continuous real-valued functions. In this sense, knowledge, whether represented as logic, rules or frames, is used to relate symbolic states in a knowledge based description in the same way as real-valued parameters and differential equation representations can be used to relate continuous real-valued variables. Both approaches attempt to model the underlying system. The essential difference being in the level of abstraction - symbols or numbers - used to represent the physical variables.

CRITIQUE OF ANALYTIC DESIGN METHODS

Conventional control system design methods assume an analytic model of the process to be controlled. This involves making basic suppositions about the behaviour of the system within the anticipated range of operating conditions. Typically, linearity and time-invariance is presumed and differential calculus used to derive a differential/difference equation representation of the system. This is a crucial, and often neglected, part of the design process. Many elaborate design methods necessitate the existence of a well-posed model and proceed from there. However, for many processes, an adequate representation is difficult, if not impossible, to obtain over the complete range of operating conditions. This is certainly true of many plants found in the process industries and furthermore is also true of some so called 'well-defined' processes. In practice, even if a process is amenable to mathematical solution, the time and effort required to extract the model coupled with the shortage and expense of the necessary expertise combine to render the analytical approach non-viable. Designs based upon an ill-posed model can result in performance that is significantly different from that predicted by the theory, resulting in a loss of confidence in the system. As a consequence, a considerable theory/applications gap has developed. Although well recognised by control practitioners, this gap has largely been ignored by control theorists.

In recent years, much research has been directed at developing adaptive control methods capable of identifying, on-line, the unknown parameters of a model of assumed structure. This estimated model then forms the basis of a real-time design procedure to produce the required control action. This approach certainly extends the range of plant operation for analytic methods. However, the performance and convergence of such systems is still heavily dependent upon design assumptions of system order, inherent time delays, etc. The effect of the adaptive loop is to shift the decision making away from the design parameters of the feedback loop to the performance parameters of the adaptive loop. In current jargon, this requires knowledge of the **control** process (meta-knowledge) rather than knowledge about the **process** itself. However, the adaptive solution is directed at processes that are difficult or ill-posed; it is knowledge about the process that is deficient. Sophisticated signal processing techniques can extend the range of acceptable performance of such systems, however, the underlying lack of knowledge about the process remains.

The human understanding of the process and its conventional mathematical description are alien and this results in a lack of an effective man-machine interface. Meaningful explanation of control decisions cannot be provided and an operator is unable to use experiential knowledge to modify the controller. In the adaptive case, this difficulty is compounded by the control 'levers' being at a further level of description removed from the feedback loop.

Based on the above comments it is tempting to conclude that many industrial processes are in fact uncontrollable. This is clearly not the case, however, as many so-called ill-posed systems are adequately, and sometimes easily, controlled using manual techniques.

COMPARISON WITH MANUAL METHODS

Manual control offers some significant advantages over conventional control methodologies, unfortunately it also has some considerable disadvantages. Paradoxically, it would now seem that the previously perceived weakness of humans is exactly what makes the human approach work. It is the intentional use of generality and vagueness that allows people to model (understand) complex processes and to infer suitable control actions. The inference procedure employed is usually symbolic in nature [1] and is only rarely numeric. That is, the observation and interpretation of the process is made on a qualitative basis, and control decisions achieved by combining qualitative judgements. Imprecision, therefore, is a necessary part of the reasoning process. The penalty of this approach is the loss of uniqueness and optimality for a particular control scheme. However, optimality in the conventional sense is defined with respect to the assumed mathematical model. If this model is ill-posed, as previously argued, sub-optimal performance results.

A disadvantage of manual methods is that they are only applicable to processes with fairly slow response times and are liable to degradation under stress . Psychological factors, such as boredom, tiredness or lack of motivation can also lead to a deterioration in performance. There is, furthermore, a need to experiment in order to learn, consequently the control policy is in continual revision, resulting in a loss of consistency. Contrary to much of the work on Expert Systems, the rationality of human thinking (in the formal sense) can be questioned, e.g. most people experience difficulty with a simple combination of logical negatives. Furthermore, human memory is prone to forgetting and distorting stored information - as a result manual methods are volatile and retraining can be expensive in time and product. Manual methods **do** work, however, and in some cases work surprisingly well. The goal for Knowledge Based Control Systems is, therefore, to combine the 'vague' symbolic processing capabilities of humans with the speed, capacity and alertness of machines.

FUZZY SETS

One way to represent vagueness is by the use of fuzzy sets [6]. Fuzzy sets are an attempt to bring the inference of computers closer to that used by people. In conventional set theory, which underpins all of the established mathematics, a set has sharp boundaries; either an element belongs to a particular set or it does not. By contrast, in fuzzy sets theory the transition from membership to non-membership is gradual rather than abrupt. The grade of membership is specified by a number usually between 1 (full member) and 0 (non-member). With fuzzy sets vague qualitative concepts like 'high', 'low', 'increasing', 'rapidly increasing' etc., can be represented in a form suitable for computers . Figure 1 shows characteristic membership functions for the values of a qualitative temperature variable. Note that an aspect of this theory is that these categories need not be mutually exclusive - a situation that is impossible in conventional 'crisp' set theory. For example, if an operator is making a decision on a given temperature being high, in crisp theory he could define all $T > 100^{\circ}C$ say, as being high. In this case $95^{\circ}C$ and $50^{\circ}C$ are both not high! By contrast, in the fuzzy sets represented below, $95^{\circ}C$ belongs to high with a membership of 0.9 and to medium with a membership of 0.1. Similarly $50^{\circ}C$ belongs to medium with a membership of 1 and to high with membership 0.2.

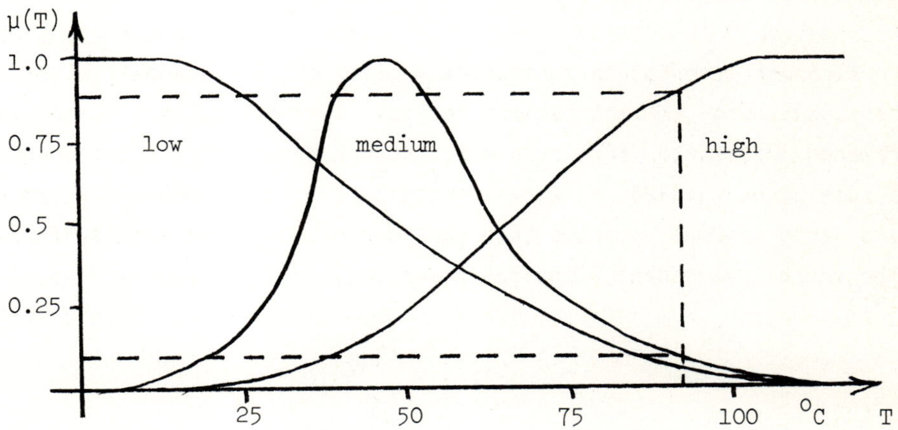

Figure 1. Characteristic Membership Functions

Basically, the grade of membership is subjective in nature and corresponds to the intentional use of vagueness in order to allow processing or reasoning about uncertain or complex systems. It is a measure of the compatability of measurement data with the category represented by the fuzzy set. The membership function is, therefore, a measure of vagueness and generality: the more vague and general the concept, the 'wider' the membership function must be, and relatively precise concepts are represented by 'thin' membership functions - approaching the characterisation of crisp sets. Fuzzy sets thus provide a way of translating the common-sense reasoning that humans tend to use into a form suitable for processing by machines.

RULE BASED CONTROL

One of the earliest, and most prominent, attempts to utilise Fuzzy Set Theory was by Mamdani and Assilian [7] in developing Fuzzy Logic Controllers (FLC), however the original idea of applying fuzzy sets to control problems is due to Zadeh . FLC is an attempt to provide a formalism for representing the control protocol of an operator. The protocol is expressed as a collection of condition/action rules, as shown in Figure 2. The 'condition' of each rule consists of the conjunction of values of the qualitative variables, the 'action' is a modification to the control inputs.

Although the FLC is conceptually quite simple, a number of technical issues concerning implementation remain. In particular, there is no established procedure for rule elicitation. While this is acceptable for small rule bases (10-20 rules), controllers for complex systems would require more formal methods than the ad hoc techniques used so far. A FLC for the control of cement kilns is commercially available.

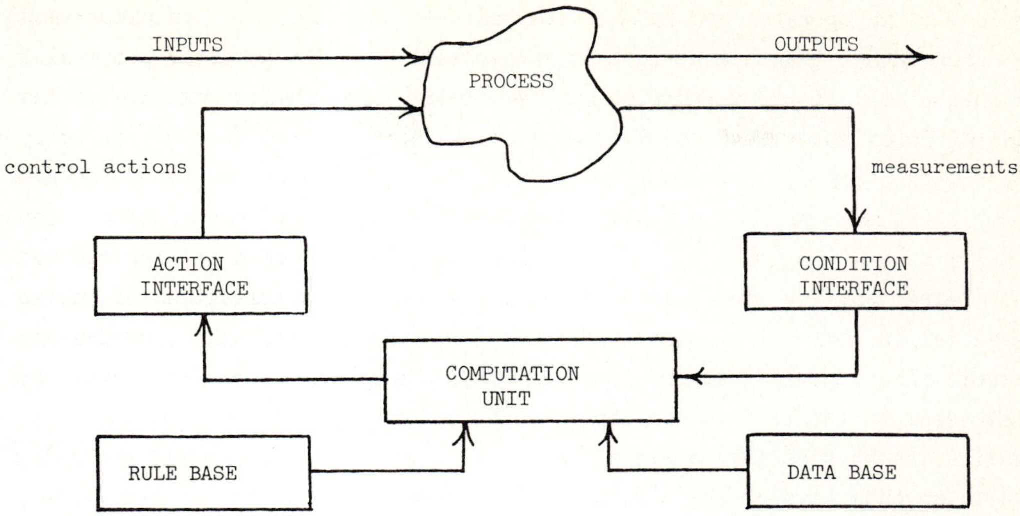

Figure 2. Rule Based Controller

EXPERT SYSTEMS

An Expert System is a knowledge-intensive program that solves problems that normally require human expertise [8]. The underlying premise of Expert Systems technology is that in certain domains there exists problem solving methods that cannot be expressed mathematically, but are nonetheless amenable to computer representation and processing. A common method of representing knowledge in current Expert Systems is the **rule-based** format. These rules have largely been used to represent empirical associations capturing problem solving expertise.

In terms of Expert System technology FLC is essentially a rule-based formalism using fuzzy logic to represent uncertainty, and a composition rule as an inference mechanism. Operator expertise is represented by empirical associations relating observations of the plant to control actions.

Compared to a human expert, today's expert system appears narrow, shallow and brittle. They do not have the same breadth of knowledge or understanding of the fundamental principles upon which their knowledge is based. In this respect their knowledge represents the 'compiled' knowledge gained through extensive experience. In these terms FLC represents the actions of an operator and not his reasoning ability i.e. it describes **what** he does rather than **how** he does it. Consequently, such systems are not able to reason from 'first principles' to derive hypotheses about situations that have not previously been compiled.

QUALITATIVE MODELLING

The ability to reason about behaviour at the qualitative level is essential in performing such tasks as designing, modelling, analysing, trouble-shooting and controlling of physical systems. This has led to a great deal of interest in developing a theory of **qualitative modelling** (e.g. qualitative physics [9], qualitative process theory [9] and causal modelling [10]. The objective of this research is to develop qualitative descriptions of the internal **causal** mechanisms within physical systems which generate the observed behaviour. By analogy with qualitative methods the causal modelling approach is similar to a qualitative state-space approach i.e. is derived from system structure. Whereas the rule-based approach represents an input/output description. In this respect, physical systems have the distinct advantage that the underlying structure and mechanism is known, even although it may be difficult to analyse. In more traditional applications of knowledge-based system technology (e.g. infectious diseases) the detailed causal mechanism is unknown; only empirical associations are known.

In conventional analysis, modelling is in terms of variables which can potentially take any real value. Qualitative modelling also uses the notion of a variable, but describes each variable with a small finite number of distinctions. For example, the concept of the derivative, which plays a crucial role in modelling, is usually described in qualitative analysis by three values, "+", "0", "-", or, in linguistic terms, "increasing", "steady" and "decreasing", respectively. The qualitative values define a set of open regions (intervals) separated by boundaries. For a particular domain, the construction of a set of qualitative values can be viewed as a mapping between continuous functions of time to a set of open intervals.

A qualitative calculus can also be defined using the qualitative values defined above. In particular for two qualitative values A and B, negation, addition and multiplication are defined as shown below:

NEGATION

A	-A
+	-
0	0
-	+

ADDITION

A\B	+	0	-
+	+	+	?
0	+	0	-
-	?	-	-

MULTIPLICATION

A\B	+	0	-
+	+	0	-
0	0	0	0
-	-	0	+

In the above table for addition the symbol (?) means that the result of the sum is **ambiguous**, and the sign of the sum cannot be determined without additional information.

A central influence within this approach is **qualitative simulation** (problem solving): derivation of a description of the behaviour of a mechanism from its qualitative model. A differential equation describes the physical system in terms of a set of state variables and constraints. The solution to the equation may be a function representing the behaviour of the system over time, see Fig. 3. The qualitative model is a further abstraction of the same system, and qualitative simulation is interpreted to yield a corresponding abstraction of its behaviour. The effect of this abstraction is that qualitative simulations predict **multiple** possible behaviours, called **envisioning**. The way in which these possible behaviours are determined represents one of the major differences between the alternative approaches to qualitative modelling. The other difference occurs in the way the model is obtained from the physical system i.e. their ontology.

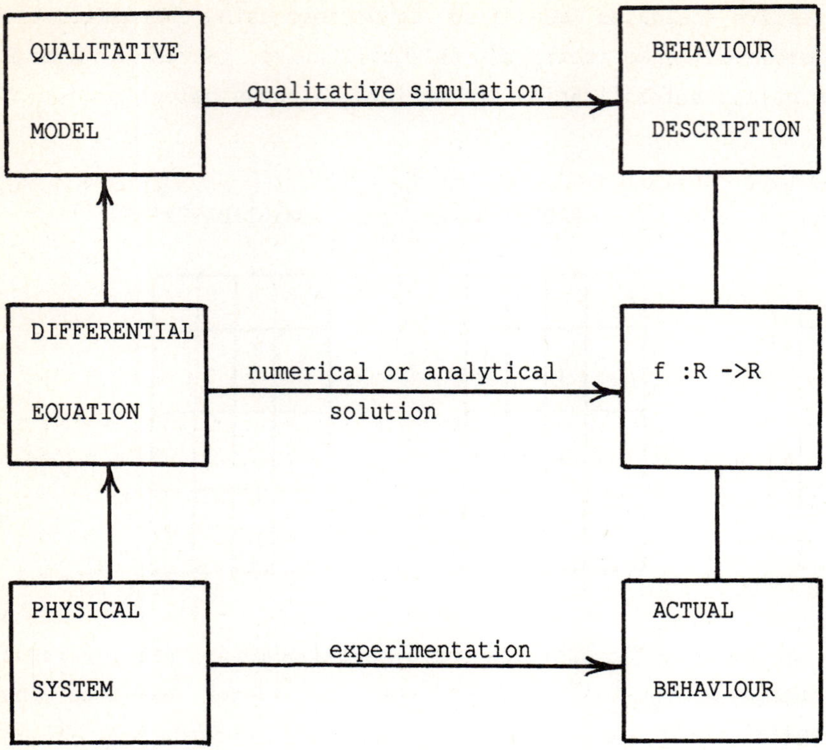

Figure 3. Qualitative models and differential equations are both abstractions of actual behaviour

The approach to qualitative modelling that is most similar to the traditional systems approach to modelling is that proposed by de Kleer[9]. A system (device) consists of physically distinct parts (components) connected together. The goal is to draw inferences about the behaviour of the composite device solely from laws governing the behaviours of its parts. It has two central characteristics: it is a physics, since it can be used to predict the qualitative behaviour of devices and it is a theory of **causality**, since it can be used to produce causal explanations acceptable to humans. The basic modelling primitive is the qualitative differential equation called a **confluence,** which acts as a constraint on the variables and derivatives associated with components. The confluence represents multiple competing influences: a change in one variable produces an immediate change in the others such that the constraint remains satisfied. The same variable can appear in other confluences describing other components. A consistent set of values is obtained by a constraint-propagation algorithm.

In the theory proposed by Kuipers[9], a qualitative description of the behaviour of a system characterised by continuous time variables, is obtained from a qualitative description of its structure. Five types of individual constraints amongst the variables are allowed: arithmetic, functional, derivative, inequality and conditional. The arithmetic constraint asserts that the values of the variables must have the indicated relationship within any time-point. The functional constraint Y = M+(X) asserts that Y is a strictly increasing (or decreasing if M-) function of X. The derivative constraint Y = dX/dt asserts that at any time-point, Y is the rate of change of X. The inequality and conditional constraints specify conditions under which some other constraint holds. Application of these functional constraints results in a "causal structure description" which is a qualitative equivalent of the standard block diagram of quantitative methods. To solve this model a qualitative simulation is done by propagating the qualitative values using constraint propagation. Additional analysis rules are required to determine the behaviour of the situation over time, but no augmentation of the model is required to do so.

ARTIFACT: A REAL TIME SHELL FOR INTELLIGENT FEEDBACK CONTROL

In this section an outline of a control scheme based on a causal model will be given. Further details can be found in [10,11]. The ARTIFACT shell was developed as a research tool for real-time knowledge based process control. It uses domain dependent knowledge to model the process under control and a continuous inference mechanism to deduce an appropriate control action such that a prescribed control objective is satisfied. The PROLOG language has been used for a large part of the shell, with the real-time aspects being handled by the language 'C'.

In conventional control systems, a data value is taken to denote a number corresponding to a measured variable from a sensor or transducer. In knowledge based control a more general notion is provided that includes the explicit input of qualitative data. This permits reasoning with non-measurable qualitative data, such as colour and texture.
Formally, for a system consisting of j-outputs and k-inputs a data tuple, d(n), is defined at observation time nT by

$$d(n) = (y_1(n), \ldots, y_j(n), u_1(n), \ldots, u_k(n))$$

Where the input and output values are either numerical or qualitative items. The **observation matrix** $\psi(n)$ is defined in terms of data tuples by

$$\psi(n) = \begin{matrix} y_1(n), \ldots\ldots, y_j(n), u_1(n), \ldots\ldots, u_k(n) \\ \vdots \quad \vdots \quad \vdots \quad \vdots \\ y_1(n-h+1), \ldots, y_j(n-h+1), u_1(n-h+1), \ldots, u_k(n-h+1) \end{matrix}$$

The number h, called the **time history number** of the system, gives the number of past data tuples that need to be stored in order to infer the state of the system at time nT. Since an entry in the observation matrix can contain qualitative data, the set Ψ of observation matrices need not form a vector space.

As ARTIFACT uses a qualitative causal model, the input information must first be translated into a qualitative description of the plant. This is achieved by defining domain dependent predicates. A two-valued predicate is simply a function p, such that

$$p: \Psi \rightarrow \{true , false\}$$

Every predicate is associated with an attribute; the predicate tests to see whether the current observation matrix has the associated attribute or not. A many-valued predicate generalises this notion by allowing truth (membership) values other than true or false. Typically the range of many-valued predicates is taken to be a number between 0 and 1, and corresponds to the definition of a fuzzy set.

The causal model is expressed in a goal-driven formulation as a collection of sub-system relations. The form of each relation is

$$\text{condition} \Rightarrow (\text{subgoal} , \text{goal})$$

where => denotes logical implication and the subgoal, goal pair contains arbitrary state attributes. The interpretation of this rule is that if a specified condition occurs, or is inferred, at time t and the goal is to be true at time $t + t2$, then the subgoal should occur at time $t + t1$, where $0 < t1 < t2$. The separation of the current condition and the causal relationship increases the heuristic power of the knowledge representation formalism, and results in a considerable reduction in the search space.

At each iteration the ARTIFACT shell evaluates the 'current-condition' part of each rule and discards the rules that evaluate to false or are below a fixed truth-threshold (thus making a closed world assumption). This is achieved by forward chaining from measurement data. The consequents of the remaining rules form a set of applicable causal relations, and a planning routine operates on the resulting sub-space to find a path from the desired goal to an enforceable control action. A structure diagram of the shell, showing the main functional components, is given in Figure 4. The ARTIFACT shell has been applied to a fluid level control problem using the laboratory-scale coupled tanks apparatus shown in Figure 5.

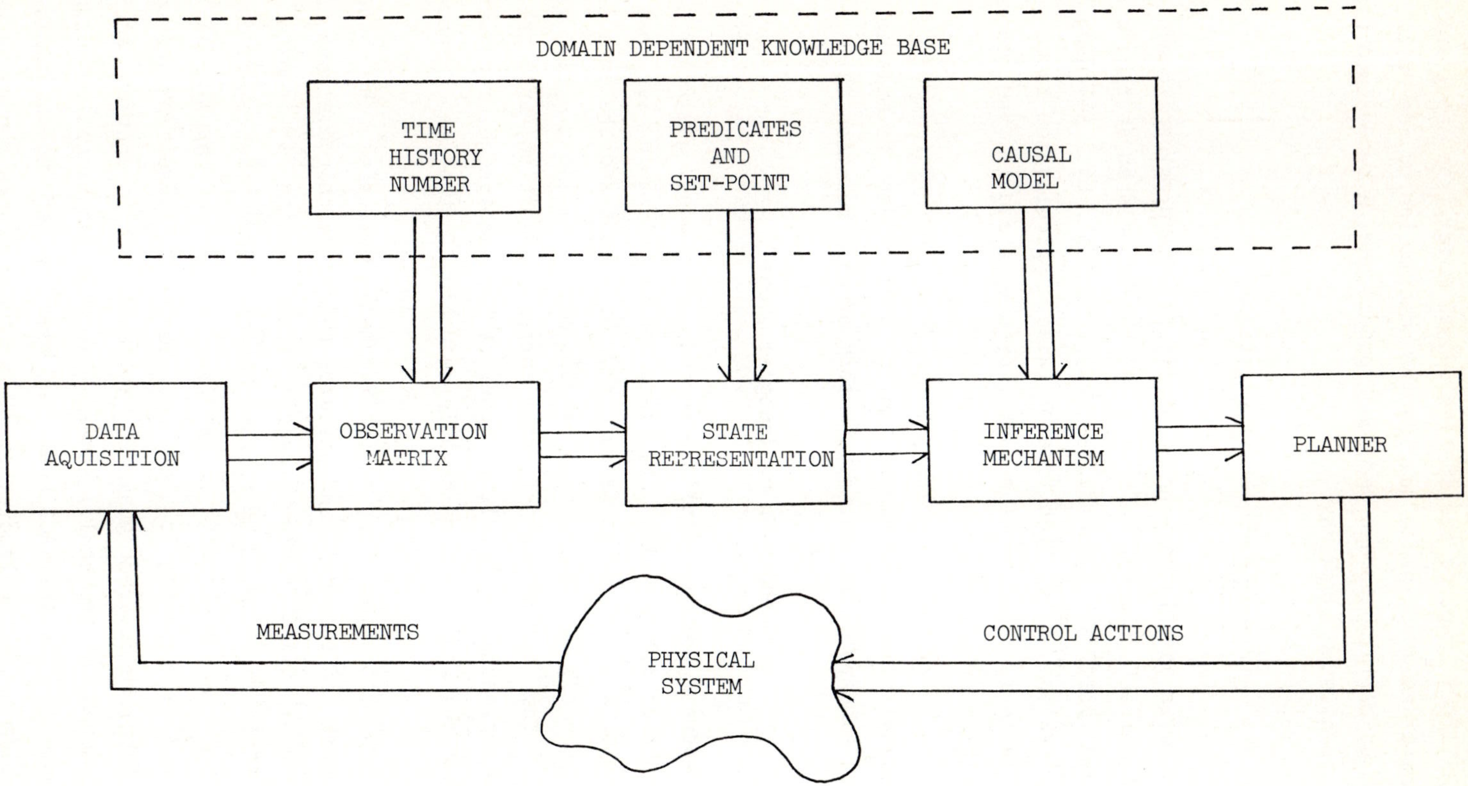

Fig. 4 Structure Diagram of ARTIFACT shell

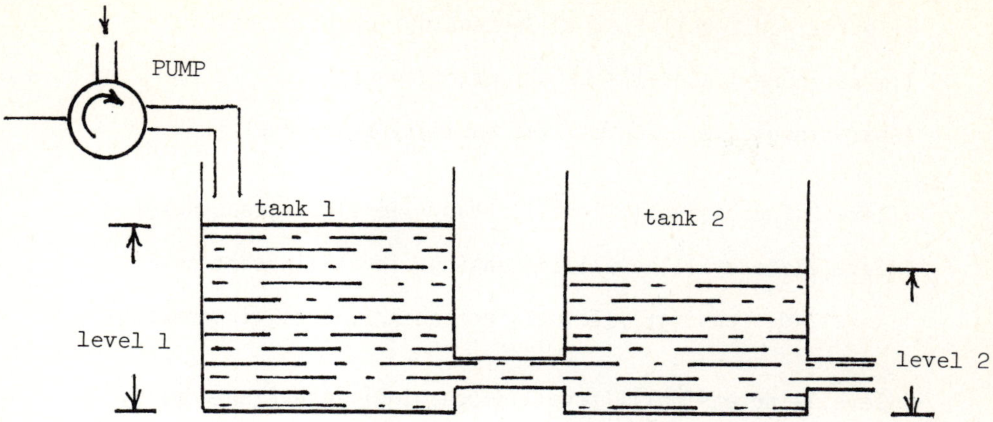

Figure 5. Coupled Tanks Apparatus.

For the case of two-valued predicates, 12 attributes were defined as follows:

level1(increasing) , level1(steady) , level1(decreasing)

level2(increasing) , level2(steady) , level2(decreasing)

error(positive) , error(none) , error(negative)

pump(increase) , pump(steady) , pump(decrease)

Where the error is defined as the difference between level2 and a specified set-point. Table 1 gives a list of 21 relationships forming a causal model of the coupled tanks apparatus.

```
[[[error(positive)],[level2(decreasing),error(none)]],
[[error(none)],[level2(steady),error(none)]],
[[error(negative)],[level2(increasing),error(none)]],

[[level2(increasing)],[level1(steady),level2(increasing)]],
[[level2(steady)],[level1(increasing),level2(increasing)]],
[[level2(decreasing)],[level1(increasing),level2(increasing)]],

[[level2(increasing)],[level1(decreasing),level2(steady)]],
[[level2(steady)],[level1(steady),level2(steady)]],
[[level2(decreasing)],[level1(increasing),level2(steady)]],

[[level2(increasing)],[level1(decreasing),level2(decreasing)]],
[[level2(steady)],[level1(decreasing),level2(decreasing)]],
[[level2(decreasing)],[level1(steady),level2(decreasing)]],

[[level1(increasing)],[pump(steady),level1(increasing)]],
[[level1(steady)],[pump(increase),level1(increasing)]],
[[level1(decreasing)],[pump(increase),level1(increasing)]],

[[level1(increasing)],[pump(decrease),level1(steady)]],
[[level1(steady)],[pump(steady),level1(steady)]],
[[level1(decreasing)],[pump(increase),level1(steady)]],

[[level1(increasing)],[pump(decrease),level1(decreasing)]],
[[level1(steady)],[pump(decrease),level1(decreasing)]],
[[level1(decreasing)],[pump(steady),level1(decreasing)]]]
```

Table 1. Causal Model of Coupled Tanks.

Consider the case of the following attributes being true at a given time:

a) level1(increasing)

b) level2(steady)

c) error(none)

The applicable relations set is obtained as:

1) [level2(steady),error(none)],

2) [level1(increasing),level2(increasing)],

3) [level1(steady),level2(steady)],

4) [level1(decreasing),level2(decreasing)],

5) [pump(steady),level1(increasing)],

6) [pump(decrease),level1(steady)],

7) [pump(decrease),level1(decreasing)].

It is readily seen that relations provide a path from the goal of error(none) to the subgoal of pump(decrease).

The above rules have been successfully used to control the coupled tanks apparatus and the performance obtained was comparable to that of a conventional control scheme using a microprocessor based 3-term controller [11]. However, the knowledge based controller has the important advantage of being readily understood by operating personnel, thereby allowing the addition and deletion of rules as experience grows or the process changes.

CONCLUSION

Within the preceding sections, three approaches to control system design have been discussed: analytic methods, rule-based controllers describing empirical associations and systems based on qualitative causal models. Each method has it's own advantages and disadvantages. In fact, using the terminology of Artificial Intelligence, each approach can be recognised as being based on a different method of representing knowledge about the process to be controlled. The decision as to which representation to use largely depends on the quality of the available information, and on the particular task for which the model will be employed. The main question is, therefore, not **how** to represent knowledge but **what** knowledge to represent.

Psychological studies have indicated that people tend to operate with multiple representations of knowledge [12], switching between representations to solve the particular task at hand, with least effort. This suggests that a necessary requirement for machines to exhibit intelligent behaviour is the use of multiple representations of knowledge and appropriate problem solving strategies. These representations may be organised in a hierarchy based upon the principle of increasing generality or vagueness [13]. In fact, this notion is very close to many definitions of human intelligence , for example, Terman: 'the capacity to form concepts and to grasp their significance'. The hierarchical organisation of knowledge is also an effective way in which to deal with complexity, since for real-time decision making the efficient organisation of knowledge is essential. Another advantage of a hierarchical organisation is that as the level of description increases within a hierarchy, the time constraints, in general, decrease. That is, the measure of real-time is dependent upon the position within the hierarchy, thereby, allowing default reasoning in cases where complete hypotheses, at a given, level cannot be formed within the time constraints. Further, the position within the hierarchy will dictate the appropriate man-machine interface; information can be input and justification of control decisions given in language appropriate to the task.

The recognition of the attributes of the different types, or level, of knowledge, is the main contribution that Artificial Intelligence brings to process control. However, the use of that knowledge in real-time control is still subject to the engineering methodology and constraints that form an essential part of the control engineer's skill. It is my belief that future developments in Knowledge Based Process Control will come from a proper synthesis of these two disciplines.

REFERENCES

1. ASTROM, K.J.,ANTON, J.J. and ARZEN, K.E.:'Expert Control',AUTOMATICA, 1986,vol.22, No.3,pp.277-286.

2. FRANCIS, J.C. and LEITCH R.R.:'Intelligent Knowledge Based Process Control',Proc. CONTROL 85, IEE Int. Conf.,1985,pp483-489.

3. ASHBY, W.R.:'An Introduction to Cybernetics',METHUEN,1965.

4. ASTROM, K.J. and WITTENMARK, B.: 'Computer Controlled Systems: theory and design',Prentice-Hall International,1984

5. NEWELL, A.,SHAW J.C. and SIMON H.A.: 'Elements of a theory of human problem solving', Psychological Review,1958,Vol.65,pp151-166.

6. Zadeh L. 'Making Computers Think Like People'. IEEE Spectrum. (1984), pp 26-32.

7. Mamdani E.H. and Assilian, S. 'An Experiment in Linguistic Synthesis with a Fuzzy Logic Controller', Int. J. Man-Machine Studies 7, (1976), pp 669-678.

8. Hayes-Roth F., Waterman D.A. and Lenat D.B. 'Building Expert Systems', Addison Wesley, (1983).

9. 'Qualitative Reasoning about Physical Systems'. Special Issue, Artificial Intelligence, Vol. 24, No. 1-3, (1984).

10. Francis J.C. and Leitch R.R.,'Intelligent Knowledge-Based Process Control'. Proc. IEE International Conference Control.85, (1985), pp 483-488.1.

11. Francis J.C. and Leitch R.R., 'ARTIFACT: A Real-Time Shell For Intelligent Feedback Control', in Research and Development in Expert Systems, Bramer M.A.(Ed.), Cambridge University Press,(1986).

12. RASMUSSEN, J,:'The role of Hierarchical Knowledge Representation in Decisionmaking and System Management',TRANS.IEEE SMC,1985,Vol.15, pp234-243.

13. SIMON, H.A.:'The Sciences of the Artificial',MIT Press,1981.

Computer Algebra Algorithms for Nonlinear Control

O. Akhrif[†] G. L. Blankenship[†]

Electrical Engineering Department
and
Systems Research Center
University of Maryland
College Park, Maryland 20742

Abstract: This paper describes some of the computer algebra algorithms for nonlinear control systems design embodied in CONDENS, a symbolic manipulation software system, which employs certain differential geometric tools for the analysis and design of control systems. Feedback equivalence among nonlinear systems is used to linearize and thereby control certain classes of nonlinear control systems. Left and right invertibility of nonlinear systems is used to solve the output tracking problem. CONDENS makes these analytical procedures available to users who may not have an extensive knowledge of differential geometry. The system consists of a main part containing functions that perform basic differential geometric computations, two modules for study and analysis of nonlinear control systems, and two packages for the design of nonlinear controllers for the output tracking problem. We shall discuss the implementation of these packages and the functions they include.

[†]This research supported in part by NSF Grant CDR-85-00108 and in part by Grant INT-84-13793 from the NSF International Programs Office. Participation by G.L. Blankenship in the NATO Advanced Study Institute was supported by NATO.

NATO ASI Series, Vol. F47
Advanced Computing Concepts and Techniques
in Control Engineering
Edited by M. J. Denham and A. J. Laub

1 Introduction

In this paper, we shall describe CONDENS, a software system written in the computer algebra system MACSYMA for the analysis and design of nonlinear control systems based on differential geometric methods. Differential geometry provides a very useful framework for the analysis of nonlinear control systems. It permits the generalization of many constructions for linear control systems to the nonlinear case. However, differential geometric objects (Lie brackets, Lie derivatives, ..., etc.) are not easily manipulated by hand. CONDENS employs MACSYMA as a powerful tool for manipulating differential geometric computations. CONDENS uses differential geometric tools expressed as function blocks in MACSYMA for the analysis and design of nonlinear control systems, emphasizing the design of controllers for the output tracking problem. Our objective here is to explain how some of these algorithms are realized as MACSYMA functions.

The programs that form CONDENS fall into three different categories:

1. A set of user-defined functions that perform certain differential geometric computations, including straightforward computations such as *Lie brackets* and *Lie derivatives* and more complex computations such as *Kronecker indices* or *Relative order* of nonlinear systems.

2. CONDENS contains more sophisticated modules that use the basic functions to address two theoretical issues in geometric control theory – *feedback equivalence* and *left-invertibility* – for nonlinear systems affine in control, i.e, systems described (in local coordinates) by:
$$\frac{dx}{dt} = f(x) + \sum_{i=1}^{m} u_i(t) g_i(x) \tag{1}$$

3. Finally, CONDENS contains two design packages. Their purpose is to use either feedback linearization or invertibility of nonlinear systems to design a control law that forces the output of a nonlinear system to follow some desired trajectory.

These design packages include the capability to automatically generate Fortran programs for the solution of problems posed in symbolic form or for simulation purposes. Thus, for a new design problem, the time involved in analyzing the analytical structure of the problem and then writing the Fortran code to execute the design algorithm is eliminated. The time required to write, test, and debug the associated Fortran code is eliminated.

The advantages of CONDENS are clear. It can make available to the design engineer design methods that rely on more or less sophisticated mathematical tools. Most importantly, the system allows the engineer to interact with the computer for design at the symbolic manipulation level. In this way, he can modify his analysis or design problem by modifying the symbolic functional form of the model. The Fortran subroutines that he might have to modify by hand to accomplish this in conventional design procedures are written automatically for him.

There are several computer algebra systems, e.g, FORMAC, MACSYMA, REDUCE, which might have been used in this work. MACSYMA appears to be the system of choice in several centers of control system work. It can implemented on computer systems supporting LISP, it is readily available and, consequently, widespread.

In the next section, we recall briefly the concepts of state and feedback equivalence among control systems and right and left invertibility of nonlinear systems. We also show how we can use these concepts to design an output tracking controller for the nonlinear system (1).

In Section 3, we present a summary of the programming work. A description of the different modules of the program is given together with examples to illustrate the performance of the system.

Acknowledgement: We would like to thank C.I. Byrnes for suggesting this research area and J.P. Quadrat for introducing us to the potential of MACSYMA. This work is part of a collaboration between a group at INRIA headed by J.P. Quadrat. We would like to thank the INRIA team for their contributions to our work.

2 Theory

2.1 Feedback Equivalence among nonlinear systems

Feedback equivalence is an equivalence relation (transitive, symmetric, and reflexive) among control systems which generalizes the concept of the linear feedback group in linear system theory (*Wonham* (1979)) leading, among other things, to the Brunovsky canonical form and the definition of controllability indices. *Krener* (1973), *Brockett* (1978), *Jacubczyck* and *Respondek* (1980), *Hunt et al* (1983) and *Su* (1982) introduced the concept of feedback linearization or feedback equivalence for nonlinear systems and gave necessary and sufficient conditions determining those nonlinear systems which are feedback equivalent to linear controllable ones.

The nonlinear system affine in control:

$$\dot{x} = f(x) + \sum_{i=1}^{m} g_i(x)u_i \qquad x \in R^n \tag{2}$$

is said to be *feedback linearizable* in a neighborhood U of the origin ($f(0) = 0$) if it can be transformed via change of coordinates in the state space, change of coordinates in the control space and feedback to a controllable linear system in Brunovsky canonical form:

$$\dot{z} = Az + Bv \qquad z \in R^n \tag{3}$$

with *Kronecker* indices $k_1, \ldots, k_m$.

Hunt, Su and *Meyer* (1983) gave necessary and sufficient conditions on the Lie structure of system (2) under which a transformation of the type we consider exists. They also gave a procedure for construction of such a transformation. This involves solving a set of partial differential equations, or equivalently n sets of n ordinary differential equations. This is not always possible or easy to do. However, in some cases–the single input case *Brockett* (1978) or systems in block triangular form *Meyer* and *Cicolani* (1980)–the construction of the transformation is simple.

CONDENS contains a module (*TRANSFORM*) that treats the feedback linearization problem. It first investigates the existence of the linearizing one to one transformation by checking the necessary and sufficient conditions. If the transformation is possible, it attempts to construct the transformation (change of coordinates and nonlinear feedback) and its inverse.

Application: Design of output tracking controllers

The design technique is to build a controller for the nonlinear system in terms of one for the equivalent linear canonical system and then inverting the transformation. The design proceeds in three steps. First, the given nonlinear system is transformed into a constant, decoupled, controllable linear representation. Second, standard linear design techniques, such as pole placement, are used to design an output tracking control for the equivalent linear system, forcing its output to follow a desired trajectory. Third, the resulting control law is transformed back out into the original coordinates to obtain the control law in terms of the original controls.

G.Meyer at NASA Ames Research Center proposed this scheme in the design of exact model following automatic pilots for vertical and short take-off aircraft and has applied it to several aircraft of increasing complexity (*Meyer* 1981).

This design procedure is performed by FENOLS, one of the design packages contained in CONDENS (see Figure 1). A description of FENOLS and examples of its functions are given in section 3.

2.2 Invertibility of nonlinear systems

In this subsection, we briefly review some of the results on left and right invertibility for nonlinear systems that are also implemented in CONDENS.

We consider systems of the following form:

$$\begin{aligned} \dot{x}(t) &= f(x(t)) + \sum_{i=1}^{m} g_i(x(t))u_i(t) \qquad x(0) = x_0 \in M \\ y(t) &= h(x(t)) \qquad u \in R^m, y \in R^m \end{aligned} \tag{4}$$

where the state space M is a connected real analytic manifold, f, g are analytic vector fields on M, $h_i, i = 1, \ldots, m$ is a real analytic mapping, and $u \in U$, the class of real analytic functions from $[0, \infty)$ into R^m.

The left-invertibility problem is the problem of injectivity of the input-output map of system (4). If the input-output map is invertible from the left, then it is possible to reconstruct uniquely the input acting on the system from knowledge of the corresponding output. The main application of this is in the output tracking problem where we try to control a system so

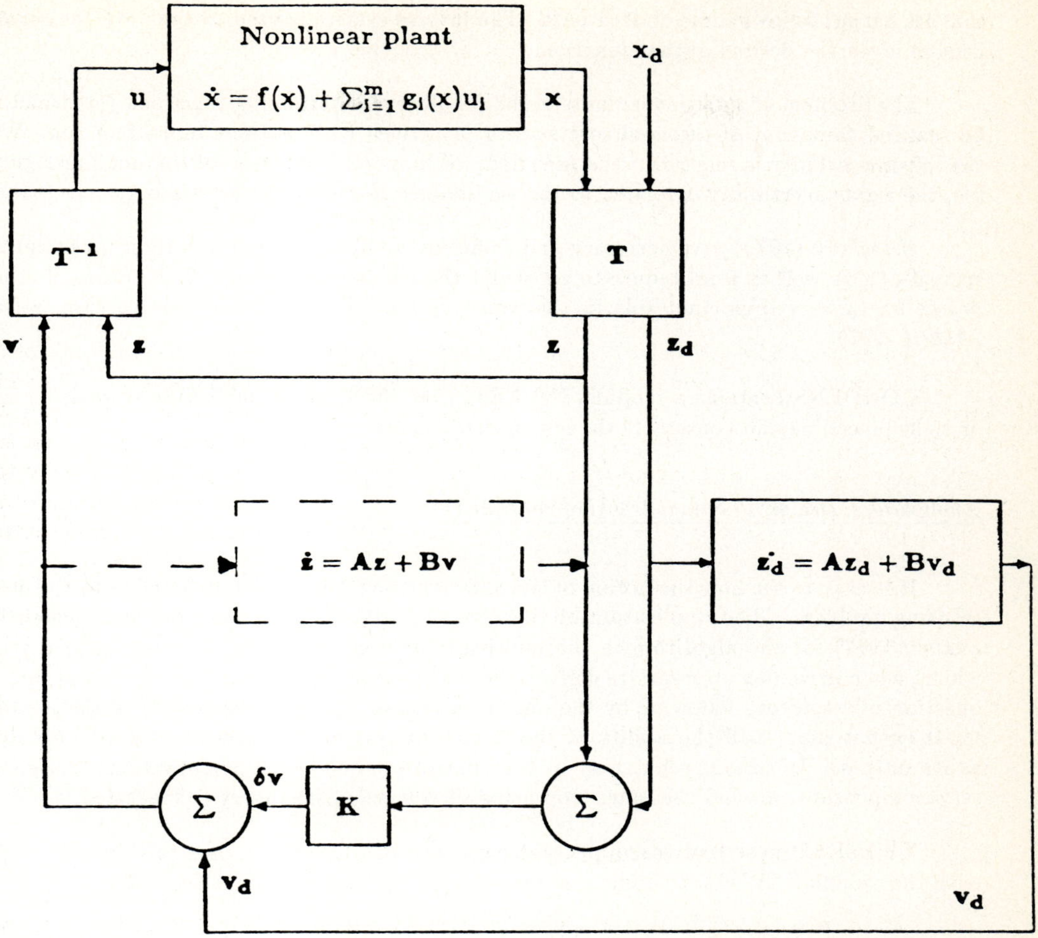

Figure 1: Design scheme used by FENOLS

that its output follows some desired path. The inverse system is used to generate the required control given the desired output function.

The problem of right-invertibility is the problem of determining functions $f(t)$ which can be realized as output of the nonlinear system (4) driven by a suitable input function. While the left-invertibility is related to the *injectivity* of Input/Output map of the nonlinear system (4), the right-invertibility is related to the *surjectivity* of the Input/Output map.

Hirschorn (1979) gave necessary and sufficient conditions of left and right invertibility of system (4), as well as a procedure to construct the left-inverse system. The notion of *relative degree* has a very important role in this construction. For the multivariable case, refer to (*Akhrif* 1987).

CONDENS contains a module, INVERT, that checks if the nonlinear system (4) is left or right invertible and constructs the left inverse system.

Application: Design of output tracking controllers

It is easy to see how the notion of left and right invertibility can be applied to the output tracking problem. The application of the inversion algorithm to the nonlinear system (see (*Akhrif* 1987) for the algorithm in the multivariable case) gives rise to a left-inverse system which, when driven by appropriate derivatives of the output y, produces $u(\cdot)$ as its output. The question of trajectory following by the output is related to right-invertibility of the nonlinear input-output map, and the ability of the nonlinear system to reproduce the reference path as its output. To obtain robustness in the control system under perturbations, design of a servocompensator around the inner loop using servomechanism theory is suggested.

TRACKS, the second design package contained in CONDENS, uses this procedure along with the module INVERT to design an output tracking controller (see Figure 2).

3 CONDENS: A software package using symbolic manipulations

In this section, CONDENS, a software package based on computer algebra is presented. Employing the analytical methods presented above, CONDENS can, given a nonlinear system affine in control, answer questions such as: Is this system feedback-equivalent to a controllable linear one? If so, can we construct the diffeomorphism that makes this equivalence explicit? Is the nonlinear system invertible? What is the relative order of the system? In case the system is invertible, can we construct a left-inverse for it? Given a real analytic function $y(t)$, can $y(t)$ appear as output of the nonlinear system? If so, what is the required control?

CONDENS tries to answer these questions and uses this knowledge to solve a design problem: The output tracking problem for the given nonlinear control system. If we are interested in simulation results, the system can automatically generate Fortran programs for that purpose.

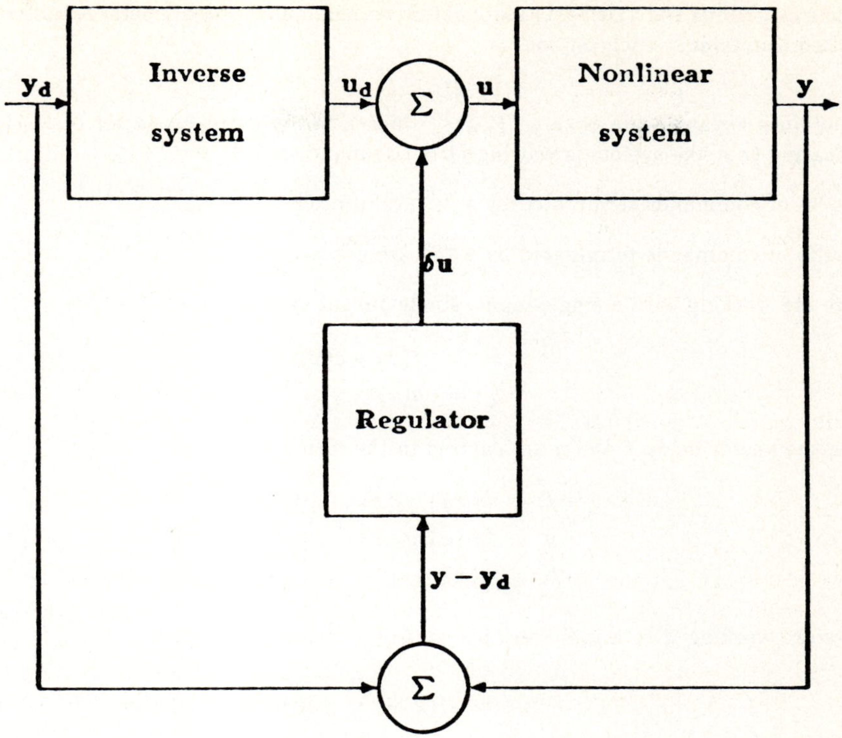

Figure 2: Design scheme using TRACKS

Before describing CONDENS in more detail, we shall make some general remarks relevant to all worked examples in each section.

1. Input lines always begin with a "(C_n)," which is the prompt character of MACSYMA, indicating that the system is waiting for a command.
2. Results of commands terminated by a ";" are printed.
3. Results of commands terminated by a "$" are not printed.
4. If we are working with a single-input, single-output system:

$$\begin{aligned} \dot{x} &= f(x) + g(x)u \\ y &= h(x) \end{aligned}$$

then the vector fields f and g are entered in the form of lists:

$$\begin{aligned} f &: [f_1(x), f_2(x), \ldots, f_n(x)]; \\ g &: [g_1(x), g_2(x), \ldots, g_n(x)]; \\ h &: h(x); \end{aligned}$$

5. If we are working with a multivariable control system:

$$\begin{aligned} \dot{x} &= f(x) + \sum_{i=1}^{m} g_i(x)u_i \\ y &= h(x) \end{aligned}$$

$x \in R^n$, $u \in R^m$, $y \in R^m$ then the vecter field f is entered in the form:

$$f : [f_1(x), \ldots, f_n(x)];$$

but the m vector fields $g_1, g_2, \ldots, g_m$ are entered in the compact form:

$$g : [g_1, \ldots, g_m]; \quad \text{where each } g_i \text{ is} \quad g_i : [g_{i1}, \ldots, g_{in}];$$

3.1 Package facilities

CONDENS is formed by a set of large functions, which, in turn are built up of smaller subfunctions. In order to save time, we would like, before using a certain function, to be able to automatically load into MACSYMA only the subfunctions that this specific function uses. This way, we will not have to load the entire package into MACSYMA each time we want to use it.

For this, the **`setup-autoload`** MACSYMA command is used. All the functions in CONDENS are collected in a special file (`''initfile.mac''`) with file addresses for all the functions. Once this file is loaded into MACSYMA, a CONDENS function, not previously defined, will be

```
/*-*-*-*-*-*-*-*-*-*-*-*-*-*-*-*-*-*-*-*-*-*-*-*-*-*-*-*-*-*-*-*-*-*
This is an initialization file for the package

       CONDENS: CONTROL DESIGN OF NONLINEAR SYSTEMS.

This is the only file that has to be loaded in MACSYMA by the user,
all the other files will be automatically loaded as required.
To get started, the user has to type: 'load(''initfile.mac'')',
followed by the command 'condens();' which will give him some hints
on how to use the package.
-*-*-*-*-*-*-*-*-*-*-*-*-*-*-*-*-*-*-*-*-*-*-*-*-*-*-*-*-*-*-*-*-*-*

loadprint:false$
setup_autoload("CONDENS/menu.mac",menu)$
setup_autoload("CONDENS/geofun1.mac",jacob,pjacob,lie,plie,adj)$
setup_autoload("CONDENS/geofun2.mac",lidev,plidev,nlidev,relord)$
setup_autoload("CONDENS/transform.mac",transform)$
setup_autoload("CONDENS/fenols.mac",fenols)$
setup_autoload("CONDENS/tracks.mac",tracks)$
setup_autoload("CONDENS/invert.mac",invert)$
setup_autoload("CONDENS/btriang.mac",btriang)$
setup_autoload("CONDENS/simu.mac",ff)$
setup_autoload("CONDENS/condens.mac",condens)$
print_true:false$
help(arg):=arg(help)$
```

Figure 3: Text of the initialization file ''initfile.mac''.

automatically loaded into MACSYMA when it is called. After entering MACSYMA, the user begins a session with CONDENS by typing `'load ''initfile.mac'';'` to load the login file. He can then start the package using the command `''condens();''` which will give him some hints on how to use CONDENS. The file `''initfile.mac''` is listed in Figure 3.

CONDENS also contains some help facilities. The standard help functions in MACSYMA are `DISPLAY(''`*fct-name*`'')` and `APROPOS(''`*fct-name*`'')`. We added two *help* functions: `MENU` and `HELP`. `MENU` will display the list of all the functions the package contains. `HELP(''`*fct-name*`'')` returns an information text describing the function, its syntax, how to enter its arguments, and an example. In Figure 4 we show how to start a session with CONDENS.

Many of the functions not explicitly described in this section are available in MACSYMA. We recommend that the reader consult the *MACSYMA Reference Manual* for details.

3.2 Differential geometric tools in CONDENS:

In this subsection, we describe the basic functions available in CONDENS. Since we are using concepts from differential geometry, we have to manipulate vector fields and differential forms. Manipulations with vector fields and forms are systematic and well suited for symbolic computations. The functions described in this section perform some straightforward differential geometric computations such as *Lie brackets* and *Lie derivatives*, and more complex computations such as *Kronecker indices* or *relative degree* of nonlinear systems.

JACOB(f): computes the *Jacobian* of f. That is, JACOB(f) returns the matrix:

$$\begin{pmatrix} \frac{\partial f_1}{\partial x_1} & \cdots & \frac{\partial f_1}{\partial x_n} \\ \vdots & \cdots & \vdots \\ \frac{\partial f_n}{\partial x_1} & \cdots & \frac{\partial f_n}{\partial x_n} \end{pmatrix}$$

Example: For the following nonlinear system:

$$\begin{aligned} \begin{pmatrix} \dot{x}_1 \\ \dot{x}_2 \end{pmatrix} &= \begin{pmatrix} x_1^2 \\ x_1 x_2 \end{pmatrix} + \begin{pmatrix} 2x_1 \\ 1 \end{pmatrix} \\ y &= x_1^2 + x_2^2 \end{aligned} \tag{5}$$

```
(c1) f:[x1**2,x1*x2];
                                   2
(d1)                           [x1 , x1 x2]
(c2) jacob(f);
                          df    [ 2 x1   0  ]
(d2)                      -- =  [           ]
                          dx    [  x2   x1  ]
```

The code for this function, which illustrates some of the basic features of MACSYMA, is shown in Figure 5.

```
(c1) load("initfile.mac");
Batching the file initfile.mac
(d1)                              initfile.mac
(c2) condens();
**********************************************************************
*                                                                    *
*  Hello ! WELCOME to CONDENS :  CONtrol DEsign of Nonlinear Systems *
*                                                                    *
*  a MACSYMA package for the design of controllers for the output    *
*  tracking problem using differential geometric concepts.           *
*                                                                    *
*      by:               OUASSIMA AKHRIF                             *
*                                                                    *
*      under the supervision of  Professor Gilmer Blankenship        *
*                                                                    *
**********************************************************************

        f,g and h, whenever mentioned, always designate the nonlinear
        dynamics of the nonlinear system affine in control:
                     dx/dt = f(x) + g(x) u
                        y  = h(x)
        Type  MENU(), to get a list of all functions available in CONDENS.
(d2)                               done
(c3) menu();

HELP(fun-name): generates an information text describing the function.
JACOB(f): computes the Jacobian matrix of f.
LIE(f,g): computes the Lie brackets of the vector fields f and g.
ADJ(f,g,k): computes the k-th adjoint of f and g.
LIDEV(f,h): computes the Lie derivative of the real valued function h
            along the direction defined by the vector field f.
NLIDEV(f,h,k): computes the k-th Lie derivative of h along f.
BTRIANG(f,g): checks if the nonlinear system is in block triangular form
RELORD(f,g,h): computes the relative order of the scalar control system
               dx/dt=f(x)+g(x)u, y=h(x).
TRANSFORM(f,g): treats the feedback linearization problem,that is checks
                if the system is linearizable and solves for the nonlinear
                transformation.
FENOLS(): module that treats the output tracking problem using feedback
          linearization.
INVERT(f,g,h): finds the left inverse to the original nonlinear system.
TRACKS(): module that treats the output tracking problem using the left-
          inverse system.

(d3)                               done

(c4) help(relord);

 relord(f,g,h) : computes the relative order of the scalar nonlinear
                 system:   dx/dt = f(x) + g(x) u
                                y   = h(x)
                 f is entered in the form   f:[f1(x),....,fn(x)],
                 example:(if n=2)         f:[x1,x2**2],
                 g is entered in the form   g:[g1(x),....,gn(x)],
                 example:(if n=2)         g:[x2,x1*x2],
                 h is entered in the form   h:h(x),
                 example:                 h:x1+x2,

(d4)                               done
```

Figure 4: How to start a session in CONDENS.

```
jacob(arg1):=block([p,temp1,temp2,temp3],
          if arg1=help then return((
print(" jacob(f) : computes the Jacobian matrix of f.  "),
print("            f is entered in the form   f:[f1(x),...,fn(x)],      "),
print("            example:(if n=2)  f:[x1**2,x2],       "),done)),
          p:length(arg1),temp1:[],
          for i1:1 thru p do
          (temp2:makelist(diff(arg1[i1],concat('x,j1)),j1,1,p),
          temp1:endcons(temp2,temp1)),
          temp3:apply(matrix,temp1),
          'df/dx= temp3)$
```

Figure 5: MACSYMA code for computation of the Jacobian.

LIE(f,g): computes the *Lie brackets* of the vector fields f and g:

$$[f,g] = \frac{\partial g}{\partial x}f - \frac{\partial f}{\partial x}g$$

ADJ(f,g,k): computes the kth adjoint of f and g:

$$\begin{aligned} ad_f^k g &= [f, ad_f^{k-1} g] \\ ad_f^0 g &= g \end{aligned}$$

Example: For system (5):

```
(c3) g:[2*x1,1];
(d3)                                    [2 x1, 1]
(c4) lie(f,g);
                                                 2
(d4)                          [f, g] = [- 2 x1 , - 2 x1 x2 - x1]
(c5) adj(f,g,2);
                                      2
(d5)            ad (f, g) = [0, x1  (- 2 x2 - 1) - x1 (- 2 x1 x2 - x1)]
                  2
```

LIDEV(f,h): computes the *Lie derivative* of the real valued function h along the direction defined by the vector field f: $\mathbf{L_f h}$.

NLIDEV(f,h,k): computes the kth *Lie derivative* of h along f:

$$\begin{aligned} L_f^k h &= L_f\left(L_f^{k-1} h\right) \\ L_f^0 h &= h \end{aligned}$$

Example: For system (5):

```
(c6) lidev(f,h);
```

```
                                       2       3
(d6)                      lf(h) = 2 x1 x2  + 2 x1
(c7) nlidev(f,h,3);
           2            2       2               2       3          3   2
(d7) lf (h) = x1  (2 x1 (2 x2  + 6 x1 ) + 8 x1 x2  + 12 x1 ) + 12 x1  x2
       3
(c8) ratsimp(%);
                                     3   2          5
(d8)                    lf (h) = 24 x1  x2  + 24 x1
                          3
```

Next, we have three more complicated functions. They use the "sub-functions" LIE, ADJ,...etc to compute quantities such as *Kronecker* indices or *relative degree* of nonlinear systems.

KRONECK(f,g): used for multivariable systems,where g represents the set of vector fields $g_1, \ldots, g_m$

This function is useful when the nonlinear system is transformable to a linear controllable system in Brunovsky canonical form. It computes the *Kronecker indices* of the linear system equivalent to the original nonlinear system. It returns a set of numbers:

$$k_1 \geq k_2 \geq \cdots \geq k_m$$

$$k_1 + k_2 + \cdots + k_m = n$$

BTRIANG(f,g): This function checks if the nonlinear system $\dot{x} = f(x) + \sum_{i=1}^{m} g_i(x)u_i$ is in *block triangular form* (see (*Meyer* 1981) for a definition). The argument g of the function represents the m vector fields $g_1, \ldots, g_m$. This function is perhaps the most "complicated" program included in CONDENS. The algorithm is sketched in Figure 6; the code is in Figure 7.

RELORD(f,g,h): computes the *relative order* (see [3] for definition) of the single-input, single-output nonlinear system:

$$\begin{aligned} \dot{x} &= f(x) + g(x)u \\ y &= h(x) \end{aligned}$$

<u>Example:</u> For system (5):

```
(c7) h:x1**2+x2**2;
                                    2    2
(d7)                              x2  + x1
(c8) relord(f,g,h);
relative order of system = 1

(d8)                              done
```

3.3 Analysis tools in CONDENS

Using the above functions, CONDENS contains two independent modules for analysis of nonlinear control systems: TRANSFORM and INVERT. They treat the two problems: feedback linearization and invertibility of nonlinear control systems, respectively.

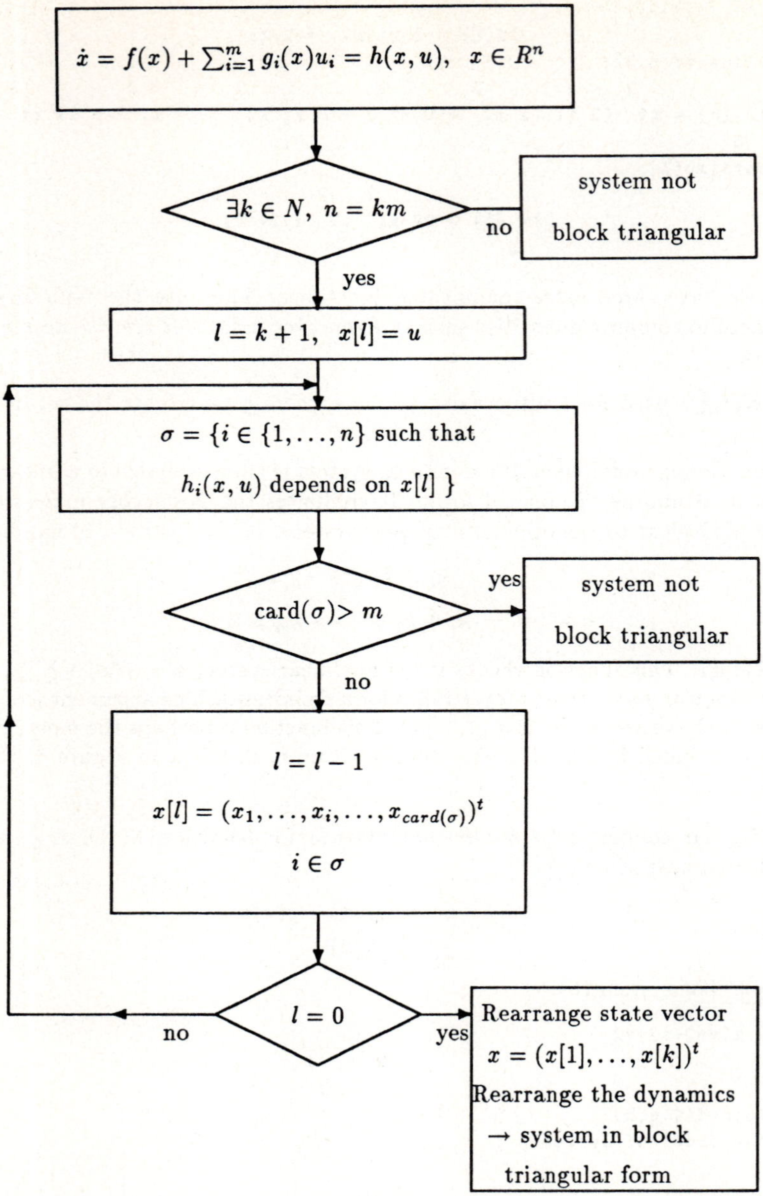

Figure 6: Flow diagram for `btriang(f,g)`.

```
/*****************************************************************************/
/*     This function checks if the system is in block triangular form       */
/*****************************************************************************/

btriang(arg1,[arg2]):=block([x,xl,u,ind],
                        if arg1=help then return((
print("Btriang(f,g) :   checks if the multivarible system dx/dt=f(x)+g(x)u is "),
print("                 in block triangular form.  "),
print("                 f and g are entered in the form f:[f1(x),...,fn(x)],   "),
print("                                                 g:[[g1(x)],..[gm(x)]], "),
print("                 Example:(if n=2,m=2)            f:[x1,x2**2],          "),
print("                                                 g:[[x1,1],[x2,x1]],   ")
,done)),
                        f:arg1,g:part(arg2,1),n:length(f),m:length(g),
                        x:makelist(concat('x,i),i,1,n),
                        ind:makelist(concat('ind,i),i,1,n),
                        ind[1]:0,d:1,p1:1,it:n,
                        u:makelist(concat('u,i),i,1,m),
                        xl:makelist([],i,1,n),xl[p1]:u,
                        for i:1 thru n do
                        equ[i]:f[i]+sum(g[j][i]*u[j],j,1,m),
                        for p:1 thru n/m do(l:0,los:0,
                            for i:n thru 1 step -1 do (
                        for id :1 thru d do
                        if i=ind[id] then (if i>1 then i:i-1
                        else (los:1,id:d)),
                        if los #1 then for r:1 thru length(xl[p]) do
                              if equ[i]#subst(0,xl[p][r],equ[i]) then (
                              l:l+1,cx[it]:x[i],it:it-1,
                              xl[p+1]:endcons(x[i],xl[p+1]),
                              ind[d]:i,if d=n then (i:1,p:n/m),d:d+1,
                              r:length(xl[p]))),
                        if l#m then
                        (print("system is not in block triangular form"),
                        resul:0,p:n/m)
                        else resul:1 ),if resul=1 then
                        print("system in block triangular form"),done)$
```

Figure 7: MACSYMA code for checking block triangular form.

TRANSFORM(f,g): TRANSFORM treats the feedback linearization problem presented in Section 2.1.

Given the nonlinear control system:

$$\dot{x} = f(x) + \sum_{i=1}^{m} g_i(x) u_i \tag{6}$$

TRANSFORM proceeds in the following manner:

- Using LIE, ADJ and KRONECK, it computes the *Kronecker* indices and checks the three necessary and sufficient conditions (see (*Hunt, Su, Meyer* 1983)) for feedback linearizability.
- If the the system is feedback linearizable, TRANSFORM investigates properties of the nonlinear system which simplify computation of the transformation. For example, TRANSFORM checks to whether the system is in block triangular form (see (*Meyer* 1981)) (using the function BTRIANG), or if it is scalar and satisfies the more restrictive conditions presented in (*Brockett* 1978).
- In the more general case, TRANSFORM proceeds to solve for the transformation (change of coordinates and feedback), say T and its inverse T^{-1} by using the MACSYMA function "`desolve`". This function tries to solve the set of n ordinary differential equations defining the transformation (see (*Hunt, Su, Meyer* 1983).[1]

A flow chart setting out the implementation of `transform(f,g)` is shown in Figure 8. Refering to the figure, there are five major operations that take place in `transform`.

Block 1: The function `btriang(f,g)` is called; if successful, it rearranges the system into the ordered form

$$\begin{aligned} \dot{x}_1 &= f_1(x_1, x_2) \\ \dot{x}_2 &= f_2(x_1, x_2, x_3) \\ &\vdots \\ \dot{x}_k &= f_k(x_1, \ldots, x_k) + \sum_{i=1}^{m} g_{ki}(x) u_i \end{aligned} \tag{7}$$

where $k = n/m$, and $x_1, \ldots, x_k$ are m-dimensional sub-vectors.

Block 2: In case the system is in block triangular form, the function `feed(f,g)` is called to construct the transformation T and the feedback pair α, β. `feed(f,g)` constructs the following transformation:

$$z_1 = x_1$$

[1] If the MACSYMA function "`desolve`" fails, then TRANSFORM fails. It would be possible to augment this function to solve the ODE's numerically; however, we have not implemented these enhancements.

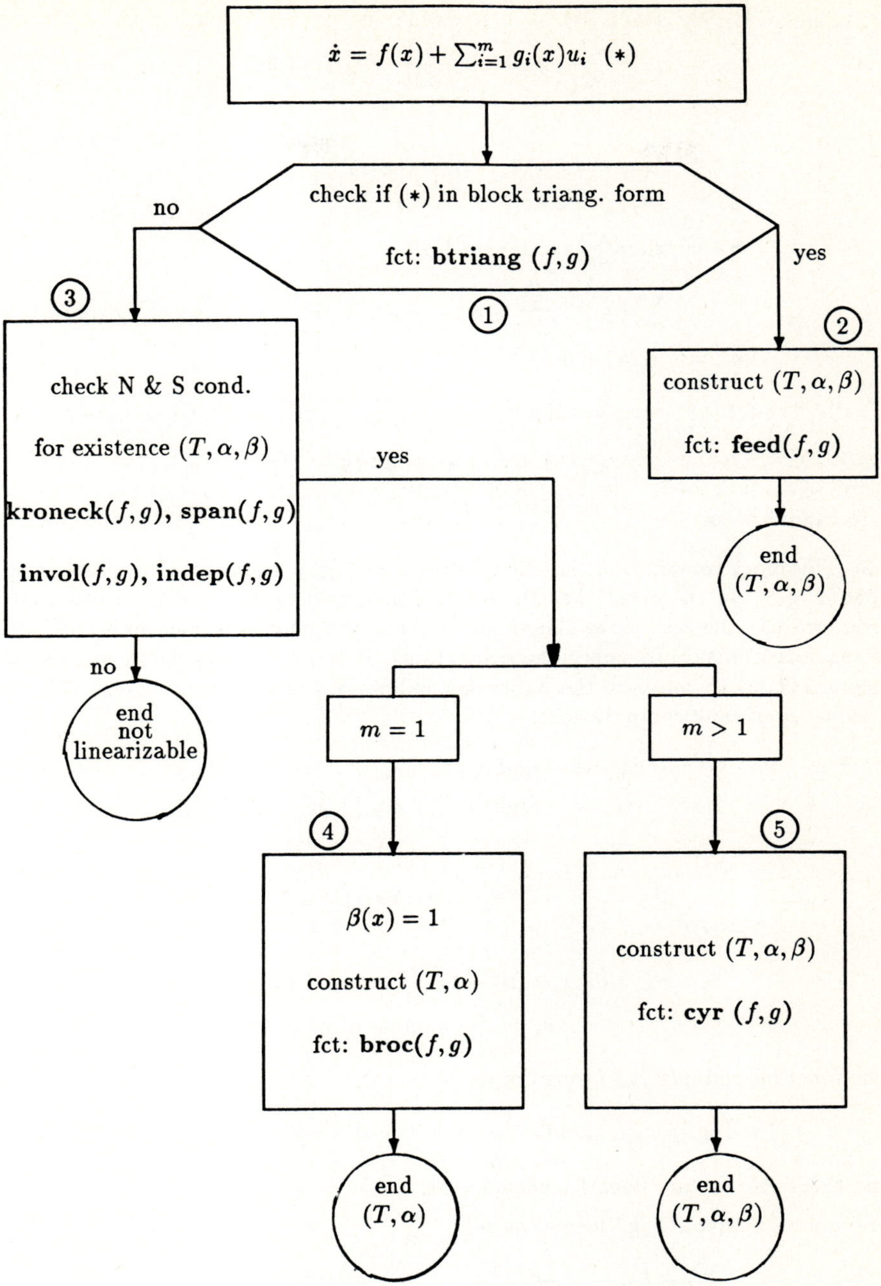

Figure 8: Flow chart of the function `transform(f,g)`.

$$\begin{aligned}
z_2 &= f_1(x_1, x_2) \\
z_3 &= \dot{z}_2 = \frac{\partial f_1}{\partial x_1} \cdot f_1(x_1, x_2) \frac{\partial f_1}{\partial x_2} \cdot f_2(x_1, x_2, x_3) \\
&\vdots \\
z_{k-1} &= \dot{z}_{k-2} = F_{k-1}(x_1, x_2, \ldots, x_{k-1}) \\
z_k &= \dot{z}_{k-1} = F_k(x_1, x_2, \ldots, x_k) \\
v &= \dot{z}_k = \sum i = 1^k \frac{\partial F_k}{\partial x_i} \dot{x}_i \\
&= \sum i = 1^k \frac{\partial F_k}{\partial x_i} \cdot f_i(x_1, \ldots, x_{i+1}) + \frac{\partial F_k}{\partial x_k} \cdot \sum i = 1^m g_{ki}(x) u_i \\
v &= \alpha_1(x) + \beta_1(x) u
\end{aligned}$$

Finally, `feed(f,g)` arranges the vector $z = [z_1, \ldots, z_k]^T$ so that the system

$$\dot{z} = Az + Bv$$

is in canonical form.

Block 3: This block contains four functions: `kroneck(f,g)`, `indep(f,g)`, `invol(f,g)`, and `span(f.g,)`. In the event that the original nonlinear system is not in block triangular form, we use the last three functions to check the necessary and sufficient conditions of tranformability. To apply these functions, it is first necessary to use the function `kroneck(f,g)` to compute the Kronecker indices of the nonlinear system. This involves a sequence of rank computations:

$$\begin{aligned}
\alpha_0 &= \text{rank}\{g_1, \ldots, g_m\} \\
\alpha_1 &= \text{rank}\{g_1, \ldots, g_m, [f, g_1], \ldots, [f, g_m]\} \\
&\vdots \\
\alpha_{n-1} &= \text{rank}\{g_1, \ldots, g_m, [f, g_1], \ldots, ad_f^{n-1} g\}
\end{aligned}$$

$$r_0 = \alpha_0, r_1 = \alpha_1 - \alpha_0, \ldots, r_{n-1} = \alpha_{n-1} - \alpha_{n-2}$$

$$k_i = \text{ the number of } r_j \geq i$$

The function `indep(f,g)` forms the set of vectors

$$C = \{g_1, [f, g_1], \ldots, ad_f^{k_1-1} g_1, g_2, \ldots, ad_f^{k_2-1} g_2, \ldots, g_m, \ldots, ad_f^{k_m-1} g_m\}$$

and checks if they are linearly independent.
The function `invol(f,g)` forms the sets

$$C_j = \{g_1, \ldots, ad_f^{k_j-2} g_1, \ldots, g_m, ad_f^{k_j-2} g_m\} \quad j = 1, \ldots, m$$

and checks to see if each is involutive. This is done by checking the linear dependence of

$$C_j \cup \{[X, Y]\} \quad \forall X, Y \in C_j$$

```
invol(f,g,w):=block([dep1,dep2,dep3,dep4,l,m1,r1,r2],dep3:[],
                for d:1 thru m do
                (dep1:[],for i5:1 thru m do(
                dep2:g[i5],dep1:endcons(dep2,dep1),
                if k[d]>w then for j5:1 thru k[d]-w do
                (dep2:plie(f,dep2),dep1:endcons(dep2,dep1))),
                l:apply(matrix,dep1),if (m*(k[d]-1)+1)>n
                then dep4:[concat(c,d),"involutive"]
                else for j6:1 thru m*(k[d]-1)-1 do
                for j7:j6+1 thru m*(k[d]-1) do
                (r1:makelist(l[j6,t],t,1,n),
                 r2:makelist(l[j7,t],t,1,n),
                m1:addrow(l,plie(r1,r2)),
                if rank(m1)<(m*(k[d]-1)+1)
                then dep4:[concat(c,d),"is involutive"]
                else(dep4:[concat(c,d),"not invol"],resul:1)),
                dep3:endcons(dep4,dep3)),dep3)$
```

Figure 9: MACSYMA code to test the involutivity of a set of vector fields.

The function `span(f,g)` forms the sets $C, C_j, j = 1, \ldots, m$ as above and checks that

$$\text{span } (C_j \cap C) = \text{span } C_j \quad \forall j = 1, \ldots, m$$

Since $(C_j \cap C) \subset C$, we only have to check

$$\text{rank } (C_j \cap C) = \text{rank } C \quad \forall j = 1, \ldots, m$$

In each of the functions **kroneck, indep, invol, span** we essentially check either the linear dependence or the linear independence of a set of vectors. The associated code makes heavy use of the MACSYMA function RANK(A) which computes the rank of a matrix A. Since the vectors we are dealing with are actually vector fields, the ranks we compute in this way will not be constant over the manifold M, the state space of the original nonlinear system. However, by smoothness, they will be constant over a dense submanifold $M' \subset M$. Thus, M' is the effective state space of the system.

The code for `invol` is shown in Figure 9. The other functions are similar. Note the use of the internal MACSYMA function RANK.

Block 4: This block constructs the feedback linearization in the single-input case using a function `broc(f,g)`[2] This function works as follows: Since the transformation $T : R^n \to R^n$ satisfies

$$\begin{aligned} < dT_1, g > &= 0 \\ < dT_1, [f, g] > &= 0 \\ &\vdots \end{aligned}$$

[2]Named for Roger Brockett.

$$\begin{aligned} < dT_1, ad_f^{n-2} g > &= 0 \\ < dT_1, ad_f^{n-1} g > &= \beta \neq 0 \end{aligned}$$

$$v =< dT_n, f(x) + g(x)u >= \alpha(x) + \beta(x)u$$

if we restrict attention to the case $\beta(x) \equiv 1$, then the PDE's can be rewritten as

$$\begin{aligned} \frac{\partial T_1}{\partial x_1} g_1(x) + \dots \frac{\partial T_1}{\partial x_n} g_n(x) &= 0 \\ &\vdots \\ \frac{\partial T_1}{\partial x_1} ad_f^{n-2} g_1(x) + \dots \frac{\partial T_1}{\partial x_n} ad_f^{n-2} g_n(x) &= 0 \\ \frac{\partial T_1}{\partial x_1} ad_f^{n-1} g_1(x) + \dots \frac{\partial T_1}{\partial x_n} ad_f^{n-1} g_n(x) &= 1 \end{aligned}$$

and so, we have the vector equation

$$\left(\frac{\partial T_1}{\partial x}\right)^T = \left[g, ad_f g, \dots, ad_f^{n-1} g\right]^{-1} \cdot \begin{pmatrix} 0 \\ 0 \\ \vdots 0 \\ 1 \end{pmatrix}$$

This expression is "integrated" to find T_1, and the remaining elements $T_2, \dots, T_n$ are found from this. The code for the function **broc** is given in Figure 10. Note the use of the MACSYMA functions INTEGRATE and SOLVE. If these internal functions fail in a specific case, then the design cannot be completed.

Block 5: This block, called `cyr(f,g)` in our code, constructs the feedback linearizing transform (T, α, β) in the general case by implementing the algorithm of Hunt, Su, and Meyer converting the set of PDE's for the transformation into a n-sets of ODE's. This is a complex block, and we shall omit its construction here. (See [14] for details.)

A key element of its implementation is use of the MACSYMA function DESOLVE to solve the set of ODE's generated in the Hunt–Su–Meyer algorithm. DESOLVE implements several methods for first order ODE's which are tested in the following order: linear, separable, exact (perhaps requiring an integrating factor), homogeneous, Bernoulli's equation, and a generalized homogeneous method. If DESOLVE cannot obtain a solution, it returns the value "false." One way to extend the capabilities of the function **transform** would be to add methods to DESOLVE applicable to the cases encountered in control system design. Alternately, one could easily add a facility which would integrate the ODE's of the Hunt–Su–Meyer algorithm numerically, by "automatically" generating a Fortran program for this purpose, similar to the simulation codes generated by the design packages FENOLS and TRACKS. The function `cyr(f,g)` also uses the MACSYMA function SOLVE to invert the Hunt–Su–Meyer ODE's to produce the transformation T from the solutions of the ODE's. If the ODE's were solved numerically, this function would have to be implemented in a recursive algorithm.

```
broc(f,g):=block([x,v,s,a,b,T,u,eqa],x:makelist(concat('x,i),i,1,n),
                    u:[concat('u,1)],
                    zr:makelist(concat('z,i),i,1,n),dep:g[1],
                    for j:1 thru n do s[1,j]:dep[j],
                    if n>=2 then for i:2 thru n do ( dep:plie(f,dep),for j:1
                    thru n do
                    s[i,j]:dep[j]),a:genmatrix(s,n,n),
                    b:a^^-1,for i:1 thru n do T[i]:b[i,n],
                    zr[1]:integrate(T[1],x[1],0,x[1]),for i:2 thru n do (
                         for j:1 thru i-1 do T[i]:subst(0,x[j],T[i]),
                         zr[1]:zr[1]+integrate(T[i],x[i],0,x[i])),
                    for i:2 thru n do zr[i]:makelist(diff(zr[i-1],x[j]),j,1,n).
                                             transpose(f+g[1]*u[1]),
                    vr[1]:makelist(diff(zr[1][n],x[j]),j,1,n).transpose
                    (f+g[1]*u[1]),
                    print("The new state variables are:"),
                    for i:1 thru n do print ("z[",i,"]=",zr[i]),
                    print("The new control variables are:"),
                    for i:1 thru m do print ("v[",i,"]=",vr[i]),
                    eqa:concat('v,1)=vr[1],
                    dd2[1]:rhs(solve(eqa,u[1])[1]))$
```

Figure 10: MACSYMA code for the function broc.

Example: Check if the following nonlinear system is feedback-linearizable to a controllable linear system and if so, find the F-transformation.

$$\begin{pmatrix} \dot{x_1} \\ \dot{x_2} \\ \dot{x_3} \\ \dot{x_4} \\ \dot{x_5} \end{pmatrix} = \begin{pmatrix} \sin(x_2) \\ \sin(x_3) \\ x_4^3 \\ x_5 + x_4^3 - x_1^{10} \\ 0 \end{pmatrix} + \begin{pmatrix} 0 & 0 \\ 0 & 0 \\ 1 & 0 \\ 0 & 0 \\ 0 & 1 \end{pmatrix} \begin{pmatrix} u_1 \\ u_2 \end{pmatrix}$$

The results of a session with CONDENS illustrating the use of **transform(f,g)** are shown in Figure 11.

INVERT(f,g,h): Given the nonlinear control system:

$$\begin{aligned} \dot{x} &= f(x) + \sum_{i=1}^{m} g_i(x) u_i \\ y &= h(x) \end{aligned} \tag{8}$$

INVERT tries to check if the system is strongly invertible by investigating its relative order. In case it is invertible (that is in case the relative order is finite), INVERT returns the relative order, furthermore, it computes the left-inverse to system (7) and returns:

```
(c2) load("initfile.mac");
(d2)                          initfile.mac
(c3) f:[sin(x2),sin(x3),x4**3,x5+x4**3-x1**10,0];
                                         3        3      10
(d3)           [sin(x2), sin(x3), x4 , x5 + x4   - x1  , 0]
(c4) g:[[0,0,1,0,0],[0,0,0,0,1]];
(d4)            [[0, 0, 1, 0, 0], [0, 0, 0, 0, 1]]
(c5) transform(f,g);

Hello,TRANSFORM tries to solve the problem:
           Given the non linear system:
                        dx
                        -- = f(x) + u  g (x) + u  g (x)
                        dt           2 2        1 1
find a non singular transformation that takes this system to a controllable
linear system:
                              dz
                              -- = b . v + a . z
                              dt
checking if the system is in block triangular form....
system not in block triangular form ==> trying the general method...
the system is multi-input ==> computing first the Kronecker indices of the
equivalent controllable linear system
Kronecker indices are:
k[ 1 ]= 3
k[ 2 ]= 2
checking the first condition of transformability.....
checking the second condition of transformability...
checking the third condition of transformability....
span of c1 =span of c1 /C
span of c2 =span of c2 /C
All the conditions are satisfied
trying to construct (if possible) the transformation....
The new state variables are :
z[ 1 ]= x1
z[ 2 ]= sin(x2)
z[ 3 ]= cos(x2) sin(x3)
z[ 4 ]= - x4
                3     10
z[ 5 ]= - x5 - x4  + x1
The new control variables are :
                             3                2
v[ 1 ]= cos(x2) cos(x3) (x4  + u1) - sin(x2) sin (x3)
             2        3     10          9
v[ 2 ]= - 3 x4  (x5 + x4  - x1  ) + 10 x1  sin(x2) - u2

(d5)                              done
```

Figure 11: Illustrating the behavior of `transform(f,g)`.

$A_{inv}(x), B_{inv}(x), C_{inv}(x), D_{inv}(x)$ where:

$$\begin{aligned} \dot{x} &= A_{inv}(x) + B_{inv}(x)\hat{y} \\ u &= C_{inv}(x) + D_{inv}(x)\hat{y} \end{aligned}$$

For an example of the use of INVERT, see the example on the design package TRACKS. We shall omit the details of the realization of INVERT[3]

3.4 Control design packages in CONDENS

FENOLS and TRACKS are two design packages contained in CONDENS. Their purpose is essentially to use, respectively, the two modules TRANSFORM and INVERT to design a control law that forces the output of a nonlinear system to follow some desired trajectory. They are both "user-friendly," asking progressively for all the data they need. They are capable of automatically generating, upon request, Fortran codes for simulation purposes.

FENOLS (): This package implements the design scheme of Figure 1, that is:

- It takes as input the nonlinear dynamics $f, g_1, \ldots, g_m$, the desired path $x_d(t)$ and the desired eigenvalues for the linear regulator.
- It uses the package TRANSFORM to transform the system into a controllable linear one, then FENOLS uses the desired eigenvalues (entered as inputs) to design a linear feedback control (by pole placement). This control solves the output tracking problem for the equivalent linear system.
- If the user is interested in simulation results, all he has to do is answer "yes" to a question posed by FENOLS at the end: "Are you interested in simulation results?".

Example: Tracking control for an industrial robot

This is the example of a *basic industrial robot*. It has one rotational joint and a translational joint in the (x, y) plane. Using FENOLS, we try to design the control or the values of the torques that make the joint angles follow some desired trajectories. The system can be described by a state space representation of the form:

$$\begin{pmatrix} \dot{x}_1 \\ \dot{x}_2 \\ \dot{x}_3 \\ \dot{x}_4 \end{pmatrix} = \begin{pmatrix} x_2 \\ f_1(\mathbf{x}) \\ x_4 \\ f_2(\mathbf{x}) \end{pmatrix} + \begin{pmatrix} 0 & 0 \\ \frac{1}{m_R+m_L} & 0 \\ 0 & 0 \\ 0 & \frac{1}{k-m_R\ell x_1+(m_R+m_L)x_1^2} \end{pmatrix} \begin{pmatrix} u_1 \\ u_2 \end{pmatrix}$$

[3]Which is being revised at the time of this writing.

where

$$f_1(\mathbf{x}) = x_1 x_4^2 - \frac{m_R \ell}{2(m_R + m_L)} x_4^2$$

$$f_2(\mathbf{x}) = \frac{-2[(m_R + m_L)x_1 - m_R \frac{\ell}{2}]}{k - m_R \ell x_1 + (m_R + m_L) x_1^2} x_2 x_4$$

$$y_1 = x_1$$

$$y_2 = x_3$$

with $m_R = m_L = \ell = 2, k = 5$

We are interested in designing a tracking controller to track the trajectories:

$$y_1(t) = exp(t)$$
$$y_2(t) = t$$

```
(c2) load("initfile.mac");
(d2)   initfile.mac
(c3) fenols();

Hello,FENOLS tries to solve the problem:                                    #
        Given the non linear system:
                                  m
                                 ====
                          dx     \
                          -- =   >     u  g (x) + f(x)
                          dt     /      i  i
                                 ====
                                 i = 1
find a non singular transformation that takes this system to a controllable li#
near system:
                               dz
                               -- = b . v + a . z
                               dt
enter dimension of the state space
4;
enter dimension of the control space
2;
enter the values of f in the form[f1(x),f2(x),....,fn(x)] followed by ,
[x2,x1*x4**2-x4**2/2,x4,(4*x2*x4-8*x1*x2*x4)/(4*x1**2-4*x1+5)];
enter g (x) in the form [ g    (x) .... g    (x) ]
       1                   1, 1          1, 4
[0,1,0,0];
enter g (x) in the form [ g    (x) .... g    (x) ]
       2                   1, 1          1, 4
[0,0,0,4/(4*x1**2-4*x1+5)];

Entering TRANSFORM :                                                        #
checking if the system is in block triangular form....
system in block triangular form
the transformation is easy to construct
```

```
The new state variables are:
z[ 1 ]= x1
z[ 2 ]= x3
z[ 3 ]= x2
z[ 4 ]= x4
The new control variables are:
                        2
        2         x4
v[ 1 ]= x1 x4   - --- + u1
                   2
        4 x2 x4 - 8 x1 x2 x4          4 u2
v[ 2 ]= -------------------- + ----------------
            2                      2
        4 x1  - 4 x1 + 5     4 x1  - 4 x1 + 5
enter the desired trajectory in the form [xd(1),...xd(n)]
[exp(t),exp(t),t,1];
enter the desired eigenvalues of the linear controller in the form [delta(1),.#
..,delta(n)] followed by a ,
[-2,-2,-2,-2];
The tracking controller is :
                                2                          t
                   (2 x1 - 1) x4  - 2 (- 4 x3 - 4 x1 + 5 %e  + 4 t)
u(x,xd)[ 1 ]= - -------------------------------------------------
                                         2

                                          2                     t
u(x,xd)[ 2 ]= ((8 x1 x2 - 4 x2) x4 + 4 x1  (- 4 x4 - 4 x2 + 4 %e  + 5)

     t     t
       - 4 x1 (- 4 x4 - 4 x2 + 4 %e  + 5) + 5 (- 4 x4 - 4 x2 + 4 %e  + 5))/4

Are you interested in simulation results?(answer y or n)
y;
enter filename of fortran code(without adding'.f')
examp1;
enter initial time you would like the simulation to start from
0.0;
enter final time tf
5.0;
enter step size h
0.01;
enter initial condition in the form[xo[1],...xo[n]]
[0,0,0,0];

      dimension x( 4 ),dx( 4 ),datad(1000, 4 ),data(1000,
     1 4 ),u( 2 ),y( 4 )
c     set no of equations
      n= 4
      m= 2
               .        ||
               .        ||    Fortran code generated
               .        \/    and written to a file.
```

```
(d3)                          done
```

TRACKS (): This package follows the design scheme in Figure 2. It uses the module INVERT to generate the output tracking controller. It presents the same interaction facilities as FENOLS, that is, takes as input the nonlinear dynamics $f, g_1, \ldots, g_m, h_1, \ldots, h_m$ and the desired trajectory $y_d(t)$ and generates upon request a Fortran program for simulation purposes.

Example: For the example of the basic industrial robot, applying this second design method, we obtain:

```
(c2) load("initfile.mac");
(d2)    initfile.mac
(c3) tracks();
Hello,TRACKS tries to solve the problem:
          .    ||
          .    ||    "Entering Input data"
          .    \/

enter the desired trajectory you want the output of your system to track
yd[ 1 ](t)=
exp(t);
yd[ 2 ](t)=
t;

Entering INVERT...
relative order of system = 2
The inverse system to our non linear control system is :
                              dx
                              -- = binv(x) . yi(t) + ainv(x)
                              dt
                             u(t) = Cinv(x) + Dinv(x) . yi(t)


                                2      2
                             d y1   d y2
where         yi(t) = [----, ----]
                                  2      2
                              dt     dt

              [ x2 ]
              [    ]                          [              2       2  ]
              [ 0  ]                          [      2 x1 x4    - x4    ]
 Ainv(x) = [    ]                   Cinv(x) = [ - -------------- ]
              [ x4 ]                          [            2            ]
              [    ]                          [                          ]
              [ 0  ]                          [ 2 x1 x2 x4 - x2 x4 ]

              [ 0   0 ]                       [ 1           0            ]
              [       ]                       [                          ]
Binv(x) =  [ 1   0 ]                  Dinv(x) = [           2              ]
              [       ]                       [      4 x1    - 4 x1 + 5 ]
              [ 0   0 ]                       [ 0   ---------------- ]
```

```
         [      ]                           [              4       ]
         [ 0  1 ]

Checking if yd(t) is trackable by our system (right-invertibility)
yd(t) can be tracked by the output y(t)

 The control ud(t) that makes y(t) track yd(t) is:
                                      2      2         t
                            2 x1 x4   - x4   - 2 %e
    ud(t)[ 1 ]= - ----------------------
                                         2
    ud(t)[ 2 ]= 2 x1 x2 x4 - x2 x4

Are you interested in simulation results? (Answer y or n)
y;
enter filename of fortran code
          .   ||
          .   ||  generating the fortran code
          .   \/

(d3)                     done
```

4 Conclusions

Differential geometric tools have proved to be very useful for analysis of nonlinear systems. Manipulations of vector fields and forms are essentially straightforward; however, doing the computations by hand, in particular on higher-dimensional spaces, is hard since the amount of labour grows rapidly and becomes rather tedious. Moreover, there is every chance that errors will be introduced. By using a computer algebra system like CONDENS, much valuable time can be saved and design methods that rely on rather sophisticated mathematical tools can be made available to a wider class of users.

If a reader is interested in using CONDENS, he or she should contact one of the authors.

References

[1] Boothby, W. M., 1975, *An Introduction to Differentiable Manifolds and Riemannian Geometry*, Academic Press, New York.

[2] Brockett, R. W., 1978, *Feedback Invariants for Nonlinear Systems*, IFAC Congress, Helsinki. p. 1115-1120.

[3] Hirschorn, R. M., 1979, *Invertibility of nonlinear control systems*, SIAM J Control Optim, 17, pp. 289-297.

[4] Hunt, R. L., Su, R., and Meyer, G., 1983, *Design for multi-input systems*, Differential Geometric Control Theory, edited by R. Brockett, R. Millman and H. J. Sussman, Birkhauser, Boston, vol. 27, pp. 268-298.

[5] HUNT, R. L., SU, R., and MEYER, G., 1983, *Global Transformations of Nonlinear Systems*, IEEE Trans Aut Control, AC-28, No. 1, pp. 24-31.

[6] JAKUBCZYK, B., and RESPONDEK, W., 1980, *On linearization of control systems*, Bull Acad Polon Sci, Ser Sci Math Astronom Phys, 28, pp.517-522.

[7] KRENER, A. J., 1973, *On the equivalence of control systems and the linearization of nonlinear systems*, SIAM J Control, 11, pp. 670-676.

[8] MEYER, G., and CICOLANI, L., 1980, *Application of nonlinear system inverses to automatic flight control design-system concepts and flight evaluations*, AGARDograph 251 on Theory and Applications of Optimal Control in Aerospace Systems, P. Kent, ed., reprinted by NATO.

[9] MEYER, G., 1981, *The design of exact nonlinear model followers*, Proceedings of Joint Automatic Control Conference, FA3A.

[10] SU, R., 1982, *On the linear equivalents of nonlinear systems*, Systems and Control Letters, 2, No 1, pp. 48-52.

[11] THE MATLAB GROUP LABORATORY FOR COMPUTER SCIENCE, 1983, *MACSYMA Reference Manual*, M.I.T., Cambridge, Mass.

[12] WONHAM, W. M., 1979, *Linear Multivariable Control: A Geometric Approach*, (New York: Springer-Verlag).

[13] AKHRIF, O., and BLANKENSHIP, G.L., 1987, *Using Computer Algebra for Design of Nonlinear Control Systems*, Invited paper, ACC 87.

[14] AKHRIF, O., and BLANKENSHIP, G.L., 1987, *Computer Algebra for Analysis and Design of Nonlinear Control Systems*, SRC TR-87-52.

PANDORE

J.P.Chancelier
C.Gomez
J.P.Quadrat
A.Sulem

INRIA
Domaine de Voluceau
78153 Le Chesnay
France

Abstract

The languages Lisp, Macsyma and Prolog provides a powerful environment for manipulating programs and reports. Macrofort is a language for interfacing Macsyma and Fortran. MacroTex does the samething for Latex. Using these facilities Pandore generates complete reports in Latex about studies in stochastic control . The user has only to specify its model in term of control of diffusion processes. Pandore does everthing else. The pandore paper in these proceedings has been completely generated by the system.

Introduction | Macsyma | Prolog | Macrofort | Macrotex | Pandore | References

Introduction

- **Pandore** — Pandore is a system able to make complete engineering studies in optimization of dynamic sytem.
- **Macsyma** — It makes an intensive use of the three languages Macsyma, Prolog and Lisp. The complete
- **Prolog** — environment is Lisp since Macsyma and the Prolog used are written in Lisp.
- **Macrofort** — To facilitate the fortran generation from macsyma one uses a language called Macrofort.
- **Macrotex** — For Tex report generation one uses another language called Macrotex.

Introduction | Macsyma | Prolog | Macrofort | Macrotex | Pandore | References

M a c s y m a

- **Formal calulus systems**
- **Trees**
- **Demos**
- **Interpret.**
- **Simplific.**
- **Evaluation**

Macsyma is one of the most famous formal calculus system.

Because it is written in Lisp it can manipulate easily trees representing algebraic formulas or typed expressions like rationals.

Mainly, the interpretor makes a "read,eval-simplify,print" loop, the simplifier and the evaluator calling each other recursively.

Most of the Macsyma power comes from its simplification possibilities which confer it surprising possibilities in partial evaluation.

Introduction | Macsyma | Prolog | Macrofort | Macrotex | Pandore | References

Macsyma demos

The following simple demos show some algebraic possibilities of macsyma:

- rational calculus,
- differential calculus,
- matricial calculus,
- progamming language,
- equations solver.

Introduction | Macsyma | Prolog | Macrofort | Macrotex | Pandore | References

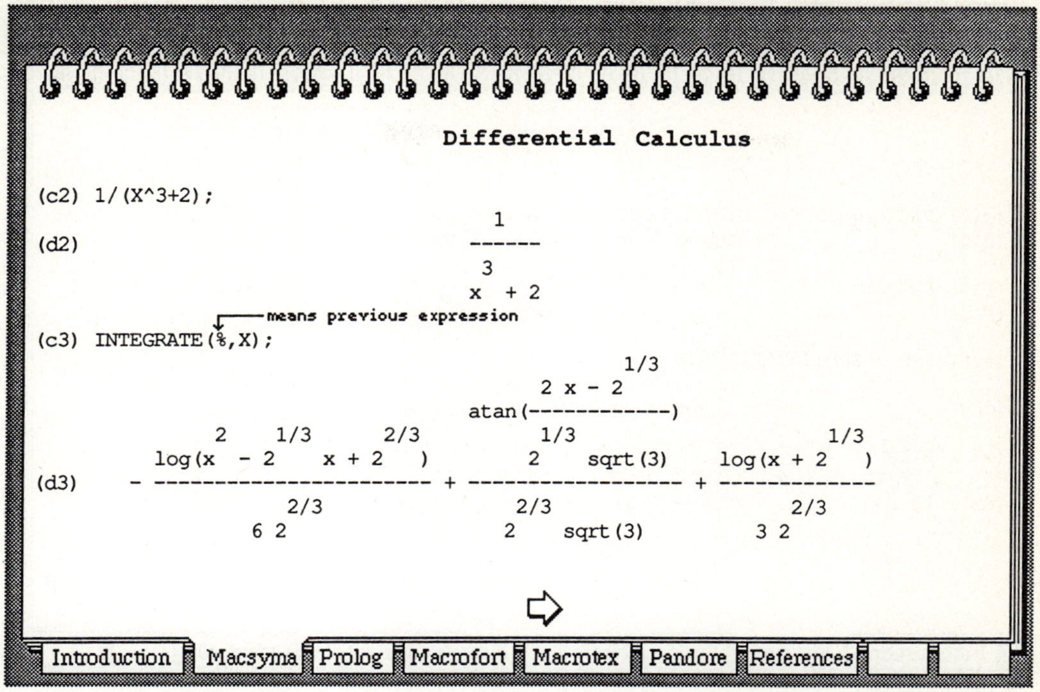
Differential Calculus
(c2) 1/(X^3+2);
(d2) 1 / (x^3 + 2)
means previous expression
(c3) INTEGRATE(%,X);
(d3) - log(x^2 - 2^(1/3) x + 2^(2/3)) / (6 2^(2/3)) + atan((2 x - 2^(1/3)) / (2^(1/3) sqrt(3))) / (2^(2/3) sqrt(3)) + log(x + 2^(1/3)) / (3 2^(2/3))
Introduction
Macsyma
Prolog
Macrofort
Macrotex
Pandore
References

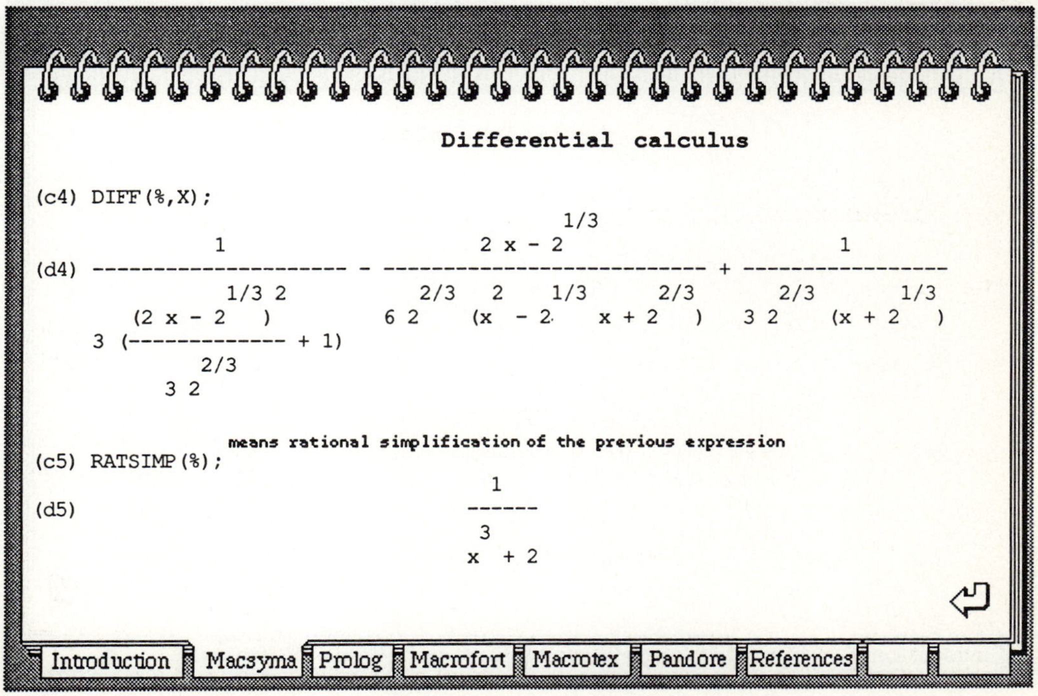
Differential calculus
(c4) DIFF(%,X);
(d4) 1 / (3 ((2 x - 2^(1/3))^2 / (3 2^(2/3)) + 1)) - (2 x - 2^(1/3)) / (6 2^(2/3) (x^2 - 2^(1/3) x + 2^(2/3))) + 1 / (3 2^(2/3) (x + 2^(1/3)))
means rational simplification of the previous expression
(c5) RATSIMP(%);
(d5) 1 / (x^3 + 2)
Introduction
Macsyma
Prolog
Macrofort
Macrotex
Pandore
References

Programming language

```
(c18) FAC(N):=IF N=0 THEN 1 ELSE N*FAC(N-1);
(d18)              fac(n) := IF n = 0 THEN 1 ELSE n fac(n - 1)

(c19) FAC(5);
(d19)                             120

(c20) G(N):=SUM(I*X^I,I,0,N);
                                         i
(d20)                     g(n) := sum(i x , i, 0, n)

(c21) G(10);
          10       9       8       7       6       5       4       3       2
(d21) 10 x    + 9 x  + 8 x  + 7 x  + 6 x  + 5 x  + 4 x  + 3 x  + 2 x  + x
```

Introduction | Macsyma | Prolog | Macrofort | Macrotex | Pandore | References

Matricial Calculus

```
(c16) MAT:MATRIX([A,B,C],[D,E,F],[G,H,I]);
                                   [ a  b  c ]
                                   [         ]
(d16)                              [ d  e  f ]
                                   [         ]
                                   [ g  h  i ]
(c17) %^^2;
              [                2                                   ]
              [ c g + b d + a     c h + b e + a b   c i + b f + a c ]
              [                                                     ]
(d17)/R/      [                          2                          ]
              [ f g + d e + a d   f h + e  + b d    f i + e f + c d ]
              [                                                     ]
              [                                       2             ]
              [ g i + d h + a g   h i + e h + b g   i  + f h + c g  ]
```

Introduction | Macsyma | Prolog | Macrofort | Macrotex | Pandore | References

Equation solver

```
(c13) SOLVE(X^6-1);
          sqrt(3) %i + 1      sqrt(3) %i - 1                     sqrt(3) %i + 1
(d13) [x = --------------, x = --------------, x = - 1, x = - --------------,
                 2                   2                              2

                                                       sqrt(3) %i - 1
                                               x = - --------------, x = 1]
                                                            2
(c14) X^6-1,%[1];
                                               6
                               (sqrt(3) %i + 1)
(d14)                          ----------------- - 1
                                      64
(c15) EXPAND(%);
(d15)                                  0
```

Introduction | Macsyma | Prolog | Macrofort | Macrotex | Pandore | References

Rational Calculus

```
(c8) (X+3)^20;
                                           20
(d8)                                (x + 3)

(c9) RAT(%);
         20        19          18            17              16               15
(d9)/R/ x   + 60 x   + 1710 x   + 30780 x   + 392445 x   + 3767472 x

               14              13              12                11
 + 28256040 x   + 169536240 x   + 826489170 x   + 3305956680 x

                 10                 9                8                  7
 + 10909657044 x   + 29753610120 x   + 66945622770 x   + 123591918960 x

                  6                  5                  4                  3
 + 185387878440 x   + 222465454128 x   + 208561363245 x   + 147219785820 x

                2
 + 73609892910 x   + 23245229340 x + 3486784401
```

Introduction | Macsyma | Prolog | Macrofort | Macrotex | Pandore | References

Rational Calculus

```
(c10) DIFF(%,X);
               19         18            17              16                15
(d10)/R/ 20 x    + 1140 x    + 30780 x    + 523260 x    + 6279120 x

               14              13                12                 11
 + 56512080 x    + 395584560 x    + 2203971120 x    + 9917870040 x

                  10                9                  8                   7
 + 36365523480 x    + 109096570440 x  + 267782491080 x  + 535564982160 x

                   6                 5                   4                  3
 + 865143432720 x   + 1112327270640 x  + 1112327270640 x  + 834245452980 x

                   2
 + 441659357460 x   + 147219785820 x + 23245229340

(c11) FACTOR(%);
                                                 19
(d11)                                20 (x + 3)
```

Introduction | Macsyma | Prolog | Macrofort | Macrotex | Pandore | References

Interpretation

Macsyma objects

Almost all of the objects of macsyma are internally lists or typed lists.

Typed list

Typed lists are lists which have a first element specifying the type for example:
- the macsyma list [1,2] is represented by ((mlist) 1 2)
- the macsyma matrix "matrix([1,2])" is represented by

The other lists are similar to the lisp ones for example "a+b" is represented by ((mplus) $a $b)

Introduction | Macsyma | Prolog | Macrofort | Macrotex | Pandore | References

Interpretation

"read, eval-simp, print" loop

Macsyma manipulates these objects. Its interpretor does a "read eval-simp display loop".

For example suppose we want compute the derivative of (x+1)^3. The interpretor

1) reads "diff((x+1)^3,x)"
2) parses the expression to obtain:
   ((\$diff) ((mexpt) ((mplus) \$x 1) 3) \$x)
3) evaluates and simplifies the expression
   ((mtimes simp) 3
   ((mexpt simp) ((mplus simp) \$x 1) 2))
4) 2D-prints the result: $3(x+1)^2$

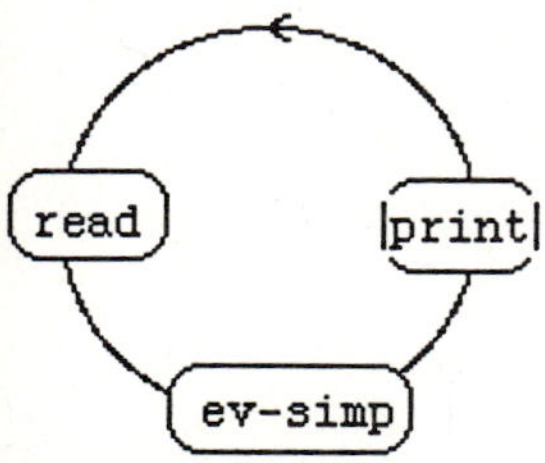

Introduction | Macsyma | Prolog | Macrofort | Macrotex | Pandore | References

Trees

Very often formulas are represented by trees in Macsyma. Some typed expressions have a specific internal representation like rational expressions, Taylor series etc...

Trees are represented easily by list in Lisp. For example the macsyma expression x^2+3x+1 is written in Lisp by '(+ (* x x) (* 3 x) 1) which is the tree given by the picture.

The rationals have a specialized internal representation nevertheless they have variable lengths and thus can be easily represented by lists.

Introduction | Macsyma | Prolog | Macrofort | Macrotex | Pandore | References

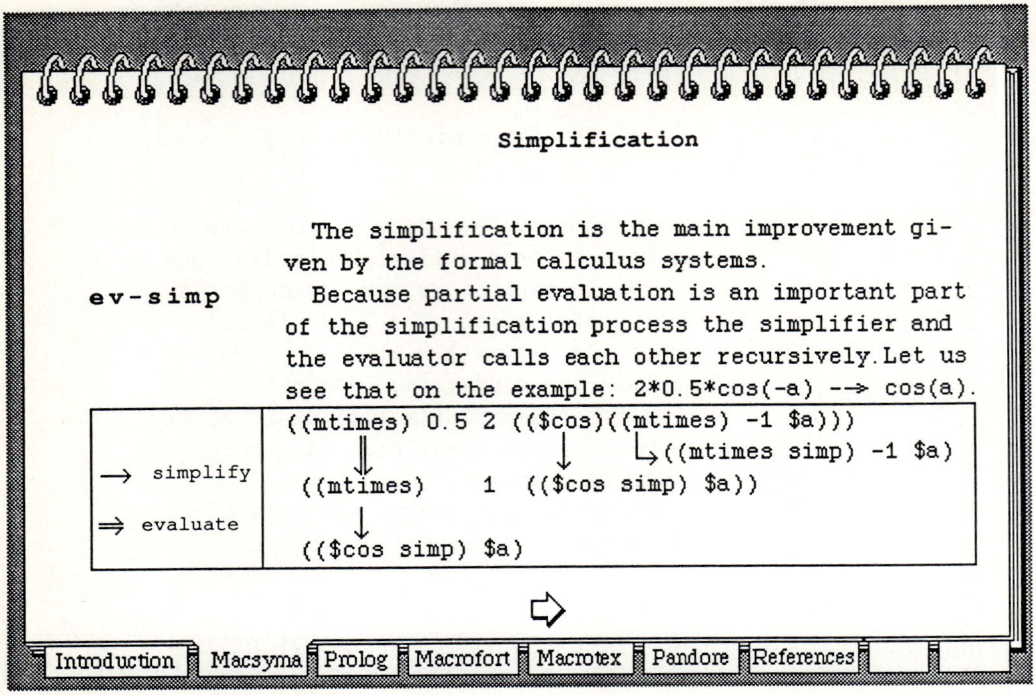
Simplification
The simplification is the main improvement given by the formal calculus systems.
ev-simp
Because partial evaluation is an important part of the simplification process the simplifier and the evaluator calls each other recursively. Let us see that on the example: 2*0.5*cos(-a) --> cos(a).
→ simplify
⇒ evaluate
((mtimes) 0.5 2 (($cos)((mtimes) -1 $a)))
((mtimes simp) -1 $a)
((mtimes) 1 (($cos simp) $a))
(($cos simp) $a)
Introduction
Macsyma
Prolog
Macrofort
Macrotex
Pandore
References

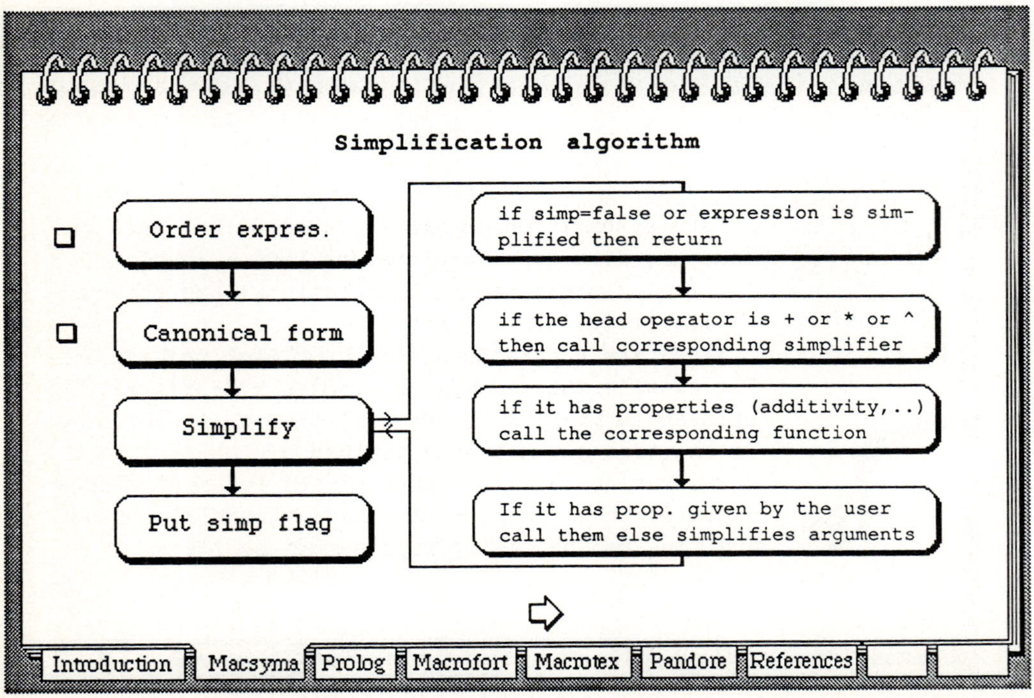
Simplification algorithm
Order expres.
Canonical form
Simplify
Put simp flag
if simp=false or expression is simplified then return
if the head operator is + or * or ^ then call corresponding simplifier
if it has properties (additivity,..) call the corresponding function
If it has prop. given by the user call them else simplifies arguments
Introduction
Macsyma
Prolog
Macrofort
Macrotex
Pandore
References

Customizing the Simplification

The user can customize the simplification of macsyma for that it has 3 possibilities:

1) puts preexisting properties on an operator:

declare
```
(declare(f,additive),f(x+y)) --> f(x)+f(y),
```

2) defines new rules for the simplifier:

tellsimp
```
(matchdeclare(a,freeof(y)),
 tellsimp(f(a+y),f(a)),f(b+c+y)) --> f(b+c),
```

3) defines its own simplifier:

defrule
apply1
```
(matchdeclare(x,true),defrule(r1,f(f(x)),g(x)),
apply1(f(f(f(x))),r1)) --> g(f(x)) .
```

Introduction | Macsyma | Prolog | Macrofort | Macrotex | Pandore | References

ordering expression, canonical form

ordering expression Macsyma order expression in the inverse lexicographic order for example:

```
%pi-3+a+x  ---> x+a+%pi-3
```

canonical form Macsyma puts the intered expressions in canonical forms. In this canonical form the operators ""-" and "/" do not exist. For example a-b+1/c has the internal representation :

```
((mplus simp) ((mexpt simp) c -1)
              b
              ((mtimes simp) a -1)))
```

Introduction | Macsyma | Prolog | Macrofort | Macrotex | Pandore | References

Partial Evaluation

"foo"--->	foo
2	2
a	a
a:2	2
a	2
'a	a
a^3+b	8 + b
f(b)	f(b)
f(a)	f(2)
f(x):=2*x	f(x):=2x
f(b)	2 b
f(a)	4
a=b	2 = b

diff(x^2,x) --------->	2 x
'diff(x^2,x)	$\frac{d(x^2)}{d\,x}$
declare(diff,noun)	done
diff(x^2,x)	$\frac{d(x^2)}{d\,x}$
remove(diff,noun)	done
diff(x^2,x)	2 x
sum(concat(x,i),i,1,2)	x1+x2
x1+x2,x1=2	x2+2

The simplification abilites of macsyma allows it to make good partial evaluation and to give an interesting answer to any syntactly correct expression.

Introduction | Macsyma | Prolog | Macrofort | Macrotex | Pandore | References

Some Formal Calculus Systems

Macsyma Macsyma comes from MIT (Engelman, Martin, Moses and their students..). The project started in 1969. Its size is about 300.000 lines of Lisp. It represents more than 100 engineer-years of designing and 600 engineer-years of testing. The Lisp used are Maclisp, Zetalisp Franzlisp and Commonlisp.

Reduce Reduce started in 1969. Its author is Hearn. It works with Cambridge-Lisp, Standard-Lisp, Maclisp, Franzlisp, Le_lisp. It is almost as powerful as Macsyma now. It is the most available system.

Introduction | Macsyma | Prolog | Macrofort | Macrotex | Pandore | References

Some Formal Calculus Systems

Scratchpad Scratchpad is the formal calculus system of IBM it is written in Lisp S/360, Lisp 1.5. It started in 1968. Two mains contributors are Griesmer and Jenks. Its new version has certainly the most beautiful design of all the systems.

Maple Maple comes from the university of Waterloo. Its authors are Char, Geddes and Gonnet. The project started in 1983. Maple is written in C. It is also almost as powerful as Macsyma. Its algorithms are more recent, thus sometimes it is much faster than Macsyma.

Introduction | Macsyma | Prolog | Macrofort | Macrotex | Pandore | References

P r o l o g

Prolog allows to express easily relations between trees. It is very useful to build small data bases on such objects, with a powerful interrogating language.

- **Oblogis**
- **Clauses**
- **Unificat.**
- **And-or trees**

We use a Prolog written in Lisp called Oblogis. In this language Prolog is seen as two Lisp Macros, the first for installing the Prolog clauses, the second for asking questions.

Prolog adds two functionalities to Lisp: unification and exploration of and-or-trees.

Introduction | Macsyma | Prolog | Macrofort | Macrotex | Pandore | References

Clauses

Prolog make inference on clauses of one of the following kind:

(1) hyp_1 & ... & hyp_n ==> concl

(2) ==> concl

❑ **literal** where hyp_i and concl are literals

The second kind is called a fact and is a particular case of the first kind.

❑ **variable** One asks some questions by asking a literal to be proved. That is finding the substitutions of the variables such that the litteral becomes true. True means proved or infered in this universe.

Introduction | Macsyma | Prolog | Macrofort | Macrotex | Pandore | References

Literal

A literal is a tree without logic connectors, the leaves of which being of two kinds the variables and the constants.

variable The variables are universally quantified that means that each clauses where appears the variable is true for any substitution of the variable by a constant.

constant The constants are any symbol which is not a variable. Some syntactic sign permits the discrimination between the two kinds of symbols.

Introduction | Macsyma | Prolog | Macrofort | Macrotex | Pandore | References

Exploration of and-or trees

Given a literal c to be proved. For all (or) the clauses for which the conclusion is c prolog tries to find all the substitutions which make the hypotheses true simultaneously (and).

Each hypothesis constrainted by the previous substitutions becomes a new literal to be proved.

Prolog makes the exploration by a depth first stategy that is at each level of its demonstration it starts by proving the first hypothesis. Only when it has achieved to prove it, it starts the second one.

Introduction | Macsyma | Prolog | Macrofort | Macrotex | Pandore | References

Unification

Definition:

Two literals a and b are unifiable if it exists a substitution of the variables which appears in them denoted by s such that:

s(a)=s(b)

❑ **Oblogis**

Example using Oblogis syntax:

(+ !x z) and (+ y !x) are unifiable by the substitution:

```
s:{ y ---> !x
  { z ---> !x.
```

Introduction | Macsyma | Prolog | Macrofort | Macrotex | Pandore | References

Oblogis

❑ **example**

It is prolog written in Lisp by P.Gloess.
The syntax of the clauses are :

```
(1) (<- (concl) (hyp_1) ... (hyp _n))
(2) (<- (concl))
```

The litterals are lisp lists.
The constant are preceded by the macro character "!".
We ask the question "concl" by :

```
(? (concl))
```

Oblogis is able to manipulate structured objects but we don't discuss this feature here.

Introduction | Macsyma | Prolog | Macrofort | Macrotex | Pandore | References

Oblogis example

```
(<- (hamiltonien b c (+ (* b !p) c)))
    (drift b)(cost c))
```

means that the expression (+ (* b p) c) is an hamiltonien if b is a drift and c is a cost.

```
(<- (drift (+ (* 2 !x) !u)))
(<- (cost (+ (* !x !x)(* 3 !u !u))))
```

are two facts which give a drift and a cost.
Then to the question (? (hamiltonien b c hm)) "what are the hamiltonians" the answer is:

```
b=(+ (* 2 x) u), c=(+ (* x x)( 3 u u)),
hm=(+  (* (+ (* 2 x) u) p)  (+ (* x x)(* 3 u u)))
```

Introduction | Macsyma | Prolog | Macrofort | Macrotex | Pandore | References

M a c r o f o r t

- ❑ Elem. Inst.
- ❑ Macros
- ❑ Stack Inst
- ❑ Fortran Generation
- ❑ Example

On one hand Macrofort is a Pascal like language written in Macsyma, on the other hand it is a set of facilities added to Macsyma to generate Fortran. He takes sense only in the latter case.

There are three kinds of instructions: -the elementary instructions, -the macro instructions, -the stack manipulation instructions.

We use Macsyma to make the algebraic computations and the macrofort program manipulations the result is a Macsyma list the evaluation of which produce the fortran program.

Introduction | Macsyma | Prolog | Macrofort | Macrotex | Pandore | References

Macsyma function	Fortran generated
equalf(niv,var,exp)	var=exp
stopf(niv)	STOP
returnf(niv)	RETURN
endf(niv)	END
programf(niv,nom)	PROGRAM nom
readf(niv,fich,etiq,[liste])	READ (fich,etiq)liste
writef(niv,fich,etiq,[liste])	WRITE (fich,etiq)liste
formatf(niv,etiq,[liste])	etiq FORMAT (liste)
gotof(niv,etiq)	GOTO etiq
ifgotof(niv,cond,etiq)	IF cond GOTO etiq
dof(niv,etiq,ind,fin)	DO etiq ind=1,fin
continuef(niv,etiq)	etiq CONTINUE
subroutinef(niv,nom,[liste])	SUBROUTINE nom(liste)
functionf(niv,type,nom,[liste])	typ FUNCTION nom(list)
callf(niv,nom,[liste])	COMMON /nom/liste
externalf(niv,liste)	EXTERNAL liste
declaref(niv,type,[liste])	type liste

Elementary Instructions

The elementary instructions are macsyma functions which prints one fortran instruction each.
"niv" is an indentation level.

Introduction | Macsyma | Prolog | Macrofort | Macrotex | Pandore | References

Macsyma functions	Fortran generated
fairem(ind,max,prog)	DO etiq ind=1,max prog etiq CONTINUE
if_then_elsem(cond,then,else)	IF.NOT.cond GOTO etiq then GOTO fin etiq else fin CONTINUE
casem([cond1,prog1], ... [condn,progn], else)	IF.NOT.cond1 GOTO etiq1 prog1 GOTO fin etiq1 IF.NOT.cond2 GOTO etiq2 ...... etiqn-1 IF.NOT.condn GOTO etiq progn GOTO fin etiq else fin CONTINUE

Macro Instructions

The macro instructions are macsyma functions which return a set of elementary instructions. Then elementary instructions generate fortran code.

Introduction | Macsyma | Prolog | Macrofort | Macrotex | Pandore | References

Macsyma functions	Fortran generated
untilm(cond,init,until)	init nuntil=0 deb CONTINUE until IF cond GOTO fin IF nuntil.GT.mxntl GOTO max nuntil=nuntil+1 GOTO deb max CONTINUE ierr=numuntil WRITE(out,format) format FORMAT("until overflow") fin CONTINUE numuntil=numuntil+1
writem([liste],[format],fich)	READ (fich,etiq)liste etiq FORMAT(format)
readm([liste],[format],fich)	READ (fich,etiq)liste etiq FORMAT(format)

Macro Instructions

"prog","init", "until" etc... mean a list of elementary instructions in a list form*. "cond" is a condition in macsyma syntax.

Introduction | Macsyma | Prolog | Macrofort | Macrotex | Pandore | References

Macsyma functions	Fortran generated
writetabm(tab,list,fich)	WRITE(out,etiq)((tab(i1,.,in), i1,list[1]),i2...) etiq FORMAT(............ (k2-1) ('[',(f12.5,','),f12.5,']')..)
mainm(prog,nom)	real integer commons init prog formats END
subroutinem(name,[param],prog)	SUBROUTINE name(param,arg,error) real integer commons init prog formats END

Macro Instructions

"real","integer","commons", "formats","arg" are updated by stack instructions*.
The function "writetabm" is used to print in a file an array readable by macsyma.

Introduction | Macsyma | Prolog | Macrofort | Macrotex | Pandore | References

Macsyma functions	Fortran generated
functionm(type,name,[param],prog)	type FUNCTION name(param,arg,error) real integer commons init prog formats END

Macro Instructions

It exists two other utilitaries:

-declarelispm(namef,namel,dim) which gives the access to the fortran array nomf of dimension dim by namel (only on Symbolics Lisp Machine).
-commentf(list) which displays in 2D the comments in "list" as a fortran comment.

Introduction | Macsyma | Prolog | Macrofort | Macrotex | Pandore | References

Macsyma functions	Updated stack
realm(name)	**dimension_reel**
realdpm(name)	**dimension_double_precision_reel**
intergerm(name)	**dimension_entier**
commonm([name,[list]])	**commons**
commentairem(string)	**commentaires**
formatm(form)	**formats**
externalm(name)	**externals**
initm(inst)	**init**
progm(inst)	**prog**
prog1m(inst)	**prog1**
prog2m(inst)	**prog2**
argumentm(name)	**arguments**

General stack manipulation

push(val,pile)	**add val on the top of pile**
pushe(val,pile)	**add val on the bottom of pile**
pop(pile)	**pop from the top of pile**
pope(pile)	**pop from the bottom of pile**

Stack Instruction

Macros use stacks to generate the fortran code.
These stacks are updated by special instructions. All these stacks are initialized by the call: "contexte()".

Introduction Macsyma Prolog Macrofort Macrotex Pandore References

Fortran Code Generation

Instruct. in the list form

To generate fortran code one build a macsyma list (prog) of elementary instructions in the list form, that is the macsyma list composed of the name of the function followed by its arguments without the first one: "niv". For example [gotof,100] is the list form of gotof(niv,100).

aplat

The evaluation of this macrofort program is obtained by calling aplat on this: list aplat(prog). The result is the display of the corresponding fortran program. We are urged to use only macro-instructions to build prog.

Introduction Macsyma Prolog Macrofort Macrotex Pandore References

Example of a Macrofort use

```
gradient_const_step(f,vars):=block(
 [dim:length(vars),var],contexte(),
 for i:1 thru dim do ( var:vars[i],
  initm([equalf,var,0]),
  proglm([equalf,concat(var,n),
                 var-ro*diff(f,var)]),
  prog2m([equalf,var,concat(var,n)]) ),

aplat(
 subroutinem(gradient,[ro,eps],
   untilm(error<eps,[equalf,error,100],
    [progl,
     [equalf,error,sum((concat(vars[i],n)
                  -vars[i])^2,i,1,dim],
     prog2
     [writem,8,vars,[concat(dim,f12\.5)]]]
))))$
```

```
      SUBROUTINE gradient(ro,eps,ierr,maxuntil0)
        x=0
        y=0
        ierr=0
c-until error < eps faire liste_until
c-initialisation
           nuntil0=0
           error=100
c-debut-d'iteration-d'until
1000       CONTINUE
           nuntil0=nuntil0+1
c-debut-liste_until
             xn=x-2*ro*x
             yn=y-2*ro*y
             error=(xn-x)^2+(yn-y)^2
             x=xn
             y=yn
```

Introduction | Macsyma | Prolog | Macrofort | Macrotex | Pandore | References

Example of use of Macrofort

The fortran program has been obtained by the call: gradient_const_step(x^2+y^2, [x,y]).

The macsyma generator is more compact and much more general than the FORTRAN program generated. The latter is very similar to a hand written one.

```
             WRITE(8,1003) x,y
c-fin-liste_until
c-tests-de-sortied'until
         IF (error.LT.eps)GOTO 1002
         IF (nuntil0.GT.maxuntil0) GOTO 1001
c-reiterer-until
         GOTO 1000
c-sortie-d'until-depassement-du-max-d'iter
 1001 CONTINUE
         WRITE(9,1004)
         ierr=1
 1002 CONTINUE
c-fin-d'until
 1003 FORMAT( 2 f12.5)
 1004 FORMAT( ' maxuntil0')
         END
```

Introduction | Macsyma | Prolog | Macrofort | Macrotex | Pandore | References

M a c r o T e x

- Elem. Inst.
- Macros
- Stack Inst
- Latex Generation
- Example

On one hand MacroTex is a Latex like language written in Lisp, on the other hand it is a set of facilities added to Macsyma to generate Latex. He takes sense only in the latter case.

There are three kinds of instructions: -the elementary instructions, -the macro instructions, -the stack manipulation instructions.

We use Macsyma to make the algebraic computations and the MacroTex program manipulations the result is a Macsyma list the evaluation of which produce the Latex program.

Introduction | Macsyma | Prolog | Macrofort | Macrotex | Pandore | References

Macsyma function	Text generated
tex(exp)	exp in math mode
exptxt(exp)	exp in text math mode
format(string,arg1,...argn)	string(arg1..argn) following the directives:
~k	-tex(argi)
~l	-exptxt(argi)
~n	-argi (no display math)
~y	-quotation argi
~z	-reference argi
decoupet(exp,[e1..en],[op1..])	exp = e1 op1 e2
documenstylet(string)	\documenstyle{string}
labelt(string)	\label{string}
reft(string)	\ref{string}
citet(string)	\cite{string}
....	....
titlet(string,arg1,..argn)	\title{string(arg1..argn)}
sectiont(string,arg1..argn)	\section{string(arg1..argn)}
authort([name1,ad1]..)	\author{name1\\ad1}

Elementary Instructions

The elementary instructions are macsyma functions which prints one latex instruction each.

They have other optional arguments*.

Introduction | Macsyma | Prolog | Macrofort | Macrotex | Pandore | References

Macsyma function	Latex instructions
decoupem(exp)	cut exp in subexpressions having algebraic meaning and generate corresponding latex code.
abstractm(prog)	\begin{abstract} prog \end{abstract}
itemizem(item1,..)	\begin{itemize} \item item1 .... \end{itemize}
enumeratem(prog1,..)	similar to itemize
stylem(doc,font,height,width, top,odd,even,par)	\documenstyle[font]{doc} \textheight=height pt \textwidth=width pt \topmargin=top pt \oddsimarginde=odd pt \evensidemargin=even pt \marginparwidth=par pt

Macro Instructions

The macro instructions are macsyma functions which generate a list of elementary macrotex instructions in the list form* and update or use some stacks.

Introduction | Macsyma | Prolog | Macrofort | Macrotex | Pandore | References

Macsyma function	Latex instructions
citem([key1,source1],..)	\cite{key1} update the biblography with source1 .....
notationm()	\section{notation} \begin{itemize} notation1 .. \end{itemize}
bibliom()	\begin{thebibliography}{3} \bibitem{key1} source1 .... \end{thebibliography}
reportm(title,authors,abstact, chapters)	style title authors \begin{document} \maketitle \tableofcontents abstract notations chapters biblio \end{document}

Macro Instructions

The function contextt() initialize the global variables and the stack used. It must be called before starting the generation.

Introduction | Macsyma | Prolog | Macrofort | Macrotex | Pandore | References

Stack Manipulation Instructions

The two stacks : list_references and list_notations are used by the macro reportm.

They are updated by the functions :

- add_referencet(key,source),
- add_notationt(notation).

Then the references can be invoked using "key" in the format instruction with the directive ~y.
For example :

```
add_referencet([Flem,"FLEMING-RISHEL...."]),
formatt("The optimal cost satisfies the dynamic programming equation ~y.",Flem).
```

Introduction | Macsyma | Prolog | Macrofort | Macrotex | Pandore | References

Latex Generation

To generate Latex code one build a macsyma list (prog) of elementary instructions in the list form, that is the macsyma list composed of the name of the function followed by its arguments. For example [authort,["Pandore","Inria"]] is the list form of authort(["Pandore",Inria"]).

list form

execute

The evaluation of this macrotex program is obtained by calling "execute" on this list:

execute(prog).

The result is the display of the corresponding Latex program.

Introduction | Macsyma | Prolog | Macrofort | Macrotex | Pandore | References

Example of MacroTex use

```
gradient_const_step(f,vars):=block(
 [dim:length(vars),var],contextt(),
 for i:1 thru dim do ( var:vars[i],
  initm([tex,var=0]),
  prog1m([tex,concat(var,n)=
                var-ro*diff(f,var)]),
  prog2m([tex,var=concat(var,n)]) ),
execute(
 rapportm([titlet,"Gradient method"],
    [authort,["Pandore","Inria"]],
    [format,"Until error<eps do"]
    [enumeratem,
     [prog1,
      [tex,error=sum((concat(vars[i],n)
                -vars[i])^2,i,1,dim],
      prog2]])))$
```

Introduction | Macsyma | Prolog | Macrofort | Macrotex | Pandore | References

P a n d o r e

- Costo
- Idsto
- Basile
- Objectives

Pandore is an expert system on identificaton (idsto) and control (costo) of stochastic processes of diffusion type. It is oriented towards the resolution of nonlinear systems. The linear part is very rudimentary. In the future an interface with Basile (matlab based system for classical automatic control) will be done.

The objective is to automatize all the engineering tasks in these domains. The part achieved corresponds to the mouse sensitive blocks in the objective diagram.

Introduction | Macsyma | Prolog | Macrofort | Macrotex | Pandore | References

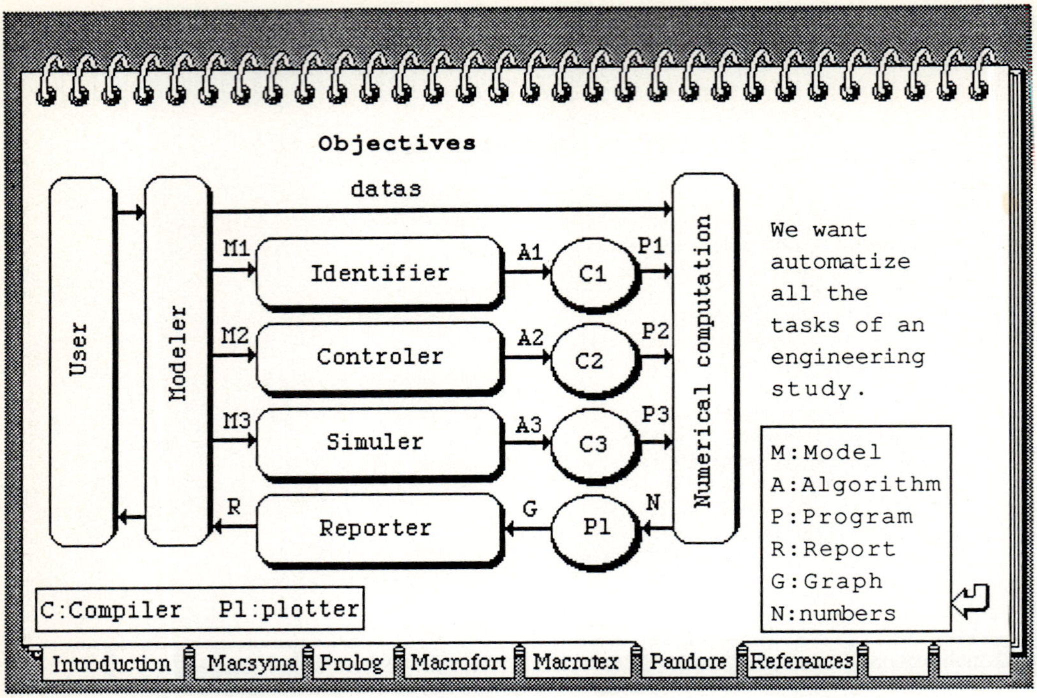

Modeler

- **pandore editor**
- **control commands**
- **identif. commands**

Now the modeler is only an interface with the user. The latter enter the model using a specialized editor, by filling some named slots. The slots appearing being not fixed but depending of the previous answers.

Then a specialized command language permit to ask questions about the model entered. Some of these commands are of very high level. They are partitioned according the problem to solve : control, identification, or simulation.

Introduction | Macsyma | Prolog | Macrofort | Macrotex | Pandore | References

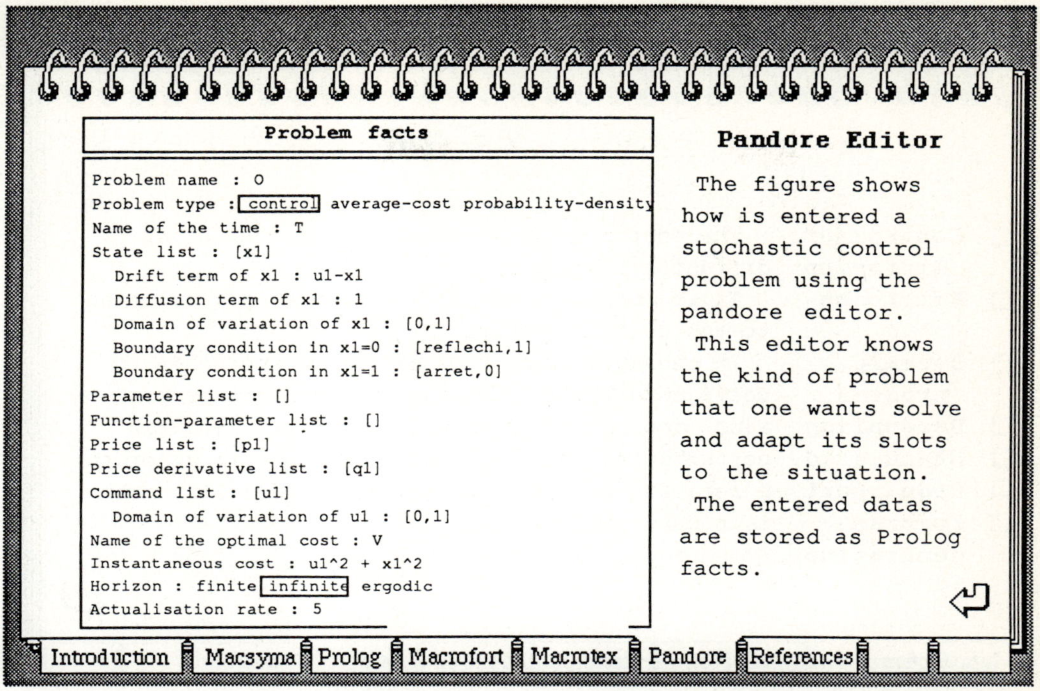
Problem facts
Problem name : O
Problem type : control average-cost probability-density
Name of the time : T
State list : [x1]
Drift term of x1 : u1-x1
Diffusion term of x1 : 1
Domain of variation of x1 : [0,1]
Boundary condition in x1=0 : [reflechi,1]
Boundary condition in x1=1 : [arret,0]
Parameter list : []
Function-parameter list : []
Price list : [p1]
Price derivative list : [q1]
Command list : [u1]
Domain of variation of u1 : [0,1]
Name of the optimal cost : V
Instantaneous cost : u1^2 + x1^2
Horizon : finite infinite ergodic
Actualisation rate : 5
Pandore Editor
The figure shows how is entered a stochastic control problem using the pandore editor.
This editor knows the kind of problem that one wants solve and adapt its slots to the situation.
The entered datas are stored as Prolog facts.
Introduction
Macsyma
Prolog
Macrofort
Macrotex
Pandore
References

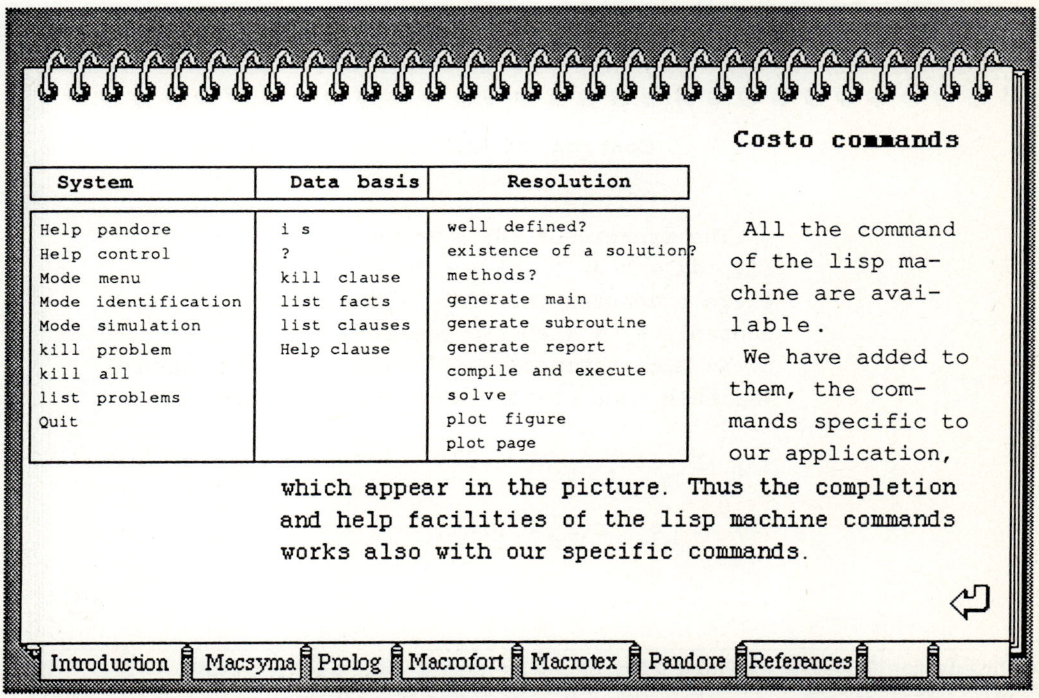
Costo commands
System
Data basis
Resolution
Help pandore
Help control
Mode menu
Mode identification
Mode simulation
kill problem
kill all
list problems
Quit
i s
?
kill clause
list facts
list clauses
Help clause
well defined?
existence of a solution?
methods?
generate main
generate subroutine
generate report
compile and execute
solve
plot figure
plot page
All the command of the lisp machine are available.
We have added to them, the commands specific to our application, which appear in the picture. Thus the completion and help facilities of the lisp machine commands works also with our specific commands.
Introduction
Macsyma
Prolog
Macrofort
Macrotex
Pandore
References

Controler

- **Control of diffusion**
- **Existence**
- **Dynamic program.**
- **Decoupling**
- **Stoch.grad.**
- **Regul.pert.**
- **Fortran generation**

The controler is a specialist of control of diffusion processes.

It is able to study the existence of a solution to the control problem.

It choses between four different approches to solve numerically the problem : dynamic programming, decoupling, stochastic gradient, regular perturbation. For each one it is able to generate a specific fortran routine and to call it to obtain numerical results.

Introduction | Macsyma | Prolog | Macrofort | Macrotex | Pandore | References

Control of diffusion processes

The dynamic of the systems is:

(1) $dX_t = b(X_t, U_t)dt + \sigma dW_t$

where t denotes the time, X_t the state, U_t the control, W_t a Wiener process.

We want minimize, with respect to the control variable one of the following criterium :

(2) $E \int_0^T c(X_t, U_t)dt$

(3) $E \int_0^\infty e^{-lt} c(X_t, U_t)dt$

Introduction | Macsyma | Prolog | Macrofort | Macrotex | Pandore | References

Dynamic programming method

The otimal cost V(t,x) of the stochastic control problem satisfies the parabolic partial differential equation :

$$D_tV + \text{Min}\ (b(x,u).D_xV + c(x,u)) + tr(aD_{xx}V)=0$$

for the criterium (1) and the elliptic equation :

$$-lV + \text{Min}\ (b(x,u).D_xV + c(x,u)) + tr(aD_{xx}V)=0$$

for the criterium (2).

These equations are solved numerically after discretization by finite difference methods.

A complete description of the method is given in the generated report.

Introduction | Macsyma | Prolog | Macrofort | Macrotex | Pandore | References

Decoupling method

In the case where the dynamic is uncoupled and the criterium is a general function of a sum of local cost we can optimize in the class of local feedbacks which does not change the coupling structure of the system.

Then we have to solve a control of the system of partial differential equations which describes the probability density of the state. Indeed in this case this density is the product of the density of each subsystem. The complete method is described in the generated report.

Introduction | Macsyma | Prolog | Macrofort | Macrotex | Pandore | References

Stochastic gradient method

The stochastic gradient method consists in

1) parametrizing the feedback law,

2) optimizing the parameter a, seeing it as open loop control, by a stochastic iteration.

In the stochastic iteration we first compute the gradient $J(\omega)$ of the deterministic control obtained by simulating a trajectory of the noise and considering it as a known parameter, then we do the iteration:

$a \longrightarrow a - r_n J(\omega)$ with r_n satisfying

$\Sigma r_n = \infty,\ r_n > 0,\ r_n \longrightarrow 0$

Introduction | Macsyma | Prolog | Macrofort | Macrotex | Pandore | References

Regular perturbation method

When the intensity of the noise σ is small we know that the affine feedback :

$u(t,x) = u_0(t) + K(t)(x - x_0(t))$,

(where : $u_0(t)$ and $x_0(t)$ are respectively the optimal control and trajectory of the deterministic control problem obtained by annulating the noise, and $k(t)$ the gain of the linear quadratic control problem osculator to the optimal deterministic trajectory) leads to a cost optimal up to σ^4.

Thus the method consists in computing this affine Feedback.

Introduction | Macsyma | Prolog | Macrofort | Macrotex | Pandore | References

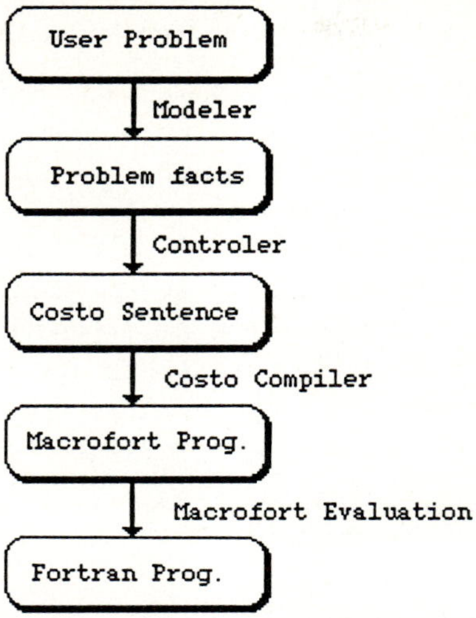

Generation of numerical routines

The steps to generate a numerical routine to solve the control problem are given in the diagram.

The controler choses a method makes the algorithmic choices and expresses its choices in a language called Costo. A compiler written in Mac-syma produces the fortran code using the Macrofort facilities

Introduction | Macsyma | Prolog | Macrofort | Macrotex | Pandore | References

Existence of a solution

The system is able to prove existence result of the solution of the the dynamic programming equation.

It verifies the hypotheses of the Lax-Milgram* theorem or those of the monotony theorem* of J.L. Lions. For that it builds the variational formulation of the problem and proves some coercivity.

To be able to do that some estimation facilities and a function able to apply the Green formula to a differential operator have been to be added to Macsyma.

Introduction | Macsyma | Prolog | Macrofort | Macrotex | Pandore | References

[1]J.P.AUBIN, Approximation of elliptic Boundary-value Problems. Wiley Inter-science, New-York, 1972.

[2]F.BASKET,M.CHANDY,R.MUNTZ,J.PALACIOS Open Closed Mixed Network Of Queus With Different Class of Customers, J.A.C.M n°22, p. 248-260, 1975.

[3]R.BELLMAN, Dynamic Programming,Princeton University Press,1957.

[4]A.BENSOUSSAN, Stochastic Control by Functional Analysis Method, North-Holland 1982.

[5]A.BENSOUSSAN, Méthodes de Perturbartion en Contrôle Optimal,1985.

[6]A.BENSOUSSAN,J.L.LIONS, Applications des inéquations variationnelles en contrôle stochastique, Dunod,1978.

[7]P.F.CIARLET, The finite element method for elliptic problems, North Holland, 1978.

[8]D.CLAUDE, Decoupling of Nonlinear Systems, Systems Control Letters n°1, p242-248, 1982.

[9]A.etC.COLMERAUER, Prolog en 10 figures, La Recherche 1985.

[10]J.B.CRUZ, Feedback Systems, McGraw-Hill, 1972.

[11]F.DELEBECQUE,J.P.QUADRAT,

-Sur l'Estimation des Caractéristiques Locales d'un Processus de diffusion avec Sauts, Rapport INRIA 1978.

-Contribution of Stochastic Control Singular Perturbation Averaging and Team Theories to an Example of Large-Scale System: Management of Hydropower Production, IEEE AC23 n°2, p209-221 1978.

⇨

Introduction | Macsyma | Prolog | Macrofort | Macrotex | Pandore | References

[12]J.C.DODU,M.GOURSAT,A.HERTZ,J.P.QUADRAT,M.VIOT Méthodes de gradient stochastique pour l'optimisation des investissements dans un réseau électrique, EDF Bulletin Serie C, n°2, 1981

[13]I.EKLAND,R.TEMAM, Analyse convexe et problèmes variationnels, Dunod, 1974.

[14]W.H.FLEMING, Control for small noise intensities, SIAM J.Control, vol.9, n°3, 1971.

[15]W.H.FLEMING-R.RISHEL, Optimal Deterministic and Stochastic Control, Springer-Verlag, 1975.

[16]F.GEROMEL,J.LEVINE,P.WILLIS, A fast Algorithm for Systems Decoupling using Formal Calculus, L.N.C.I.S n°63, Springer-Verlag 1984.

[17]P.GLOESS, Logis User's manual, Université de Compiegne, janvier 1984.

[18]P.GLOESS, Understanding Expert Systems,Université de Compiegne, janvier 1984.

[19]C.GOMEZ,J.P.QUADRAT,A.SULEM, Towards an Expert System in Stochastic Control : the Hamilton-Jacobi equation Part, L.N.C.I.S n°63, Springer-Verlag 1984.

[20]C.GOMEZ,J.P.QUADRAT,A.SULEM, Towards an Expert System in Stochastic Control : the Local-feedback Part, Congrés Rome sur le contrôle Stochastique, L.N.C.I.S Springer-Verlag 1985.

[21]THEOSYS Numerical methods in stochastic control., RAIRO Automatique 1983.

[22]M.GOURSAT,J.P.QUADRAT,

- Analyse numériques d'inéquations quasi variationnelles elliptiques associées à des problèmes de contrôle impulsionnel, IRIA Rapport,1975.

- Analyse numériques d'inéquations variationnelles elliptiques associées à des problèmes de temps d'arrêt optimaux, IRIA Rapport,1975.

⇨

Introduction | Macsyma | Prolog | Macrofort | Macrotex | Pandore | References

[23]J.JACOD, Calcul Stochastique et Problèmes de Martingale, Springer-Verlag, 1979.
[24]E.KIEFER,J.WOLFOWITZ, Stochastic estimation of the maximum of a regression function, Ann. Math. Statistic, 23,n°3,1952.
[25]H.J.KUSHNER, Probabilty methods in stochastic control and for elliptic equations, Academic Press, 1977.
[26]H.J.KUSHNER,D.S.CLARK, Stochastic approximation methods for constrained and unconstrained systems,Springer Verlag,1978.

[27]P.L.LIONS,B.MERCIER, Approximation numérique des équations de Jacobi-Bellman, RAIRO,14 , pp.369-393, 1980.
[28]P.LIPCER-A.SHIRIAEV Statistique des Processus Stochastiques, Presse Universitaire Moscou, 1974.
[29]LISP Machine Lisp Manual, MIT Press,1982.

Introduction | Macsyma | Prolog | Macrofort | Macrotex | Pandore | References

[30]MACSYMA Manual,MIT Press,1983.
[31]B.T.POLYAK, Convergence and convergence rate of iterative stochastic algorithms. Automatica i Telemekhanika, Vol.12,1976,pp.83-94.
[32]B.T.POLYAK, Subgradient methods a survey of soviet research in nonsmooth optimization, C. Lemarechal and K.Mifflin eds.Pergamon Press, 1978.
[33]B.T.POLYAK,Y.Z.TSYPKIN, Pseudogradient adaptation and training algorithms.Automatica i Telemekhanika,Vol.3,1973.
[34]J.P.QUADRAT, Existence de solution et algorithme de résolutions numériques de problèmes stochastiques dégénérés ou non, SIAM Journal of Control, mars 1980.
[35]J.P.QUADRAT, Analyse numérique de l'équation de Bellman stochastique, IRIA Report,1975.
[36]J.P.QUADRAT, On optimal stochastic control problem of large systems, Advances in filtering and optimal stochastic control, Lecture Notes in Control and Computer Science n° 42, Springer-Verlag,1982.
[37]J.P.QUADRAT,M.VIOT, Product form and optimal local feedback for multi-index Markov chains, Allerton Conference,1980.
[38]C.QUEINNEC, LISP Langage d'un autre type, Eyrolles, 1983.
[39]H.ROBBINS,S.MONRO, A stochastic approximation method. Ann. Math. Statist., 22, pp400-407, 1951.

Introduction | Macsyma | Prolog | Macrofort | Macrotex | Pandore | References

[37]J.P.QUADRAT,M.VIOT, Product form and optimal local feedback for multi-index Markov chains, Allerton Conference,1980.

[38]C.QUEINNEC, LISP Langage d'un autre type, Eyrolles, 1983.

[39]H.ROBBINS,S.MONRO, A stochastic approximation method. Ann. Math. Statist., 22, pp400-407, 1951.

[40]F.STROOCK,S.R.S.VARADHAN, Multidimensional Diffusion Processes, Springer-Verlag,1979.

[41]TORRION, Differentes méthodes d'optimisation appliquées à la gestion annuelle du système offre-demande français. Note EDF, EEG 1985.

[42]TURGEON, Optimal operation of multi-reservoir power system with stochastic inflows.Water resource resarch. April 1980

[43]CHANCELIER-GOMEZ-QUADRAT-SULEM Un systeme expert pour l'optimisation de système dynamique, Congres d'Analyse Numerique INRIA Versailles Decembre 85, North Holland.

Introduction | Macsyma | Prolog | Macrofort | Macrotex | Pandore | References

APPENDIX: Example of an output from the Pandore system

A DISCOUNTED 2 -DIMENSIONAL STOCHASTIC CONTROL PROBLEM

T. PANDORE
INRIA Domaine de Voluceau
BP 105 Rocquencourt
78153 Le Chesnay France

September 8, 1987

Abstract

We consider a stochastic control problem. The dynamic of the system is described by a 2 -dimensional diffusion process defined on $\Omega = [0,1] \times [0,1]$. The purpose is to minimize the expected discounted cost which includes an integral cost and reflecting costs. The optimal cost satisfies a Bellman equation derived from the optimal principle of Dynamic Programming .

The Dynamic Programming equation has a unique solution in $H^1(\Omega)$ which is solved numerically .

1 Notation

We use the following notation:

State variables: x_1, x_2
Control variables: u_1, u_2
Time variable: t
Optimal cost: V
State dimension: n
ith state variable: x_i
Derivative operator with respect to the time variable: ∂_0
Derivative operator with respect to x_i : ∂_i .

2 Evolution equation of the system

We consider the control diffusion process defined by the dynamic differential equation :

$$\begin{aligned} dx_{1t} &= -dZ^1_{1_t} + dZ^0_{1_t} + (u_2 - u_1)dt + 0.632dW_{1_t} \\ dx_{2t} &= -dZ^1_{2_t} + dZ^0_{2_t} + (0.5 - x_2)dt + \sqrt{2}dW_{2_t} \end{aligned} \quad (1)$$

where

NATO ASI Series, Vol. F47
Advanced Computing Concepts and Techniques in Control Engineering
Edited by M. J. Denham and A. J. Laub

$x_1 \in [0,1]$,
$x_2 \in [0,1]$,
$u_1 \in [0,1]$,
$u_2 \in [0,1]$,
$Z^j_{i\,t}$ is an increasing process, strictly increasing when x_{i_t} is on the boundary $x_i = j$.
x_{1t} is reflected on the boundary $x_1 = 0$.
x_{1t} is reflected on the boundary $x_1 = 1$.
x_{2t} is reflected on the boundary $x_2 = 0$.
x_{2t} is reflected on the boundary $x_2 = 1$.
W_{i_t} is a Wiener process, i.e. a continuous gaussian process with independent increments.

This process is well defined [9] . It is the limit when the time step h goes to 0 of a Markovian discrete process X^h_n which satisfies:

$$(i)\, E(\, X^h_{n+1} - X^h_n \mid \mathcal{F}_n\,) = h \begin{pmatrix} u_2 - u_1 \\ 0.5 - x_2 \end{pmatrix} + o(h)$$

$$(ii)\, E(\, \left(X^h_{n+1} - X^h_n\right)^{\otimes 2} \mid \mathcal{F}_n\,) = h \begin{pmatrix} 0.2 & 0 \\ 0 & 1 \end{pmatrix} + o(h)$$

(iii) A uniform integrability condition of the increment $X^h_{n+1} - X^h_n$ where $\mathcal{F}_n$ denotes the sigma-algebra generated by X_0 ,X_1 ,. . .,X_n .

3 Value function

The problem is to minimize the expectation of the function :

$$\begin{aligned} J(S) \;&=\; \int_0^{+\infty} e^{-5t}(-0.1 - 1.2x_2)\, dZ^0_{1\,t} \\ &+ \int_0^{+\infty} e^{-5t}((1.2u_2 - u_1)x_2)_t\, dt \end{aligned} \tag{2}$$

in the feedback class, i.e. the applications $S : [x_1, x_2] \mapsto [u_1, u_2]$.

4 Optimality conditions

The Bellman function V is defined by:

$$V(y_1, y_2) = \min_S \left(E\,[\, J(S) \mid x_{10} = y_1, x_{20} = y_2 \,]\right) \tag{3}$$

V satisfies the Dynamic Programming equation [2] [1] :

$$\min_{u_1,u_2} \left(A(u_1,u_2)V + C(u_1,u_2)\right) - 5V = 0 \tag{4}$$

with:

$$A(u_1, u_2)V = \partial_2^2 V + 0.2\partial_1^2 V + (0.5 - x_2)\partial_2 V + (u_2 - u_1)\partial_1 V \tag{5}$$

$$\begin{aligned} C(u_1, u_2) &= (1.2u_2 - u_1)x_2 \\ \partial_1 V(0, x_2) &= 1.2x_2 + 0.1 \\ \partial_1 V(1, x_2) &= 0 \\ \partial_2 V(x_1, 0) &= -0 \\ \partial_2 V(x_1, 1) &= 0 \end{aligned} \tag{6}$$

5 Existence of a solution of Bellman equation

Using the Maximum Principle, P.L.Lions has proved the existence of a solution of Bellman equation [6] .The above discretization methods are also based on the Maximum Priciple and its probabilistic interpretation.

Furthermore, using Monotonous operator theorem of J.L.Lions [5] [1] , one can easily show the convergence of discretization methods of finite-element type. These numerical methods will not be exposed here for the sake of simplicity.

On the other hand, we use the monotony method to prove the existence of a solution of the Bellman equation [8] . For this purpose, we compute the variational formulation of equation (4) in the space $H^1(\Omega)$. ($\Omega = [0,1]^2$):

$$\begin{aligned} F(V,W) &= H_1(V,W) - \int_\Omega H_2(V) W \, d\omega \\ &= \int_{\Gamma_{11}} -\frac{(12X_2 + 1)W}{10} \, d\gamma \end{aligned} \tag{7}$$

where $H_1(V,W)$ is the linear part of the operator:

$$H_1(V,W) = \int_\Omega \frac{dV}{dX_2}\frac{dW}{dX_2} \, d\omega + 0.2\int_\Omega \frac{dV}{dX_1}\frac{dW}{dX_1} \, d\omega + \int_\Omega (X_2 - 0.5)\frac{dV}{dX_2} W \, d\omega + 5\int_\Omega VW \, d\omega \tag{8}$$

and $H_2(V)$ is the non linear part (called hamiltonian).

$$H_2(V) = \min_{u_1,u_2} \left(\frac{\left((6u_2 - 5u_1)x_2 + (5u_2 - 5u_1)\frac{dV}{dx_1}\right)}{5} \right) \tag{9}$$

- The linear part is coercive in the space $H^1(\Omega)$:

$$H_1(V,V) \geq \frac{(-379\theta - 15.6\theta^2)\|V\|^2_{L^2(\Omega)}}{(-1.04 - 79\theta)} + 0.05(3.95 - 3.95\theta)\|V\|^2_{H^1(\Omega)} \tag{10}$$

with

$$0 < \theta < 1$$

- The hamiltonian is lipschitzian with a coefficient equal to $\sqrt{2}$; because the coefficients $u_2 - u_1, 0$ of the derivatives of V are bounded by 1 .

- F is bounded in the space $H^1(\Omega)$ because the cost $\frac{6u_2x_2}{5} - u_1x_2$ which appear in the hamiltonian is bounded by 1.2 .

- F is hemicontinuous in $H^1(\Omega)$ because the hamiltonian is lipschitzian.

- F is monotonous in $H^1(\Omega)$ when

$$0.0152 < \theta < 0.468$$

- F is coercive in $H^1(\Omega)$ when

$$0.00845 < \theta < 0.702$$

We infer from Theorem 2.1 Chap. 2 [5] *the existence of a unique solution of Bellman equation in the space* $H^1(\Omega)$.

6 Dynamic Programming method

Our purpose is to solve the Bellman equation (4) after discretization [7] [8] [3] [4] . This is possible because the state dimension is small.

6.1 Discretization:

we denote:

- h_i : i-th space variable discretization step.

We define the following operators:

$S_i \, : V(x_1,\ldots,x_i,\ldots,x_n) \mapsto V(x_1,\ldots,x_i+h_i,\ldots,x_n)$; $i = 1,\ldots,n$

$$\partial_i^{h+} = \frac{(S_i - 1)}{h_i}$$

$$\partial_i^h = \frac{\left(S_i - S_i^{-1}\right)}{2h_i}$$

$$\Delta_i^h = \frac{\left(S_i^{-1} + S_i - 2\right)}{h_i^2}$$

We thus approximate:

$\partial_i^2 V$ by $\Delta_i^h V$

$\partial_1 V$ by $\partial_1^h V$

$\partial_2 V$ by $\partial_2^h V$

The discretized Bellman equation is:

$$\min_{u_1,u_2}\left(C(u_1,u_2) + A^h(u_1,u_2)V^h\right) - 5V^h = 0 \tag{11}$$

with:

$$A^h(u_1,u_2)V = (0.5 - x_2)\partial_2^h V + (u_2 - u_1)\partial_1^h V + \Delta_2^h V + 0.2\Delta_1^h V \tag{12}$$

Reordering with respect to S_i , we get:

$$\begin{aligned} A^h(u_1,u_2)V &= (1.2u_2 - u_1)x_2 \\ & -\frac{S_1 V\left(-\frac{10u_2}{h_1} + \frac{10u_1}{h_1} - 4{h_1}^{-2}\right)}{20} \\ & -\frac{S_1^{-1} V\left(\frac{10u_2}{h_1} - \frac{10u_1}{h_1} - 4{h_1}^{-2}\right)}{20} \\ & -\frac{S_2^{-1} V\left(-\frac{10x_2}{h_2} + \frac{5}{h_2} - 20{h_2}^{-2}\right)}{20} \\ & -\frac{S_2 V\left(\frac{10x_2}{h_2} - \frac{5}{h_2} - 20{h_2}^{-2}\right)}{20} \\ & -\frac{\left(40{h_2}^{-2} + 8{h_1}^{-2}\right)V}{20} \end{aligned} \tag{13}$$

6.2 Probabilistic Interpretation of the discretized equation :

The discretization of the Bellman equation

$$\min_{u_1,u_2} \left(A(u_1,u_2)V + C(u_1,u_2)\right) - \lambda V = 0 \tag{14}$$

can be interpreted as a control problem of Markov chain with discount factor λk and instantaneous cost kC. The associated cost function is

$$\sum_{n=0}^{\infty} k(\lambda k + 1)^{-1-n} C(X_n, U_n) \tag{15}$$

and the Markov matrix M can be written:

$$M = kA + I$$

where I is the Identity matrix, λ the discount factor and k the inverse of the maximum of the diagonal of A.

M is given by:

$$\begin{pmatrix} Init-pt & Final-pt & Transition-probability \\ [x_1,x_2] & [x_1,x_2] & 0 \\ [x_1,x_2] & [x_1+h_1,x_2] & \frac{\left(\frac{5u_2}{h_1}-\frac{5u_1}{h_1}+2h_1{}^{-2}\right)}{2\left(10h_2{}^{-2}+2h_1{}^{-2}\right)} \\ [x_1,x_2] & [x_1,x_2+h_2] & -\frac{5\left(\frac{2x_2}{h_2}-\frac{1}{h_2}-4h_2{}^{-2}\right)}{4\left(10h_2{}^{-2}+2h_1{}^{-2}\right)} \\ [x_1,x_2] & [x_1-h_1,x_2] & -\frac{\left(\frac{5u_2}{h_1}-\frac{5u_1}{h_1}-2h_1{}^{-2}\right)}{2\left(10h_2{}^{-2}+2h_1{}^{-2}\right)} \\ [x_1,x_2] & [x_1,x_2-h_2] & \frac{5\left(\frac{2x_2}{h_2}-\frac{1}{h_2}+4h_2{}^{-2}\right)}{4\left(10h_2{}^{-2}+2h_1{}^{-2}\right)} \\ [0,x_2] & [0,x_2] & 0 \\ [0,x_2] & [h_1,x_2] & \frac{h_2{}^2}{\left(h_2{}^2+5h_1{}^2\right)} \\ [1,x_2] & [1,x_2] & 0 \\ [1,x_2] & [1-h_1,x_2] & \frac{h_2{}^2}{\left(h_2{}^2+5h_1{}^2\right)} \\ [x_1,0] & [x_1,0] & 0 \\ [x_1,0] & [x_1,h_2] & \frac{5h_1{}^2}{\left(h_2{}^2+5h_1{}^2\right)} \\ [x_1,1] & [x_1,1] & 0 \\ [x_1,1] & [x_1,1-h_2] & \frac{5h_1{}^2}{\left(h_2{}^2+5h_1{}^2\right)} \end{pmatrix}$$

Indeed if the following conditions are satisfied:

$$\begin{array}{c} h_1 \leq \max_{u_1,u_2}\left(\frac{0.4}{|u_2-u_1|}\right) \\ h_2 \leq \max_{x_2}\left(\frac{2}{|x_2-0.5|}\right) \end{array} \tag{16}$$

the matrix coefficients are positive and the sum of the coefficients on a same line is equal to 1. The matrix M is thus a transition matrix of a Markov chain. Moreover the optimal cost obeys:

$$(\lambda k+1)V^h = \min_{u_1,u_2}\left(M(u_1,u_2)V^h + kC(u_1,u_2)\right) \tag{17}$$

Thus we can use the contraction iteration:

$$V^h_{n+1} = \frac{\min_{u_1,u_2}\left(M(u_1,u_2)V^h_n + kC(u_1,u_2)\right)}{(\lambda k+1)} \tag{18}$$

6.3 Optimization of the Hamiltonian

The problem is to minimize the control-dependent part $\mathcal{H}^h$ of H^h :

$$\mathcal{H}^h = \frac{\left(6u_2x_2 - 5u_1x_2 + 5\partial_1^h V u_2 - 5\partial_1^h V u_1\right)}{5} \tag{19}$$

$\mathcal{H}^h$ is minimized with respect to $U = [u_1, u_2]$ by a projected gradient method:

$$U_{n+1} = \mathcal{P}_{[-1,1]\otimes[-1,1]}(U_n - \rho \frac{d\mathcal{H}^h(U_n)}{dU_n}) \tag{20}$$

that is:

$$\begin{pmatrix} u_{1n+1} = \mathcal{P}_{[0,1]}(u_{1n} - \frac{\rho\left(-5\partial_1^h V - 5x_2\right)}{5}) \\ u_{2n+1} = \mathcal{P}_{[0,1]}(u_{2n} - \frac{\rho\left(6x_2 + 5\partial_1^h V\right)}{5}) \end{pmatrix} \tag{21}$$

This algorithm converges when the step ρ satisfies:

$$0 < \rho < 2kK^{-2} \tag{22}$$

with:

$$k|V|^2 \leq D_U^2 \mathcal{H}^h(V)V \leq K|V|^2 \tag{23}$$

6.4 Numerical results

We perform a numerical test with:

number of space discretization points: $[13, 13]$
precision required for the implicit resolution: 0.0069
step for the implicit resolution: 0.00408
maximal number of iterations for the implicit resolution: 451
gradient algorithm step: $\rho = 1.0\,10^7$
maximal number of iterations for the optimization of the hamiltonian: 40
convergence tests
precision required: 0.01
The functions V and their contour-lines are displayed on the following figures.

References

[1] BENSOUSSAN,A.-LIONS,J.L.(1978). Applications des inéquations variationnelles en contrôle stochastique. Dunod.

[2] FLEMING,W.H.-RISHEL,R.(1975). Deterministic and Stochastic Optimal Control. Springer Verlag, New York.

[3] GOURSAT,M.-QUADRAT,J.P.(1975). Analyse numérique d'inéquations quasi variationnelles elliptiques associées à des problèmes de contrôle impulsionnel. IRIA Report.

[4] KUSHNER,H.J.(1977). Probabilty methods in stochastic control and for elliptic equations. Academic Press.

[5] LIONS,J.L. (1969). Quelques méthodes de résolution des problèmes aux limites non linéaires. Dunod Gauthier-Villars. Paris.

[6] LIONS,P.L.(1982).Generalized Solutions of Hamilton-Jacobi Equations. Research Notes in Mathematics, 69 , Pitman.

[7] QUADRAT,J.P.(1980). Existence de solution et algorithme de résolutions numériques de problèmes stochastiques dégénérés ou non. SIAM Journal of Control.

[8] QUADRAT,J.P.(1975).Analyse numérique de l'équation de Bellman stochastique. IRIA Report.

[9] STROOCK,F.-VARADHAN,S.R.S.(1979). Multidimensional Diffusion Process. Springer Verlag.

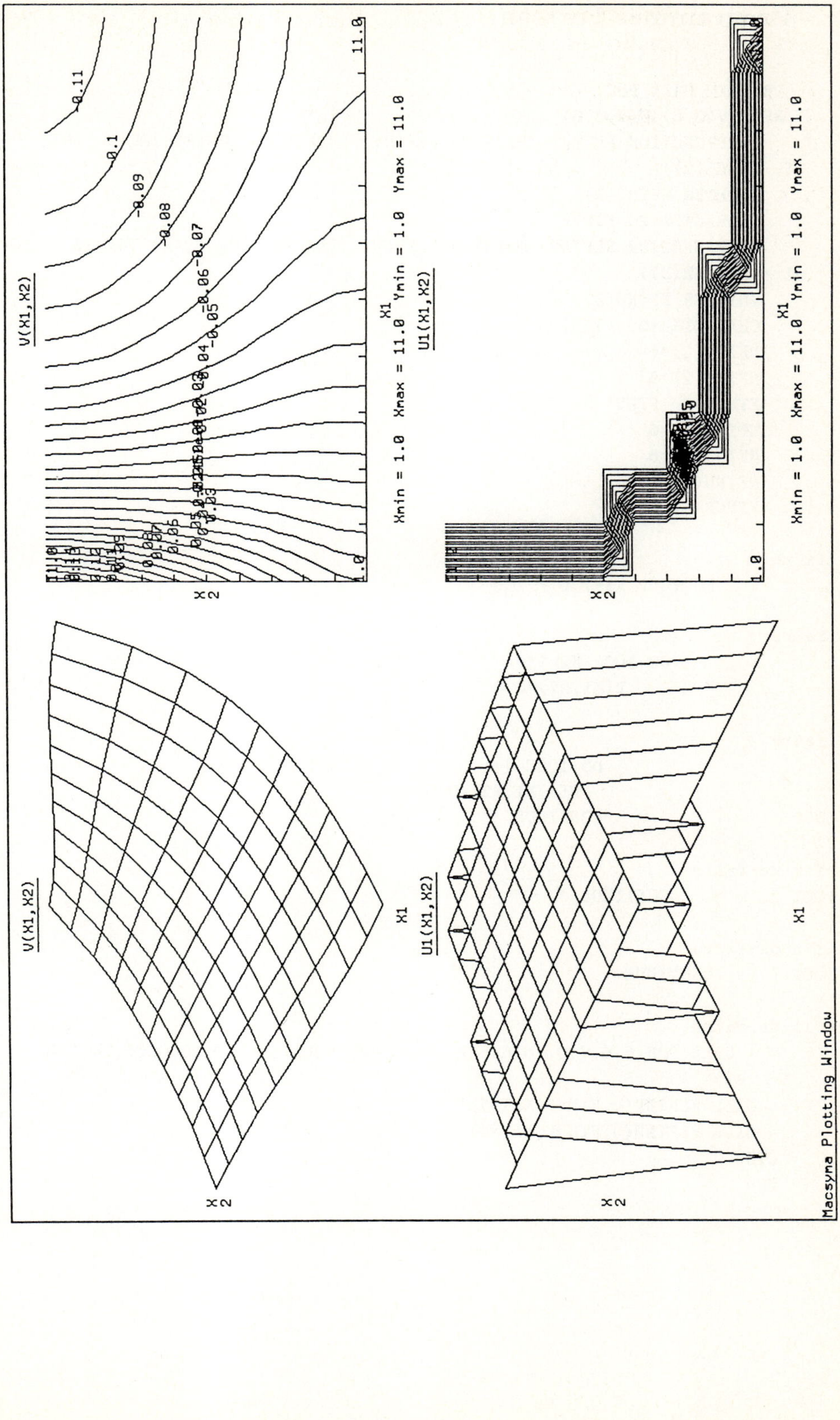

Macsyma Plotting Window
V(X1,X2)
X1
X2
V(X1,X2)
Xmin = 1.0 Xmax = 11.0 Ymin = 1.0 Ymax = 11.0
X1
X2
-0.11
-0.1
-0.09
-0.08
-0.07
-0.06
-0.05
11.0
1.0
U1(X1,X2)
X1
X2
U1(X1,X2)
Xmin = 1.0 Xmax = 11.0 Ymin = 1.0 Ymax = 11.0
X1
X2
1.0

7 Programmes Fortran

```
      PROGRAM MAIN_PPD
      REAL V(6,6),U(2,6,6)
        LISPFUNCTION FLFTNV 'MACSYMA:FLFTNV' (REAL(6,6),CHARACTER*20,INT
     1  EGER(2))
        INTEGER NFTNV(2)
        CHARACTER*20 FTNVF
        LISPFUNCTION FLFTNU 'MACSYMA:FLFTNU' (REAL(2,6,6),CHARACTER*20,I
     1  NTEGER(3))
        INTEGER NFTNU(3)
        CHARACTER*20 FTNUF
        NFTNV(1)=6
        NFTNV(2)=6
        FTNVF=  'FTNV'
        NFTNU(1)=6
        NFTNU(2)=6
        NFTNU(3)=2
        FTNUF=  'FTNU'

c-faire
            DO 1002 K1=1,6

c-faire
                DO 1001 K2=1,6
                    V(K1,K2)=0.0

c-faire
                      DO 1000 J=1,2
                        U(J,K1,K2)=0.0
 1000                 CONTINUE

c-fin-de-faire.
 1001           CONTINUE

c-fin-de-faire.
 1002       CONTINUE

c-fin-de-faire.
          CALL PPD(6,6,V,0.16363637,101,0.045454547,U,0.010000001,24,1.0
     1    e7)
        CALL FLFTNV(V,FTNVF,NFTNV)
        CALL FLFTNU(U,FTNUF,NFTNU)
      STOP
      END
```

```
      SUBROUTINE PPD(N1,N2,V,EPSIMP,IMPMAX,RO,U,EPS,NMAX,ROG)
      DIMENSION V(N1,N2),U(2,N1,N2)
C        Resolution de l equation de Kolmogorov dans le cas ou:
C              Les parametres sont
C              L etats-temps est:  X1 X2
C              La dynamique du systeme est decrite par l operateur
C        plus( - P2 X2 + Q2 + 0.2 Q1 + 0.5 P2 , Minu(
C                          (6 U2 - 5 U1) X2 + 5 P1 U2 - 5 P1 U1
C                          ------------------------------------ ) )
C                                          5
C              ou v(..) et w designe  le cout
C              ou pi designe sa derivee premiere par rapport a xi
C              ou qi designe sa derivee seconde par rapport a xi
C              Le probleme est statique
C              Les conditions aux limites sont:
C                   X2 = 0    p2 = 0.0
C                   X2 = 1    -p2 = 0.0
C                   X1 = 0    p1 = - 1.2 X2 - 0.1
C                   X1 = 1    -p1 = 0
C        Les nombres de points de discretisation sont: N1 N2
C                   X2 = 1 correspond a I2 = N2 - 1
C                   X2 = 0 correspond a I2 = 2
C                   X1 = 1 correspond a I1 = N1 - 1
C                   X1 = 0 correspond a I1 = 2
C        Le taux d actualisation vaut: 1
C        impmax designe le nbre maxi d iterations du systeme implicite
C        epsimp designe l erreur de convergence du systeme implicite
C        ro designe le pas de la resolution du systeme implicite
C                                  par une methode iterative
C        P2 est discretise par difference divise  symetrique
C        P1 est discretise par difference divise  symetrique
C        Minimisation par la methode de  gradient avec projection
C                                                de l'Hamiltonien:
C              (6 U2 - 5 U1) X2 + 5 P1 U2 - 5 P1 U1
C              ------------------------------------
C                                5
C        contraintes sur le controle:
C              0.0 =< U2 =< 1.0
C              0.0 =< U1 =< 1.0
C        nmax designe le nombre maxi d iteration de la methode de
C                                          gradient avec projection
C        eps designe l erreur de convergence de la methode de
C                                          gradient avec projection
C        rog designe le pas, qui est constant, dans la methode de gradi#
C       ent
      H2=0.999999/(N2-3)
      H1=0.999999/(N1-3)
      U2=U(2,1,1)
```

```
      U1=U(1,1,1)
      HIH2=H2**2
      HIH1=H1**2
      H22=2*H2
      H21=2*H1
      NM2=N2-1
      NM1=N1-1
      do 1017 I2=1,N2,1
      do 1017 I1=1,N1,1
      V(I1,I2)=0.0
 1017 CONTINUE
      IMITER=1
 1013 CONTINUE
      ERIMP=0
      do 1011 I1=1,N1,1
      X1=H1*(I1-2)
      V(I1,N2)=V(I1,N2-2)
      V(I1,1)=V(I1,3)
 1011 CONTINUE
      do 1009 I2=2,NM2,1
      X2=H2*(I2-2)
      V(N1,I2)=V(N1-2,I2)
      V(1,I2)=V(3,I2)-2*H1*(-1.2*X2-0.1)
 1010 CONTINUE
      do 1009 I1=2,NM1,1
      X1=H1*(I1-2)
      Q2=(V(I1,I2+1)-2*V(I1,I2)+V(I1,I2-1))/HIH2
      Q1=(V(I1+1,I2)-2*V(I1,I2)+V(I1-1,I2))/HIH1
      P2=(V(I1,I2+1)-V(I1,I2-1))/H22
      P1=(V(I1+1,I2)-V(I1-1,I2))/H21
      W=V(I1,I2)
      NITER=0
      WO=-1.0e20
 1001 CONTINUE
      NITER=NITER+1
      if (NITER-NMAX) 1002,1002,1003
 1003 CONTINUE
      WRITE(8,1801) I1,I2
 1801 FORMAT( ' descente n a pas converge' , 2i3 )
      GOTO 1004
 1002 CONTINUE
      UN1=ROG*X2+U1+P1*ROG
      UN2=-(6*ROG*X2-5*U2+5*P1*ROG)/5.0
      U1=UN1
      U2=UN2
      U1=AMAX1(U1,0.0)
      U1=AMIN1(U1,1.0)
      U2=AMAX1(U2,0.0)
```

```
      U2=AMIN1(U2,1.0)
      WW=((6*U2-5*U1)*X2+5*P1*U2-5*P1*U1)/5.0
      ER=ABS(WW-WO)
      if (ER-EPS) 1004,1004,1005
 1005 CONTINUE
      WO=WW
      GOTO 1001
 1004 CONTINUE
      U(1,I1,I2)=U1
      U(2,I1,I2)=U2
      WO=WW
      W1=-P2*X2+Q2+0.2*Q1+0.5*P2
      WO=W1+WO
      WO=WO-V(I1,I2)
      VNEW=RO*WO+V(I1,I2)
      V(I1,I2)=VNEW
      ERIMP=ABS(WO)+ERIMP
 1009 CONTINUE
      IMITER=IMITER+1
      if (IMITER-IMPMAX) 1016,1015,1015
 1016 CONTINUE
      if (EPSIMP-ERIMP) 1013,1012,1012
 1015 CONTINUE
      WRITE(8,1807)
 1807 FORMAT( ' schema implicite n a pas converge' )
 1012 CONTINUE
      RETURN
      END
```

II. DISCRETE-EVENT AND DISTRIBUTED SYSTEMS

A Control Theory for Discrete-Event Systems

W.M. Wonham
Systems Control Group
Department of Electrical Engineering
University of Toronto
Toronto, Ontario, Canada M5S 1A4

KEYWORDS/ABSTRACT: control theory / discrete-event systems / formal languages / automata

A discrete-event system is modelled as a controlled state-machine (generator) of a formal language. Concepts of controllability, observability, and decentralized and hierarchical control architecture are defined and explored. A guide is provided to the software package TCT for controller synthesis of small systems.

Contents

1. Introduction

Discrete-event systems encompass a wide variety of physical systems that arise in technology. These include manufacturing systems, traffic systems, office information systems, data-base management systems, communication protocols and computer networks. Typically the processes associated with these systems may be thought of as discrete (in time and in state space), asynchronous (not necessarily clock-driven) and in some sense generative (or nondeterministic). The underlying primitive concepts include events, conditions and signals.

This article will survey a control theory for discrete-event systems that originated with the thesis of P.J. Ramadge (see Sect. 7) and that has since been developed further by Ramadge and the author with others, including principally F. Lin. Along with the main results, we provide a user's guide to the computing program TCT, written by the author with P.A. Iglesias. TCT is suitable for developing small-scale examples of discrete-event control system synthesis on a personal computer.

In this theory our objective has been, so far, to define and explore such control-theoretic ideas as controllability, observability, and decentralized and hierarchical control from a purely qualitative point of view. To this end, and in view of the general lack of any alternative and comprehensive control theory of discrete-event systems, we have selected the simplest possible class of structures as our basic paradigm. While this class certainly does not support all the features that a complete theory of discrete-event systems should incorporate, it has proved to be both tractable and (we think)

NATO ASI Series, Vol. F47
Advanced Computing Concepts and Techniques
in Control Engineering
Edited by M.J. Denham and A.J. Laub

conceptually rich. It is also capable of natural extension in directions of greater realism.

Briefly stated, our approach is to regard the discrete-event system to be controlled as the generator of a formal language. By adjoining control structure, it is possible to vary the language generated by the system within certain limits. The desired performance of such a controlled generator is specified by stating that its generated language must belong to some specification language. It is often possible to meet this specification in an 'optimal', that is, minimally restrictive, fashion. The control problem is considered solved when a controller that causes the specification to be met has been shown to exist and to be constructible. In accordance with widely accepted control methodology, we take the state description of a system (and, in this case, a language) as fundamental.

The present article is based on lectures presented at the NATO Advanced Study Institute on *The Application of Advanced Computing Concepts and Techniques in Control Engineering*, Il Ciocco, Italy, 14-25 September 1987. The version of TCT to which we refer is 870907.

2. Representation of DES

In the control theory we shall develop in this article, a discrete-event system will be represented by its state description, or *transition structure*. Theoretically there is no compelling reason to restrict the state set to be finite; but TCT has no means of storing 'equations' for infinite transition structures. The example shown below represents a simple 'machine' with 3 states, labelled *I*, *W*, *D* for 'idle', 'working' and 'broken down'.

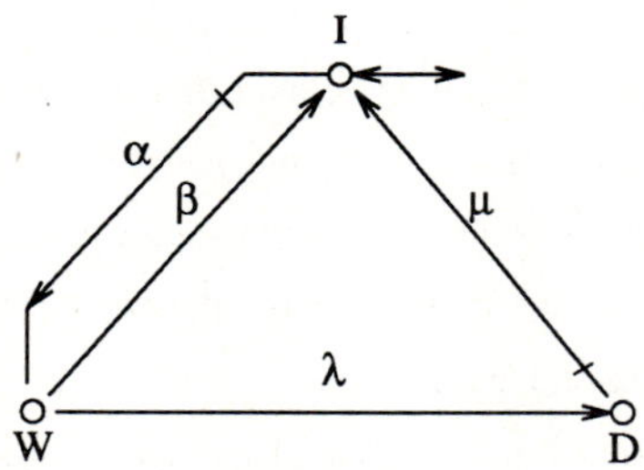

In general the *initial state* is labelled with an entering arrow (—→o), while a state labelled with an exiting arrow (o—→) will be called a *marker state*. If the initial state is also a marker state, it may be labelled with a double arrow (o↔). *Events*, or transitions, are either *uncontrollable* (o$\overset{\sigma}{\longrightarrow}$o) or *controllable* (o $\overset{\sigma}{\longmapsto}$o) (the tick indicating event controllability will be optional). Formally an event is a triple of the form (exit_state, event_label, entrance_state), for instance (I, α, W). The set of event_labels of the DES is its *alphabet* Σ, in this case $\{\alpha, \beta, \lambda, \mu\}$.

Operation of the 'machine' described above may be pictured as follows. Starting from state *I*, the machine executes a sequence of events in accordance with its transition graph. Each event is instantaneous in time. The events occur at quasi-random time instants. Upon occurrence of an event, the event_label is 'emitted' to some external agency. In this way the machine generates a string of event_labels, or a *word*, over the alphabet Σ. At a state such as W from which more than one event may occur, the machine will be thought of as selecting just one of the possibilities, in accordance with some mechanism that is hidden from the system analyst and therefore unmodelled. In this sense, machine operation is nondeterministic. However, it will be assumed that event labelling is 'deterministic' in the sense that distinct events exiting from a given state will always carry distinct labels. In general it may happen that two or more events exiting from distinct states may carry the same label. The marker states serve to distinguish those words that have some special significance, for instance represent a completed work cycle or sequence of work cycles. The controllable event labels, in this case $\{\alpha,\mu\}$, label transitions that can be enabled or disabled by an external agent. A controllable event can occur only if it is enabled. Thus if the event (labelled) α is enabled, but not otherwise, the machine can execute the transition to W from I; if the machine is at D, enablement of μ could be interpreted as the condition that the machine is under repair, and so may (eventually) execute the transition (D,μ,I). In general we shall often refer to 'the event σ', meaning any or all events (transitions) that happen to be labelled by σ.

The TCT procedure **create** allows the user to create and file a new DES. In response to the prompt, the user enters the DES name, number of states or *size*, the list of marker states and list of transitions. The TCT standard state set is the set {0,1,...,size-1}, with 0 the initial state. Event_labels must be entered as integers (between 0 and 99), where controllable events are odd and uncontrollable events are even. The procedure **show** (DES_name) displays an existing DES. For instance, the machine above could be coded and then shown as displayed below.

Example (TCT procedure **create**):

Name? mach

States? 3

TCT selects standard state set {0,1,2}

Marker State(s)? 0

User selects event labels {0,1,2,3} (events labelled 1 or 3 are controllable)

Transitions?

0 1 1
1 0 0
1 2 2
2 3 0

□

In general a DES is formally a 5-tuple

$$\mathbf{G}=(\Sigma,Q,\delta,q_o,Q_m)$$

Here Σ is the (finite) set of event labels, with $\Sigma=\Sigma_u\dot{\cup}\Sigma_c$, where Σ_u (resp. Σ_c) is the subset of uncontrollable (resp. controllable) events; Q is the (possibly infinite) state set; $\delta:\Sigma\times Q\rightarrow Q$ is the transition function (a partial function or *pfn*: we write $\delta(\sigma,q)!$ to mean that $\delta(\sigma,q)$ is defined); q_o is the initial state and Q_m is the subset of marker states.

Let Σ^* be the set of all finite strings of elements from Σ, plus the empty string, denoted ε, where $\varepsilon \notin \Sigma$. The transition function is inductively extended to strings by the recipe: $\delta(\varepsilon,q)=q$, and

$$\delta(s\sigma,q)=\delta(\sigma,(\delta(s,q))$$

whenever $q' := \delta(s,q)!$ and $\delta(\sigma,q')!$. The *languages associated with* **G** are the *closed behavior* of **G**:

$$L(\mathbf{G})=\{s\in\Sigma^* \mid \delta(s,q_o)!\}$$

and the *marked behavior* of **G**:

$$L_m(\mathbf{G})=\{s\in L(\mathbf{G}) \mid \delta(s,q_o)\in Q_m\}$$

The *reachable state subset* of **G** is

$$Q_r=\{q\in Q \mid (\exists s\in\Sigma^*)\delta(s,q_o)=q\}$$

G is *reachable* if $Q_r=Q$. The *coreachable state subset* is

$$Q_{cr}=\{q\in Q \mid (\exists s\in\Sigma^*)\delta(s,q)\in Q_m\}$$

G is *coreachable* if $Q_{cr}=Q$. **G** is *trim* if it is reachable and coreachable. Every transition structure has a trim substructure with maximal state set, obtained by replacing Q by $Q_{\text{new}} := Q_r \cap Q_{cr}$, Q_m by $Q_{m,\text{new}} := Q_m \cap Q_{\text{new}}$, and δ by $\delta_{\text{new}} := \delta \mid \Sigma \times Q_{\text{new}}$. The TCT procedure **trim** returns the trimmed version of its argument:

$$\mathbf{DES_{new}}=\mathbf{trim(DES)}$$

as illustrated below.

Example (TCT procedure **trim**):

DES$_{\text{new}}$ = trim(DES)

DES

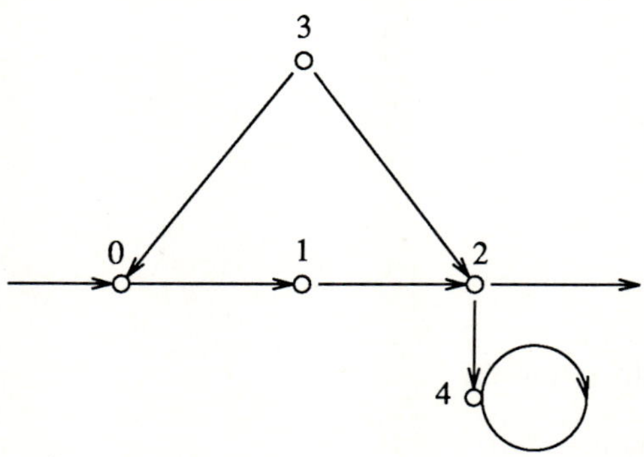

$Q_r=\{0,1,2,4\}$, $Q_{cr}=\{0,1,2,3\}$, $Q_{\text{new}}=Q_r\cap Q_{cr}$

DES$_{new}$

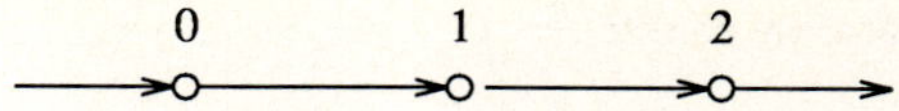

□

Let $L \subseteq \Sigma^*$ be an arbitrary language. The *prefix closure* of L is the language

$$\bar{L} = \{s \in \Sigma^* \mid (\exists t \in \Sigma^*)\ st \in L\}$$

namely the set of prefixes of strings in L. For a DES **G**, every reachable state of **G** is coreachable iff $L(\mathbf{G}) = \bar{L}_m(\mathbf{G})$. In particular this is true if **G** is trim. Because some marker state can then be reached from any reachable state, such **G** is said to be *nonblocking*. In modelling a DES it is often natural to take it to be trim, hence nonblocking.

If K is a language, we shall say that a DES **G** *represents* K if **G** is nonblocking and $L_m(\mathbf{G}) = K$. Thus the marked behavior of **G** is K and the closed behavior is $\bar{K}$.

Let L_1 and L_2 be languages over disjoint alphabets Σ_1 and Σ_2. The *shuffle language* $L_3 = L_1 \| L_2$ over $\Sigma_1 \dot{\cup} \Sigma_2$ is the set of all strings that are interleavings of strings in L_1 with strings in L_2. Suppose $L_1 = L_m(\mathbf{G}_1)$ and $L_2 = L_m(\mathbf{G}_2)$. The counterpart operation on transition structures is accomplished by the TCT procedure **shuffle**, according to

$$\mathbf{G}_3 = \mathbf{shuffle}(\mathbf{G}_1, \mathbf{G}_2)$$

which yields $L_m(\mathbf{G}_3) = L_m(\mathbf{G}_1) \| L_m(\mathbf{G}_2)$ and $L(\mathbf{G}_3) = L(\mathbf{G}_1) \| L(\mathbf{G}_2)$. Intuitively the shuffle DES $\mathbf{G}_3$ represents the concurrent, asynchronous behavior of $\mathbf{G}_1$ and $\mathbf{G}_2$; namely $\mathbf{G}_1$ and $\mathbf{G}_2$ are thought of as operating independently, with supposedly 'zero probability' of an event occurring simultaneously in each. **Shuffle** is useful in building up complex DES from simpler ones; but it should be recognized that the state size of the result goes up exponentially with the number of DES shuffled.

Example (TCT procedure **shuffle**):

G1

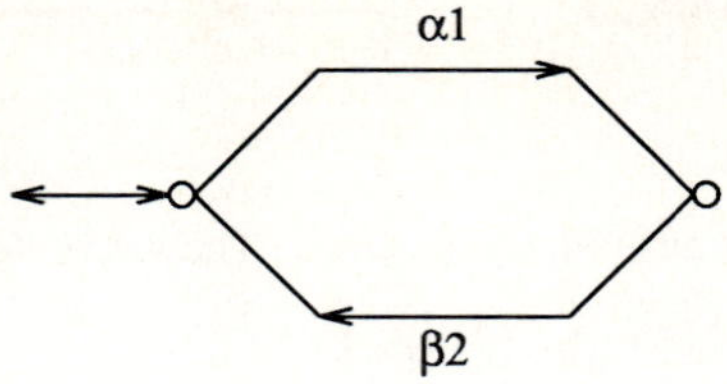

G2

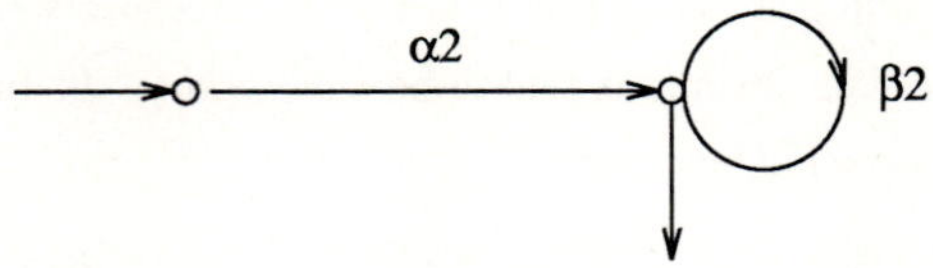

G3

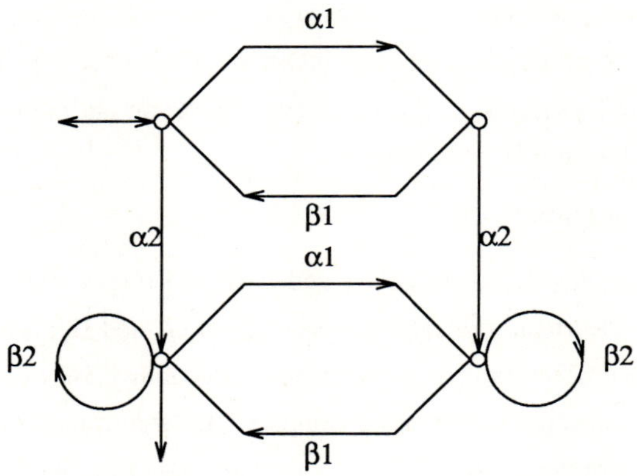

□

When a DES **G** is modelled specifically as an artifact to generate some language of interest (e.g. a specification language), it may be necessary to augment $L(\mathbf{G})$ by shuffling it with all the strings of some disjoint alphabet, say Σ_{ext}; for instance Σ_{ext} might consist of event labels from other, possibly unspecified, DES in a total system. The TCT procedure **selfloop** performs the corresponding operation on transition structures:

$$\mathbf{DES_{new} = selfloop(DES, EVENT\_LIST)}$$

Thus $L(\mathbf{DES_{new}}) = L(\mathbf{DES}) \| \Sigma_{ext}^*$. An illustration is the following.

Example (TCT procedure **selfloop**):

DES$_{\mathbf{new}}$ = selfloop(DES,EVENT_LIST)

DES

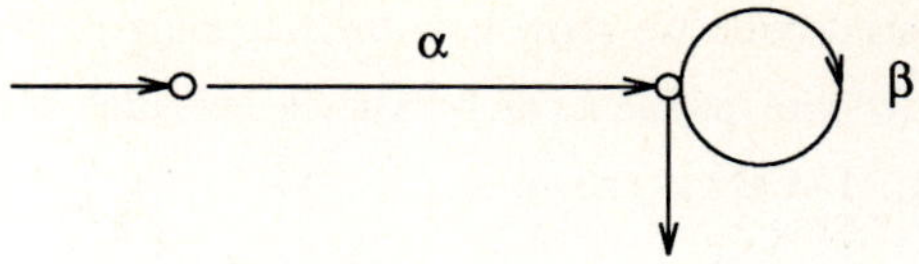

EVENT_LIST = {γ,λ}

DES_{new}

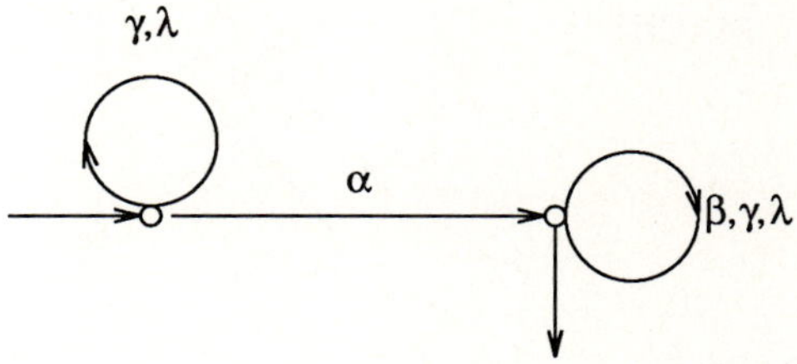

□

It is often useful to form the intersection of languages. For this let $\mathbf{G}_1$, $\mathbf{G}_2$ be DES. The TCT procedure **meet** determines $\mathbf{G}_3$=**meet**($\mathbf{G}_1$,$\mathbf{G}_2$) such that

(i) $L_m(\mathbf{G}_3)=L_m(\mathbf{G}_1)\cap L_m(\mathbf{G}_2)$

(ii) $\mathbf{G}_3$ is trim

In general it can be inferred (for the closed behaviors) only that $L(\mathbf{G}_3)\subseteq L(\mathbf{G}_1)\cap L(\mathbf{G}_2)$, even when each of $\mathbf{G}_1$ and $\mathbf{G}_2$ is trim. **Meet** is illustrated in the example to follow.

Example (TCT procedure **meet**):

G1

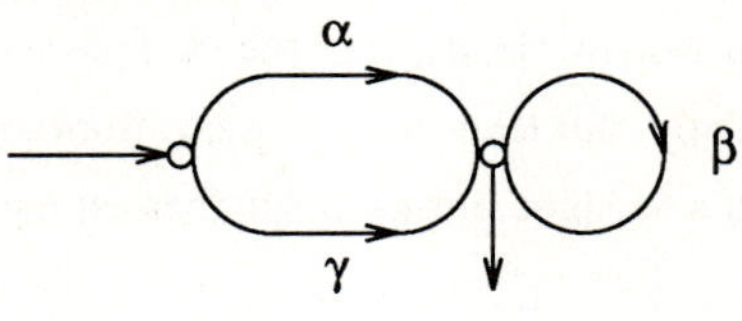

G2

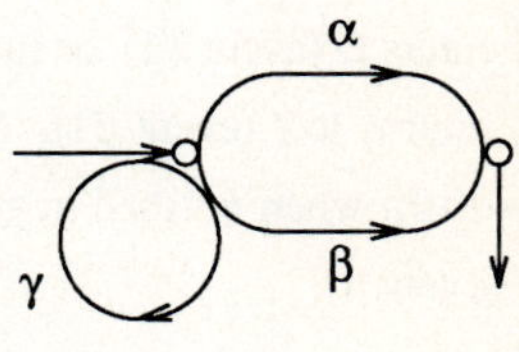

G3

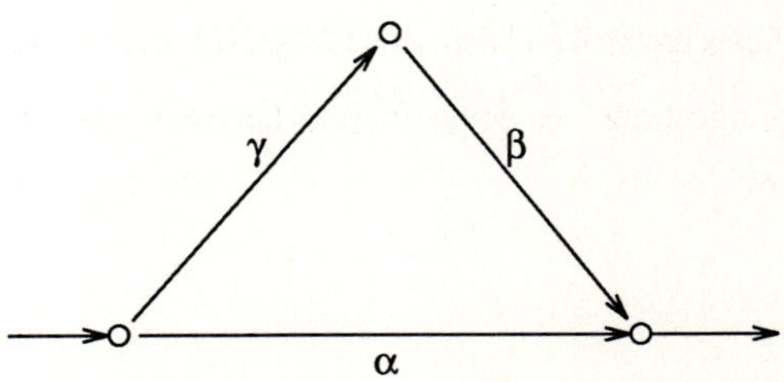

□

To conclude this section we show how the foregoing procedures can be used to build up the specifications for a control problem, to be known hereafter as Small Factory. We bring in two 'machines' **MACH1, MACH2** as shown.

MACHi

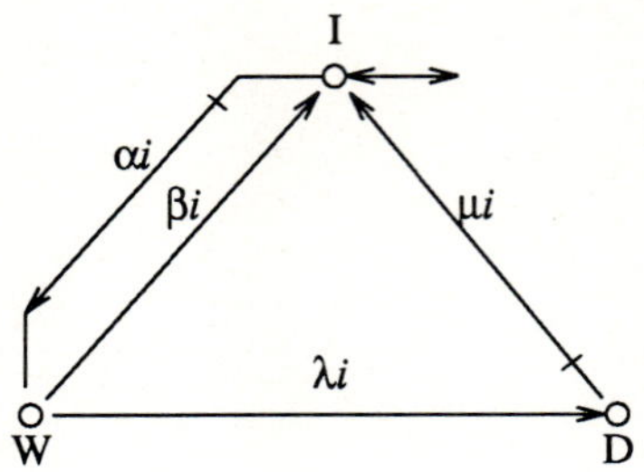

Define **MACH = shuffle (MACH1, MACH2)**. Small Factory consists of the arrangement shown below, where **BUF** denotes a buffer with one slot.

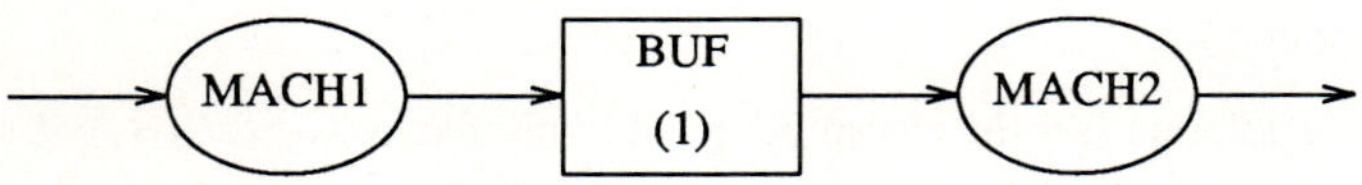

Small Factory operates as follows. Initially the buffer is empty. With the event α1, **MACH1** takes a workpiece from an infinite input bin and enters *W*. Subsequently **MACH1** either breaks down and enters *D* (event λ1), or successfully completes its work cycle, deposits the workpiece in the buffer, and returns to *I* (event β1). **MACH2** operates similarly, but takes its workpiece from the buffer and deposits it when finished in an infinite output bin. If a machine breaks down, then on repair it returns to *I* (event μ).

The informal specifications for admissible operation are the following:

1. The buffer must not overflow or underflow.

2. If both machines are broken down, then **MACH2** must be repaired before **MACH1**.

To formalize these specifications we bring in two language generators as the DES **SPEC1** and **SPEC2**, as shown below.

SPEC1

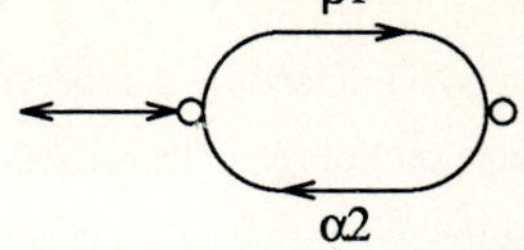

selfloop {α1,λ1,μ1,β2,λ2,μ2}

SPEC2

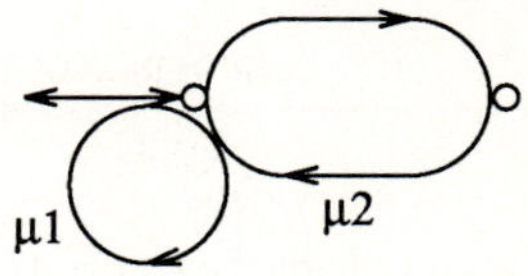

selfloop {α1,β1,λ1,α2,β2}

L(**SPEC1**) expresses the requirement that β1 and α2 must occur alternately, with β1 occurring first, while L(**SPEC2**) requires that if λ2 occurs then μ1 must not occur until μ2 occurs. The assignment of the initial state as a marker state in each of these DES is a matter of convenience. In each case selfloops must be adjoined to account for all events that are irrelevant to the specification but which may be executed in the plant. For the combined specification we form

SPEC = meet (SPEC1, SPEC2)

It is clear that **SPEC** is trim.

Temporarily denote by **G** the (as yet unknown) DES that would represent '**MACH** under control'. In general, for a given DES **G** and given specification DES **SPEC** as above, we shall say that **G** *satisfies* **SPEC** if

$$L_m(\mathbf{G}) \subseteq L_m(\mathbf{SPEC})$$

Typically **G** and **SPEC** will both be trim; and then it follows on taking closures,

$$L(\mathbf{G}) \subseteq L(\mathbf{SPEC})$$

The first condition can be checked in TCT by computing **COSPEC = complement** (SPEC,–), and then verifying that **meet(G,COSPEC) = EMPTY**, where **EMPTY** is the DES with # states = 0. The results will be illustrated for Small Factory in Sects. 3 and 4.

3. Controllability and Supervision of DES

Control of a DES **G** can be exercised by a supervisor **S** consisting of an automaton $\mathbf{S_{aut}}$ equipped with a state-determined output function ψ. The closed-loop structure is again a DES, denoted by **S/G**, the *action* of **S** on **G**. The feedback configuration is shown below.

S/G with $\mathbf{S_{aut}}$ in state x:

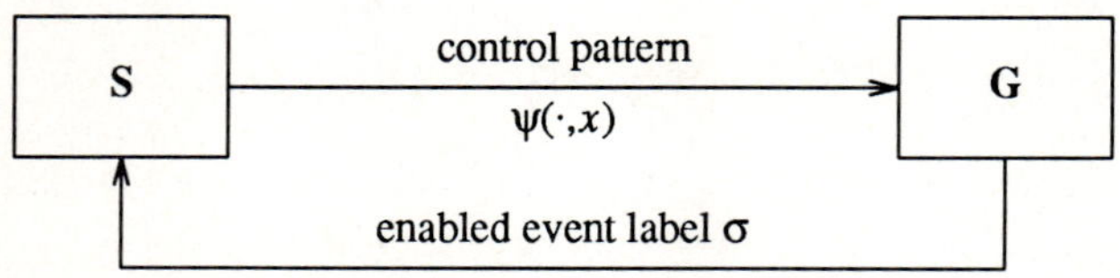

The automaton $\mathbf{S_{aut}}$ executes state transitions in response to the stream of event labels output by **G**. With $\mathbf{S_{aut}}$ in state x, the control pattern $\psi(\cdot,x)$ is imposed on the controllable transitions of **G**. In this way **S/G** generates a sublanguage of $L(\mathbf{G})$, the closed behavior *synthesized* by **S**. Formally

$$\mathbf{G}=(\Sigma,Q,\delta,q_o,Q_m)$$

$$\mathbf{S}=(\mathbf{S_{aut}},\psi)$$

$$\mathbf{S_{aut}}=(\Sigma,X,\xi,x_o,X_m)$$

$$\psi:\Sigma\times X\rightarrow\{0,1\}:$$

$$\psi(\sigma,x)=\begin{cases}0 \text{ or } 1 & \text{if } \sigma\in\Sigma_c\\ 1 & \text{if } \sigma\in\Sigma_u\end{cases}$$

When $\mathbf{S_{aut}}$ is in x, an event labelled σ is enabled (disabled) according as $\psi(\sigma,x)=1(0)$. The transitions of **S/G** are given by

$$(\sigma,q,x)\longmapsto\begin{cases}(\delta(\sigma,q),\xi(\sigma,x)) & \text{if } \psi(\sigma,x)=1\\ \text{undefined} & \text{otherwise}\end{cases}$$

A closed-loop transition is defined only if it is defined both for the supervisor and for the DES under supervision.

A supervisor **S** for **G** can be represented by its transition structure alone (without explicit mention of the control function ψ) provided $\mathbf{S_{aut}}$ is *neat*: i.e. the transition function $\xi(\sigma,x)!$ only for events σ at x such that $\psi(\sigma,x)=1$, namely σ is enabled. It is also convenient to arrange that $\mathbf{S_{aut}}$ be **G**-*complete*: i.e. $\xi(\sigma,x)!$ whenever there is some word s in $L(\mathbf{S/G})$ with $\xi(s,x_o)=x$, $q:=\delta(s,q_o)!$, $\delta(\sigma,q)!$ and $\psi(\sigma,x)=1$. A supervisor $\mathbf{S}=(\mathbf{S_{aut}},\psi)$ that is trim, neat and **G**-complete will be called *standard* (for **G**). In that case we identify $\mathbf{S}=\mathbf{S_{aut}}$ and omit reference to ψ. From now on it will be arranged or tacitly assumed that supervisors are standard.

The closed behavior of S/G, namely L(S/G), is the set of strings in Σ^* that are generated by S/G, while the marked behavior of S/G, namely L_m(S/G), is the set of strings marked by both S and G, i.e. the strings marked by G that survive under supervision and marking by S. We have

$$L(\mathbf{S/G}) = L(\mathbf{S}) \cap L(\mathbf{G}), \quad L_m(\mathbf{S/G}) = L_m(\mathbf{S}) \cap L_m(\mathbf{G})$$

A (standard) supervisor **S** is *proper* (for **G**) if **S/G** is nonblocking, i.e.

$$\bar{L}_m(\mathbf{S/G}) = L(\mathbf{S/G})$$

Since L_m(**S/G**) is represented by **meet (S,G)**, when **S** is proper we have

$$\mathbf{S/G} = \mathbf{meet}\,(\mathbf{S},\mathbf{G})$$

The properness of **S** is a nontrivial property that is not, for instance, guaranteed even if **S** and **G** are both trim. An example of improper **S** is displayed in the example below.

Example (Improper supervisory control):

G

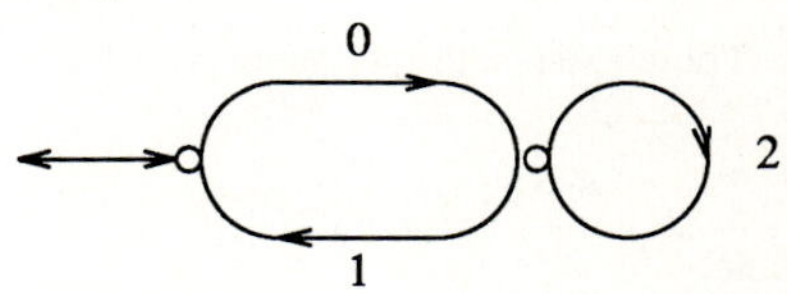

S

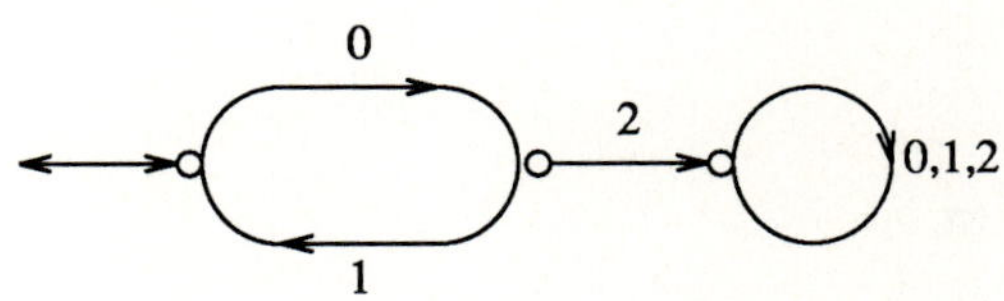

S/G

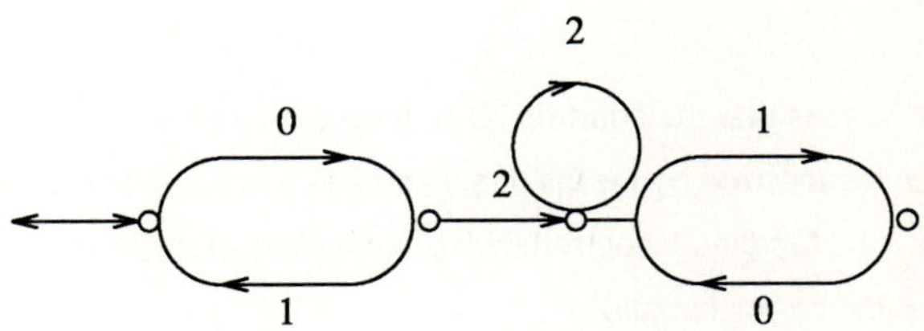

meet(S,G)

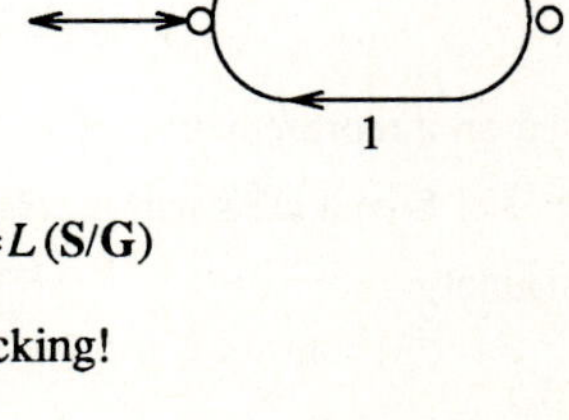

$$\mathbf{meet}\,(\mathbf{S},\mathbf{G}) \neq \mathbf{S/G}, \qquad \bar{L}_m(\mathbf{S/G}) \neq L(\mathbf{S/G})$$

S is *not* proper, in fact **S/G** fails to be nonblocking!

□

We now investigate those sublanguages of $L(\mathbf{G})$ that can be generated by $\mathbf{S/G}$ for some supervisor $\mathbf{S}$. Let $\mathbf{G}=(\Sigma,\_,\_,\_,\_)$, with $\Sigma=\Sigma_u \cup \Sigma_c$. A language $K \subseteq \Sigma^*$ is *controllable (with respect to* $\mathbf{G}$) if

$$\overline{K}\Sigma_u \cap L(\mathbf{G}) \subseteq \overline{K}$$

That is, if s is a word of $\overline{K}$ (i.e. a prefix of K) and $s\sigma$ is a word of $L(\mathbf{G})$ with σ an uncontrollable event, then $s\sigma$ must also be a word of $\overline{K}$. For a language K to be controllable, an uncontrollable event in $\mathbf{G}$ must never cause a string to exit from the prefix closure $\overline{K}$. In the example above, $\mathbf{S}$ and $\mathbf{S/G}$ represent controllable languages, but **meet(S,G)** represents an uncontrollable language. Notice that the languages $\varnothing$, $L(\mathbf{G})$ and Σ^* are always controllable.

Let the DES $\mathbf{G}$ be fixed. Our main results on controllability (with respect to $\mathbf{G}$) are the following.

Theorem 3.1

Let $K \subseteq L_m(\mathbf{G})$, $K \neq \varnothing$. There exists a proper supervisor $\mathbf{S}$ such that

$$L_m(\mathbf{S/G})=K$$

if and only if K is controllable.

□

Theorem 3.2

Let $E \subseteq \Sigma^*$ and let

$$\mathbf{C}(E)=\{K \subseteq E \mid K \text{ is controllable}\}$$

Then $\sup \mathbf{C}(E)$ exists.

□

Theorem 3.1 states that the controllability property characterizes the nonempty sublanguages of $L_m(\mathbf{G})$ that can be synthesized by means of supervisory control; while Theorem 3.2 states that there is a unique largest, or 'supremal' controllable sublanguage of a given language (however, this sublanguage could be the empty language).

Now let B be the language $B=E \cap L_m(\mathbf{G})$. By Theorems 3.1 and 3.2 there is a supervisor $\mathbf{S}$ that synthesizes the unique 'optimal' (i.e. supremal) controllable approximation to B, i.e. the language

$$K=\sup \mathbf{C}(B)$$

In fact $\mathbf{S}$ can always be based on a representation of K. That is, $\mathbf{S}$ can be constructed once the transition structure of K is known. Let $\mathbf{E}$ be a DES with $L_m(\mathbf{E})=E$. The TCT procedure **supcon** computes a trim representation $\mathbf{K}$ of K, namely

$$\mathbf{K}=\mathbf{supcon}\,(\mathbf{G},\mathbf{E})$$

To complete the description of **S**, the TCT procedure **condat** computes the control pattern (specifically, the controllable events that must be disabled) at each state of **K**:

K_dat = condat (G,K)

In the regular (finite-state) case, **supcon** works as follows. Assume $L_m(\mathbf{G})$ and L are regular and denote by $\|L\|$ the number of states in a canonical (minimal-state) representation of a language L. Let $\mathbf{P}(\Sigma^*)$ be the power set of Σ^*, i.e. the set of all languages over Σ. Define the operator

$$\Omega : \mathbf{P}(\Sigma^*) \longrightarrow \mathbf{P}(\Sigma^*)$$

according to

$$\Omega(K) = B \cap \sup \{T \mid T \subseteq \Sigma^*,\ T = \bar{T},\ T\Sigma_u \cap L(\mathbf{G}) \subseteq \bar{K}\}$$

It can be shown that sup $\mathbf{C}(B)$ is the largest fixpoint of Ω. Let $K_o = B$, $K_{j+1} = \Omega(K_j)$ $(j = 0, 1, 2, ...)$. It turns out that

$$\sup \mathbf{C}(B) = \lim K_j, \quad j \to \infty$$

Furthermore the limit is attained after a finite number of steps that is of worst-case order $\|L_m(\mathbf{G})\|\ \|E\|$. In TCT the operator Ω is implemented by a simple backtracking operation on the transition structure of B.

As an example, we consider Small Factory. The result for

SUPER := supcon (MACH,SPEC)

is displayed below. By tracing through the transition graph the reader may convince himself that the specifications are satisfied; and the theory guarantees that **SUPER** represents the freest possible behavior of **MACH** under the stated constraints.

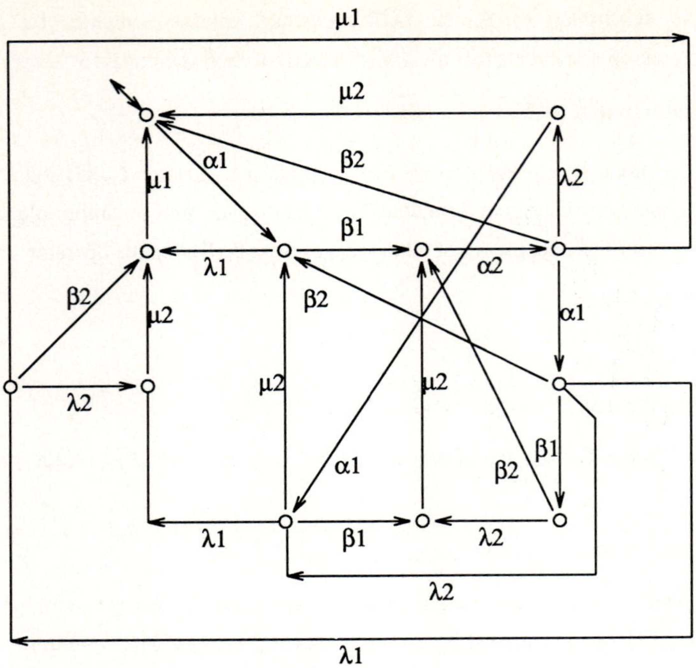

In practice a simpler supervisor that enforces the specifications can sometimes be constructed directly by intuition. For Small Factory we construct **SIMSUP** as shown below. It will be checked later that **SIMSUP** is equivalent to **SUPER** in the sense that

$$L(\mathbf{SIMSUP/MACH}) = L(\mathbf{SUPER/MACH}) \qquad (=L(\mathbf{SUPER}))$$

and

$$L_m(\mathbf{SIMSUP/MACH}) = L_m(\mathbf{SUPER/MACH}) \qquad (=L_m(\mathbf{SUPER}))$$

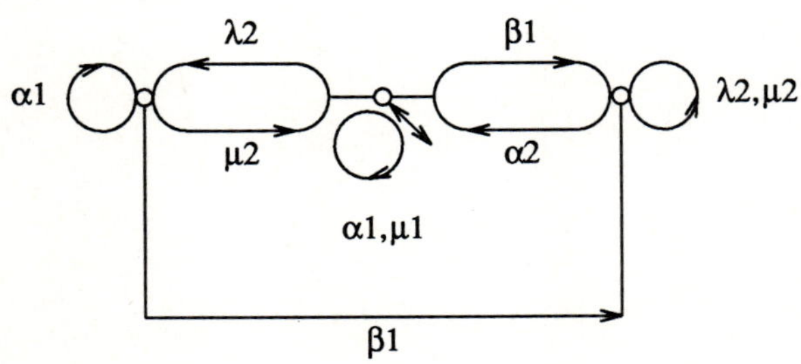

selfloop {λ1,β2}

□

To conclude this section we briefly discuss the problem of *mutual exclusion*. Many problems of resource sharing impose a requirement of mutual exclusion of various 'agents' (component DES) from certain critical state combinations: e.g. a section of guideway might be prohibited to more than one vehicle at a time. Mutual exclusion is an instance of a common type of control problem: construct a controller that holds the state of the controlled object in a given subset T of its state set. Usually this amounts to constructing a suitable subset $R \subseteq T$ with the property that R can be made *invariant* under the transitions of the system state: namely if the system starts in R then it can be controlled to remain in R thereafter. To allow the system maximal freedom of behavior one may try to arrange that R be as large as possible subject to these requirements.

The TCT procedure **mutex** can be used to construct maximal invariant sublanguages and to solve mutual exclusion problems:

G3 = mutex (G1, G2, EXCLUDED_STATE_PAIRS)

G3 represents the largest sublanguage generated by **shuffle(G1,G2)** subject to the requirement that state pairs in the **EXCLUDED_STATE_PAIRS** list are not reachable (with suitable controllable event disablements). **G3** is trim, but need not be controllable (although it often is). In general the mutual exclusion problem is solved by taking **G3** as specification and using **supcon**:

MUTEXSUP = supcon (shuffle(G1,G2),G3)

In particular suppose

G2 = DUMMY: $\longrightarrow o \longrightarrow$

EXCLUDED_STATE_PAIRS = [(i 1,0),(i 2,0),...]

Then **G3** represents the maximal language generated by **G1** with controlled exclusion from the states [i1,i2,...].

A simple illustration of the use of **mutex** is displayed below.

Example (TCT procedure **mutex**)

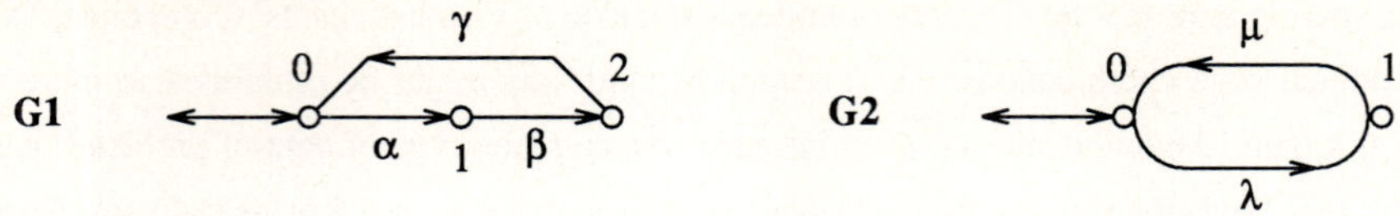

EXCLUDED_STATE_PAIRS list: [(2,1)]

G3=mutex(**G1**,**G2**,[(2,1)])

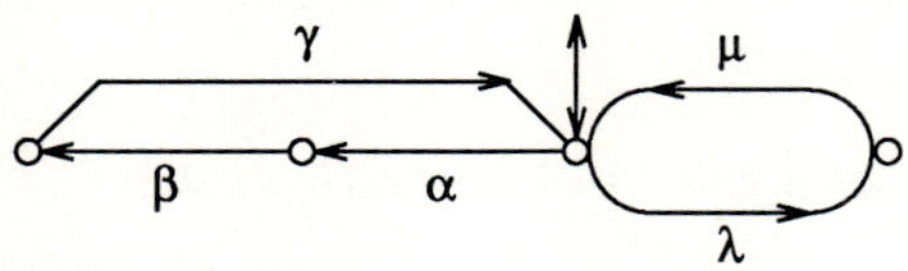

G3 is trim and, in this case, controllable.

□

As a more elaborate example of mutual exclusion, consider the following. In a certain Zoo, Tiger and Goat share a compound consisting of five pens.

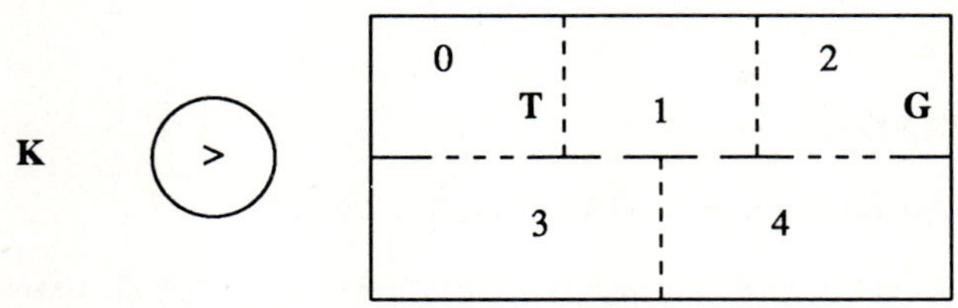

The inmates' home pens are 0 and 2, as shown. Each animal may pass from a pen to one or more adjoining pens by its own one-way gates, of which some are always open but others can be switched open or shut from outside the compound by a vigilant Keeper. The Keeper wishes to allow his charges maximum freedom to wander within the compound, subject to the restrictions that they are never to be found within the same pen at once, and they can always make their way home by a suitable sequence of moves. Thus the Keeper requires a rule for switching the gates as a function of the animals' current locations.

TIGER

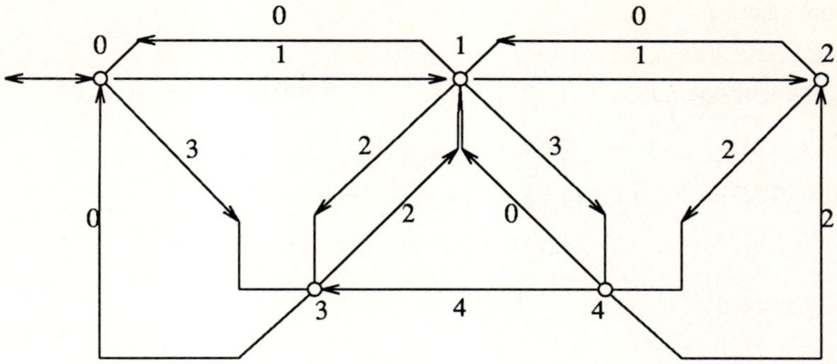

GOAT

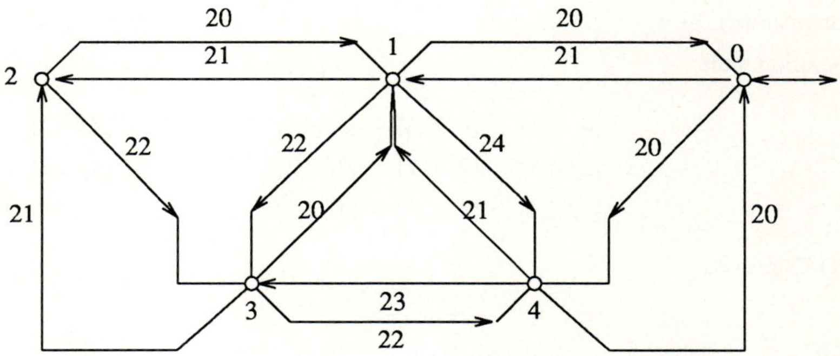

The Keeper's problem is solved by computing

ZOO = **mutex** (**TIGER**, **GOAT**, [(0,2),(1,1),(2,0),(3,3),(4,4)])

TI_GOAT = **shuffle** (**TIGER**, **GOAT**)

ZOODAT = **condat** (**TI_GOAT**, **ZOO**)

The **ZOODAT** table shows that **ZOO** is controllable. So **ZOO** is the supremal controllable sublanguage of **TI_GOAT** subject to the stated requirement of mutual exclusion. **ZOO** is tabulated below, along with the Keeper's supervisory control. One can verify that Goat may visit all pens except Tiger's home, while Tiger may only visit pens 1 and 3.

TIGER # states: 5 state set: 0...4 initial state: 0

marker states: 0

transitions: 13

transition table:

[0,1,1]	[0,3,3]	[1,0,0]	[1,1,2]	[1,2,3]
[1,3,4]	[2,0,1]	[2,2,4]	[3,0,0]	[3,2,1]
[4,0,1]	[4,2,2]	[4,4,3]		

TIGER printed.

GOAT # states: 5 state set: 0...4 initial state: 0

marker states: 0

transitions: 14

transition table:

[0,21,1]	[0,20,4]	[1,20,0]	[1,21,2]	[1,22,3]
[1,24,4]	[2,20,1]	[2,22,3]	[3,20,1]	[3,21,2]
[3,22,4]	[4,20,0]	[4,21,1]	[4,23,3]	

GOAT printed.

ZOO # states: 8 state set: 0...7 initial state: 0

marker states: 0

transitions: 26

transition table:

[0,20,1]	[0,21,3]	[0, 3,5]	[0, 1,6]	[1,20,0]
[1,23,2]	[1,21,3]	[1, 3,4]	[1, 1,7]	[2,22,1]
[2,20,3]	[3,20,0]	[3,24,1]	[3,22,2]	[4, 0,1]
[4,20,5]	[4, 2,7]	[5, 0,0]	[5,20,4]	[5, 2,6]
[6, 0,0]	[4, 2,5]	[5,20,7]	[7, 0,1]	[7, 2,4]
[7,20,6]				

ZOO printed.

ZOODAT

Control Data are displayed by listing the supervisor states where disabling occurs, together with the events that must be disabled there.

Control Data:

2:	21	1	3		3:	21	1	3	
4:	21	23			5:	21			
6:	21	1	3		7:	21	23	1	3

ZOODAT printed.

□

4. Modular Supervision of DES

The key to proper supervision and modularity is the concept of nonconflicting languages. Let $L_1, L_2 \subseteq \Sigma^*$. Automatically

$$\overline{L_1 \cap L_2} \subseteq \bar{L}_1 \cap \bar{L}_2$$

namely a prefix of any word common to L_1 and L_2 is a common prefix of L_1 and L_2. The languages L_1 and L_2 are *nonconflicting* if equality holds:

$$\overline{L_1 \cap L_2} = \bar{L}_1 \cap \bar{L}_2$$

namely every common prefix can be completed to a common word. Observe that two closed languages are always nonconflicting.

Example (Conflicting and nonconflicting languages)

$\Sigma = \{\alpha, \beta, \gamma\}$

$L_1 = \{\alpha\beta\}$ 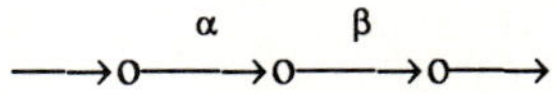

$L_2 = \{\alpha\gamma\}$ 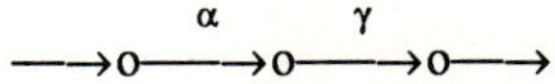

$\bar{L}_1 \cap \bar{L}_2 = \{\varepsilon, \alpha\}$, $\overline{L_1 \cap L_2} = \varnothing$

L_1, L_2 are conflicting!

Suppose instead

$L_1 = \{\alpha, \alpha\beta\}$ 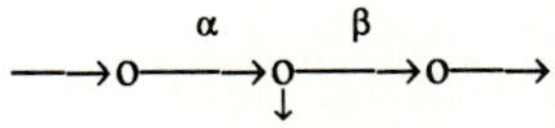

$L_2 = \{\alpha, \alpha\gamma\}$ 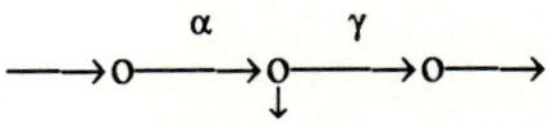

$\bar{L}_1 \cap \bar{L}_2 = \{\varepsilon, \alpha\} = \overline{L_1 \cap L_2}$

Now L_1, L_2 are nonconflicting.

□

As an application suppose that **G** is a trim DES to be supervised. Under what conditions is a given DES **S** a proper supervisor for **G**?

Theorem 4.1

Assume that **S** is trim (in particular $\bar{L}_m(\mathbf{S}) = L(\mathbf{S})$), that $L_m(\mathbf{S})$ is controllable with respect to **G**, and that $L_m(\mathbf{S})$ and $L_m(\mathbf{G})$ are nonconflicting. Then **S** is a proper supervisor for **G**, and

$$\mathbf{S/G} = \mathbf{meet}(\mathbf{S}, \mathbf{G})$$

□

The requirement that $L_m(\mathbf{S})$ and $L_m(\mathbf{G})$ be nonconflicting is necessary to ensure that **S** is proper, i.e. that **S/G** is nonblocking. In case $L_m(\mathbf{S}) \subseteq L_m(\mathbf{G})$, the 'nonconflicting' condition is satisfied automatically.

Example (Blocking due to conflict)

$$\Sigma = \{\alpha, \beta, \gamma, \delta\}, \quad \Sigma_c = \{\beta, \delta\}$$

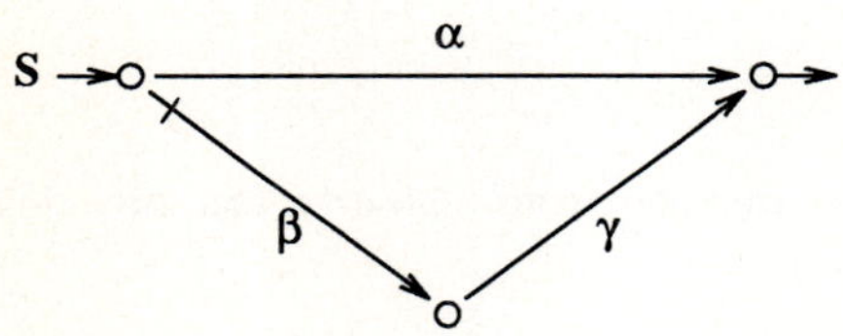

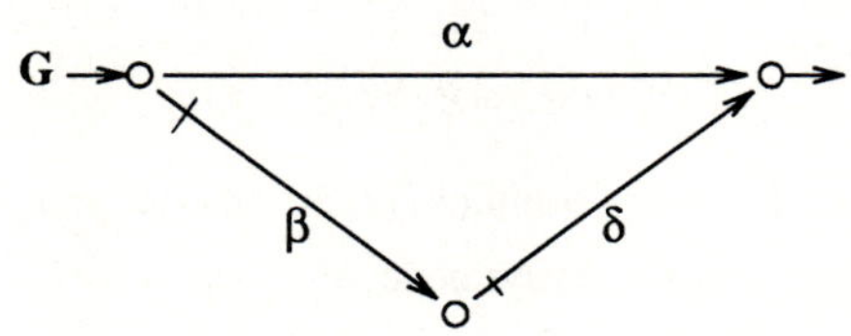

S and **G** are trim; $L_m(\mathbf{S})$ is controllable; however, $L_m(\mathbf{S})$, $L_m(\mathbf{G})$ conflict...

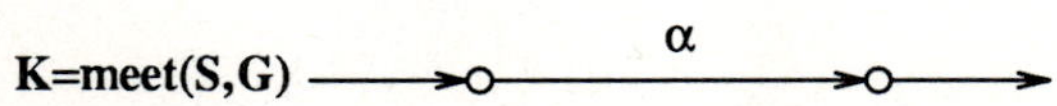

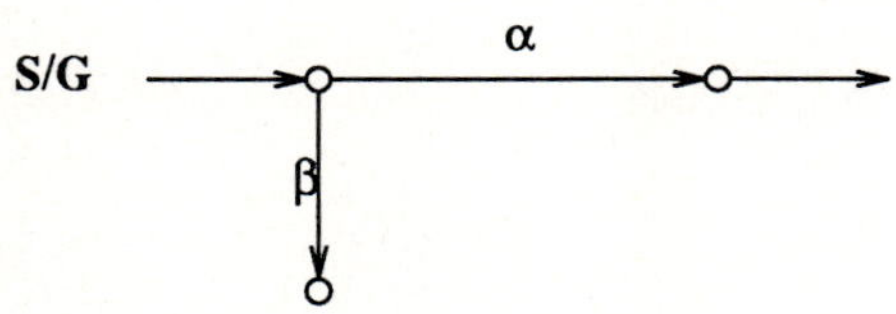

$$\bar{L}_m(\mathbf{S/G}) = \{\varepsilon, \alpha\} \underset{\neq}{\subset} \{\varepsilon, \alpha, \beta\} = L(\mathbf{S/G})$$

S is not proper!

□

The TCT procedure **nonconflict(G1,G2)** checks whether DES **G1** and **G2** are nonconflicting, by testing whether all reachable states of the product transition structure are coreachable.

As an application of these ideas we consider the simplified supervisor **SIMSUP** proposed in Sect. 2 for Small Factory. We claim that **SIMSUP** is proper and is equivalent to **SUPER** (in the sense described in Sect. 2). By inspection, **SIMSUP** is trim. Then **condat(MACH,SIMSUP)** shows that **SIMSUP** is controllable with respect to **MACH**. Next, **nonconflict(MACH,SIMSUP)** shows that L_m(**MACH**) and L_m(**SIMSUP**) are nonconflicting. By Theorem 4.1, **SIMSUP** is a proper supervisor for **MACH**, and

SIMSUP/MACH = meet (MACH,SIMSUP) = SSM (say)

A call to **isomorph(SSM,SUPER)** shows that

L_m(**SSM**) = L_m(**SUPER**)

which verifies the claim.

We shall introduce modular supervision again by way of Small Factory. As displayed below, introduce supervisors **BUFSUP** and **BRSUP** to enforce the buffer and the breakdown/repair specifications independently. It is easily verified that **BUFSUP** and **BRSUP** each satisfies the conditions of Theorem 4.1, and thus each is a proper supervisor for **MACH**. Is the combination of **BUFSUP** and **BRSUP** working concurrently also a proper supervisor for **MACH**, or could blocking occur?

BUFSUP

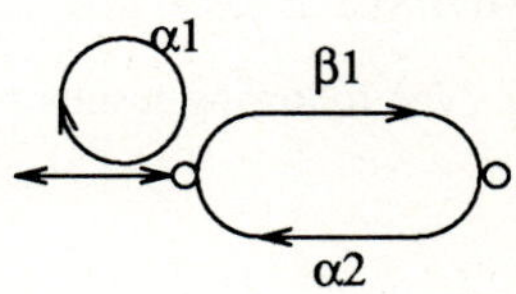

selfloop{λ1,μ1,β2,λ2,μ2}

BRSUP

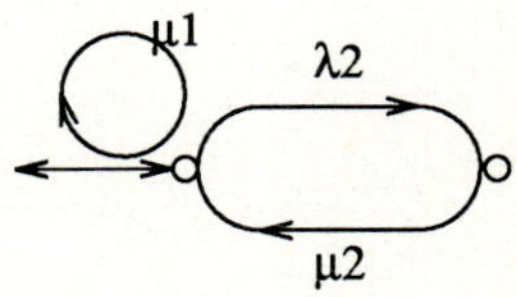

selfloop{α1,β1,λ1,α2,β2}

In general, let S_1 and S_2 be proper supervisors for a trim DES **G**. With

$$S_i = (\Sigma, X_i, \xi_i, x_{oi}, X_{mi})$$

define the conjunction

$$S_1 \wedge S_2 = \mathbf{Rch}(\Sigma, X, \xi, x_o, X_m)$$

by: $X = X_1 \times X_2$, $x_o = (x_{o1}, x_{o2})$, $X_m = X_{m1} \times X_{m2}$, and

$$\xi(\sigma,(x_1,x_2))=(\xi_1(\sigma,x_1),\xi_2(\sigma,x_2));$$

then **Rch** extracts the reachable substructure. We have

Theorem 4.2

$$L_m((\mathbf{S}_1 \wedge \mathbf{S}_2)/\mathbf{G})=L_m(\mathbf{S}_1/\mathbf{G}) \cap L_m(\mathbf{S}_2/\mathbf{G})$$

Furthermore $\mathbf{S}_1 \wedge \mathbf{S}_2$ is a proper supervisor for **G** iff it is trim, and $L_m(\mathbf{S}_1/\mathbf{G})$ and $L_m(\mathbf{S}_2/\mathbf{G})$ are nonconflicting.

□

Returning to Small Factory, and taking **BUFSUP** and **BRSUP** as modular component supervisors, let

MODSUP = BUFSUP ∧ BRSUP

By inspection,

MODSUP = meet(BUFSUP,BRSUP)

Next **nonconflict (MACH,MODSUP)** verifies the nonconflicting property, so that **MODSUP** is proper. Finally, one verifies that

meet(MACH,MODSUP) = SUPER

namely L_m(**MODSUP/MACH**)$=L_m$(**SUPER**), so **MODSUP** is optimal. **MODSUP** has 4 states as compared to 3 for **SIMSUP**, but **MODSUP** is easier to design and to comprehend than **SIMSUP**.

Let the DES **G** be arbitrary. The following results provide additional insight when exploiting modularity.

Theorem 4.3

Let $K_1,K_2 \subseteq \Sigma^*$ be controllable with respect to **G**. If K_1 and K_2 are nonconflicting then $K_1 \cap K_2$ is controllable with respect to **G**.

□

Theorem 4.4

Let $E_1,E_2 \subseteq \Sigma^*$. If sup **C**($\mathbf{E}_1$) and sup **C**($\mathbf{E}_2$) are nonconflicting then

$$\sup \mathbf{C}(E_1 \cap E_2)=\sup \mathbf{C}(E_1) \cap \sup \mathbf{C}(E_2)$$

□

Now let **G = MACH**. Theorem 4.3 holds with

$K_1=L_m$(**BUFSUP**), $K_2=L_m$(**BRSUP**)

while Theorem 4.4 holds with

$$E_1 = L_m(\mathbf{SPEC1}), \qquad E_2 = L_m(\mathbf{SPEC2})$$

As an example of the foregoing ideas we consider Big Factory, as described below. Two machines as before operate in parallel to feed a buffer with capacity 3; a third machine empties the buffer.

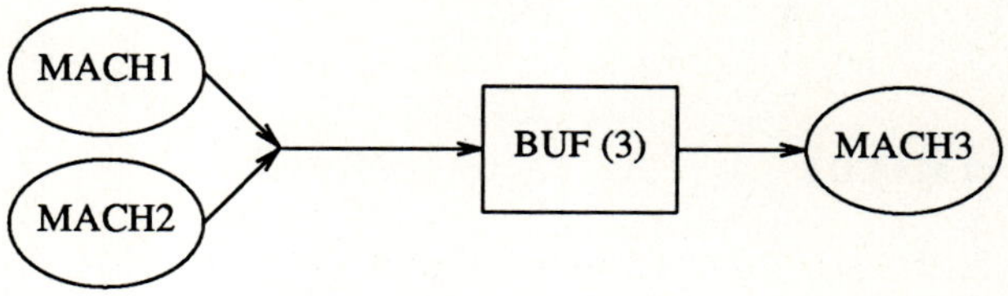

MACHi

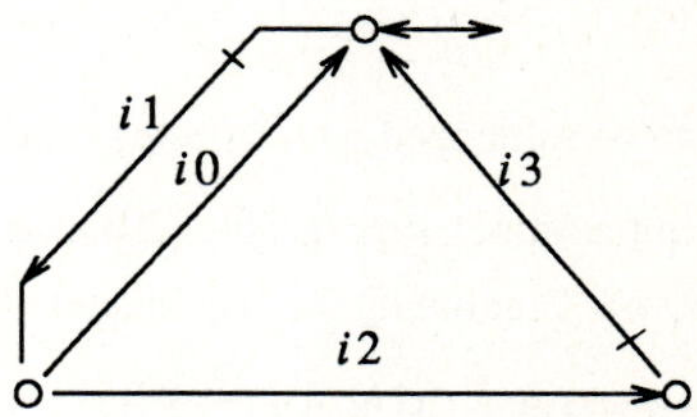

The informal specifications are:

1. Buffer must not overflow or underflow.
2. **MACH1** and **MACH2** are repaired in order of breakdown.
3. **MACH3** has priority of repair over **MACH1** and **MACH2**.

As the plant we take

BIGMACH = shuffle(shuffle(MACH1,MACH2),MACH3)

To formalize the specifications we construct the DES shown below:

Buffer overflow/underflow:

BUF

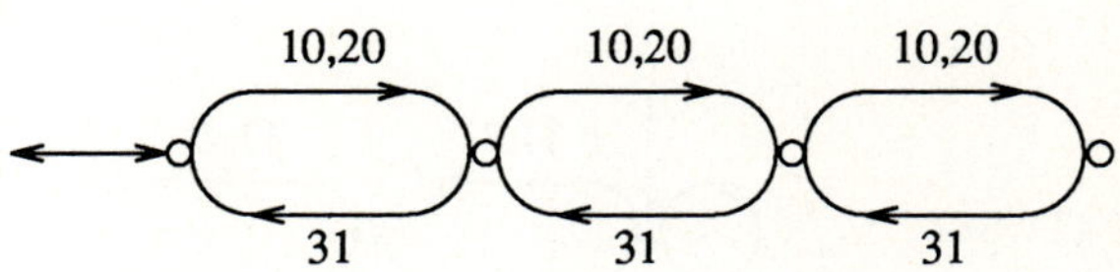

Breakdown/repair of **MACH1, MACH2**:

BR12

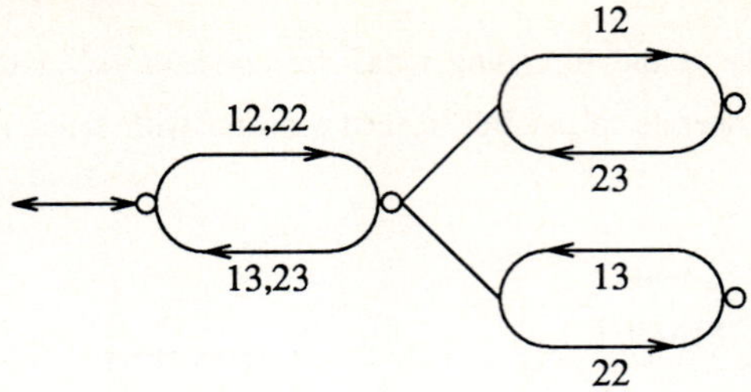

Breakdown/repair of **MACH3**:

BR3

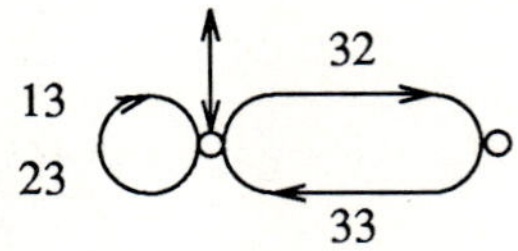

Each DES is understood to be selflooped with its complementary subalphabet.

We first consider 'monolithic' supervision. **BIGMACH** turns out to have 27 states and 108 transitions (written (27,108)). Combining the specification languages into their intersection, we define

BIGSPEC = meet(meet(BUF,BR12),BR3) (32,248)

For the 'monolithic' supervisor we then obtain

BIGSUP = supcon(BIGMACH,BIGSPEC) (96,302)

By the theory, the transition structure of the DES **BIGSUP** is that of the supremal controllable sublanguage of L_m(**BIGMACH**) that is contained in the specification language L_m(**BIGSPEC**). Thus **BIGSUP** is guaranteed to be the optimal (i.e. minimally restrictive) proper supervisor that controls **BIGMACH** subject to the three legal specifications. Nevertheless, **BIGSUP** is a rather cumbersome structure to implement directly, and it makes sense to consider a modular approach.

For prevention of buffer overflow alone, we compute

status1 = # empty buffer slots - # feeder machines at work

STATUS1

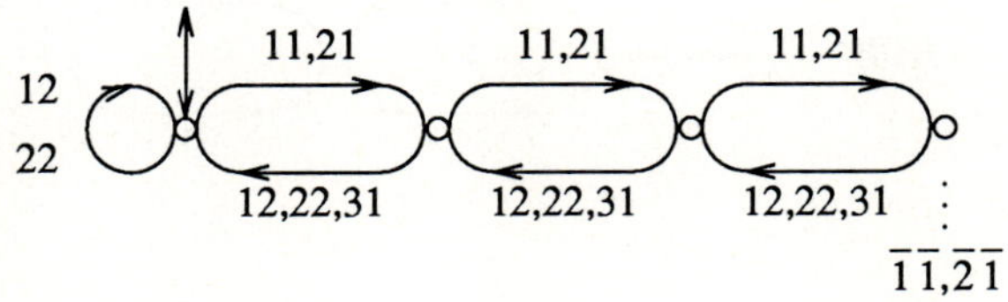

STATUS1 disables *11* and *21* when status1 = 0, and is a proper supervisor for **BIGMACH**. For prevention of buffer underflow alone, we compute

status2 = # full slots in buffer

STATUS2

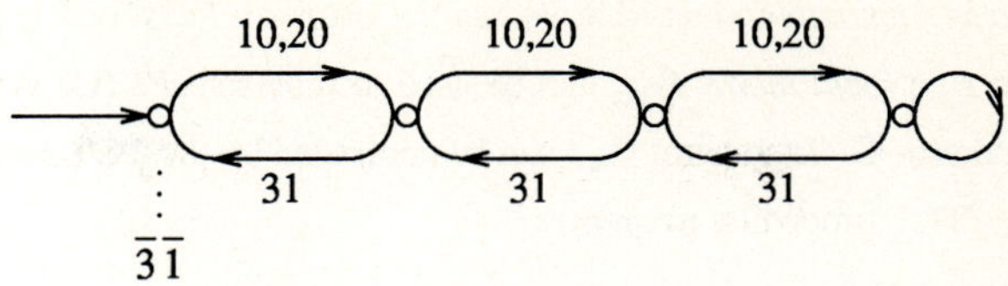

STATUS2 disables 31 when status2 = 0 and is also proper. For control of breakdown/repair, **BR12** and **BR3** are themselves proper supervisors. It can be verified that optimal (and proper) supervision of the buffer is enforced by

$$\mathbf{STATUS} = \mathbf{STATUS1} \wedge \mathbf{STATUS2}$$

while optimal (and proper) supervision of breakdown/repair is enformed by

$$\mathbf{BR} = \mathbf{BR12} \wedge \mathbf{BR3}$$

Finally, optimal (and proper) supervision with respect to all the legal specifications is enforced by

$$\mathbf{BIGMDSUP} = \mathbf{STATUS} \wedge \mathbf{BR}$$

Obviously **BIGMDSUP** is much simpler to implement than **BIGSUP**, to which it is equivalent in supervisory action.

As an exercise, the reader might like to construct a 9-state supervisor that is equivalent to **STATUS**.

5. Decentralized and Hierarchical Supervision of DES

Decentralized and hierarchical supervision are based on the concepts of local or partial information. For this we need various notions of projection of languages, which can be used to describe a local 'agent', i.e. observer and/or supervisor, as pictured below.

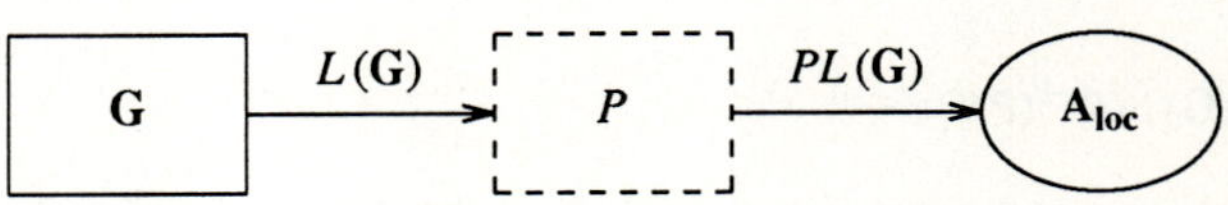

For a DES **G**, over the alphabet Σ, assume that 'locally' only events with labels in a subalphabet $\Sigma_{loc} \subseteq \Sigma$ can be observed. The simplest type of projection simply erases the unobservable event labels in the complementary subalphabet $\Sigma - \Sigma_{loc}$:

$$P : \Sigma^* \rightarrow \Sigma_{loc}^*$$

$$P\varepsilon = \varepsilon$$

$$P\sigma = \varepsilon, \qquad \sigma \in \Sigma - \Sigma_{loc}$$

$$P\sigma = \sigma, \qquad \sigma \in \Sigma_{loc}$$

$$P(s\sigma) = (Ps)(P\sigma), \quad s \in \Sigma^*, \quad \sigma \in \Sigma$$

A projection P with these properties will be called the *natural projection* from Σ^* to Σ_{loc}^*. A natural projection is thus *catenative*: that is, has the property $P(ss')=P(s)P(s')$. For a DES **G** and natural projection P, a *local model* $\mathbf{G_{loc}}$ can be used to represent $PL(\mathbf{G})$, namely the local observer's picture of the generator **G**. In general $\mathbf{G_{loc}}$ can be computed by the TCT procedure **project**.

Example (TCT procedure **project**)

G

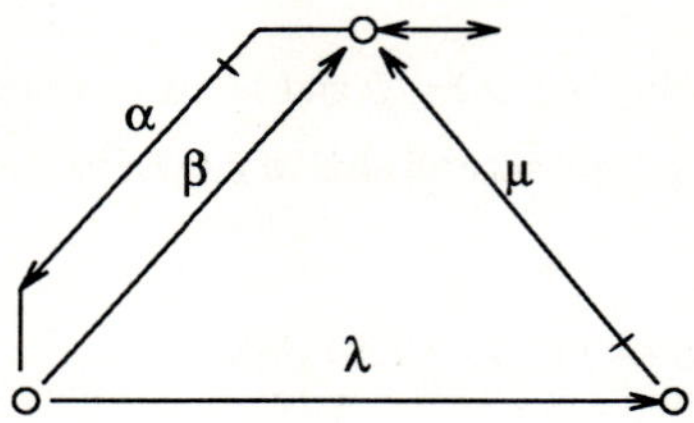

$$\Sigma=\{\alpha,\beta,\lambda,\mu\}, \quad \Sigma_{loc}=\{\alpha,\beta\}$$

$\mathbf{G_{loc}}$

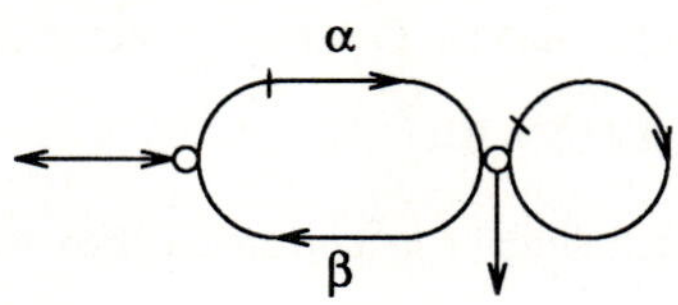

$$P(L_m(\mathbf{G}))=L_m(\mathbf{G_{loc}}), \quad P(L(\mathbf{G}))=L(\mathbf{G_{loc}})$$

$$\mathbf{G_{loc}}=\mathbf{project}\,(\mathbf{G},[\lambda,\mu])$$

As an exercise the reader might like to compute $\mathbf{G_{loc}}$ if $\Sigma_{loc}=\{\lambda,\mu\}$.

□

Closely associated with natural projections is the pleasant algebraic property of normality. Suppose **G** is a DES over Σ, $\Sigma_{loc}\subseteq\Sigma$, and $P:\Sigma^*\rightarrow\Sigma_{loc}^*$ is natural. A language $K\subseteq\Sigma^*$ is $(P,L(\mathbf{G}))$*–normal* if

$$K=L(\mathbf{G})\cap P^{-1}(PK)$$

If P and **G** are fixed, we just say that K is *normal*. K is $(P,L(\mathbf{G}))$-normal if and only if it is the largest sublanguage of $L(\mathbf{G})$ having PK as its projection; thus K is determined uniquely by its projection and the constraints imposed by the structure of $L(\mathbf{G})$. As boundary cases, the languages $\varnothing$ and $L(\mathbf{G})$ are normal. For any language $K_o\subseteq\Sigma_{loc}^*$, the language

$$L(\mathbf{G})\cap P^{-1}(K_o)$$

is normal. The family of $(P,L(\mathbf{G}))$-normal languages contained in a given sublanguage of Σ^* is closed under union and intersection. Thus the normal languages are algebraically well-behaved.

Proposition 5.1

A sublanguage $B \subseteq L(\mathbf{G})$ is normal only if

$$B(\Sigma - \Sigma_{loc}) \cap L(\mathbf{G}) \subseteq B$$

□

Thus B is normal only if it is 'invariant under the occurrence of unobservable events'.

Example (Normality)

For the previous machine $\mathbf{G}$, and

$$\Sigma_{loc} = \{\alpha, \beta\}, \quad \text{i.e. } P\lambda = P\mu = \varepsilon,$$

the sublanguage $H = \overline{(\alpha\beta)^*} \subseteq L(\mathbf{G})$ generated by

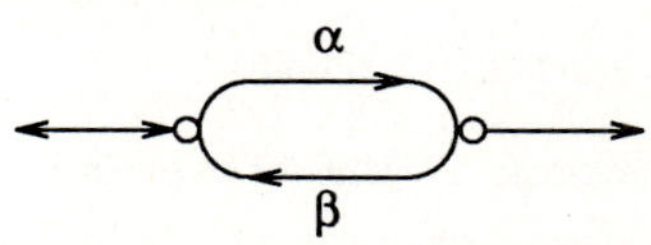

is not $(P, L(\mathbf{G}))$-normal: e.g. the word $\alpha\beta\alpha$ "escapes" from H on execution of the unobservable event λ.

By contrast, the same language H is $(Q, L(\mathbf{G}))$-normal, where Q nulls only μ.

□

Next we turn to a discussion of local versus global supervision. Given the DES $\mathbf{G}$ over Σ, together with the local alphabet $\Sigma_{loc} \subseteq \Sigma$ and the natural projection $P: \Sigma^* \rightarrow \Sigma_{loc}^*$, define

$$\Sigma_{u,loc} = \Sigma_u \cap \Sigma_{loc}, \quad \Sigma_{c,loc} = \Sigma_c \cap \Sigma_{loc}$$

Let the local model $\mathbf{G_{loc}}$ represent $P(L_m(\mathbf{G}))$. Assume also that a 'local legal specification' is given as the closed language $E_{loc} \subseteq \Sigma_{loc}^*$. The corresponding global legal specification is the closed language

$$E = L(\mathbf{G}) \cap P^{-1}(E_{loc})$$

Thus the globally optimal supervisor $\mathbf{S}$ will satisfy:

$$L_m(\mathbf{S}/\mathbf{G}) = K = \sup \mathbf{C}(E) \subseteq \Sigma^*$$

Now for the locally optimal supervisor $\mathbf{S_{loc}}$ we have:

$$\begin{aligned} L_m(\mathbf{S_{loc}}/\mathbf{G_{loc}}) &= K_{loc} \\ &= \sup \mathbf{C}_{loc}(L(\mathbf{G_{loc}}) \cap E_{loc}) \\ &\subseteq \Sigma_{loc}^* \end{aligned}$$

The corresponding global controlled behavior is easily seen to be:

$$\hat{K} = L(\mathbf{G}) \cap P^{-1} K_{loc} \subseteq \Sigma^*$$

And so the basic problem of local versus global supervision can be stated as: Is $\hat{K}$ globally optimal, i.e. $\hat{K}=K$?

Theorem 5.1

Local supervision is globally optimal (i.e. $\hat{K}=K$) if and only if K is normal. In that case, events in $\Sigma_c-\Sigma_{c,loc}$ never need to be disabled.

□

In order to identify situations in which the condition of Theorem 5.1 may be achieved, we bring in the following

Main Condition (MC)

$$(\forall w,\sigma) w\in K_{loc} \wedge \sigma\in\Sigma_{c,loc} \wedge w\sigma\in PL(\mathbf{G}) \wedge w\sigma\notin K_{loc} \Rightarrow$$

$$(\forall s) s\in K \wedge Ps=w \wedge s\sigma\in L(\mathbf{G}) \Rightarrow s\sigma\notin K \,.$$

MC may be paraphrased by the statement: "Whenever an event $\sigma\in\Sigma_{c,loc}$ must be disabled by $\mathbf{S_{loc}}$ then it must also be disabled by **S**."

Theorem 5.2

Assume $K_{loc}\neq\emptyset$. Then $\hat{K}=K$ if and only if MC.

□

In practice, we will probably wish to avoid computing the global object K. Sometimes this can be done on the basis of sufficient conditions derived from MC which do not require K to be known explicitly, but we shall not develop this idea here.

The following is an example where MC, and therefore the optimality of local supervision, fails.

Example (Local supervision: nonoptimality)

G

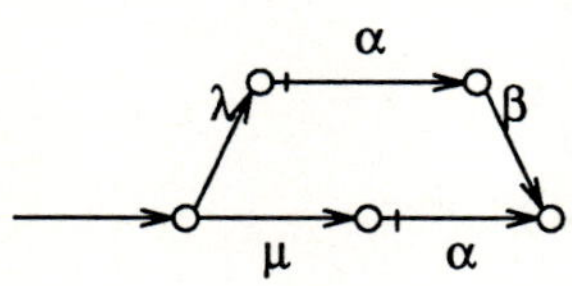

(all states marked)

$$\Sigma=\{\alpha,\beta,\lambda,\mu\}, \quad \Sigma_{loc}=\{\alpha,\beta\}, \quad \Sigma_{c,loc}=\alpha$$

G_{loc}

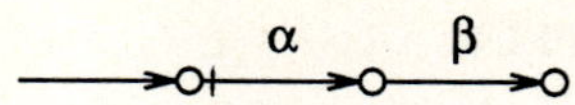

(all states marked)

$$E_{loc}=\alpha^*, \quad L(G_{loc})\cap E_{loc}=\{\varepsilon,\alpha\}, \quad K_{loc}=\varepsilon.$$

$$P^{-1}(K_{loc})=\{\lambda,\mu\}^*, \quad \hat{K}=\{\varepsilon,\lambda,\mu\}$$

$$P^{-1}(E_{loc})=P^{-1}\alpha^*=\{\alpha,\lambda,\mu\}^*$$

$$L(G)\cap P^{-1}(E_{loc})=\{\varepsilon,\lambda,\lambda\alpha,\mu,\mu\alpha\}$$

$$K=\{\varepsilon,\lambda,\mu,\mu\alpha\}$$

The conclusion from these calculations is that $K \underset{\neq}{\supset} \hat{K}$, namely locally optimal supervision is not globally optimal.

□

For a positive illustration let us consider local supervision of Small Factory. Here

$$\Sigma=\{\alpha i,\beta i,\lambda i,\mu i \mid i=1,2\}$$

Suppose the local supervisor views only normal operation:

$$\Sigma_{loc}=\{\alpha i,\beta i \mid i=1,2\}, \quad \Sigma_{nul}=\Sigma-\Sigma_{loc}$$

Define

MACHLOC = **project(MACH,** $[\Sigma_{null}]$**)**

= **shuffle(MACH1LOC,MACH2LOC)**,

MACHiLOC = **project(MACHi,** $[\lambda i,\mu i]$**)**

MACHiLOC

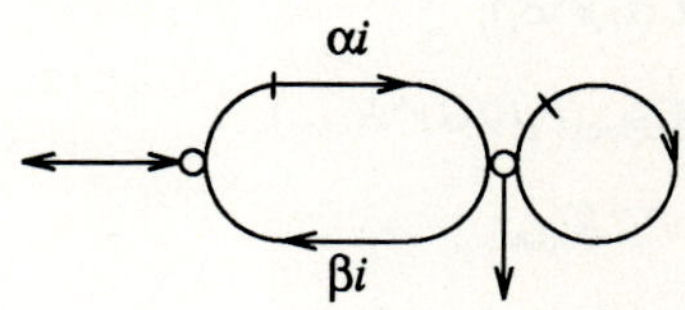

SPECLOC

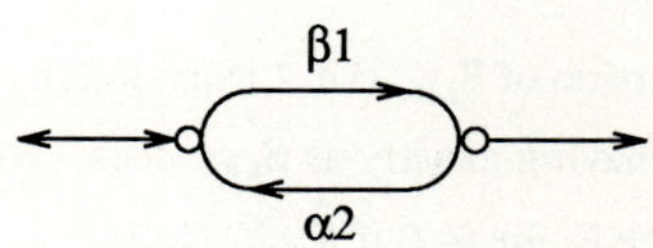

selfloop $\{\alpha 1,\beta 2\}$

KLOC = **supcon(MACHLOC,SPECLOC)**

KLOC ^= **meet(MACH,** *selfloop* **(KLOC,** $[\Sigma_{null}]$**))**

SPEC = **selfloop(SPECLOC,** $[\Sigma_{null}]$**)**

$$\mathbf{K} = \mathbf{supcon(MACH,SPEC)}$$

It can be verified that **isomorph(K,minstate(KLOC^))**. Thus locally optimal supervision is indeed globally optimal.

□

Suppose now that our DES **G** is to be supervised by several local agents, indexed by $i \in I$. Let the local subalphabets be $\Sigma_i \subseteq \Sigma$, and the corresponding natural projections be $P_i : \Sigma^* \rightarrow \Sigma_i^*$. For the local legal specifications we take closed languages $E_{i,loc} \subseteq \Sigma_i^*$; the corresponding global legal specifications are then $E_i = P_i^{-1} E_{i,loc} \subseteq \Sigma^*$, with global supervisors, say, $\mathbf{S_i}$. The overall global legal specification is then $E = \cap E_i$. On this basis we have

Theorem 5.3

If each $\mathbf{S_i}$ is optimal for the corresponding legal specification, i.e.

$$(\forall i) L(\mathbf{S_i}/\mathbf{G}) = \sup \mathbf{C}(L(\mathbf{G}) \cap E_i)$$

then

$$L((\wedge \mathbf{S_i})/\mathbf{G}) = \sup \mathbf{C}(L(\mathbf{G}) \cap E)$$

□

The result may be paraphrased by the statement: "The concurrent operation of globally optimal decentralized supervisors is globally optimal." In this formulation the issue of blocking does not arise because, for simplicity, the relevant languages are declared to be closed.

We turn now to decentralized supervision. For this let

$$K_i = \sup \mathbf{C}(L(\mathbf{G}) \cap E_i)$$

$$K_{i,loc} = \sup \mathbf{C}_{i,loc}(P_i L(\mathbf{G}) \cap E_{i,loc})$$

$$\hat{K}_i = L(\mathbf{G}) \cap P_i^{-1} K_{i,loc}$$

$$L(\mathbf{S_{i,loc}}/\mathbf{G_{i,loc}}) = K_{i,loc}$$

Let $\hat{\mathbf{S}}_\mathbf{i}$ be the global version of $\mathbf{S_{i,loc}}$ (i.e. $L(\hat{\mathbf{S}}_\mathbf{i}/\mathbf{G}) = \hat{K}_i$): thus $\hat{\mathbf{S}}_i$ permanently enables all events not seen by $\mathbf{S_{i,loc}}$, otherwise behaving exactly as $\mathbf{S_{i,loc}}$ does. We shall say that the DES **G** is *locally controllable (with respect to the* E_i *for* $i \in I$*)* if $\hat{K}_i = K_i$ $(i \in I)$.

Theorem 5.4

Let **G** be locally controllable. Then

$$L((\wedge \hat{\mathbf{S}}_\mathbf{i})/\mathbf{G}) = \sup \mathbf{C}(L(\mathbf{G}) \cap E)$$

□

The result may be paraphrased by the statement: "If **G** is locally controllable, then decentralized supervision of **G** is globally optimal."

As an illustration of the foregoing ideas, consider the following scenario involving machines and buffers. Given $\mathbf{MACH_i}$ ($i \in \mathbf{N}$) as before, define **MACH** to be the **shuffle** of $\mathbf{MACH_1}, \ldots, \mathbf{MACH_N}$. Let

$$\Sigma = \{\alpha_i, \beta_i, \lambda_i, \mu_i \mid i \in \mathbf{N}\},$$

$$\Sigma_c = \{\alpha_i, \mu_i \mid i \in \mathbf{N}\}$$

Consider local supervisory control of breakdown and repair:

$$\Sigma_{br} = \{\lambda_i, \mu_i \mid i \in \mathbf{N}\}, \quad E_{br} \subseteq \Sigma_{br}^*$$

An example of a legal specification E_{br} governing breakdown and repair might be that machines are to be repaired in order of breakdown. On E_{br} we shall impose only the mild general condition

$$E_{br}\lambda_i \subseteq E_{br} \qquad (i \in \mathbf{N})$$

which states that the occurrence of breakdown is never illegal. We shall use the following.

Proposition 5.2

If $PL(\mathbf{G}) \cap E_{loc}$ is locally controllable, then $\hat{K} = K$.

□

It can now be readily verified that $\hat{K} = K$, i.e. optimal local and optimal global control achieve the same behavior.

As a second illustration we consider decentralized supervision of a production line consisting of machines and buffers in the linear array displayed below.

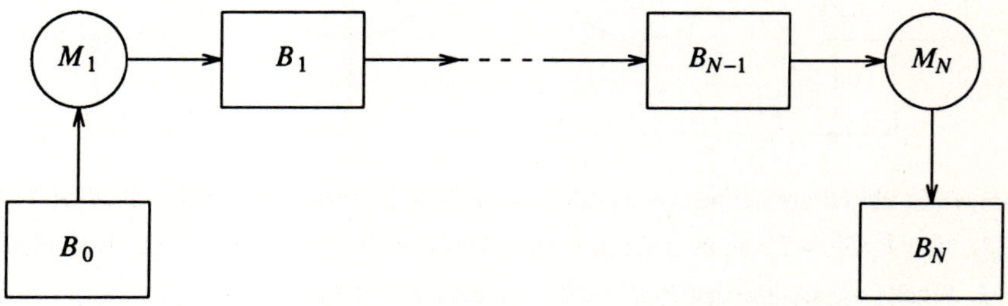

The buffers $B_1, \ldots, B_{N-1}$ are assumed to have finite capacity, while B_0, B_N are infinite. Initially the machines $\mathbf{M_i}$ are all idle, B_0 is full, and $B_1, \ldots, B_{N-1}$ are empty. For normal operation we have the subalphabet $\Sigma_{no} = \{\alpha_i, \beta_i \mid i \in \mathbf{N}\}$ together with the legal specification $E_{no} \subseteq \Sigma_{no}^*$: "No buffer shall overflow or underflow." As usual we take the plant to be the shuffle **M** of $\mathbf{M_1}, \ldots, \mathbf{M_N}$. Thus

$$L(\mathbf{M}) = L(\mathbf{M_1}) \,\|\, \cdots \,\|\, L(\mathbf{M_N})$$

$$PL(\mathbf{M}) = PL(\mathbf{M_1}) \parallel \cdots \parallel PL(\mathbf{M_N})$$

For decentralized supervision the locally optimal supervisor is constructed according to

$$\mathbf{S_{no}} : L(\mathbf{S_{no}}) = \sup \mathbf{C}_{no}(PL(\mathbf{M}) \cap E_{no})$$

MC may be used to verify that $\hat{\mathbf{S}}_{\mathbf{no}}$ is globally optimal. Combining this fact with the result from the previous example, we now know that **M** is locally controllable with respect to the specification languages $\{E_{br}, E_{no}\}$. Define

$$E = P_{br}^{-1} E_{br} \cap P_{no}^{-1} E_{no}$$

By Theorem 5.4

$$L((\hat{\mathbf{S}}_{\mathbf{br}} \wedge \hat{\mathbf{S}}_{\mathbf{no}})/\mathbf{M}) = \sup \mathbf{C}(L(\mathbf{M}) \cap E)$$

Thus we may conclude that optimal supervision of breakdown and repair may be carried out independently of optimal supervision of normal operation. It turns out that nonblocking may also be proved under the assumptions that $\mathbf{S_{br}}$ and $\mathbf{S_{no}}$ are nonblocking, plus a mild supplementary assumption on E_{br}.

To conclude this section we present a brief introduction to hierarchical control in the context of a simple example. The diagram below displays information flow in a two-level hierarchy with two 'low-level' DES **G1, G2** and their local or 'low-level' supervisors **S1,S2**, coordinated by a single 'high-level' supervisor **S0**.

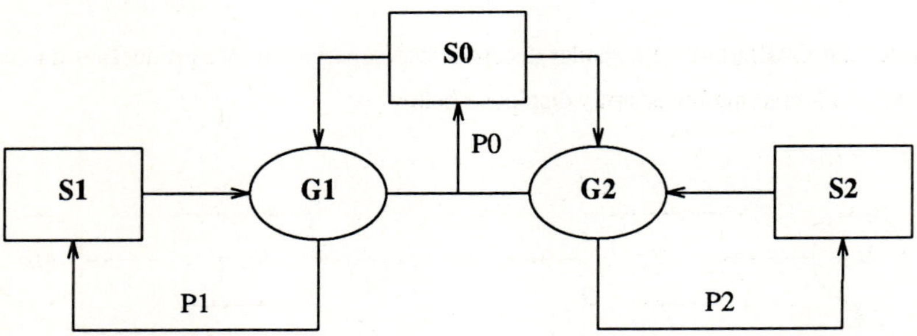

We assume that the **Gi** are defined over disjoint alphabets Σ_i, and consider **G = shuffle(G1,G2)** over $\Sigma = \Sigma_1 \cup \Sigma_2$. Let $P_i : \Sigma^* \to \Sigma_i^*$ be the natural projections, and let $E_i \subseteq \Sigma_i^*$ $(i=1,2)$ be closed specification languages. Finally we assume that $P_0 : \Sigma^* \to \Sigma_0^*$ is a map with the properties:

$$P_0(\varepsilon) = \varepsilon$$

$$(\forall s \in \Sigma^*)(\forall \sigma \in \Sigma) P_0(s\sigma) = \begin{cases} \text{either} & P_0(s) \\ \text{or}\ P_0(s)\sigma_0, & \text{some } \sigma_0 \in \Sigma_0 \end{cases}$$

Thus P_0 is *prefix-preserving* ($s \le s'$ implies $P_0(s) \le P_0(s')$) but need not be catenative.

The coordination specification is a language $E_0 \subseteq \Sigma_0^*$. Thus the complete specification language is, in low-level terms, the closed language $E \subseteq \Sigma^*$ defined by

$$E = P_1^{-1} E_1 \cap P_2^{-1} E_2 \cap P_0^{-1} E_0$$

Stated informally, the fundamental problem of hierarchical control is the following: Given a specification language $E \subseteq \Sigma^*$, construct 'minimal' (P_0, E_0), and then the corresponding coordinating supervisor **S0**. Roughly speaking, 'minimal' will mean that the information about low-level behavior extracted by P_0 is no more than is required for control (by **S0**) of the logical interaction specified for **G1** and **G2**.

As an illustration we consider a fable entitled 'Reds and Whites Together'. In a certain City, factions known as Reds and Whites must share a communal Bathtub. The structures for a typical Red and a typical White are displayed below.

Red

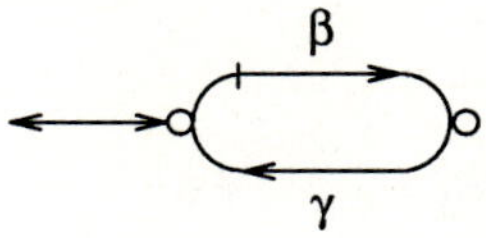

$\mathbf{R} = \|_{i=1}^{\infty} \mathbf{R}_i$

White

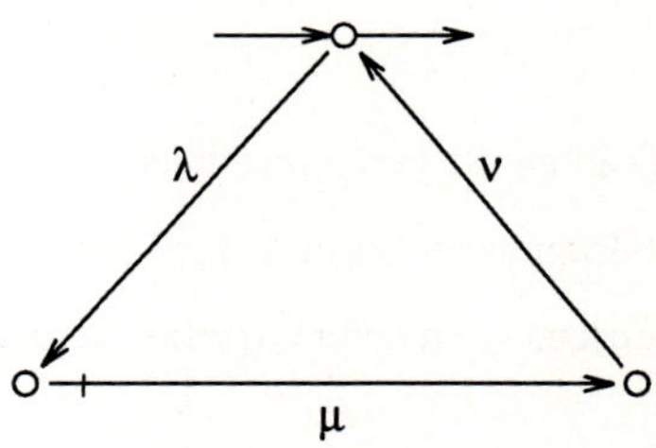

$\mathbf{W} = \|_{i=1}^{\infty} \mathbf{W}_i$

Let **G** be the shuffle (in an obvious sense) of **R** and **W**. The legal specifications (hallowed by tradition) are the following:

1. At most 10 Reds are allowed in the tub at a time.
2. At most 1 White is allowed in the tub at a time.
3. Reds and Whites never bathe together.
4. Reds enter at will, but Whites enter first come, first served.
5. Whites have overall priority over Reds.

It is pretty clear that local supervisors can be constructed based on a counter for the Reds (# in tub ≤ 10) and a queue for the Whites (fcfs). The coordinating supervisor must enforce mutual exclusion. For this, fairly obviously, we must construct a coordinating alphabet to mark the start and finish (relative to the tub) of strings of Reds and Whites. Thus let

$$\Sigma_0 = \{\rho_0, \rho_1, \tau_0, \tau_1, \tau_2\}$$

and define a noncatenative projection $P_0 : \Sigma^* \to \Sigma_0^*$ as follows.

$$\begin{aligned}
P_0(s\beta_i) &= P_0(s)\rho_0, & & \\
P_0(s\gamma_i) &= P_0(s)\rho_1 & & \text{if } \gamma_i \text{ ends a Red string} \\
P_0(s\lambda_i) &= P_0(s)\tau_0 & & \text{if } W_i \text{ heads queue in a White string} \\
P_0(s\mu_i) &= P(s)\tau_1 & & \text{if } W_i \text{ first to bathe in a White string} \\
P_0(s\nu_i) &= P(s)\tau_2 & & \text{if } W_i \text{ ends a White string} \\
P_0(s\sigma) &= P_0(s) & & \text{otherwise}
\end{aligned}$$

A generator **G0** for $P_0(L_m(\mathbf{G}))$ must be defined over Σ_0^*: let **G0** = shuffle(**R0**,**W0**), with **R0** and **W0** as shown.

R0

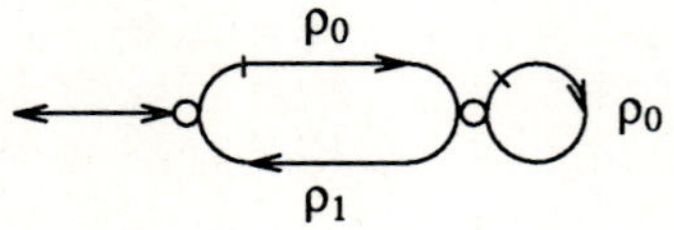

W0

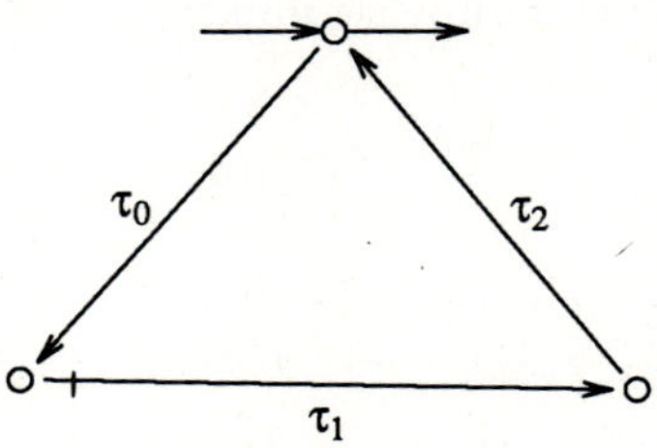

At this level the appropriate legal specifications are easily seen to be the following.

1. If τ_0 occurs then ρ_0 must not occur again before τ_2 next occurs.
2. If ρ_0 occurs then τ_1 must not occur again before ρ_1 next occurs.

These are formalized in the DES below.

L1

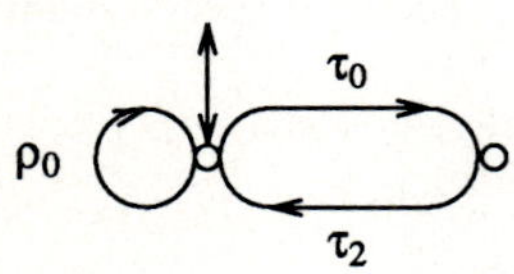

L2

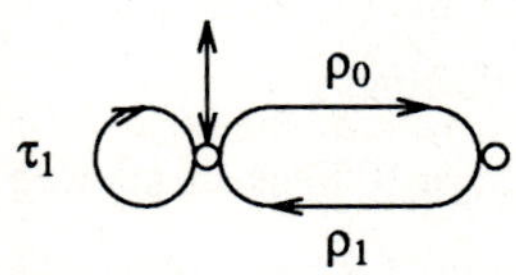

The high-level coordination problem can now be solved by setting

L0 = meet (L1,L2)

and then computing

S0 = supcon (G0,L0)

The result for the coordinating supervisor is displayed below:

S0

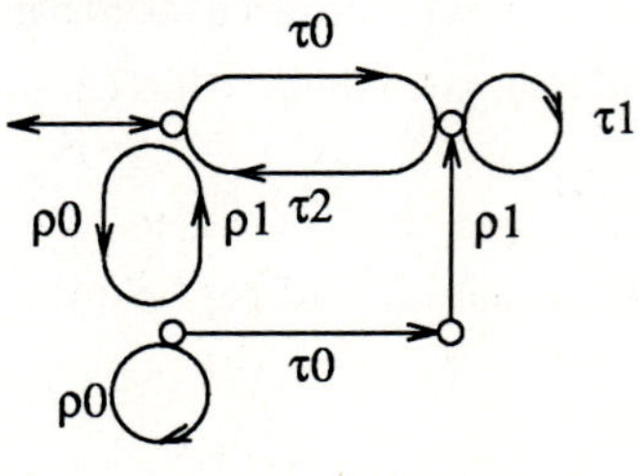

The algebraic relationship between the two levels of the hierarchy can be summarized by the commutative diagram:

$$
\begin{array}{cccccc}
\hat{\mathbf{S}}_0 & \Sigma\times X & \longrightarrow & X & & \text{low level} \\
 & \Big\downarrow \theta\times\pi & & \Big\downarrow \pi & & \\
\mathbf{S}_0 & \tilde{\Sigma}_0\times X_0 & \longrightarrow & X_0 & & \text{high level}
\end{array}
$$

Here $\mathbf{S}_0$ is the high-level coordinating supervisor, while $\hat{\mathbf{S}}_0$ is an equivalent but fictitious low-level coordinating supervisor. The augmented alphabet $\tilde{\Sigma}_0=\Sigma_0\cup\{\text{null}\}$, where $(\text{null},x_o)\vdash\!\!\rightarrow x_o$ for all $x_o\in X_0$. The map $\pi:X\rightarrow X_0$ is surjective and simply lumps states in X to form states in X_0, while $\theta:\Sigma\times X\rightarrow\tilde{\Sigma}_0$ is an output map in the sense of Mealy automata. The maps π and θ are determined by $\mathbf{S}_0$ and P_0.

6. Observability and Supervision of DES

In this final section we introduce into our framework for control of DES a version of the concept of observability, or more precisely the problem of supervision in the presence of partial observations of the system behavior. Unlike the situation with decentralized control, we shall not assume any particular relation between the events that can be observed and the events that can be controlled.

Let $\mathbf{G}$ be a DES over $\Sigma=\Sigma_u\cup\Sigma_c$, and let $\Sigma_o\subseteq\Sigma$. The events with labels in Σ_o will be those that are directly observed by the supervisor; there is no special relation between Σ_o and Σ_c. Let $P:\Sigma^*\rightarrow\Sigma_o^*$ be the natural projection, and let $\ker P$ be the equivalence kernel of P: that is, $(s,s')\in\ker P$ (or s,s' are equivalent mod P) if $Ps=Ps'$.

Let K be a closed sublanguage of $L(\mathbf{G})$. We define the binary relation act_K on Σ^* as follows. Let $s,s'\in\Sigma^*$. The pair $(s,s')\in\text{act}_K$ if

$$
\begin{aligned}
&s\in K\wedge s'\in K\\
&\qquad\Rightarrow \textit{not}\ (\exists\sigma)((s\sigma\in K\wedge s'\sigma\in L(\mathbf{G})-K)\vee(s\sigma\in L(\mathbf{G})-K\wedge s'\sigma\in K))
\end{aligned}
$$

Thus $(s,s')\in\text{act}_K$ if all the one-step continuations of s and s' in $L(\mathbf{G})$ yield the same result with respect to membership in K. The relation act_K is a *tolerance* relation on Σ^*; that is, it is reflexive and symmetric, but not in general transitive. Now we can make the definition: the (closed) language $K\subseteq\Sigma^*$ is $(P,L(\mathbf{G}))$*–observable* if $\ker P\leq\text{act}_K$, namely

$$(\forall s,s'\in\Sigma^*)(s,s')\in\ker P\Rightarrow(s,s')\in\text{act}_K$$

The definition states that the projection P retains sufficient information to decide whether or not, when some event occurs, the resulting string is still a word of K.

Example (Observability)

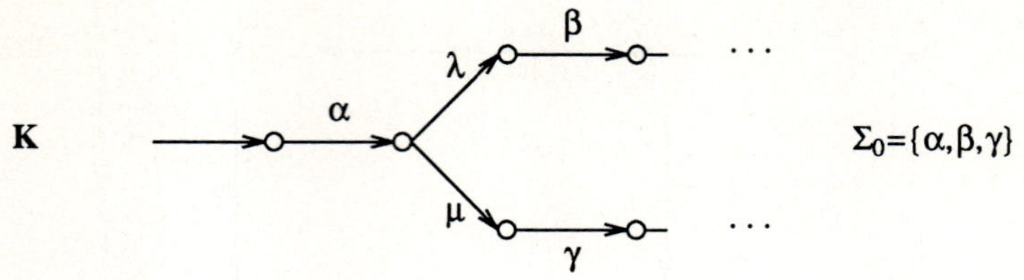

Assume $L(\mathbf{G}) \supset \{K, \alpha\mu\beta\}$. We have

$$P(\alpha\lambda)=P(\alpha\mu), \quad \alpha\lambda\beta\in K, \quad \alpha\mu\beta\in L(\mathbf{G})-K$$

and therefore K is not observable!

□

We shall say that a supervisor **S** for **G** is *feasible* if it sees only $PL(\mathbf{G})$; that is, if $\xi:\Sigma\times X\rightarrow X$ is the transition function of **S**, $\sigma\in\Sigma-\Sigma_o$, and $\xi(\sigma,x)!$, then $\xi(\sigma,x)=x$. We now have the following.

Theorem 6.1

Let $K\subseteq L(\mathbf{G})$ be closed. There exists a proper feasible supervisor **S** for **G** such that $L(\mathbf{S}/\mathbf{G})=K$ if and only if K is controllable and observable.

□

Let $Q:\Sigma^*\rightarrow(\Sigma_c\cup\Sigma_o)^*$ be the natural projection, let A and E be closed sublanguages of $(\Sigma_c\cup\Sigma_o)^*$ with $A\subseteq E\subseteq QL(\mathbf{G})$, and let

$$\mathbf{O}(A)=\{B\subseteq\Sigma^* \mid B\supseteq A\wedge B \text{ observable}\}$$

Proposition 6.1

$\mathbf{O}(A)$ is closed under intersection; $\inf\mathbf{O}(A)$ exists.

□

The proposition states that there is a smallest observable sublanguage of Σ^* containing a given sublanguage. Our main result on supervision in the presence of partial observations is now the following.

Theorem 6.2

There exists a proper feasible supervisor **S** for **G** such that $A\subseteq QL(\mathbf{S}/\mathbf{G})\subseteq E$ if and only if

$$\inf\mathbf{O}(A)\subseteq\sup\mathbf{C}(E)$$

□

While the foregoing results exhibit a pleasing duality between controllability and observability, from a practical point of view they have the deficiency that, in a given situation, a unique maximal controllable and observable language in general does not exist. In order to obtain a solution to the problem of feasible supervision that allows as much freedom of behavior to the supervised system as seems practical to determine, we consider again the property of normality that was exploited in Sect. 5.

Let K be a closed sublanguage of $L(\mathbf{G})$, and recall that K is *normal* if $K = L(\mathbf{G}) \cap P^{-1}(PK)$. Then we have

Proposition 6.2

If K is normal then K is observable.

□

The normal languages form a subclass of the observable languages that is closed under union. In particular sup $\mathbf{N}(B)$, the largest normal sublanguage of a closed sublanguage $B \subseteq L(\mathbf{G})$, always exists. Thus the *supremal controllable normal sublanguage* sup $\mathbf{CN}(E)$ of a closed sublanguage $E \subseteq L(\mathbf{G})$ always exists, and so provides a natural quasi-optimal solution to the problem of feasible supervision. In principle this solution may be computed by means of the following formulas.

Theorem 6.3

$$\sup \mathbf{CN}(E) = L(\mathbf{G}) \cap P^{-1} \sup \mathbf{C}_{loc}(P \sup \mathbf{N}(E))$$

$$\sup \mathbf{N}(B) = L(\mathbf{G}) \cap P^{-1}(PB - P(L(\mathbf{G}) \cap P^{-1}PB - B))$$

□

The following example illustrates the foregoing ideas in an intuitively simple setting, as well as the computation of the various DES involved using TCT. Of course, this 'universal' approach to the example problem is far from being the most efficient; furthermore, the piecemeal TCT computations could certainly be combined into higher-level procedures if desired.

Example

Stations A and B on a guideway are connected by a single one-way track from A to B. The track consists of 4 sections, with stoplights (*) and detectors (!) installed at various section junctions.

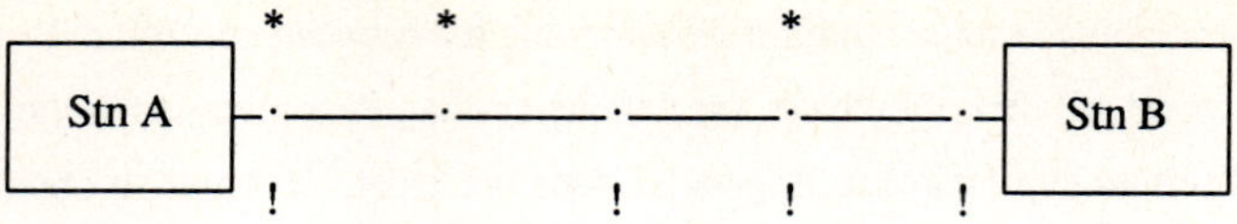

Two vehicles **V1, V2** use the guideway simultaneously. **V_** is in state 0 (at A), state i (while travelling in section i, i=1,...,4), or state 5 (at B).

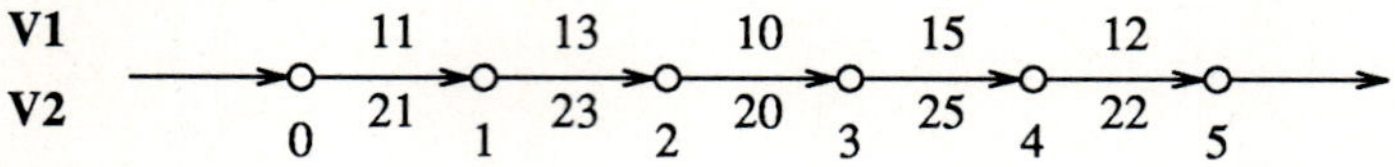

To avoid collision, control of the stoplights must ensure that **V1** and **V2** never travel on the same section of track simultaneously: i.e. the **V**'s are subject to mutual exclusion of the state pairs (i,i), i=1,...,4. Controllable events are odd-numbered; the unobservable events are $\{13,23\}$.

By use of existing TCT procedures the solution can be carried out as follows. Bracketed numbers (m,n) report the state size m and number of transitions n of the corresponding DES.

Steps 1 to 13 compute the legal specification language **LE**, then its supremal normal sublanguage **SUPNLE**:

1.	**V = shuffle(V1,V2)**	(36,60)
2.	**LE = mutex(V1,V2,**[(1,1),(2,2),(3,3),(4,4)]**)**	(30,40)
3.	**NULL** = [13,23]	{null events of P}
4.	**PLE = project(LE,NULL)**	(24,32)
5.	**PPLE = selfloop(PLE,NULL)**	(24,80)
6.	**VPPLE = meet(V,PPLE)**	(38,56)
7.	**COLE = complement(LE,-)**	(31,310)
8.	**CVPPLE = meet(COLE,VPPLE)**	(38,54)
9.	**PCVPPLE = project(CVPPLE,NULL)**	(26,32)
10.	**COPCV = complement(PCVPPLE,-)**	(27,216)
11.	**PLECOPCV = meet(PLE,COPCV)**	(20,24)
12.	**PPLECO = selfloop(PLECOPCV,NULL)**	(20,64)
13.	**SUPNLE = meet(V,PPLECO)**	(26,32)

The transition tables verify that **LE** is not normal, as indicated by the reduction in size from **LE** to **SUPNLE**. Starting from **SUPNLE**, steps 14 to 18 compute the supremal controllable normal sublanguage **SUPCNLE** of **LE**.

14. **PSUPNLE = project(SUPNLE,NULL)** (20,24)
15. **PV = project(V,NULL)** (25,40)
16. **CSUPNLE = supcon(PV,PSUPNLE)** (20,24)
17. **PCSUPNLE = selfloop(CSUPNLE,NULL)** (20,64)
18. **SUPCNLE = meet(V,PCSUPNLE)** (26,32)

The supervisory action of **SUPCNLE** can be read from the tabulated transition structure or from the transition graph and is the following (where tsi stands for 'track section i'): If **V2** starts first (event *21*) it must enter ts4 before **V1** may start (event *11*: disabled by light #1). **V1** may then continue into ts3 (event *10*), but may not enter ts4 (event *15*: disabled by light #3) until **V2** enters Stn B (event *22*). Light #2 is not used. If all events were observable, supervision could be based on **LE**, allowing **V1** to start when **V2** has entered ts2. But then **V1** must halt at light #2 until **V2** has entered ts4.

Transition graph for **SUPCNLE** when **V2** starts first (for **LE** adjoin events shown dashed):

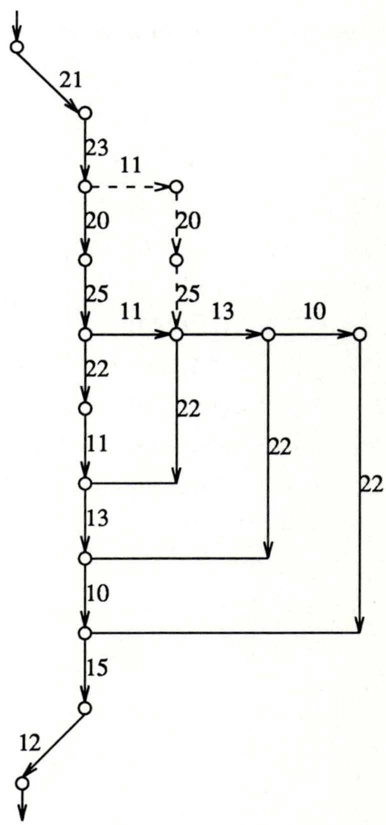

SUPCNLE # states: 26 state set: 0...25 initial state: 0

marker states: 10

transitions: 32

transition table:

[0,11, 1]	[0,21,14]	[1,13, 2]	[2,10, 3]	[3,15, 4]
[4,12, 5]	[4,21,11]	[5,21, 6]	[6,23, 7]	[7,20, 8]
[8,25, 9]	[9,22,10]	[11,12, 6]	[11,23,12]	[12,12, 7]
[12,20,13]	[13,12, 8]	[14,23,15]	[15,20,16]	[16,25,17]
[17,11,18]	[17,22,25]	[18,13,19]	[18,22,24]	[19,10,20]
[19,22,23}	[20,22,21]	[21,15,22]	[22,12,10]	[23,10,21]
[24,13,23]	[25,11,24]			

SUPCNLE printed.

LE # states: 30 state set: 0...29 initial state: 0

marker states: 10

transitions: 40

transition table:

[0,21, 1]	[0,11,16]	[1,23, 2]	[2,20, 3]	[2,11,15]
[3,25, 4]	[3,11,14]	[4,22, 5]	[4,11,11]	[5,11, 6]
[6,13, 7]	[7,10, 8]	[8,15, 9]	[9,12,10]	[11,22, 6]
[11,13,12]	[12,22, 7]	[12,10,13]	[13,22, 8]	[14,25,11]
[15,20,14]	[16,13,17]	[17,21,18]	[17,10,27]	[18,10,19]
[19,15,20]	[20,23,21]	[20,12,26]	[21,20,22]	[21,12,25]
[22,12,23]	[23,25,24]	[24,22,10]	[25,20,23]	[26,23,25]
[27,21,19]	[27,15,28]	[28,21,20]	[28,12,29]	[29,21,26]

LE printed.

LEDAT

Control Data are displayed by listing the supervisor states where disabling occurs, together with the events that must be disabled there.

Control Data:

1:	11	13:	15
14:	13	15:	13
16:	21	18:	23
19:	23	22:	25

LEDAT printed.

□

7. Selected References

1. P.J. Ramadge. Control and Supervision of Discrete Event Processes. Ph.D. Thesis, Dept. of Electl. Engrg., University of Toronto, May 1983.
2. P.A. Iglesias. Computational Methods for the Control of Discrete Event Systems. Divn. of Engrg. Science, University of Toronto, April 1987.
3. P.J. Ramadge, W.M. Wonham. Supervisory control of a class of discrete event processes. SIAM J. Control and Optimization **25** (1), 1987, pp. 206-230.
4. W.M. Wonham, P.J. Ramadge. On the supremal controllable sublanguage of a given language. SIAM J. Control and Optimization **25** (3), 1987, pp. 637-659.
5. F. Lin, W.M. Wonham. Decentralized supervisory control of discrete-event systems. Information Sciences **44** (2), 1988, to appear.
6. F. Lin, W.M. Wonham. On observability of discrete-event systems. Information Sciences **44** (2), 1988, to appear.
7. W.M. Wonham, P.J. Ramadge. Modular supervisory control of discrete event systems. Maths. of Control, Signals and Systems **1** (1), 1988, pp. 13-30.
8. TCT Software: diskette for IBM/PC/XT/AT; current version available from the author on request.

The Complexity of Some Basic Control Problems for Discrete Event Systems

Peter J. Ramadge †

Department of Electrical Engineering
Princeton University, Princeton NJ 08544.

Abstract

We present a summary of some of the known results on the complexity of supervisory control of discrete event systems in the framework of [RW1]. In addition some recent results on the controllability of sequential behaviors and on the complexity of basic coordination problems are presented.

Keywords: discrete event systems, supervisory control, complexity.

1 Introduction

There have been a variety of models proposed for the study of the control of discrete event systems (DES). We shall be concerned in this paper with the model proposed by Ramadge and Wonham in [RW1]. The reader is assumed to be familiar with this model and the associated framework. For background please refer to the paper by W.M. Wonham in these proceedings.

Control problems for discrete event systems (DES) typically require the coordination of component subsystems in order to achieve some prescribed orderly behavior. One of the principal difficulties in this task is that of the computational complexity of the associated decision and synthesis problems. In [RW3] it was reported that the basic decision and synthesis problems arising in the supervisory control problems examined in [RW1] are of polynomial complexity when measured in terms of the number of system states. However, there are classes of DES of interest where these results are not strong enough to ensure computational tractability. For example, in systems composed of a finite set of n asynchronous components the number of system states is usually exponential in n. Such systems arise naturally when modeling the concurrent operation of several asynchronous, or partially synchronous discrete dynamical systems. For example, in modeling certain aspects of communication protocols, operating systems, or the online control of flexible manufacturing systems. It has been shown in [R] that for a certain

† Research partially supported by the National Science Foundation through grant ECS-8715217 and by an IBM Faculty Development Award.

NATO ASI Series, Vol. F47
Advanced Computing Concepts and Techniques
in Control Engineering
Edited by M. J. Denham and A. J. Laub

class of such DES several interesting coordination problems are polynomially solvable. This paper gives a summary of the above results.

We restrict attention to problems with complete information. There have been extensions of the framework of [RW1] to include the possibility of partial observations ([CDFV],[LW1],[LW2]) and the complexity of some of the problems that arise in the extension are examined in [T].

The remainder of the paper consists of two main parts. In the first part the necessary notation and terminology is introduced and qualitative aspects of control and coordination are discussed. The second part deals with the complexity of basic coordination problems for product systems.

Part I

2 Discrete Event Systems

Intuitively a DES consists of a set of elementary events together with a specification of their possible order and time of occurrence. We shall make the simplifying assumption that we are primarily interested in the order of events. In this case we model a DES by specifying its set of elementary events and the possible order in which these events can occur.

Let Σ denote a finite set of events, and Σ^* denote the set of all finite strings of elements of the set Σ, including the empty string 1. A string $u \in \Sigma^*$ is a *prefix* of a string $v \in \Sigma^*$ if for some $w \in \Sigma^*$, $v = wu$. The *prefix closure* of $L \subseteq \Sigma^*$ is the subset $\bar{L} \subset \Sigma^*$ defined by

$$\bar{L} = \{u \colon uv \in L \text{ for some } v \in \Sigma^*\}$$

i.e., $\bar{L}$ is the set of all prefixes of words in L, and L is *prefix closed* if $\bar{L} = L$.

We model the behavior of a DES as a language $L \subseteq \Sigma^*$. Since L is intended to represent the set of all possible generated sequences of events it is natural to require that L be prefix closed.

Example 2.1. (Prefix closure)

(1) If $K = (ab)^* = \{1, ab, abab, ababab, \ldots\}$ then the prefix closure of K is

$$\bar{K} = \{1, a, ab, aba, abab, ababa, \ldots\}$$

In a DES with behavior $L = \bar{K}$ the events a and b occur in strict alternation with a occurring first.

(2) If $K = \{1, ab, aabb, aaabbb, \ldots\}$, then the prefix closure of K is

$$\bar{K} = \{1, a, ab, aa, aab, aabb, aaa, aaab, aaabb, aaabbb, \ldots\}$$

In a DES with behavior $L = \bar{K}$ a finite number of occurrences of event a are followed (eventually) by an equal number of occurrences of event b.

(3) Let $|w|_a$ denote the number of occurrences of a in the string w, and set

$$L = \{w \in \Sigma^*: \text{ for each prefix } u \prec w, |u|_a \leq |u|_b\}$$

Then L is prefix closed. In a DES with behavior L the number of occurrences of event a is always less than or equal to the number of occurrences of event b. This might be the case, for example, if in a manufacturing system a corresponds to taking a part from a buffer and b corresponds to placing a part in the buffer.

A natural extension of the above model is to consider infinite sequences of events rather than, or in addition to, finite strings. To set up such a model we will need the following additional concepts and notation.

Let N denote the set of natural numbers, and Σ^ω denote the set of all sequences of elements of Σ. A subset $B \subseteq \Sigma^\omega$ is termed an *ω-language* over Σ. For an element $e \in \Sigma^\omega$ let e^j denote the string consisting of its first j elements.

The *prefix* of $B \subseteq \Sigma^\omega$ is the set $pr(B) \subseteq \Sigma^*$ defined by

$$pr(B) = \{1\} \cup \{e^j: j \geq 1, e \in B\}$$

i.e., $pr(B)$ is the set consisting of the empty string together with each string in Σ^* that is a prefix of a sequence in B. Note that pr is a mapping from the set of subsets of Σ^ω into the set of subsets of Σ^*, i.e., $pr: 2^{\Sigma^\omega} \to 2^{\Sigma^*}$.

Example 2.2. (Prefix operator)

(1) Set $S = a^*ba^\omega = \{baaa..., abaaa..., aabaaa..., ...\}$. Then the prefix of S is

$$pr(S) = a^* + a^*ba^* = \{1, b, a, ba, ab, aa, baa, aba, aab, aaa, baaa, ...\}$$

(2) Let $S = a^\omega + a^*b^\omega = \{aaaa..., bbbb..., abbbb..., aabbbb..., ...\}$. Then

$$pr(S) = a^* + a^*b^* = \{1, a, b, aa, ab, bb, aaa, aab, abb, bbb, ...\}$$

The *adherence* or *limit* of $L \subseteq \Sigma^*$ is the ω-language

$$L^\infty = \{e: e \in \Sigma^\omega \ \& \ e^j \in L \text{ for infinitely many } j \in N\}$$

Note that the adherence operator is a mapping from the subsets of Σ^* into the subsets of Σ^∞, i.e., $\infty: 2^{\Sigma^*} \to 2^{\Sigma^\omega}$.

Example 2.3. (Adherence operator)

(1) If $L = a^* = \{1, a, aa, aaa, \ldots\}$, then $L^\infty = a^\omega = \{aaaa\ldots\}$.

(2) If $L = \overline{ab^*a} = \{1, a, aa, ab, aba, aba, abb, \ldots\}$, then $L^\infty = ab^\omega = \{abbbbbb...\}$.

(3) If $L = \{a^n b^n, n \geq 1\} = \{1, ab, aabb, aaabbb, \ldots\}$, then $L^\infty = \emptyset$.

There is a natural topology on Σ^ω that matches the notions of prefix and adherence. The easiest way to introduce this topology is through the following metric:

$$\rho(e_1, e_2) = \begin{cases} 1/n, & \text{if } e_1^{n-1} = e_2^{n-1} \text{ and } e_1^n \neq e_2^n; \\ 0, & \text{if } e_1 = e_2. \end{cases}$$

For $S \subseteq \Sigma^\omega$ we let $\bar{S}$ denote the closure of S in the metric topology. Note we are using the bar notation for two distinct operations: prefix closure of string languages, and topological closure of ω-languages. The operations of prefix, adherence and topological closure are related by

$$\bar{S} = pr(S)^\infty$$

Example 2.4. (Topological closure)

(1) If $S = a^*b^\omega$, then $\bar{S} = a^\omega + a^*b^\omega$.

(2) If $S = aaba^\omega$, then $\bar{S} = S$, i.e., S is (topologically) closed.

We now extend our DES model as follows: A DES is a pair $A = (L, S)$ where L is a prefix closed subset of Σ^* and S is a subset of L^∞.

In general it need not be the case that $pr(S) = L$. Equality implies that every string in L is a prefix of some sequence in S. Roughly this can be interpreted to mean that the system is never permanently 'blocked' from producing a string in S. When $pr(S) = L$ we say that A is *nonblocking.*

Example 2.5. (Discrete event systems)

(i) Let $\Sigma = \{a, b\}$. Set $A = (L, S)$ with

$$L = (a + b)^*$$
$$S = (a + b)^* a^\omega$$

S is the set of all sequences over Σ containing a finite number of occurrences of the event b. Clearly $S \subseteq L^\infty$, and $pr(S) = L$, so that A is nonblocking.

(2) Let $\Sigma = \{a, b\}$. Set $A = (L, S)$ with

$$L = a^* + a^*ba^*$$
$$S = a^\omega + a^*ba^\omega$$

S is the set of all sequences over Σ with at most one occurrence of the event b. In this case $pr(S) = L$, so that A is nonblocking, and $S = L^\infty$.

2.1 Control

In order to control the evolution of the DES in time we assume that certain events can be disabled (i.e., prevented from occurring) when desired. Since it may be the case that not all events have this property we partition Σ into *uncontrollable* and *controllable* events: $\Sigma = \Sigma_u \cup \Sigma_c$. An uncontrollable event models the occurrence of events over which the controlling agent has no influence, e.g., machine breakdown in a manufacturing system, loss of a packet in a communication channel, etc.

An admissible input for G consists of a subset $\gamma \subseteq \Sigma$ satisfying $\Sigma_u \subseteq \gamma$. Let $\Gamma \subseteq 2^{\Sigma}$ denote the set of admissible inputs. If $\gamma \in \Gamma$ and $\sigma \in \gamma$, then we say σ is *enabled* by γ, otherwise we say σ is *disabled* by γ. Disabled events are prevented from occurring while enabled events can occur when permitted by the prescribed dynamics.

Control of a DES A with control set Γ consists of selecting a pattern of enabled an disabled events, i.e., an element $\gamma \in \Gamma$, based on the previously generated events. Thus a *supervisor* for the controlled DES (CDES) $A = (L, S)$ is a map

$$f : L \to \Gamma$$

specifying for each possible (finite) string of generated events the next input to be applied.

The closed loop DES consisting of f supervising A is denoted by (A, f), and the closed loop behaviors L_f and S_f are defined as follows:

(i) $1 \in L_f$; and

(ii) $w\sigma \in L_f$ if and only if $w \in L_f$ & $\sigma \in f(w)$.

(iii) $S_f = L_f^{\infty} \cap S$

It is readily verified that $L_f \subseteq L$, and $\overline{L_f} = L_f$.

It is always the case that $pr(S_f) \subseteq L_f$. However, in general there need not be equality in this expression, i.e., (A, f) need not be nonblocking. When $pr(S_f) = L_f$ we say that f is *nonblocking* for A.

Example 2.6. (Supervision)

Let $\Sigma = \{a, b\}$, $\Sigma_c = \{b\}$, and A be the DES with

$$L = a^* + a^*ba^*$$

$$S = a^{\omega} + a^*ba^{\omega}$$

S is the subset of Σ^{ω} consisting of the sequences that contain at most one occurrence of the event b. Let $n \geq 0$ be a fixed integer and define a supervisor f for A by

$$f(w) = \begin{cases} \{a\}, & \text{if } w = a^n; \\ \{a, b\}, & \text{otherwise;} \end{cases}$$

Then (A, f) is the DES with

$$L_f = a^k + a^k ba^* \quad 0 \leq k \leq n$$
$$S_f = a^k ba^\omega \quad 0 \leq k \leq n$$

S_f is the subset of S consisting of the sequences with exactly one occurrence of the event b after at most n initial occurrences of the event a.

3 Controllable Languages

3.1 Some basic results

The basic problem in supervisory control is to modify the open loop behavior of a given DES A so that it lies (as a set) within some prescribed desirable range. This desirable range may be specified by actually giving the desired closed loop behavior, by giving a behavior within which the closed loop behavior must be contained, or by specifying such sets indirectly through other qualitative performance objectives. A basic problem to solve is thus: given a DES $A = (L, S)$ what closed loop behaviors $(\hat{L}, \hat{S})$ can be achieved by supervision ?

It can be shown [RW1] (1) that for nonempty $K \subseteq L$ there exists a supervisor f such that $L_f = K$ if and only if K is prefix closed and controllable, i.e., if and only if

$$\bar{K} = K \quad \text{and} \quad \bar{K}\Sigma_u \cap L \subseteq \bar{K}$$

It can be further shown ([RW1],[WR1]) that the family of closed and controllable sublanguages of L is closed under set union and set intersection, and for any closed $K \subseteq L$ there exists a unique largest closed and controllable language $K^\uparrow$ and a unique smallest closed and controllable language $K^\downarrow$ such that $K^\uparrow \subseteq K \subseteq K^\downarrow$.

3.2 The Controllability of Sequential Behaviors

Similar results to those above hold in the case of sequential behaviors:

Proposition 3.1.

If $B \subseteq S$ is nonempty, then there exists a nonblocking supervisor f such that $S_f = B$ if and only if

(1) $pr(B)$ is controllable, i.e., $pr(B)\Sigma_u \cap L \subseteq pr(B)$; and

(2) $\bar{B} \cap S = B$.

(1) See the paper at this conference by W.M. Wonham.

Proof.

See [R]. ◊

An ω-language B is said to be *closed relative to* S if $\bar{B} \cap S = B$. In view of the above result it is natural to say that $B \subseteq S$ is a *controllable sequential behavior* if $pr(B)$ is controllable and B is (topologically) closed relative to S.

It is readily shown that the set of controllable ω-languages is closed under finite unions. However, as shown in the second example below the set of controllable ω-languages is not closed under countable unions.

Example 3.1. (Controllability)

Consider the DES $A = (L, S)$ over the event set $\Sigma = \{a, b\}$ with

$$S = a^\omega + a^* b a^\omega$$
$$L = a^* + a^* b a^* \quad (= \bar{S})$$

and $\Sigma_c = \{a\}$. Let $B = a^* b a^\omega \subseteq S$. Since $pr(B) = L$ it is clear that $pr(B)$ is controllable. Thus there exists a supervisor f such that $L_f = pr(B)$. However,

$$S_f = pr(B)^\infty \cap S = S \supset B$$

Thus the controllability of $pr(B)$ is not sufficient for the existence of a nonblocking supervisor f with $S_f = B$.

Example 3.2. (Closure under countable set union)

Consider the DES of the previous example. For each fixed $n \geq 0$ the regular ω-language

$$B_n = \{a^k b a^\omega, 0 \leq k \leq n\} \subseteq S$$

is controllable. Indeed the nonblocking supervisor

$$f_n(w) = \begin{cases} \{b\}, & \text{if } w = a^n; \\ \{a, b\}, & \text{otherwise.} \end{cases}$$

will ensure that $L_{f_n} = pr(B_n)$ and $S_{f_n} = B_n$. On the other hand as shown in the previous example

$$\cup_{n=0}^{\infty} B_n = a^* b a^\omega$$

is not a controllable ω-language . Thus the family of controllable ω-languages is not closed under countable unions.

3.3 Supremal Controllable Sequential Behaviors

Despite the fact that the family of controllable sublanguages of B need not be closed under arbitrary set unions, there still may exist a supremal controllable sublanguage contained in B. For this we impose the additional constraint that B be closed relative to S.

Proposition 3.2.

Let $A = (L, S)$ and $B \subseteq S$ be closed relative to S, i.e., $\bar{B} \cap S = B$. Then there exists a unique maximal controllable sequential behavior $B^{\dagger}$ contained in B.

Proof.

See [R]. ◇

Example 3.3. (Supremal controllable sublanguage)

Consider the DES of the Example 3.1. In this case the language B is not closed relative to S so the conditions of the previous proposition are not satisfied. The ω-languages

$$B_n = \{a^k ba^\omega, 0 \leq k \leq n\} \subseteq S \quad n \geq 0$$

are controllable, and $B_0 \subset B_1 \subset B_2 \subset \ldots \subset B$. Thus if B contains a unique maximal controllable sublanguage, say S_0, then this sublanguage must contain each of the B_n, and hence must contain their union. But

$$\cup_{n=0}^{\infty} B_n = a^* ba^\omega = B$$

Hence $S_0 = B$ is the only possibility, and since B is not controllable, this is impossible. This shows that in general if a language B is not closed relative to S, then it need not contain a supremal controllable sublanguage.

4 DES Representations

The above language based model for a DES is representation independent. By this we mean that the languages L and S could be represented by any of a number of different schemes, e.g., Turing machines, pushdown automata, Petri nets, temporal logic formulas, recursive fixpoint equations, etc. The above results apply to all representations. A particular representation has to be selected at some stage either to narrow the scope of the languages under study or in order to carry out computation. We shall restrict attention here to the simplest representation: deterministic finite automata.

A *generator* G is a dynamic system consisting of a state set Q, an initial state q_0, and transition function $\delta : \Sigma \times Q \rightarrow Q$ (in general a partial function). Without loss of generality we assume that every state of G is reachable from the initial state, i.e., that G is accessible. δ is extended to a (partial) function on $\Sigma^* \times Q$ in the standard fashion [HU, p.17]. We write $\delta(w, q)!$ as an abbreviation for the phrase '$\delta(w, q)$ is defined'. Then the language *generated* by G is defined to be the subset

$$L(G) = \{w : w \in \Sigma^* \ \& \ \delta(w, q_0)!\}$$

Every closed language $L \subseteq$ has such a realization. However, as is well known, L has a finite state realization if and only if it is a closed regular language.

Example 4.1. (Generators)

(1) A generator can be visualized as a labelled directed graph. The generator below is a representation of the string language $L = (ab)^*$.

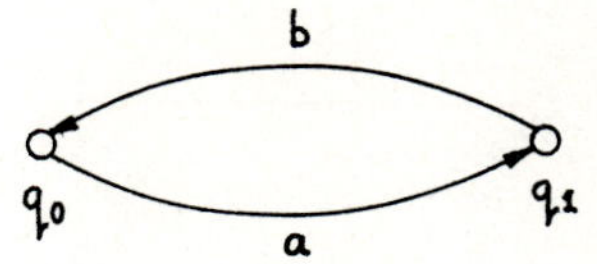

(2) The following generator is a representation of the string language $L = a^* + a^*ba^*$.

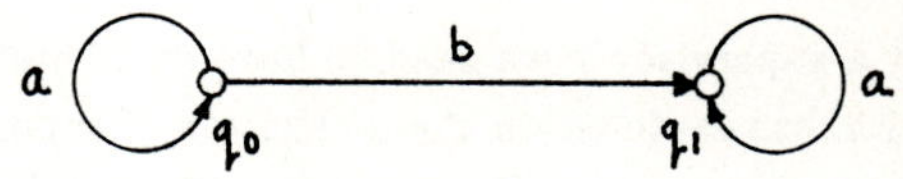

The limit behavior S can be specified as follows. Adjoin to $G = (\Sigma, Q, \delta, q_0)$ a subset of states $Q_m \subseteq Q$ To each sequence of events $e \in L(G)^\infty$ there corresponds a unique state trajectory $s_e: N \to Q$ satisfying

$$s_e(j) = \delta(e^j, q_0)$$

The sequence e and trajectory s_e are said to be *admissible* if s_e visits the set Q_m infinitely often. The set of event sequences generated by $G = (\Sigma, Q, \delta, q_0, Q_m)$ is then defined to be [(2)]

$$S(G) = \{e: e \in L(G)^\infty, \text{ and } s_e \text{ is admissible } \}$$

It is clear that

$$S(G) \subseteq L(G)^\infty \tag{G1}$$

with equality if $Q_m = Q$; and that

$$pr(S(G)) \subseteq L(G) \tag{G2}$$

In general we need not have equality in (G2) since there may be strings in $L(G)$ that cannot be extended to an admissible event sequence, i.e., G may not be nonblocking.

---

(2) This is a deterministic Büchi automaton [B].

Example 4.2. (Generators)

(1) The DES $A = (L, S)$ with $L = a^* + a^*ba^*$ and $S = a^\omega + a^*ba^\omega$ has the following finite state generator:

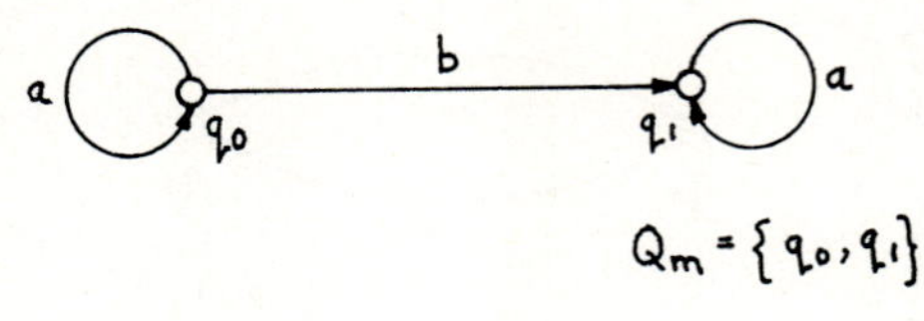

(2) The DES $A = (L, S)$ with $L = (a+b)^*$ and $S = (a+b)^*a^\omega$ does not have a deterministic finite state generator. It does, however, have a finite state nondeterministic generator.

4.1 Supervisor Realizations

To actually implement a supervisor f we need to have some finite realization for the map $f: \Sigma^* \to \Gamma$. To date attention has centered on the realization of supervisors in terms of a finite state machine together with an output map [RW1],[WR1]. While attractive for small problems it is not clear that this is the best form of realization for more complex problems or problems with variable parameters. Ideally one would like the realization to reflect any structure in the supervisor f. As yet it is not clear how this is best done.

Let $S = (\Sigma, X, \xi, x_0)$ be an automaton and $\phi : X \to \Gamma$. We say that the pair (S, ϕ) *realizes* the supervisor f if for each $w \in L_f$

$$\phi(\xi(w, x_0)) = f(w)$$

We interpret S as a standard automaton whose state transitions are driven by the events in Σ. In turn the state feedback map ϕ determines the input for G as a function of the state of S. Every supervisor has such a realization. We say that f is a *finite state supervisor* if it has a finite state realization.

Example 4.3. (Supervisor realization)

The supervisor f in Example 2.6 has the following finite state realization.

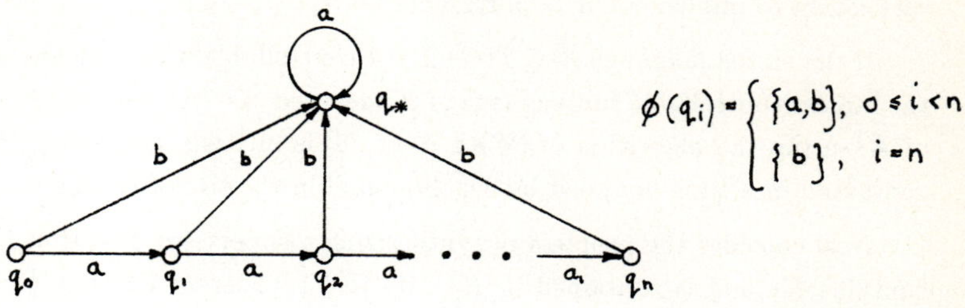

5 Some Complexity Issues

5.1 Controllability and supervisor synthesis

We say that supervisor synthesis problem is *polynomially decidable* if given an instance of the problem it is possible to decide in a time bounded by a polynomial in the size of the instance whether or not it is solvable. Similarly we shall say that the problem is *polynomially solvable* if given a solvable instance of the problem, it is possible to synthesize a solution in a time bounded by a polynomial in the size of the particular instance.

The basic supervisory control problems introduced in [RW1] are phrased in terms of a generator G and one or more sublanguages of $L(G)$. Assume that G is finite state with n_0 states, and that all other languages involved in the synthesis problem have finite state realizations. We measure the size of a problem instance in terms of n_0 and the number of states required in a finite state realization for other language involved in the problem. Under this scheme the supervisor synthesis problems of [RW1] were reported in [WR2] to be polynomially decidable and polynomially solvable. These results are quickly reviewed below.

The controllability of a prescribed $K \subseteq L(G)$ is polynomially decidable. Indeed a polynomial upper bound for the time complexity of this decision problem is easily derived from the algorithm for computing supremal controllable sublanguages given in [WR1, Sect. 6]. Let K be specified by deterministic finite state generator with n_1 states. Then according to the above algorithm (in the worst case) one first takes the intersection of the generator for K with G, an operation of complexity $O(|\Sigma|n_0n_1)$. This yields a generator with at most n_0n_1 states. Then for each state of this generator one checks a subset inclusion; each check being a computation of complexity $O(|\Sigma||\Sigma_u|)$. Thus the time complexity of the decision problem is certainly $O(n_0n_1|\Sigma_u||\Sigma|)$.

The problem of synthesizing a supervisor so that the closed loop behavior is a prescribed language K is both polynomially decidable and solvable. As shown in [RW1, Prop. 5.1] there exists a supervisor f for G that $L(G,f) = K$ if and only if K is closed and controllable. Hence it follows from the results of the previous paragraph that the problem is polynomially decidable. Now assume that K is a closed and controllable sublanguage of $L(G)$. Following the scheme outlined in [RW1, Prop. 5.1], one sees that the time complexity of the synthesis of a supervisor realization to implement K is polynomial in n_1.

If the closed language $K \subseteq L(G)$ is not controllable, then one has the option of computing the largest controllable sublanguage of K, denoted $K^{\uparrow}$. As reported in [WR2] the computation of $K^{\uparrow}$ using the algorithm of [WR1, Sect. 6] is of time complexity $O(n_0^2 n_1^2)$. So $K^{\uparrow}$ can be computed in a time bounded by a polynomial in the size of G and K.

Now consider the problem of synthesizing a supervisor f so that the closed loop behavior contains K_1 and is contained in K_2. By [RW1,Theorem 7.1] this problem is solvable if and only if the largest controllable sublanguage of K_2 contains K_1. Let K_1 and K_2 be specified by deterministic finite state generators with n_1 and n_2 states, respectively. Then $K_2^{\uparrow}$ can be computed in a time bounded by a polynomial in n_0 and n_2, and the inclusion of K_1 in $K_2^{\uparrow}$ can be checked in a time bounded by a polynomial in n_1 and n_2. Thus the synthesis problem is polynomially decidable. Furthermore, when solvable it reduces to the first synthesis problem with $K = K_2^{\uparrow}$. Hence it is polynomially solvable.

5.2 Minimal supervisor realizations

In realizing a supervisor in terms of a finite state automaton it is natural to consider the possibility of seeking a realization with the least number of states. Rather than consider this general problem it is sufficient for our purposes to restrict attention to the following simplified problem:

SUPERVISOR MINIMIZATION (SM): *Let $f : K \to \Gamma$ be a finite state supervisor for the closed and controllable regular language K. Find a minimal realization for f.*

This problem is certainly no harder than the general problem, and in what follows that will be all that matters.

We assume that one finite state realization of f, say (S, ϕ), is given. The minimal realization problem is then equivalent to the much studied problem of the minimization of a partially specified sequential machine. There are two approaches to the problem. First, as explored in [RW1], one can look for congruences on the state set X of S finer than *ker* ϕ. A *congruence* is an equivalence relation on X with the property that for each $\sigma \in \Sigma$ if $x \equiv y$ and $\xi(\sigma, x)$ and $\xi(\sigma, y)$ are both defined, then

$$\xi(\sigma, x) \equiv \xi(\sigma, y)$$

Conceptually this is a very useful construction. However, except in special cases, it is unlikely to be computationally feasible. Indeed if we translate, with some minor modifications, the results of Pfleeger [P] into the current setting we have:

Proposition 5.1.

Let (S, ϕ) be a finite state supervisor realization. The decision problem: *does there exist a congruence on S finer than ker ϕ and having at most k equivalences classes*, is NP-complete.

Proof.

See [P]. ◇

A second, and more general approach to the problem involves the use of invariant covers. This was explored in [VW]. It can be shown that the minimal realizations of f can be obtained from any given realization (S, ϕ) by examining the invariant covers of S finer than ϕ. This fact together with a simple extension of Pfleeger's previous result can be used to prove:

Proposition 5.2.

Let $f : K \rightarrow \Gamma$ be a finite state supervisor. The decision problem: *does there exist a finite state realization* (S, ϕ) *of* f *that has at most* k *states,* is NP-complete.

Proof.

See [P].

These results indicate that minimal supervisor realization is unlikely to be feasible for problems of significant size (assuming P $\neq$ NP). While it is important to be aware of this fact it is also important to keep it in perspective. In many supervisory control problems minimal supervisor realizations are of questionable merit. Indeed given the complexity of most discrete event systems a structured (nonminimal) solution is usually more desirable than an unstructured minimal solution.

There are special cases where supervisor simplification is a polynomial time problem. For example, if the dynamic part S of a supervisor realization is required to be a generator for the closed loop language K, then a minimal state realization exists and can be constructed in polynomial time. Such an approach might be desired, for example, if, in addition to computing the desired control action, the supervisor realization is required to detect erroneous behavior.

Part II

6 Product Systems

In the second part of the paper we restrict attention to DES composed of a finite set of asynchronous components. Such product systems arise naturally when modeling the concurrent operation of several asynchronous, or partially synchronous discrete dynamical systems. For example, in modeling certain aspects of communication protocols, operating systems, or the online control of flexible manufacturing systems.

One of the principal difficulties in dealing with product systems is that the number of states increases exponentially with the number of components. Thus synthesis methods based on searching over the product state space are unlikely to be computationally feasible. For example, the general supervisor synthesis problems posed and solved in [RW1], [WR1], although known to be of polynomial complexity when the size of a problem instance is measured in terms of the number of system states (Part I), cannot be regarded as computationally tractable for product systems.

We regard a decision or synthesis problem for a product system as computationally feasible if it can be solved in a time bounded by a polynomial in the size of the component subsystems n and the number of components p. Several interesting supervisory control problems for a class of product DES are computationally feasible in this sense.

Let $A_i = (L_i, S_i)$ be be p finite state DES over disjoint alphabets $\Sigma_1, \ldots, \Sigma_p$, with control partitions $\Sigma_i = \Sigma_{ci} \cup \Sigma_{ui}$. For each A_i we assume $S_i \neq \emptyset$ and $pr(S_i) = L_i$

Let $\Sigma = \cup_{i=1}^{p} \Sigma_i$, and define the projection $p_i : \Sigma^* \to \Sigma_i^*$ of Σ^* onto Σ_i^* by

$$p_i(\sigma) = \begin{cases} \sigma, & \text{if } \sigma \in \Sigma_i; \\ 1, & \text{if } \sigma \in \Sigma_j \text{ with } i \neq j. \end{cases}$$
$$p_i(w\sigma) = p_i(w)p_i(\sigma) \quad w \in \Sigma^*,\ \sigma \in \Sigma$$

The action of p_i on $w \in \Sigma^*$ is simply to extract from w the string of events in Σ_i. A sequence $e \in \Sigma^\omega$ is Σ_i-*recurrent* if $e(j) \in \Sigma_i$ infinitely often. In this case let e_i denote the unique subsequence of e consisting of the elements of Σ_i, and extend the projection p_i to a partial function $p_i : \Sigma^\omega \to \Sigma_i^\omega$ by defining

$$p_i(e) = \begin{cases} e_i, & \text{if } e \text{ is } \Sigma_i\text{-recurrent;} \\ \text{undefined}, & \text{otherwise.} \end{cases}$$

We are using the symbol p_i for two different functions but it will be clear from context what is intended.

The *product system* $A = (A_1, A_2, \ldots, A_p)$ is defined to be the DES (L, S) with

$$L = \{w: w \in \Sigma^* \ \&\ p_i(w) \in L_i, i = 1, \ldots, p\}$$

and

$$S = \{e: e \in \Sigma^\omega \ \&\ p_i(e) \in S_i,\ i = 1, \ldots, p\}$$

i.e., $e \in S$ if and only if for each i, $e(t) \in \Sigma_i$ infinitely often, and if e_i is the subsequence of e consisting of all elements in Σ_i, then $e_i \in S_i$.

6.1 Product system representation

Assume that for $i = 1, \ldots, p$ the component DES A_i has a finite state realization

$$G_i = (\Sigma_i, Q_i, \delta_i, q_{0i}, Q_{mi})$$

Let $|Q_i|$ denote the cardinality of Q_i and set $n = max\{|Q_i| : 1 \leq i \leq p\}$.

The product generator $G = ||_{i=1}^{p} G_i$ is defined to be

$$G = (\Sigma, Q, \delta, q_0)$$

with

$$\Sigma = \cup_{i=1}^{p} \Sigma_i \quad (\Sigma_c = \cup_{i=1}^{p} \Sigma_{ci})$$
$$Q = \Pi_{i=1}^{p} Q_i$$
$$q_0 = (q_{01}, \ldots, q_{0p})$$

and for $\sigma \in \Sigma_i$

$$\delta(\sigma, (q_1, \ldots, q_i, \ldots, q_p)) = (q_1, \ldots, \delta_i(\sigma, q_i), \ldots, q_p)$$

provided $\delta_i(\sigma, q_i)!$.

For $i = 1, \ldots, p$ let

$$Y_{mi} = \{q : q \in Q, q_i \in Q_{mi}\}$$

These sets will pay the role of the set Q_{mi} for the generator G_i, except for G there are p such recurrent sets - one for each of the component generators.

The language generated by G is defined in the usual fashion, i.e.,

$$L(G) = \{w : w \in \Sigma^*, \delta(w, q)!\}$$

To each event sequence $e \in L(G)^\infty$ there corresponds a unique state trajectory s_e. The sequence e and trajectory s_e are *admissible* if s_e visits each of the sets Y_{mi}, $i = 1, \ldots, p$, infinitely often. The sequential behavior of G is then defined to be the set

$$S(G) = \{e : e \in L(G)^\infty, \text{ and } s_e \text{ is admissible }\}$$

It is readily verified that for each i, $1 \leq i \leq p$,

$$L(G) = \{w : w \in \Sigma^* \ \&\ p_i(w) \in L(G_i), i = 1, \ldots, p\}$$

and that

$$S(G) = \{e : \ e \in \Sigma^\omega \ \&\ p_i(e) \in S(G_i), \ i = 1, ..., p\}$$

Thus G is a realization of the product system A.

Note that if each G_i has n states, then G has n^p states. If p is bounded, then the size of G is bounded by a polynomial in n. It follows from our opening remarks that control problems for G formulated in the framework of [RW1] are decidable in a time bounded by a polynomial in n. Here, however, we are interested in the case when both p and n are variable and both are to be taken as a measure of problem size. In this case we say that a supervision problem for a product system is *polynomially decidable* (resp. *polynomially solvable*) if the time required to determine whether or not it is solvable (resp. to synthesize a solution) is bounded by a polynomial in n and p.

6.2 Coordination

A supervisor f for the product system $A = ||_{i=1}^{p} A_i$ is a *coordinator* if for every set of p event sequences $e_1, e_2, \ldots, e_p$ with $e_i \in S_i$, $i = 1, ..., p$, there exists a sequence e in the closed loop behavior S_f such that for $i = 1, ..., p$

$$p_i(e) = e_i$$

Roughly this means that the supervisor does not modify the open loop behaviors of the individual DES; it only constrains how they interact by controlling the relative order of events. Thus if the individual DES have selected to execute the event trajectories e_i, $i = 1, ..., p$, then there exists at least one trajectory in the closed loop system in which this is possible.

7 Coordination Problems

We end by considering two interesting coordination problems for product systems.

7.1 Mutual exclusion

A subset $\bar{Q}$ of the state set of the generator G is said to be *nontransient* if there exists an admissible state trajectory for G that visits $\bar{Q}$ infinitely often.

We first consider the following standard problem:

MUTUAL EXCLUSION (MEX): *Let $\bar{Q}_i \subseteq Q_i$ be p given nontransient subsets and k be a fixed integer with $1 \leq k < p$. Synthesize (if possible) a nonblocking coordinator f for G such that for each $e \in S_f$, and each $j \geq 1$, after e^j at most k of the generators satisfy $q_i \in \bar{Q}_i$.*

The problem requires the G_i to be coordinated so that at most k of the generators are in the designated subsets of states at any one time. For $k = 1$ this is the traditional mutual exclusion problem.

Let $W(e^j)$ be the number of component DES satisfying $q_i \in \bar{Q}_i$ after e^j has been generated. Thus a sequence $e \in S$ satisfies the MEX constraint iff

$$W(e^j) \leq k \quad \text{all } j \geq 1$$

Let $B \subseteq S$ denote the set of all such sequences. Then MEX requires the synthesis of a nonblocking coordinator f such that $S_f \subseteq B$.

It is straightforward to show that B is closed relative to S. It follows that there exists a unique maximal controllable sequential behavior $B^\dagger$ contained in B, and provided $B^\dagger$ is nonempty, there exists a nonblocking supervisor f such that $S_f = B^\dagger \subseteq B$. If f is a coordinator, then obviously it is a solution to MEX. Moreover, in this case the solution is *minimally restrictive* in the sense that the resultant closed loop behavior is the largest possible behavior that does not violate the MEX constraint. On the other hand if f is not a coordinator, then MEX has no solution. This follows by noting that if g is a proposed solution to MEX, then

$$S_g \subseteq B^\dagger = S_f$$

But if f is not a coordinator, there exist $e_i \in S_i$, $i = 1, \ldots, p$, such that for no $e \in S_f$ (and hence no $e \in S_g$) do we have $p_i(e) = e_i$, $i = 1, \ldots, p$. Thus g can not be a coordinator.

We conclude from the above that if an instance of MEX is solvable, then it has a minimally restrictive solution, and that this solution implements the closed loop behavior $B^{\uparrow}$.

Our main result on the complexity of MEX is:

Theorem 7.1.

MEX is polynomially decidable. Moreover, given a solvable instance of MEX, it is possible to synthesize a minimally restrictive solution in polynomial time.

That the problem is polynomial is due to the fact that it can be decoupled and analyzed in terms of the component DES. To see this let $\Sigma_{ui} = \Sigma_i - \Sigma_{ci}$ be the set of uncontrolled events of G_i, and D_i denote the set of states of G_i from which is possible to reach $\bar{Q}_i$ via uncontrollable events:

$$D_i = \{q_i : q_i \in Q_i \ \& \ \delta_i(w, q_i) \in \bar{Q}_i, \text{ for some } w \in \Sigma_{ui}^*\}$$

It is easily seen that an instance of MEX is solvable if and only if it is solvable with the sets $\bar{Q}_i$ replaced by the sets D_i, and that f is a solution to the former problem if and only if it is a solution to the latter. Necessary and sufficient conditions for the solvability of MEX are readily determined in terms of the sets D_i:

Proposition 7.1.

MEX is solvable if and only if the following conditions are satisfied:

(1) For at most k of the G_i, $q_{0i} \in D_i$; and

(2) There exist $p - k + 1$ generators with the property that every admissible state trajectory of G_i enters $Q_i - D_i$ infinitely often.

Proof.

See [R]. ◇

The second condition of the previous proposition can be further resolved as follows:

Proposition 7.2.

Every admissible state trajectory of the generator G_i enters the set $Q_i - D_i$ infinitely often if and only if G_i has no cycles in D_i that intersect Q_{mi}.

Proof.

See [R]. ◇

Using these results it is now straightforward to prove Theorem 7.1. To determine if an instance of MEX is solvable we construct the sets D_i and test the conditions of the previous two propositions. The determination of D_i for generator G_i is a simple reachability problem, and the time required for this computation is bounded by a polynomial in n. The verification of item (i) in the first proposition requires at most p set membership tests, each with at most n comparisons. Thus the condition can be verified in polynomial time. To determine if G_i has a cycle in D_i intersecting Q_{mi} one need only check whether or not each state $q \in D_i \cap Q_{mi}$ is reachable from itself in the restricted generator $G_i|D_i$. The time required for this verification is bounded by a polynomial in n. Thus all p generators can be tested for such cycles in polynomial time. It follows that MEX is polynomially decidable. The proof that a minimally restrictive solution can be synthesized in polynomial time requires knowledge of the proof of the first proposition (see [R]).

7.2 Uncontrolled string exclusion

A string $w \in \Sigma^*$ is said to be *nontransient* for the DES $A = (L, S)$ if there exists $e \in S$ such that w occurs infinitely often as a contiguous string in e.

Our second coordination problem is:

UNCONTROLLED STRING EXCLUSION (USE): *Let $w_i \in \Sigma^+_{ui}$, $1 \le i \le p$, be p nontransient strings of uncontrolled events, and let the string w be formed from a shuffling of the w_i. Synthesize (if possible) a nonblocking coordinator f for G such that for every $e \in S_f$ and every $j \ge 1$, $e^j \ne s_1 w s_2$ for some $s_1, s_2 \in \Sigma^*$.*

The second item requires that the generated sequence of events e never contains the 'illegal string' w.

We show that there exists a polynomial transformation of any instance of USE to an instance of MEX. From this it will follow that USE is polynomially decidable and polynomially solvable.

Set

$$X_i = \{q_i \in Q_i : \delta_i(w_i, q_i)!\}$$
$$X = \Pi_{i=1}^p X_i$$

Since each w_i is nontransient so are the sets X_i. We claim that an instance of USE is solvable if and only if the instance of MEX with $k = p-1$ and $\bar{Q}_i = X_i$ is solvable. Indeed if USE is solvable then the solution , say f, must keep the state of G from reaching X since the uncontrolled string w is possible from any state in X. Conversely, if f solves MEX with $k = p - 1$ and $\bar{Q} = X$, then under f the state of G never reaches X, and since these are the only states of G from which w is possible, the illegal string never occurs. Finally, for each generator the set of states X_i can be found in a time bounded by a polynomial in n and $|w|$. We need only test the condition $\delta_i(w_i, q_i)!$ at each state of G_i. It follows that all of the X_i can be found in a time bounded by a polynomial in n and p and $|w|$.

We end by noting that we have proved:

Theorem 7.2.

USE is polynomially decidable and polynomially solvable. Moreover, given a solvable instance of USE it is possible to synthesize a minimally restrictive solution in polynomial time.

8 References.

[B] Buchi, J.R., *On a decision method in restricted second order arithmetic, International Congress Logic Methodology and Philosophy of Science,* Stanford, Calif., 1960.

[CDFV] Cieslak, R., C. Desclaux, A. Fawaz, and P. Varaiya, *Supervisory control of discrete event processes with partial observations,* Memo no. UCB/ERL M86/63, Electronics Research Lab., College of Eng., Univ. of Calf., Berkeley, 1986.

[HU] Hopcroft, J.E., and Ullman, J.D., *Introduction to Automata Theory, Languages and Computation,* Addison-Wesley Pub. Co., Reading, MA., 1979.

[LW1] Lin, F., and W.M. Wonham, *Decentralized supervisory control of discrete-event systems,* to appear: *Information Sciences,* 1987.

[LW2] Lin, F., and W.M. Wonham, *On observability of discrete-event systems,* to appear: *Information Sciences;* see also: Systems Control Group Report #8701, Department of Elect. Eng., University of Toronto, 1987.

[P] Pfleeger, C.P. *State reduction in incompletely specified finite state machines, IEEE Transactions on Computers,* C-22 (12), pp. 1099–1102, January 1979.

[R] Ramadge, P.J., *Some tractable supervisory control problems for discrete event systems,* ISS Technical Report, Department of Electrical Engineering, Princeton University, Princeton, N.J., October 1987.

[RW1] Ramadge, P.J., and W.M. Wonham, *Supervisory control of a class of discrete-event processes, SIAM J. on Contr. and Optimization,* **25** (1), pp. 206–230, January 1987.

[RW2] Ramadge, P.J., and W.M. Wonham, *Modular feedback logic for discrete event systems, SIAM J. on Contr. and Optimization,* **25** (5), pp. 1202-1218, September 1987.

[RW3] Ramadge, P.J., and W.M. Wonham, *Modular supervisory control of discrete event systems, Proc. of the Seventh International Conference on Analysis and Optimization of Systems,* Antibes, June, 1986.

[T] Tsitsiklis, J.N., *On the control of discrete event dynamical systems* Technical Report LIDS-P-1661, MIT, March 1987

[VW] Vaz, A., and W.M. Wonham, *On supervisor reduction in discrete-event systems, Int. J. Control,* **44** (2), pp. 475–491, 1986.

[WR1] Wonham, W.M., and P.J. Ramadge, *On the supremal controllable sublanguage of a given language*, *SIAM J. on Contr. and Optimization*, **25** (3), pp. 637–659, May 1987.

[WR2] Wonham, W.M., and P.J. Ramadge, *Modular supervisor control of discrete event systems*, *Mathematics of Control, Signals, and Systems*, **1** (1), 1988.

A Petri-net Approach to the Control of Discrete-event Systems

M.J. Denham
School of Computing
Kingston Polytechnic
Kingston upon Thames
Surrey, KT1 2EE
United Kingdom

Abstract - In this paper we describe a theory for the synthesis of supervisory controllers for discrete-event systems such as computing, communication and manufacturing systems. The required behaviour of the system is specified by an invariant relation which must hold during operation of the system and a controller is derived which when combined with the system will ensure the required behaviour. The theory is based on viewing the Petri-net model of the system as a two-sorted algebra and the composition operation as a natural product between two such algebras.

Keywords - discrete-event systems/ supervisory control/ Petri-nets.

1. Introduction

It is clear that there exists a substantial need for the development of a comprehensive and effective theory for the control of systems which are described by discrete-event types of behaviour. There is a wide range of systems which display such behaviour:

* computing systems
  - programs
  - operating systems
  - human/computer interfaces
* communication systems
  - networks
  - protocols
* manufacturing systems
  - batch (job-shop) production
  - flexible manufacturing systems (FMS)

NATO ASI Series, Vol. F47
Advanced Computing Concepts and Techniques
in Control Engineering
Edited by M.J. Denham and A.J. Laub

The control requirement for such systems takes a number of different forms:

* operational control - the need to maintain the behaviour of an individual subsystem or set of subsystems according to a pre-determined sequence of required operations/events,
* supervisory control - the need to maintain overall control of a number of interacting systems or subsystems to prevent conflicts and enable cooperation on shared tasks,
* scheduling control - the need to determine the order in which tasks or activities in a system are to be carried out in order to meet some specified objective.

Techniques for designing operational controllers for individual units of a communication or manufacturing system, for example, are well understood. For supervisory control, which involves specifying, analysing and verifying the behaviour of a usually large number of closely interacting subsystems and synthesising a controller to coordinate this behaviour to avoid problems such as conflicts during access to shared resources, a comprehensive and practically useful theory is still awaited. This paper is one attempt to provide the basis for the development of such a theory, and ultimately a controller design methodology.

To meet the need for such a theory, we will need to provide the following:

* a language - in which the behaviour of the system can be expressed, i.e. the means to create a model of the system,
* analysis and verification procedures - in order to determine and validate the behaviour of the system representation or model,
* a method for coping with the potential complexity of the analysis and verification procedures - e.g. the use of modular approaches
* synthesis procedures - to create new behaviours from

existing ones,

* implementation techniques - to allow the model of the required system behaviour to form a specification for the actual controller hardware or software from which the latter can be directly generated.

In this paper we will be concerned with the structural or qualitative aspects of system behaviour and supervisory control, rather than the performance or quantitative aspects. That is, we are not concerned with properties such as execution time or throughput, or with the associated task of optimising such properties. Our theory therfore only requires that we are able to model the operational behaviour of the system in terms of a sequence of states and events which determine the transition of the system from one state to another. Since the system is generally composed of a number of subsystems, and in order to cope with the complexity problem resulting from trying to handle a large number of such subsystems and their interactions, our theory also needs mechanisms for describing the concurrent and interactive operation of a number of communicating and co-operating subsystems. In summary therfore our theory needs to capture the following aspects of system behaviour:

* concurrency - the simultaneous occurrence of events in different subsystems,
* co-operation - the requirement that events in different subsystems occur simultaneously in order to carry out some jointly executed task,
* communication - the need for subsystems to exchange information, e.g. in order to carry out some jointly executed task,
* control - the ability of a controller subsystem to restrict the execution of events in other subsystems, or to initiate such events, in order to impose a required behaviour on the system.

It might perhaps be clear to those with some background in computer science that the basic need to be able to model sequences of states and events, together with the additional requirement to capture the notions of concurrency, co-operation and communication, leads one naturally to the theories which have been developed in that field in order to model exactly the same characteristics for programs and computer systems. Indeed, the pioneering work by Ramadge and Wonham (1987) in the field of supervisory control of discrete-event systems, was based on the theory of regular languages, which has been one of the foundation theories for computer science for some years. More recently, the work of Ostroff (1987) on extended state machines draws substantialy on the theory of communicating sequential processes (Hoare, 1978) and on the synthesis of communicating processes from temporal logic specifications (Manna & Wolper, 1984). Other types of model and associated theory are possible candidates for use, including flow algebras (e.g. Milner, 1980), event structures, path expressions and Petri-nets. In this paper we will consider the latter approach and use the Petri-net model as our basic formalism for describing dicrete-event system behaviour. This has the following attractions. Firstly, the theory is well-developed and continues to be an active area of research. Secondly, the notion of concurrency is captured by the model in an intuitively appealing and simple manner. Thirdly, the use of Petri-nets as a model for analysing behaviour in the important application area of manufacturing systems is of growing interest to practitioners in the field, with the ability to extend the basic model with timing information in order to assess quantitative aspects such as throughput. Also, finally, the Petri-net model is proving to be of some use in the area of specifying the behaviour of real-time control software, e.g. Bruno and Marchetto (1986), and distributed systems, e.g. Bruno and Balsamo (1986). Work in this area could provide the avenue to an implementation procedure which allows direct generation of the controller hardware and software from the specification embodied in the synthesized Petri-net model of the controller.

In the remainder of the paper, we will first review some fundamental aspects of modelling discrete-event behaviour and then consider Petri-net models in particular. In the next section we will use the results of Winskel (1985) to formulate a Petri-net as a two-sorted algebra and thus provide the formal framework in which our analysis of behaviour can be carried out. In particular, in the next section we will describe a composition operation for Petri-nets which will be used to provide the basis for our required control action. This will lead naturally to the outline formulation of a synthesis procedure for a controller which, when composed with the system, will ensure the system behaves in the specified manner. Finally, we summarise our results and propose possible ways ahead in the development of this approach.

2. Modelling Discrete-event Behaviour

As we have already said, a large number of models exist for representing systems which exhibit a discrete-event type of behaviour. Fundamental to all of these models is the notion that the behaviour of the system is described by a sequence of events. Events are generally assumed to be atomic, i.e. indivisible in respect of the level of abstraction at which the system is being modelled. Thus they can be assumed to occur instantaneously and their effect is immediately registered, although in some models there may be a time delay defined between the moment when the event is enabled, i.e. the conditions which allow the event to occur are satisfied, and the actual occurrence of the event. The set of possible events which can occur at a given moment of time is determined by:

* the sequence of events which have occurred up to that moment of time from some initial point in time, represented by the state of the system,
* the set of pre-conditions which are defined for that event and which must hold before the event can occur.

In general, events are assumed to occur asynchronously, i.e. not at specific moments of time defined by some underlying clock, although a clock subsystem may well be part of the system model and reference made to it in specifying the behaviour of the system. The occurrence of an event results in the set of post-conditions associated with that event being satisfied and in the system evolving into a new state, although this new state may possibly be indistinguishable from the previous state.

A fundamental model proposed by Winskel (1986) serves well in illustrating the basic notion of discrete-event behaviour:

Definition: An event structure consists of

* a set of events E,
* a non-empty subset F of the finite subsets of E, each member of which represents a feasible set of events, i.e. can form a history of events (a past sequence of events),
* an enabling relation $T \subseteq E \times F$ which constrains the set of events in E which can occur after a given subset of events in F.

In this model therefore, a state is a subset $X \in F$ of events which have occurred up to the present time, and all sequences of events which could have led to the state X can be determined by considering every $e \in X$ which is enabled (under T) by $Y = X - e$, $Y \in F$. The resulting chain of enablings will end with the set of events which are enabled by the empty set, i.e. require no prior events to have occurred.

As an example of a system which has a discrete-event behaviour and to which we will return again later, consider the well-known dining philosophers problem. The picture (Figure 1) is as follows. Five philosophers are sitting at a table on which there are five forks, one to the left and one to the right of each philosopher, and a bowl of pasta. Each philosopher is initially thinking and all the forks are on the table. A philosopher can then carry out the following sequence of events: pick up the fork which is to the left; pick up the fork to the right and eat the pasta; put down the forks. Each event results in the transition from one state to another, i.e. from thinking to hungry to eating and back to thinking. Note that in order to eat, a philosopher needs both forks and will not release the forks until finished eating. The model of this system needs, for example, to capture the possible simultaneous occurrence of each philosopher picking up his/her left-hand fork which will lead to a state of the system in which no further event is possible. There is also a need to introduce control to both ensure that this state is not reached and to allow each philosopher an opportunity to eat. Later we will build a Petri-net model of this system and introduce a suitable controller.

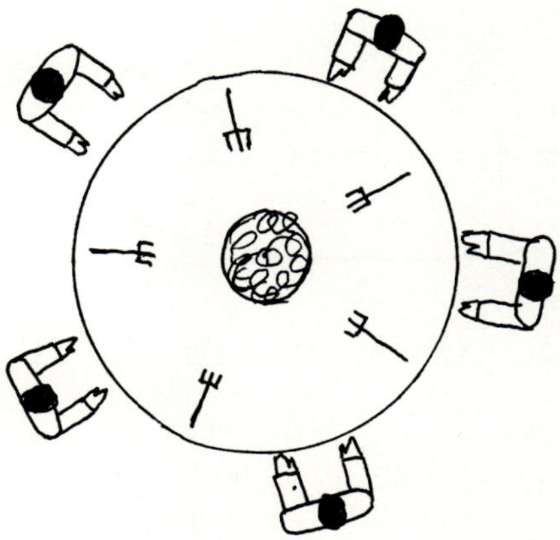

Figure 1. The dining philosophers problem

3. Petri-net Models

Definition: A Petri-net model consists of

* a finite set of places P,
* a finite set of events E,
* an initial marking m_o of the set of places, where a marking m is a function $m: P \rightarrow N$ (the set of non-negative integers),
* a transition function (partial)
  $t: [P \rightarrow N] \times [E \rightarrow N] \rightarrow [P \rightarrow N]$
  which maps a marking m into a marking m' according to
  $m' = m - W^{-}q + W^{+}q$
  where $q: E \rightarrow N$ is called a transition, and

$W^-, W^+: [E \to N] \to [P \to N]$ are called the causal dependency matrices.

The Petri-net model defined above, which is also known by other names such as place-transition net, has a simple graphical representation. For example, consider the net shown in Figure 2.

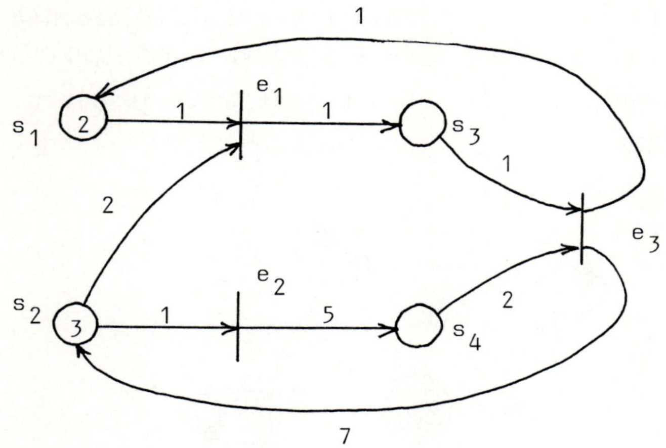

Figure 2. Example of a Petri-net

In this case, $P = \{s_1, s_2, s_3, s_4\}$, $E = \{e_1, e_2, e_3\}$

$m_o = \begin{vmatrix} 2 \\ 3 \\ 0 \\ 0 \end{vmatrix}$, and $W^+ = \begin{vmatrix} 0 & 0 & 1 \\ 0 & 0 & 7 \\ 1 & 0 & 0 \\ 0 & 5 & 0 \end{vmatrix}$, $W^- = \begin{vmatrix} 1 & 0 & 0 \\ 2 & 1 & 0 \\ 0 & 0 & 1 \\ 0 & 0 & 2 \end{vmatrix}$.

Thus, in the graphical form, places are represented as circles and events as bars. The arcs from places to events are labelled as shown by the appropriate entries in the causal dependency matrix W^-, and those from events to places by the appropriate entries in the matrix W^+. The marking of the places is shown by the numbers inserted in each of the places (empty implies the value zero). The occurrence of an event causes the value labelling each arc into that event to be subtracted from the value marking the corresponding place, and the value labelling each arc out of the event to be added to the value marking the place to which the arc goes. A transition q, which is a vector over the set of events, represents the possibly multiple occurrence of each event at a given moment in time.

Since the marking of any place must always be non-negative we have the following definition and constraint on the set of possible transitions which can occur at any time:

Definition: A transition $q{:}E \rightarrow N$ is enabled in a marking m if and only if $m \geq W^{-}q$. (Note: for two markings m,m', we have that $m \geq m'$ if and only if $m(p) \geq m'(p)$, for all $p \in P$).

Thus $q = \begin{vmatrix} 1 \\ 1 \\ 0 \end{vmatrix}$, which represents the simultaneous occurrence of the two events e_1 and e_2, is enabled in m_o since $W^{-}q \leq m_o$. Thus, from the definition of the transition function, we have that

$$m' = m_o - W^{-}q + W^{+}q$$
$$= \begin{vmatrix} 1 \\ 0 \\ 1 \\ 5 \end{vmatrix}$$

It should be clear from this example that in a Petri-net, the concurrency of events is determined by their respective pre-conditions, represented by the entries in the matrix W^{-}, being simultaneously satisfied or not. For example, if the marking of s_2 is less than 3 then it is not possible for events e_1 and e_2 to occur concurrently.

The dynamic behaviour of a Petri-net model is denoted by

$$M:\; m_o \xrightarrow{q_1} m_1 \xrightarrow{q_2} m_2 \xrightarrow{q_3} m_3 \xrightarrow{q_4} \cdots \xrightarrow{q_N} m_N$$

for sequences of markings $(m_0, m_1, m_2, \ldots, m_N)$ and of transitions $(q_1, q_2, q_3, \ldots, q_N)$. We also say that a marking m is reachable from an initial marking m_o if and only if $m = m_N$ for some pair of sequences of markings and transitions.

We now come to the important concept of a <u>net invariant</u> which we shall use in this paper to specify the required behaviour of a Petri-net model.

Theorem: Let $v{:}\ P \rightarrow Z$ (the set of integers) be such that

$vW^- = vW^+$ and m be any marking reachable from m_o. Then $vm = vm_o$.

Proof:

$$\begin{aligned} vm = vm_N &= vm_{N-1} - v(W^- - W^+)q_N \\ &= vm_{N-1} \\ &= vm_{N-2} - v(W^- - W^+)q_{N-1} \\ &= vm_{N-2} \\ &\vdots \\ &= vm_o \end{aligned}$$

As an example, $v = |1\ 0\ 1\ 0|$ is an invariant for the net shown in Figure 2.

The use of an invariant as a specification of system behaviour will become clearer later, but it should be evident now that each invariant defines a relation or predicate on the markings of the net which holds for all reachable markings, and thus for the entire dynamic behaviour of the system.

We will return to the dining philosophers problem to provide another example of a Petri-net model (Figure 3). The figure shows only a part of the total model, that for two philosophers and their adjacent forks. The model extends in the same way to the left and right and eventually loops back on itself as the left fork of one philosopher becomes the right fork of another to complete the circle.

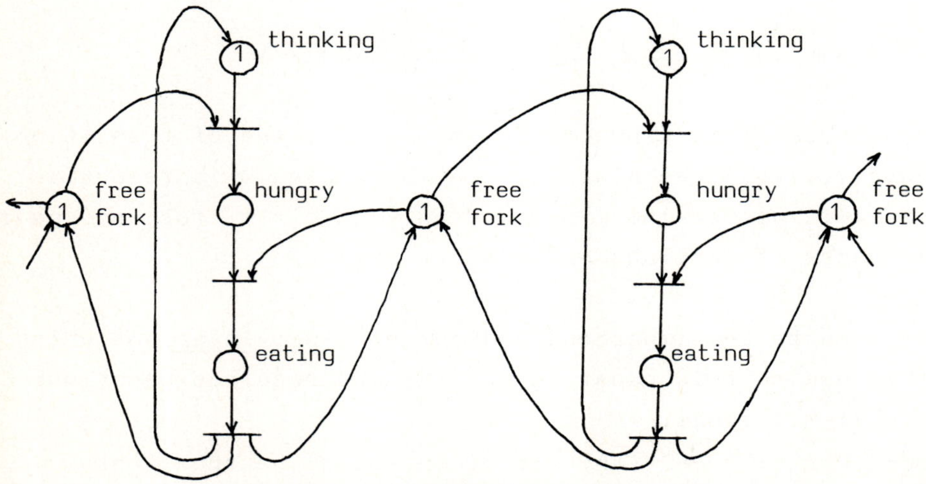

Figure 3. Dining philosophers model

We can also use this model to illustrate an important concept which will be central to our controller synthesis procedure. Further consideration of the dining philosophers model shows that it is formed by the composition of a set of models, one for each philosopher and one for each fork. The models for one philosopher and for one fork are shown in Figure 4. This also illustrates the way in which these two models are composed. Events in each model are synchronised with one another in relation to the way in which the two systems interact. Thus, for example , when a philosopher goes from hungry to eating the appropriate fork goes from free to in use. We can then make the two pairs of synchronised events each into a single event, as shown in Figure 4, and then combine the two places, eating in the philosopher model and in use in the fork model, into a single place. It is clear, and can be proven using the invariants of the model, that the markings of these two places are always equal. Composition through synchronisation of events of the five philosopher models with the five fork models will yield the required model of the overall system behaviour.

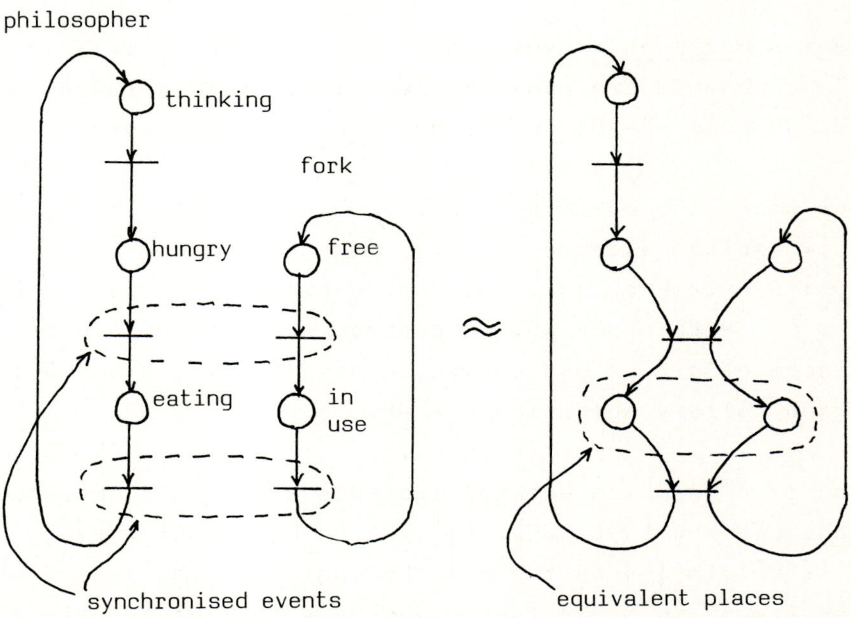

Figure 4. Composition of the philosopher and the fork models

4. Petri-nets as Two-sorted Algebras

Before we show how a Petri-net model can be formulated as an algebra, we will establish the necessary mathematical framework for our discussion.

Definition: An algebra is a system $\mathcal{A} = [\mathcal{S}, F]$ in which

- $\mathcal{S} = S_i$ is a family of non-empty sets called sorts, indexed by $i \in I$,
- $F = \{f_a\}$ is a set of finitary operations, where each f_a is a mapping

  $f_a : S_{i(1,a)} \times S_{i(2,a)} \times \ldots \times S_{i(n(a),a)} \rightarrow S_{r(a)}$

  for some $n(a) \in N$ and function $i_a: k \rightarrow i(k,a)$, $k \in \{1,\ldots,n(a)\}$, $i(k,a) \in I$, $r(a) \in I$.

Definition: Let $A = [\{S_i\}, F]$ and $B = [\{T_i\}, F]$ be two algebras belonging to the same family, i.e. they have sorts and operations having the same index set I. A morphism from A to B is a set of functions $\Phi_i : S_i \rightarrow T_i$, $i \in I$, such that for $f_a \in F$,

$$f_a \cdot (\Phi_{i(1,a)} \times \ldots \times \Phi_{i(n(a),a)}) = \Phi_{r(a)} \cdot f_a$$

Definition: A multiset v over a set X is a vector over X in which all components are non-negative, i.e. it is a function $v : X \rightarrow N$. We call μX the set or space of all multisets.

We can now state the basic definition which establishes a Petri-net as a two-sorted algebra.

Definition: A Petri-net is a two-sorted algebra $N = [\{\mu P, \mu E\}, \{W^-, W^+, m_o\}]$ with sorts μP, μE corresponding to multisets over the sets of places and events, unary operations W^-, W^+: $\mu E \rightarrow \mu P$, and nullary operation $m_o \in \mu P$.

The notion of a morphism between two nets can now be introduced.

Definition: Let $N = [\{\mu P, \mu E\}, \{W^-, W^+, m_o\}]$ and $N' = [\{\mu P', \mu E'\}, \{W^{-\prime}, W^{+\prime}, m_o'\}]$ be two nets belonging to the same family. A morphism from N to N' is then given by a pair of functions $\eta : \mu E \rightarrow \mu E'$ and $\beta : \mu P \rightarrow \mu P'$ such that $\beta m_o = m_o'$ and for all $x \in \mu E$, $W^{-\prime}\eta(x) = \beta(W^- x)$ and $W^{+\prime}\eta(x) = \beta(W^+ x)$.

An important property of morphisms between nets is that they preserve the dynamic behaviour of the nets.

Definition: Let m and m' be markings of a net N, i.e. m, m'$\in \mu P$. Let $x \in \mu E$. The dynamic behaviour of N is defined by the transition relation $N : m \xrightarrow{x} m'$ if and only if $W^- x \leq m$ and $m' = m - W^- x + W^+ x$.

Definition: Any marking $m \in \mu P$ for which there exists a sequence $(x_o, x_1, \ldots, x_{n-1})$, where $x_i \in \mu E$, such that from an initial marking m_o,

$$N : m_o \xrightarrow{x_o} m_1 \xrightarrow{x_1} \ldots \xrightarrow{x_{n-1}} m_n = m$$

is a reachable marking of N.

Theorem: Let $(\eta, \beta) : N \rightarrow N'$ be a morphism of nets. Then if $N : m \xrightarrow{x} m'$, it follows that $N' : \beta(m) \xrightarrow{\eta x} \beta(m')$.

Corollary: If m is a reachable marking of N then $\beta(m)$ is a reachable marking of N'.

Finally, we redefine the notion of an invariant of a net within this framework.

Definition: Let $N = [\{\mu P, \mu E\}, \{W^-, W^+, m_o\}]$ be a net. An invariant of N is a matrix $v : 1 \times P \rightarrow Z$ (here 1 denotes the single element set) such that $vm = vm_o$ for all reachable markings m of N. We write I_N as the set of all invariants of N.

5. Composition of Petri-nets

At the end of section 3, we saw that a possible composition operation which could be used to build systems from their component subsystems was one which involved the putting together of two nets with synchronisation between selected events. These events are subsequently combined into single events and redundant places are eliminated. We now wish to investigate the properties of this composition operator within the framework set up in the previous section. In particular, we wish to know what effect the composition has on the dynamic behaviour of the system and on the invariants of the model. Since we intend to use the invariants of the net as a specification of its behaviour, we need to see how the invariants of two nets propogate through the composition operation, i.e. if N and N' are two nets with invariants I_N and $I_{N'}$ respectively, and || denotes the composition operation, the we wish to determine the set of invariants I_S for the net defined by S = N||N'.

First, we define the composition operation in terms of two nets viewed as two-sorted algebras.

Definition: Let $N = [\{\mu P, \mu E\} , \{ W^-, W^+, m_o\}]$ and $N' = [\{\mu P', \mu E'\} , \{ W^-, W^+, m'_o \}]$ be two nets. The composition of N and N', denoted by N||N', is the concurrent operation of N and N' in which selected events, e.g. $e \in E$ and $e' \in E'$, are synchronised and form a synchronised event (e,e') such that, for all $p \in P$ and $p' \in P'$, $W^-(p,(e,e')) = W^-(p,e)$ and $W^{-}{}'(p',(e,e')) = W^{-}{}'(p',e')$, and similarly for W^+, $W^{+}{}'$.

The synchronisation of two events from N and N' is illustrated in Figure 5.

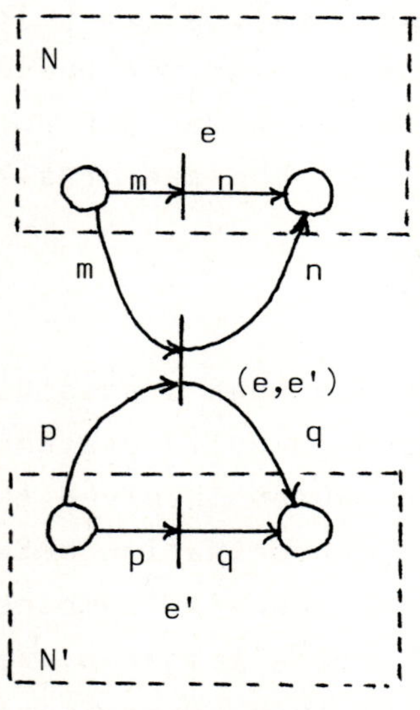

Figure 5. Synchronisation of two events

We can now consider the dynamic behaviour of the net composed in this way. Firstly, note that the composed net $S = N||N'$ has places $P_S = P \cup P'$, and events $E_S = \{(e,*) | e \in E\} \cup \{(*,e') | e' \in E'\} \cup \{(e,e') | e \in E \text{ and } e' \in E'\}$ where $(e,*)$ and $(*,e')$ denote events in N and N' respectively which are not synchronised with an event in the other net. The dynamic behaviour of the composed net is now defined in terms of the projection mappings from S into N and N' as follows.

<u>Theorem</u>: Let (π, ρ) and (π', ρ') be the projections from $S = N||N'$ to N and N' respectively, i.e. $\pi : P_S \longrightarrow P$, $\rho : E_S \longrightarrow E$, etc. Then $S : m \xrightarrow{x} m'$ if and only if $N : \rho m \xrightarrow{\pi x} \rho m'$ and $N' : \rho' m \xrightarrow{\pi' x} \rho' m'$.

<u>Corollary</u>: A marking m in $S = N||N'$ is reachable if and only if ρm is reachable in N and $\rho' m$ is reachable in N'.

We should note here that the projection mappings defined above are morphisms of nets. This allows us to use a general result concerning the propogation of net invariants through morphisms to determine the effect of the composition operator on the invariants of the component nets. This result shows that invariants are not always preserved in the direction of the morphism, i.e. if $(\eta, \beta) : N \rightarrow N'$, an invariant of N is not always mapped under β into an invariant of N'. However, invariants are preserved in the opposite direction under the dual map β^* defined by $\beta^* v = v\beta$, where $v \in I_{N'}$.

<u>Lemma</u>: Let $(\eta, \beta) : N \rightarrow N'$ be a morphism of nets. Then $v \in I_{N'}$ implies that $v\beta \in I_N$.

It follows therfore that the set of invariants I_S of the composite net is the <u>union</u> of the sets of invariants I_N and $I_{N'}$ of the component nets N and N', as illustrated below:

$$I_N \xrightarrow[\rho^*]{} I_{N||N'} \xleftarrow[\rho'^*]{} I_{N'}$$

where $\rho^* v = v\rho$ and $\rho'^* v = v\rho'$

This result provides us with a basis for a synthesis procedure for constructing a controller which, when composed with the uncontrolled system, will impose the required behaviour on the composite system, where this behaviour is expressed in the form of a set of invariants of the composite Petri-net model.

6. Synthesis of controllers

Let N_P denote the net model of the uncontrolled system and N_C denote the net model of a controller, as yet to be constructed. We propose that the two systems operate concurrently and that control action is achieved by the synchronisation of events in the uncontrolled system with events in the controller. Hence the behaviour of the initially uncontrolled system is restricted since certain selected events are prevented from occurring unless the controller is also in the state which allows the correponding synchronised event to occur.

Let N_S denote the composite net formed from the nets N_P and N_C, i.e. $N_S = N_P || N_C$. We will assume that the required behaviour of the system represented by N_S can be specified by a set of invariants for this net. The limitation imposed by this assumption will be discussed later. Bearing in mind the results described above for the propogation of net invariants through the composition operation, we propose therefore the following outline synthesis procedure for the controller net:

(i) Specify the required behaviour by defining a set of invariants for the system which embody the required behaviour,
(ii) Construct a controller net such that the invariants defined in (i) are invariants of this net,
(iii) Select events in the uncontrolled system net and in the controller net to be synchronised such that those places in the uncontrolled system and in the controller net which have the same name are made equivalent, i.e. the markings of these places are always equal,
(iv) Compose the uncontrolled system and the controller net and remove any redundant places resulting from the equivalence

procedure in (iii). The composite system net will now have the required behaviour.

To illustrate this procedure, we will consider two examples. The first is the familiar problem of controlling the access of a number of computer processes to a shared disc file for both reading and writing to the file. The net model of the uncontrolled system is shown in Figure 6. There are N processes and each process can be in one of five states: request to read (RR), request to write (RW), reading (R), writing (W) and idle (I). In the uncontrolled system there is no restriction on the number of processes which are allowed simultaneous access to the file for either reading or writing. We need to impose some control on this so that access is regulated according to the following requirements:

(i) any number of processes can read from the file simultaneously,

(ii) only one process can write to the file at any time,

(iii) if a process is writing to the file then no process may simultaneously read from the file.

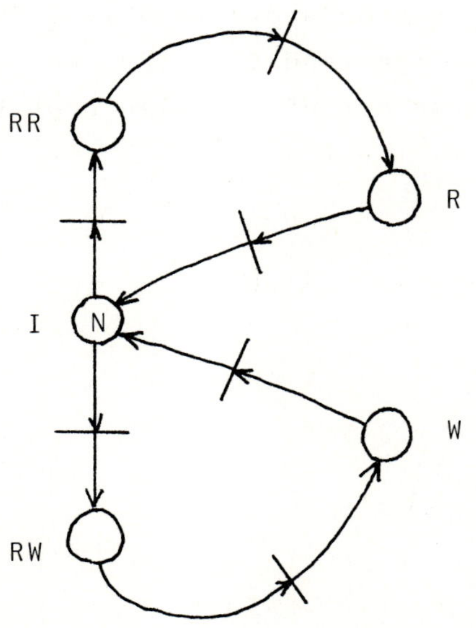

Figure 6. The reading and writing problem

Following our synthesis procedure, we first express this required behaviour in terms of a set of invariants for the system. In this case, it turns out that all three of the above requirements can be expressed in terms of a single invariant. If we introduce into the system an additional place which we will denote by SP, the required behaviour can be represented by the following invariant relation on the markings of the places SP, R, and W:

$$m(SP) + m(R) + Nm(W) = N$$

where m(.) denotes the value of the marking of the stated place. To see how this embodies the specification above, first let m(W) = 1. It follows that, since no marking can take a negative value, m(R) = 0. Hence (iii) is satisfied. Now let m(W) = 0. Then clearly m(R) can take any value between 0 and N inclusive. Thus (i) is satisfied. Finally, it is easy to see that m(W) cannot take a value greater than 1 without m(R) or m(SP) taking negative values.

The second step in our procedure is to construct a controller with an invariant corresponding to the above relation. Such a controller is shown in Figure 7. The reader is invited to check that this is so by determining the set of invariants of the net

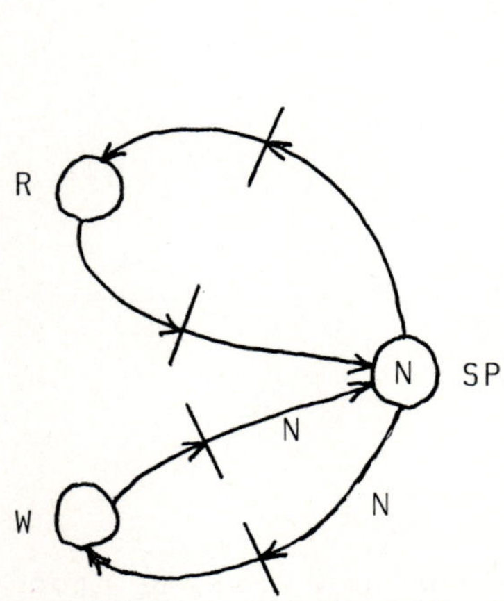

Figure 7. Controller for the reading and writing problem

from the equation $vW^- = vW^+$ and showing that for some invariant v in this set, the relation vm = vm' yields the required relation given above.

Our next step is to select the events in the uncontrolled system and the controller for synchronisation. Clearly the events to be controlled are those which result in a process going from the request to read state (RR) to the read state (R) and from the request to write state (RW) to the write state (W). In order for the controller to know when to permit a process to go from a request to read state to a read state, for example, it must also monitor the transition of processes out of the write state and out of the read state. Hence, the events to be synchronised are those which result in processes entering and leaving the read and write states, as shown in Figure 8. This choice also has the effect of making the read and write places in the system and the controller equivalent, i.e. the read place in the system and in the controller will always have markings of equal value and similarly for the write places.

Figure 8 also shows the result of the next step in the synthesis procedure, which composes the system and controller nets and removes the redundant places.

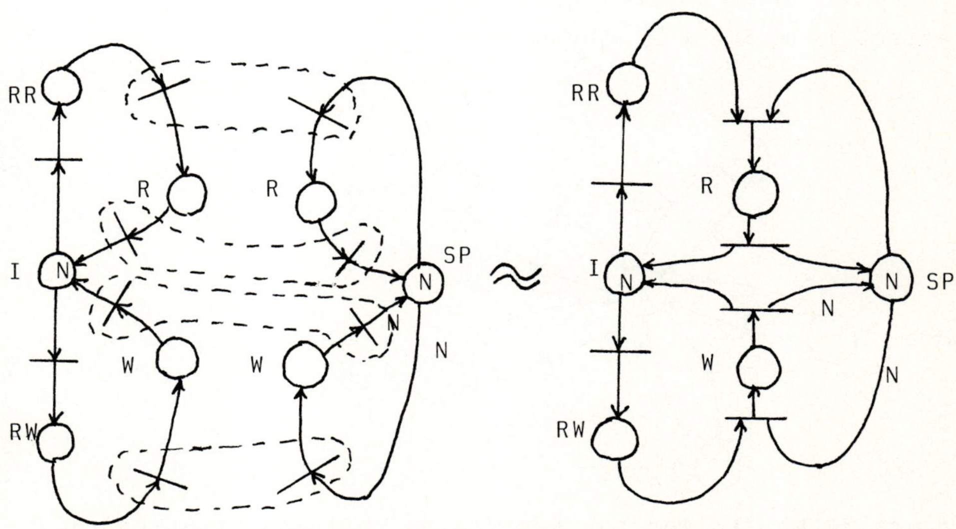

Figure 8. Composition of system and controller

That the resulting composite system has the required behaviour is easily checked by determining the invariants of the system.

For our second example we return to the dining philosophers problem. Recall that in this problem the requirement is that the system is controlled in such a way that prevents each philosopher simultaneously picking up his/her left fork and the system entering a deadlock state, i.e. when no further event can take place. We would also like to ensure that the system operates fairly in the sense that every philosopher eventually reaches the eating state.

Following our procedure we first express the requirement as an invariant of the system. As in the last example, we introduce an additional place which we will call SP. The requirement to prevent deadlock can then be expressed by the invariant relation

$$m(SP) + \sum_i m(hungry_i) = 4$$

which ensures that not all five philosophers can be simultaneously in the hungry state. A controller with this invariant is shown in Figure 9. The events to be synchronised with the system are clearly those of pick up left fork and pick up right fork for each philosopher, and the resulting composite system is shown in Figure 10, although here, for simplicity, we show it for only three philosophers. In this case the required invariant is $m(SP) + \sum_i m(hungry_i) = 2$.

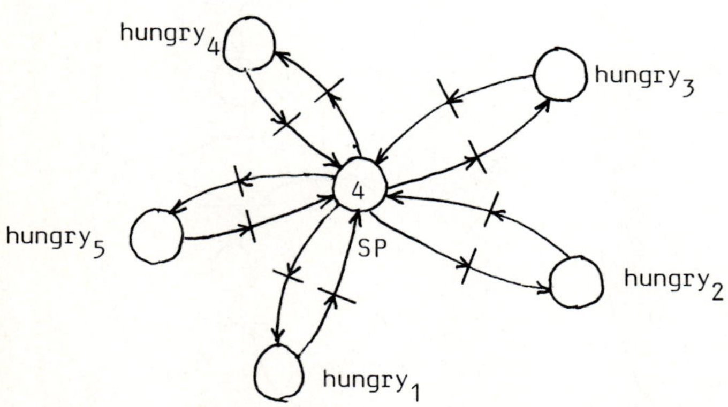

Figure 9. Controller for the dining philosophers problem

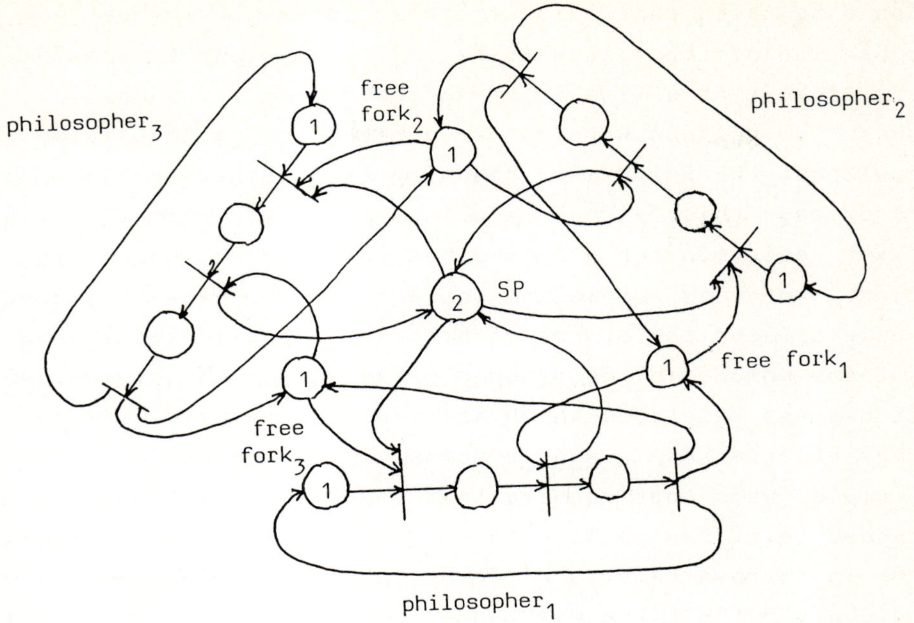

Figure 10. Controlled dining philosophers problem

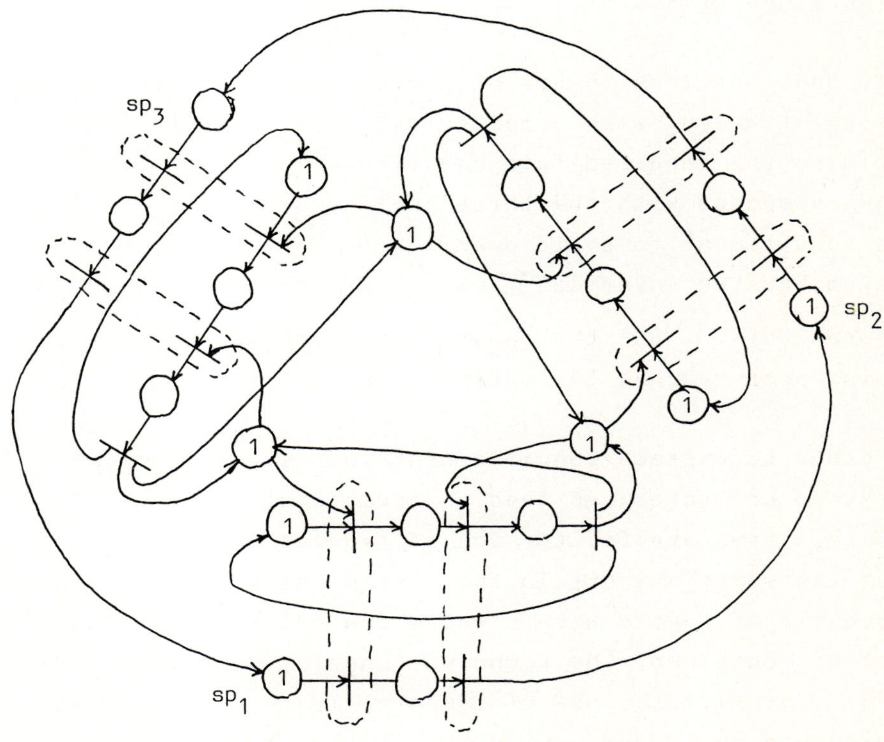

Figure 11. Alternative "fair" controller

It should be noted that this controller does not ensure that the operation of the system is fair. Clearly, any two philosophers can eat at a time in the system shown in Figure 10, but these two philosophers can keep returning to the eating state without the third ever reaching that state. In fact, we cannot express this fairness requirement in the form of an invariant realation for a simple but fundamental reason. An invariant expresses a relation between the markings of a net, and hence between the states of processes represented in the net, at any moment in the dynamic behaviour of the system. We cannot express relations which are required to hold between markings at more than a single moment in this behaviour. Thus, for example, we cannot require that whenever a certain marking is reached (e.g. two philosophers are eating), some subsequent marking is reached (e.g. a different pair of philosophers are eating) before the first marking is reached again in the dynamic behaviour of the system. In general terms, invariants cannot be used to state required behaviours which relate to sequences of markings or states.

This does not mean to say of course that a controller net does not exist which, when composed with the uncontrolled system, achieves the required fair behaviour. Such a controller is shown composed with the three philosopher system in Figure 11. The requirement to avoid deadlock is met since the composite system has the invariant $\sum_i m(sp_i) + \sum_i m(hungry_i) = 2$ which ensures this. But the controller also ensures that each philosopher reaches the eating state.

In order to express required behaviour over a sequence of markings or states, we need an additional language. Two possibilities are:(i) the use of regular language expressions over the set of events in the system which define a set of sequences of events within which the behaviour of the system must be contained, the technique used by Ramadge and Wonham (1987), or (ii) the use of temporal logic which allows formal statements to be made which are interpreted over the sequence of states or events which constitute the dynamic behaviour of the system, the technique used by Ostroff (1987). These are matters for further investigation.

7. Summary and Conclusions

We offer the following points as a summary of the results and conclusions we have arrived at in this paper:

* A Petri-net provides an abstract model of concurrent system behaviour which has an easily understandable and concisely stated semantics.
* There exists an algebraic formulation of a Petri-net and of families of Petri-nets (viewed as algebras together with the morphisms between them) which provides a basis for a theory of composition and control.
* A composition operator can be defined to express the concurrent, cooperating behaviour of a number of subsystems for which a number of important properties can be established in terms of the dynamic behaviour of the composite system.
* Invariants of Petri-nets can be used to specify certain kinds of required system behaviour and this behaviour can be propogated under the composition operator based on the fact that the space of invariants of the composition of two nets is the product of the spaces of invariants of the two nets.
* Control action is achieved by composition of a controller net with the system to be controlled, the actual control being imposed by synchronisation of events in the controller with selected events in the system.
* The required controller can be synthesised directly from the set of invariants which specify the required behaviour.
* Other work currently in progress shows that control software can be synthesised directly from Petri-net specifications of the required controlled behaviour, where, for example, events are modelled as (Ada) tasks which interact with processes through rendezvous (Bruno and Marchetto, 1986), providing the potential for an integrated controller synthesis procedure.

References

Bruno G, Balsamo A (1986) Petri net-based object-orientated modelling of distributed systems. Proc. OOPSLA '86, Sept. 1986, ACM.

Bruno G, Marchetto G (1986) Process-translatable Petri nets for the rapid prototyping of process control systems. IEEE Trans. SE-12, 2:346-357.

Hoare CAR (1978) Communicating sequential processes. Comm. ACM 21, 8:666-677.

Manna Z, Wolper P (1984) Synthesis of communicating processes from temporal logic specifications. ACM Trans. Prog. Lang. and Syst. 6, 1:68-93.

Milner R (1980) A calculus of communicating systems. Lect. Notes in Comp. Sci. 92, Springer-Verlag.

Ostroff JS (1987) Real-time computer control of discrete systems modelled by extended state machines: a temporal logic approach. PhD thesis (Rept. 8618) Dept. of Elec. Eng. Univ. of Toronto.

Ramadge PJ, Wonham WM (1987) Supervisory control of a class of discrete-event processes. SIAM J. Contr. and Optim. 25, 1: 206-230.

Winskel G (1985) Petri nets, algebras and morphisms. Tech. Rept. 79, Computer Lab., Univ. of Cambridge.

Winskel G (1986) Event structures. Tech. Rept. 95, Computer Lab., Univ. of Cambridge.

Temporal logic and extended state machines in discrete-event control

J.S. Ostroff*
Computer Science Department, York University,
4700 Keele Street, North York, Ontario, M3J 1P3.

Keywords / Abstract: Discrete-event systems / Real-time temporal logic / Extended state machines / Verification.

ESM/RTL is a framework for modelling, specifying and verifying real-time discrete event systems. Extended state machines (ESMs) are used to model the processes of the plant and the controller, and real-time temporal logic (RTL) is used as the assertion language for specifying required plant behaviour and verifying that the controller achieves the specification. A review of the framework is presented. The paper concludes with a brief discussion of current research to automate and modularize the framework.

1 Introduction

Discrete event dynamic systems (DEDS) are characterized by processes that are discrete in time and space, often asynchronous and typically nondeterministic (capable of "choices" by some mechanism unmodelled by the designer). Examples of DEDS include embedded computer systems, process control, patient monitoring systems, flexible manufacturing systems, traffic control and many other safety-critical systems. The recent position paper "Challenges to Control" [3] identifies DEDS as an important future research topic. A key problem is the lack of a suitable framework that will support modelling and analysis of DEDS, in the same way that differential or difference equations support the representation of continuous variable dynamic systems.

DEDS have also been explored in the systems and software engineering literature (e.g. [18,25,28,7]). Here also there is almost unanimous agreement that there is a major problem in the specification and design of large and complex event-driven systems. The term *reactive systems* [6,5] has been used for DEDS because, in contrast to systems that must *transform* input data into specified output, reactive systems must continually interact and react to the outside world. Reactive systems do not just compute or perform a given function, but must maintain an ongoing relationship with the environment. Features of reactive systems such as nondeterminism, concurrency, and real-time constraints all serve to add to the problem of providing a framework for modelling and analysis that is both clear and precise.

*This work was partially supported by a grant from NSERC.

NATO ASI Series, Vol. F47
Advanced Computing Concepts and Techniques
in Control Engineering
Edited by M.J. Denham and A.J. Laub

Systems with "hard" real-time constraints, requiring guaranteed response within specified time, have received increasing attention in the literature. This is because representing real-time features compounds the already difficult problem of describing concurrency [29]. Petri nets are one of the best known formalisms for dealing with real-time DEDS (e.g. see [16]). Petri-net theories have been criticised in [11] as lacking satisfactory verification methods for liveness properties and (concurrent) data structures. State-machine based requirements and specification languages are discussed in [30,1]. Real-time temporal logics have been discussed in [2,10,19], but rejected as suitable for modelling real-time programming language constructs in [11,9], on the asserted grounds that temporal logic lacks the ability to capture real-time concurrency via interleaving, and because of its lack of modularity. In fact, almost all the proposed formalisms lack satisfactory hierarchical decomposition methods thus leading to combinatorial explosion in the number of states for large systems.

Purpose of this paper:

The purpose of this paper is to describe the ESM/RTL framework (first reported in [24,21]) for modelling, specifying and verifying real-time DEDS. The framework uses logic, language and other programming methodology concepts from software engineering to pose problems of interest to the control engineer (e.g. controller correctness and synthesis).

For modelling the processes of plants and controllers, extended state machines (ESMs) having guarded local, shared and communicating events are used in a fashion similar to CSP [8,7]. In addition, time bounds on the events and a global clock allow the designer to distinguish between spontaneous and forced events in the plant, and to model real-time distributed programming constructs used in controllers. Spontaneous events represent unpredictable changes in the plant; and the guarded events allow for control without ruling out nondeterministic behaviour.

Real-time temporal logic (RTL), is used as the assertion language for specifying required plant behaviour and for verifying controller correctness. RTL is based on Manna-Pnueli temporal logic [27,17,14] and has enhancements for real-time features and the ability to refer to both states and events. A "coarse-grained" interpretation of time simplifies the representation of real-time concurrency.

In the remainder of this paper we give an overview of the ESM/RTL framework taken mainly from [22], and also from [23,21]. In section 2 we describe ESMs. In section 3 we interpret ESMs as generators of legal trajectories, an interpretation which provides a semantics for RTL (section 4). A small verification example is presented in section 5. We conclude with a brief discussion of current research involving the semi-automation of verification procedures in Prolog, and hierarchical ESMs as a first step towards modularizing the framework.

2 Extended state machines (ESMs)

An extended state machine models a process (e.g. a device, pump, valve, sensor, vehicle or task) occurring in either a plant or its controller. An ESM is described in terms of an activity variable, data variables, communication channels and guarded events.

The events comprise transitions from one activity to the next (while simultaneously changing the data variables), as well as allowing for cooperation and communication with other ESMs.

A plant-controller system $\mathbf{M}$ is denoted by

$$\mathbf{M} = \mathbf{M}_0 \parallel \mathbf{M}_1 \parallel \ldots \parallel \mathbf{M}_k \parallel \mathbf{M}_{k+1} \parallel \ldots \parallel \mathbf{M}_{k+n}$$

representing the concurrent interaction of the processes of the plant and controller. The ESM $\mathbf{M}_0$ represents a global clock, the ESMs $\mathbf{M}_1 \cdots \mathbf{M}_k$ represent plant processes, and $\mathbf{M}_{k+1} \cdots \mathbf{M}_{k+n}$ the controller processes.

This section will provide a syntactic definition of an ESM, as well as an informal account of the behaviour of a set of concurrent ESMs such as the plant-controller system. A formal operational semantics will be provided in the next section. Some examples will be given to illustrate the flexibility of ESMs for modelling plant processes. The SELECT real-time programming construct of the distributed language Conic [13, 12] will be used to illustrate the transformation of controller software constructs into ESM fragments.

Each ESM has the following components.

1. An *activity variable* x with associated *type* $\mathcal{X}$ whose elements are called activities.

2. A vector of *data variables* y. Each vector component y_i has associated type $\mathcal{Y}_i$.

3. A set of *communication channels* $\mathcal{C}$.

4. A set of *event labels* $\mathcal{L}$.

ESM operations and events are defined as follows:

operation - An ESM operation is either an *assignment*, a *send* or a *receive*.

An assignment operation is denoted by $\alpha[y : a]$ where α is an event label, y is a vector of data variables, and a is a vector of expressions. Each component expression of a must have the same type as the corresponding data variable in y. The event label may occur by itself in an operation, i.e. $[y : a]$ is optional.

A send operation is denoted by $c!m$, where c is a channel in $\mathcal{C}$ and m is a *message*. A message is either a term (i.e. an expression) in the data variables, or an event label. A receive operation is denoted $c?r$, where c is a communication channel, and r is either a data variable or an event label. A send operation $c!m$ *matches* a receive operation $c?r$ if either m and r are the same event label, or r is a data variable and m is a term with the same type as r.

event - An ESM event is a 4-tuple $(A_s, guard, operation, A_d)$, where A_s is a source activity, A_d a destination activity and *guard* is a boolean-valued expression in the data variables. The graph

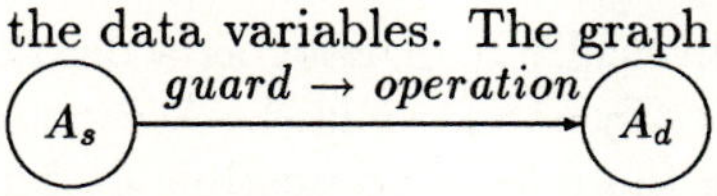

is a pictorial representation of an event. If the guard is omitted then it is assumed to be *true*.

An ESM may now be defined as a set of events. For small ESMs an equivalent representation as a transition graph (e.g. see Figure 2 later) gives the designer a clear visual picture of the plant processes being modelled.

As in CSP, we arrange for ESMs to synchronize their events when they need to interact. In this way, lower-level handshaking (e.g. via semaphores, monitors, or condition queues) may be taken for granted. Furthermore the *transition* (defined formally as a 5-tuple in the next section), consisting of the simultaneous participation of all the interacting events in the synchronized action, is then considered to be *atomic*.

If E is an event in $\mathbf{M}_i$, then (to represent cooperation or communication between event E in $\mathbf{M}_i$ and some event E_j in $\mathbf{M}_j$) define the *interaction set* $\mathcal{E}$ associated with E as follows:

cooperation - If both E and E_j have the same event label α occurring in their respective assignment operations $\alpha[y_i : a_i]$ and $\alpha[y_j : a_j]$, then $\mathcal{E} = \{(E, E_j)\}$. The cooperative action associated with the pair (E, E_j) is called a *shared transition*, which is enabled only if the guards of E and E_j are both simultaneously enabled. When the transition occurs, y_i and y_j are simultaneously assigned the values of a_i and a_j respectively. It is assumed that α does not occur in any other event of $\mathbf{M}_i$. Thus, no confusion should result if α is used to refer either to the event E or to the resulting shared transition.

In general, if there are m events (including E and m-1 other events $E_1 \cdots E_{m-1}$) in $\mathbf{M}$ with shared assignment label α, then $\mathcal{E} = \{(E, E_1, \ldots, E_{m-1})\}$. If the event label α of E is not shared by any other ESM in $\mathbf{M}$, then $\mathcal{E} = \{E\}$, and the associated action is called a *local transition.*

communication - If E has send operation $c!m$, then $\mathcal{E}$ consists of the set of all pairs (E, E_j) for which E_j has matching receive operation $c?r$. For simplicity, it is assumed that channel connections are one-to-one (i.e. each channel has a single sending ESM and a single receiving ESM) as in CSP. The synchronized action associated with each pair in $\mathcal{E}$ is called a *communicating transition*, which is enabled if the guards of each event in the pair are simultaneously enabled. If m is a term and r is a data variable, then on making the transition there is a "distributed" assignment of (the value of) the message m to r.

Each transition has a lower and upper time bound. In order to properly define the effect of these time bounds, it is assumed that each plant-controller system $\mathbf{M}$ has as one of its ESMs a (conceptual) global clock $\mathbf{M}_0$. There is a choice of whether to model the key features of the clock with a data variable t (as in Figure 1b) that is incremented by one each time the clock ticks, or with an infinite number of activity variables (as in Figure 1a). Data variables provide a concise representation of quantitative (e.g. arithmetic) features as distinguished from logical features of the process. The clock representation in Figure 1b is thus preferred.

In a distributed system a global clock is not physically realizable, as a distributed set of clocks will drift apart even if they are initially synchronized. We may take sufficiently small clock drift into account by adding the maximum possible drift to the upper time bound of transitions, and subtracting it from the lower time bound.

The following constraints impose the proper relationship between transition time bounds and clock ticks:

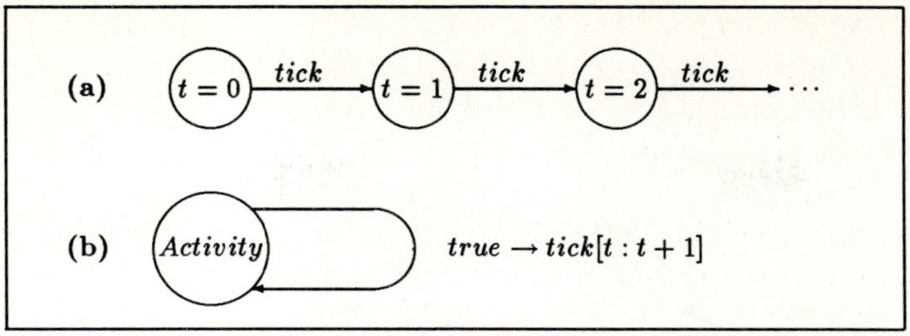

Figure 1: ESM for the clock

(a) The clock must tick infinitely often, and

(b) Let τ be any transition with lower time bound l and upper time bound u. Once τ is enabled, it is prevented from occurring for at least l ticks of the clock, but may not be continuously enabled for more than u ticks of the clock.

A *spontaneous* transition has upper time bound of infinity, i.e. a spontaneous transition is never *forced* to occur. Spontaneous transitions may be used to model unpredictable changes in the plant. Guarded events in the plant allow for control without ruling out nondeterministic behaviour because two events exiting from the same activity may have their respective guards simultaneously enabled.

The clock and time bound assumptions ((a) and (b) above) embody a "coarse-grained" interpretation of time which simplifies the modelling of real-time concurrency. For instance, it will not be necessary to update the clock by one tick every time a transition occurs; furthermore, spontaneous transitions that are continuously enabled may occur many times between any two clock ticks.

Figure 2 is a transition graph representation of the plant-controller system $\mathbf{M} = clock \parallel train \parallel gate \parallel controller$ for a railway crossing (a Petri net model of a similar system is presented in [16]). The plant consists of a *train* (with activity variable x_1) and *gate* (with activity variable x_2). The formal specifications of required plant behaviour will be given in section 4. Informally, the controller must ensure that the train is never within the crossing ($x_1 = ingate$) at the same time that the gate is up (safety), and the gate must remain up so long as the train is travelling. In order to meet the safety requirement, it will be necessary to show that the gate must be lowered within a fixed time (e.g. 10 clock ticks).

The gate has a forced event β (the matching event with operation $c!\beta$ is in the controller ESM) corresponding to the action "the controller commands over channel c that the gate be lowered". The transition β has an upper bound computed by taking into account the time to evaluate the guard ($y = approaching$), and the time to do the handshaking over the control channel. The spontaneous shared transition δ corresponds to the action "when the train leaves the railway crossing the gate is simultaneously raised and the controller is reset to its initial activity `L1`". A justification for treating δ as a shared transition is that the train remains in the *travelling* activity for some length of time. In contrast, α_2 is a local spontaneous event with lower bound l set to the minimum time it takes to reach *ingate*; and since l may be less than the time it takes for the measurement of the current train activity to reach the controller over the

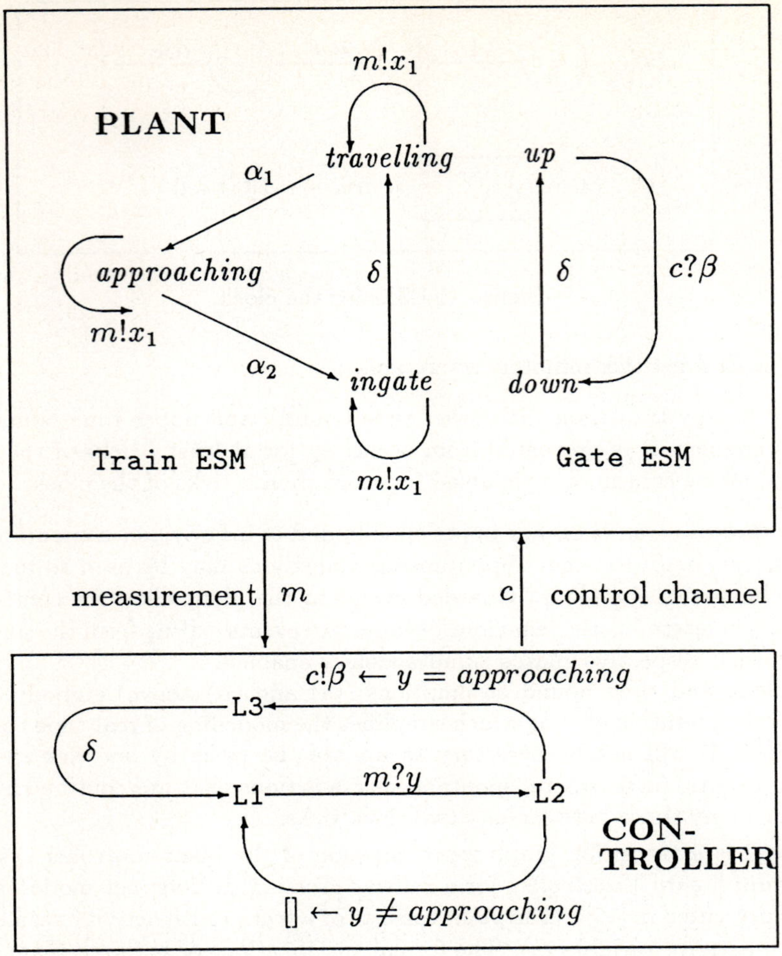

Figure 2: ESM representation of the train-gate system

channel m, we must allow for the possibility that α_2 may occur prior to the controller taking suitable action. The receive operation $m?y$ in the controller has a matching send operation $m!x_1$ at each activity of the train, corresponding to the action "send a message (the current train activity) over channel m to the controller, and put the message in the data variable y".

To illustrate the modelling flexibility provided by data variables (for representing quantitative features, e.g. liquid levels, temperatures, pressures and speed), consider a situation in which the time available to lower the gate depends on the speed of the train. In such a situation, a data variable z may be used to represent the train speed. The guarded operation given by $z \leq 95 \rightarrow \omega[z : z + 5]$ would then represent the fact that the train can increment its speed by 5 m.p.h at a time up to a maximum of 100 m.p.h., and a lower time bound could be used to model the train inertia. See the shared-track example in [23] where data variables are used for representing such quantitative features.

We conclude this section by giving transformations for some of the real-time constructs of Conic to ESM fragments, and refer the reader to [21] for a more detailed discussion.

- For the delay instruction

```
L1: DELAY(N)
L2:
```

construct

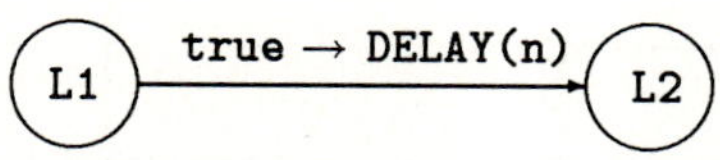

When the delay statement is executed by a given Conic task the task is delayed for n ticks of the clock before it resumes processing. In the ESM representation, the delay transition does not change any data (i.e., program) variables, and has a lower time bound set to n ticks of the clock. The instruction labels L1 and L2 correspond to ESM activities. The labels do not appear in the actual program code but are inserted to allow for reasoning about program locations.

- For the send instruction

```
L1: SEND msg TO c WAIT SIGNAL => L2:S1
    FAIL n => L3:S2 END
L4:
```

construct

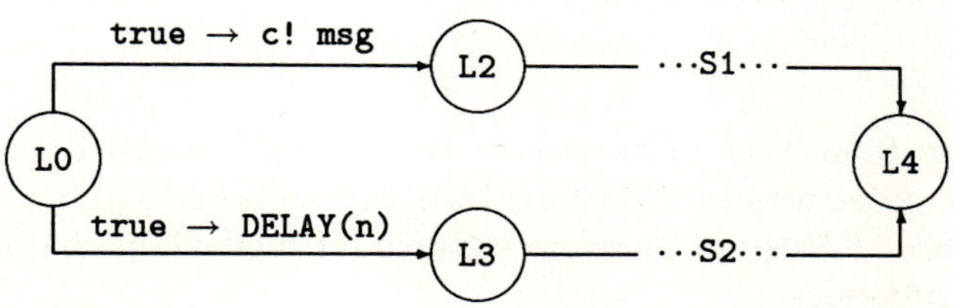

where S1 and S2 represent sequences of instructions. The WAIT SIGNAL part of the SEND instruction refers to the fact that the send and receive components of the communication are blocked (synchronized).

- For the receive instruction

```
L1: RECEIVE y FROM c REPLY SIGNAL;
L2:
```

construct

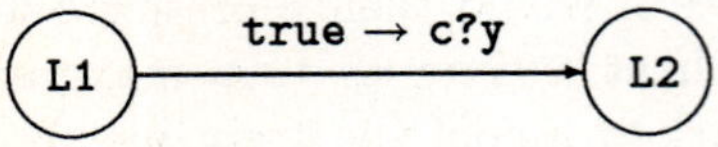

- For the select instruction

```
L0:  SELECT
       WHEN g1 RECEIVE y1 FROM c1 REPLY SIGNAL => L1:S1
       OR WHEN g2 RECEIVE y2 FROM c2 REPLY SIGNAL => L2:S2
       OR WHEN g3 TIMEOUT n => REPLY SIGNAL => L3:S3
     END
L4:
```

construct

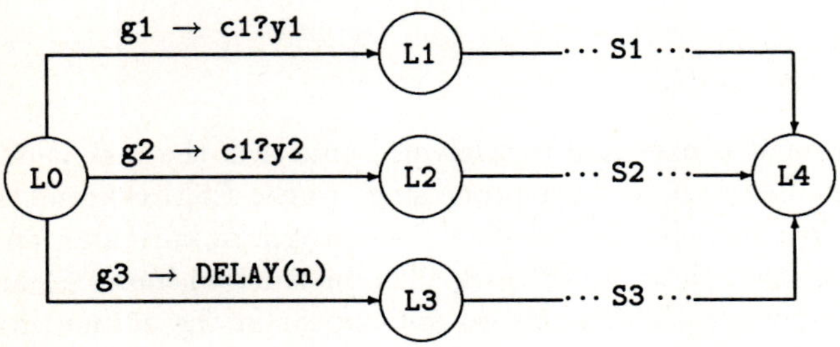

- For the composition of two instruction sequences

```
L1: S1 ; L2: S2
L3:
```

construct

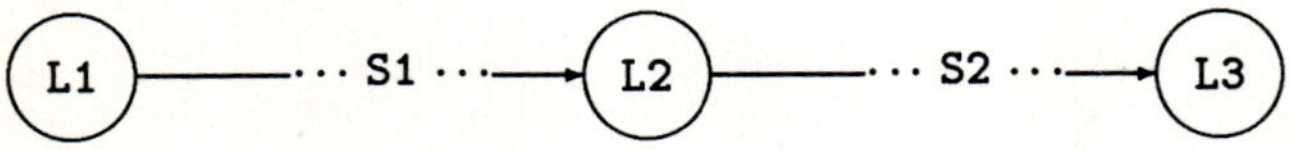

Conic tasks are built up from basic units such as the SELECT instruction. The transformation from Conic constructs to ESM fragments can be used to obtain ESM models of controller software. ESMs are therefore suitable for modelling a variety of plant as well as controller processes.

3 Legal trajectories

Intuitively a trajectory is any path, in the state-space of the plant-controller system **M**, consisting of a sequence of states and events. **M** cannot take an arbitrary path because of the constraints mentioned in the previous section (e.g. an event must be enabled if it is to occur). The paths that **M** is constrained to follow are called the *legal trajectories.* In this section we define the legal trajectories, and then interpret the plant-controller system **M** as a generator of legal trajectories.

In computing, the term *operational semantics* of a programming language construct refers to the behaviour of the computer (or an abstract model of the computer) when the construct is executed. The legal trajectories of **M** provide an abstract operational semantics that fully describes the system behaviour and provides a basis for specifying and deducing system properties.

In formal logic, the term "semantics" is used in a slightly different sense than in programming. The semantics of a well-formed formula in a given logic provides the formula's interpretation or truth value.

The legal trajectories provide not only a formal operational semantics (in the computing sense), but also a semantics for real-time temporal logic (RTL). RTL in turn will make it possible to express and deduce properties of **M**.

The operational semantics given below essentially follows [17], but with enhancements to take into account the clock $\mathbf{M}_0$, time-bounded events and additional features such as data variables for shared transitions. The designer will be able to refer to control and data variables, as well as to transitions via the distinguished variable $\mathbf{n}$ (e.g. $\mathbf{n} = \alpha$ will mean that the next transition to occur is α).

Before giving a formal definition of the legal trajectories, we first define the set of transitions of **M**. This set involves the following constituents:

$\mathcal{V}$ - The set of all activity and data variables of **M** .

$\mathcal{D}$ - The domain formed by the union of the types of all the variables in $\mathcal{V}$.

state-assignment q - A mapping $q : \mathcal{V} \rightarrow \mathcal{D}$ such that for each activity variable x of **M**, with type $\mathcal{X} \subset \mathcal{D}$, we have $q(x) \in \mathcal{X}$, and for each data variable y of **M** with type $\mathcal{Y} \subset \mathcal{D}$, $q(y) \in \mathcal{Y}$.

$\mathcal{Q}$ - The set of all state-assignments of **M** .

Θ - A set of initial state-assignments $\Theta \subset \mathcal{Q}$.

The following notation is used for state-assignment updates. Let $q \in \mathcal{Q}$ be a state-assignment, and let $q(v_1) = d$ for $v_1 \in \mathcal{V}, d \in \mathcal{D}$. We denote by $(q; v_1 : d_1)$ the state-assignment differing from q only by its value at v_1 where it has the value d_1 instead of d. Formally

$$(q; v_1 : d_1)(v) \stackrel{\text{def}}{=} \text{if } v = v_1 \text{ then } d_1 \text{ else } q(v)$$

The notation can be extended to express multiple simultaneous updates of the form $(q; v_1 : d_1, v_2 : d_2)$.

If a is a term (e.g. an arithmetic expression) then the notation $q(a)$ stands for the value of a when evaluated with state-assignment q. For a boolean-valued expression b,

b is said to be *satisfied* for state-assignment q (written: $q(b)$) if b evaluates to *true* for q. The guards on the edges of ESMs are all boolean-valued expressions in the variables of $\mathcal{V}$, and may therefore be evaluated for any state-assignment q.

For simplicity, we avoid the problem of expressions that are undefined (for some state-assignments) in the following way. An ESM event with guard g and expression a occurring in its operation, is *well-defined* if in all state-assignments q for which $q(g)$ is defined we also have that $q(a)$ is defined. For example, if expression a is $1/y_i$ then the condition $(y_i \neq 0)$ must occur in g. A *plant-controller system* **M** *is well-defined* if all the events (in the ESMs of **M**) are well-defined, and all events with a send (or receive) operation can be matched with a corresponding receive (or send) operation.

Definition 1 (Transitions) *Let* **M** *be a well-defined plant-controller system, and for each event E (in the ESM* $\mathbf{M}_i$ *of* **M***), let $\mathcal{E}$ be its interaction set. For each element in the interaction set $\mathcal{E}$, we define an associated transition $\tau = (\alpha, e, h, l, u)$, where α is the event label of E, e is the enabling condition, h is a transformation function $h : \mathcal{Q} \to \mathcal{Q}$, and l and u are the lower and upper time bounds respectively. The enabling condition and transformation function are defined as follows:*

local transitions - If $E \in \mathcal{E}$ is the event

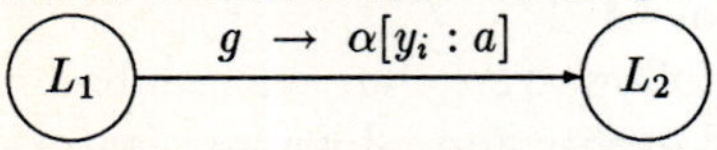

where L_1, L_2 are activities in $\mathcal{X}_i$, α is an event label (not occurring in any other event), g is a boolean-valued expression representing the guard, y_i is a data variable, and a is an expression (in the data variables), then τ has enabling condition
$e \stackrel{\text{def}}{=} (g \wedge x_i = L_1)$,
and for any state-assignment q,
$h(q) \stackrel{\text{def}}{=}$ if $q(e)$ then $(q; x_i : L_2, y_i : q(a))$ else *undefined.*

shared transitions - For each pair of events $(E, E') \in \mathcal{E}$ with E' in $\mathbf{M}_j$ given respectively by

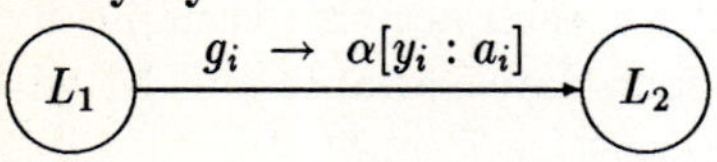

and by

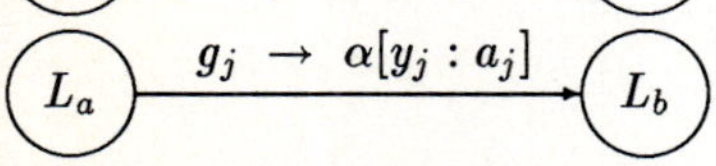

where both events have event label α, L_1 and L_2 are activities in $\mathcal{X}_i$, and L_a and L_b are activities in $\mathcal{X}_j$, the associated transition τ has enabling condition
$e \stackrel{\text{def}}{=} (g_i \wedge g_j \wedge x_i = L_1 \wedge x_j = L_a)$,
and for any state-assignment q,
$h(q) \stackrel{\text{def}}{=}$ if $q(e)$ then $(q; x_i : L_2, x_j : L_b, y_i : q(a_i), y_j : q(a_j))$ else *undefined.*
(The obvious extension to the definition is made if α is shared by more than two events.)

communicating transitions - For each $(E, E') \in \mathcal{E}$ with E' in $\mathbf{M}_j$, given respectively by

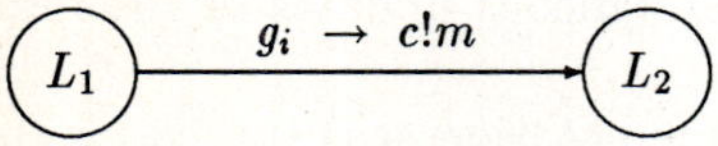

and by

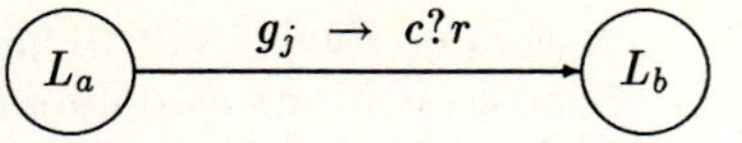

where m is the message sent on channel c, the associated transition τ has enabling condition
$e \stackrel{\text{def}}{=} (g_i \wedge g_j \wedge x_i = L_1 \wedge x_j = L_a)$,
and for any state-assignment q,

1. if m is a term and r is a data variable y_j, then
   $h(q) \stackrel{\text{def}}{=}$ if $q(e)$ then $(q; x_i : L_2, x_j : L_b, y_j : q(m))$ else *undefined*
   (as an example, m may be x_i or an arithmetic expression)

2. if m and r are both the label α, then
   $h(q) \stackrel{\text{def}}{=}$ if $q(e)$ then $(q; x_i : L_2, x_j : L_b)$ else *undefined*

For a local transition, the notation $[x_i : L_1, y_i : a]$ will be used to denote the transformation function h, and if φ is a predicate, then φ_h^v denotes $\varphi_{L_1,a}^{x_i,y_i}$, i.e, the predicate obtained from φ by simultaneously substituting every free occurrence of x_i with L_i and every free occurrence of y_i with a. Similar notations will apply for shared and communicating transitions.

The transition set $\mathcal{T}_i$ of an ESM $\mathbf{M}_i$ is defined as follows:

$$\mathcal{T}_i \stackrel{\text{def}}{=} \{\tau : \ \tau \text{ is a transition associated with an element of an interaction set } \mathcal{E} \text{ corresponding to an event } E \text{ in } \mathbf{M}_i\}$$

The set of all transitions $\mathcal{T}$ of $\mathbf{M}$ is defined as the union of the $\mathcal{T}_i$ for each machine $\mathbf{M}_i$ in $\mathbf{M}$ together with a special transition $(null, true, [], 0, 0)$ corresponding to the event of initializing or rebooting the controller. The notation [] stands for the unit transformation function that effects no change in the variables. It is possible for the same transition to occur in more than one set $\mathcal{T}_i$, so that duplicate transitions must of course be removed when computing $\mathcal{T}$.

If τ is a transition with enabling condition e, then we say that the transition τ is *enabled* in a state-assignment q if $q(e) = true$. Let the transitions in $\mathcal{T}_i$ have enabling conditions $e_1, \cdots, e_m$. Define $enabled(\mathcal{T}_i) \stackrel{\text{def}}{=} (e_1 \vee e_2 \vee \cdots \vee e_m)$. We say that $\mathcal{T}_i$ is enabled for state-assignment q if $q(enabled(\mathcal{T}_i)) = true$.

In any *interleaved* model, we must assert either a justice (weak fairness) requirement or a fairness (strong fairness) requirement to properly model concurrency. Our justice constraint will require that no enabled transition in any controller process is allowed to be postponed indefinitely. Justice is a weaker concept than fairness as justice is automatically guaranteed on a truly concurrent computer, whereas fairness can only be ensured by special scheduling software (see [4] for a complete discussion of these issues).

Many implementations of real-time languages are not fair; for example, Conic is just but not fair. For this reason, we will not insert a fairness requirement into the operational semantics, although we could do so should the need arise. The justice requirement will allow for the verification of qualitative timing properties, even in the absence of accurate time bounds on transitions.

The *Justice family* $\mathcal{J}$ is the family of sets $\mathcal{J} = \{\mathcal{T}_{k+1} \dots \mathcal{T}_{k+n}\}$. Each element in the justice family is a *justice set*. Each justice set consists of the transitions associated with a controller ESM.

Definition 2 (Initialized trajectories) *Let* **M** *be a well-defined plant-controller system with set of state-assignment* $\mathcal{Q}$, *set of transitions* $\mathcal{T}$, *set of initial state-assignments* Θ, *and justice family* $\mathcal{J}$. *An initialized trajectory* σ *is a sequence* $\sigma = q_0 \xrightarrow{\tau_0} q_1 \xrightarrow{\tau_1} q_2 \xrightarrow{\tau_2} \cdots$ *with each* $q_i \in \mathcal{Q}$ *and* $\tau_i \in \mathcal{T}$ *satisfying the following requirements:*

(a) initialization - The initial state-assignment q_0 belongs to Θ, the initial transition τ_0 is the *null* event, and there is no subsequent occurrence of *null*.

(b) succession - For each i, $q_{i+1} = h(q_i)$ where h is the transformation function of τ_i, and τ_i is enabled for the state-assignment q_i.

(c) justice - Let $\mathcal{T}_i \in \mathcal{J}$ be a justice set which is enabled at each state-assignment of σ beyond some q_j. Then some transition in $\mathcal{T}_i$ must occur at least once beyond q_j, i.e, there must be an $m \geq j$ such that $\tau_m \in \mathcal{T}_i$. Thus, transitions in $\mathcal{T}_i$ must occur infinitely often after the state-assignment q_j.

(d) ticking - The clock ticks infinitely often.
Thus, the tick transition $(tick, true, [t : t+1], 0, \infty)$ occurs an infinite number of times in the trajectory σ. This implies that all initialized trajectories are infinite sequences.

(e) real-time - Let τ be any transition in $\mathcal{T}$ with lower time bound l, upper time bound u, and enabling condition e. If τ is first enabled in q_j of σ with $q_j(t) = T$ (i.e, the clock variable reads T ticks), then for some $m \geq j$ either (τ occurs at q_m with $T + l \leq q_m(t) \leq T + u$) or ($q_m(\neg e)$ holds and $q_m(t) \leq T + u$).

The first three requirements in the above definition of initialized trajectories are similar to those found in [17]. However, the last two conditions for real-time are novel to our application.

A *state* of **M** is defined as a mapping $s : \mathcal{V} \cup \{\mathbf{n}\} \rightarrow \mathcal{D} \cup \mathcal{T}$, where $\mathbf{n}$ is the distinguished *next-transition* variable mentioned earlier ranging over the set of transitions $\mathcal{T}$. Any initialized trajectory $\sigma = q_0 \xrightarrow{\tau_0} q_1 \xrightarrow{\tau_1} q_2 \xrightarrow{\tau_2} \cdots$ may thus be written as a sequence of states $\sigma = s_0 s_1 s_2 \cdots$, where $s_i(\mathbf{n}) = \tau_i$ and $s_i(v) = q_i(v)$ for any $v \in \mathcal{V}$.

Definition 3 (Legal trajectories) *The set of legal trajectories* $\Sigma_{\mathbf{M}}$ *of the plant-controller* **M** *consists of the set of all its initialized trajectories together with all suffixes of initialized trajectories.*

In the next section, trajectories will be used to provide a semantics for RTL leading to a precise formulation of the Verification Problem.

4 Real-time Temporal Logic (RTL)

Temporal logic has been used extensively in program verification where it has proved useful in describing properties whose truth and falsity depend on time. A program, consisting of successive assignments, can be thought of as passing through a sequence of states as the assignments are executed. Assertions about the relationship among program variables will be true or false depending upon the current state. Such dynamic situations are conveniently described by Manna-Pnueli linear time temporal logic (for a complete exposition see [14]) which we use as the basis for our specification and verification language RTL.

For simplicity, we use two basic operators $\bigcirc$ (next), and $\mathcal{U}$ (until) from which we can define many other useful operators including: $\Box$ (henceforth), $\Diamond$ (eventually), U (unless), and $\mathcal{P}$ (precedes). Other operators not dealt with here such as *previous* and *since* have been used to extend the expressive power of Temporal Logic. These *past* operators have been found useful in dealing with issues of modularity.

The individual variables of the language are partitioned into *local* variables (e.g. the activity variables, the data variables and the next event variable **n**) which change from state to state, and *global variables* which do not change with time. Quantification is allowed only over global variables. A *state-formula* is any formula of first order logic in the local and global variables but not containing any of the temporal operators.

An *interpretation* for an RTL formula is a 3-tuple consisting of a trajectory, an assignment over a suitable domain to constant, function and predicate symbols, and an assignment from global variables to their domains. Since for a given system the constant, function and predicate symbols have a fixed meaning, we suppress this information, and interpret formulas with respect to state trajectories only, at the same time keeping in mind that global variables have the same value in any state of the trajectory.

For an arbitrary trajectory $\sigma = s_0 s_1 s_2 \cdots$, denote by σ^k the k-shifted trajectory $\sigma^k = s_k s_{k+1} s_{k+2} \cdots$. The following inductive definition defines the *satisfaction* relation. If w is *satisfied* in σ, then write $\models^{\sigma} w$.

Definition 4 (Satisfaction) *For temporal formulas w, w_1, w_2 and trajectory σ, the satisfaction relation is defined as follows:*

- If w is a state-formula, $\models^{\sigma} w$ iff w evaluates to *true* in initial state s_0 of σ.
- $\models^{\sigma} \bigcirc w$ iff $\models^{\sigma^1} w$.
  We may paraphrase $\bigcirc w$ as asserting that w will be true in the *next* state.
- $\models^{\sigma} w_1 \mathcal{U} w_2$
  iff $\exists k \geq 0$ such that $\models^{\sigma^k} w_2$ and $\forall i, 0 \leq i < k, \models^{\sigma^i} w_1$.
  Thus, $w_1 \mathcal{U} w_2$ can be paraphrased as: eventually w_2 will hold and *until* then w_1 holds continuously.

The other operators may now be defined as follows:

- $\Diamond w$ is an abbreviation for $(true\,\mathcal{U} w)$.
  The paraphrase of $\Diamond w$ is as follows: *eventually* w will hold true in some state.
- $\Box w$ is an abbreviation for $\neg(\Diamond(\neg w))$.
  Thus, $\Box w$ may be paraphrased as follows: *henceforth*, w holds true in all states.
- $w_1 \mathcal{P} w_2$ is an abbreviation for $(\neg((\neg w)_1 \mathcal{U} w_2))$.
  A paraphrase of $w_1 \mathcal{P} w_2$ is as follows: if w_2 eventually occurs then w_1 must *precede* w_2.

As examples of RTL formulas we offer the following:

$w_1 \wedge t = T \rightarrow \Diamond(w_2 \wedge t \leq T+5)$ - If w_1 is true now and the clock reads T ticks (T is a global variable), then within $T+5$ clock ticks w_2 must become true. Thus, once w_1 becomes true, w_2 must become true no more than 5 ticks later.

$w_1 \rightarrow (w_2 \mathcal{P} w_3)$ - If w_1 is true now, then (should w_3 occur at some future state) w_2 must precede w_3.

$\Box\Diamond(\mathbf{n} = tick)$ - The clock ticks infinitely often.

Definition 5 ($\Sigma_{\mathbf{M}}$-validity) *For any system* $\mathbf{M}$*, let* S *be a temporal formula specifying the required plant behaviour to be ensured by the controller, and let* $\Sigma_{\mathbf{M}}$ *be the set of legal trajectories. Specification* S *is* $\Sigma_{\mathbf{M}}$-valid *if it satisfies all trajectories in* $\Sigma_{\mathbf{M}}$.

Thus, a precise way of stating that a controller satisfies its specification, is to say that the specification S is $\Sigma_{\mathbf{M}}$-valid. The notation $\Sigma_{\mathbf{M}} \models$ S will be used to denote the fact that S is $\Sigma_{\mathbf{M}}$-valid.

Provided below are the temporal logic specifications for the train-gate example. Event labels are used to refer to the associated transitions, i.e. if transition τ has an associated event label α then rather than write $\mathbf{n} = \tau$ we write $\mathbf{n} = \alpha$.

(S1) - Safety $\Box\neg(x_1 = ingate \wedge x_2 = up)$
Henceforth, the gate must never be up while simultaneously the train is inside the railway crossing.

(S2) - Precedence $(\mathbf{n} = \beta) \rightarrow \bigcirc(\mathbf{n} = \alpha_1 \mathcal{P} \mathbf{n} = \beta)$
There must be no unsolicited gate lowering, i.e. once the gate has been lowered, the train must once again approach the railway crossing prior to the next gate lowering.

(S3) - Real-time liveness The gate must be lowered within 10 ticks of the train approaching the railway crossing.
$(\mathbf{n} = \alpha_1 \wedge t = T) \rightarrow \Diamond(\mathbf{n} = \beta \wedge t \leq T+10)$

Let $\mathbf{M}$ be a system given by $\mathbf{M} = \mathbf{M}_0 \| \mathbf{P} \| \mathbf{C}$ where $\mathbf{P}$ stands for the ESMs of the plant and $\mathbf{C}$ for the ESMs of the controller. The following three problems may now be posed in the RTL framework:

Verification Problem - Given a specification S for a plant $\mathbf{P}$ and a controller $\mathbf{C}$, is there a way to check for the Σ-validity of S? Is there an efficient algorithm to do this check?

Synthesis Problem - Given a plant $\mathbf{P}$ and a specification S, under what conditions does a satisfying controller $\mathbf{C}$ exist? If such a controller exists, is there an efficient algorithm to synthesize the ESMs of $\mathbf{C}$?

Modular Design Problem - What are the logics or algebras that will permit controller design in a modular fashion so as to beat combinatorial explosion?

The Verification Problem has been explored in [21]. For finite state systems, decision procedures have been developed for a small (but important) class of real-time properties. The procedures have complexity linear in the size of the global state transition graph. For (possibly infinite state) systems, a sound proof system has been developed for verifying temporal specifications. Axioms and rules in RTL are based on the operational semantics for **M** (Definition 2). Heuristics have also been developed to guide the designer in searching for correctness proofs.

In the next section we will illustrate the use of the proof system for verifying that the train-gate controller satisfies the specification (S1),(S2) and (S3).

5 Verification

In this section, the sound RTL proof system (developed in [21]) will be used to illustrate the verification of the train-gate specifications S1 (safety) and S3 (real-time liveness). An axiom A is sound for a concrete system **M** if $\Sigma_{\mathbf{M}} \models A$, and a proof rule with hypotheses H and conclusion C is sound if the truth of $\Sigma_{\mathbf{M}} \models H$ implies the truth of $\Sigma_{\mathbf{M}} \models C$. We sketch the verification method, referring the reader to [21] for more detail.

Table 1 lists the set of transitions for the train-gate example. The abbreviations are *tr* for *travelling*, *in* for *ingate* and *ap* for *approaching*. (See Section 3 for the notation used for transformation functions.)

The following notation is used in the sequel:

- $\varphi, \varphi_0, \varphi_1, \cdots$ denote state-formulas.
- For $\tau \in \mathcal{T}$, $\{\varphi_1\}\tau\{\varphi_2\}$ is an abbreviation for $(\mathbf{n} = \tau \wedge \varphi_1) \rightarrow \bigcirc \varphi_2$, and is read as "$\tau$ leads from φ_1 to φ_2" (as in Manna-Pnueli theory). The state-formula φ_1 is the *source node*, and φ_2 is the *destination node.*
- If $\mathcal{S} \subset \mathcal{T}$, then $\{\varphi_1\}\mathcal{S}\{\varphi_2\}$ abbreviates the statement that $\{\varphi_1\}\tau\{\varphi_2\}$ for each $\tau \in \mathcal{S}$.

The axiom AS may be used to deduce properties such as

$$(\mathbf{n} = \alpha_1 \wedge \psi_0) \rightarrow \bigcirc(\psi_1 \vee \psi_2) \qquad (1)$$

where $\psi_0 \cdots \psi_4$ are defined in the proof diagram of Figure 3. In AS, let $\varphi = (\psi_1 \vee \psi_2)$ and let $\tau = \alpha_1$. Recall that x_1, x_2 and x_3 are the activity variables of the *train*, *gate*

Name	Enabling condition	Transformation	Lower	Upper
α_1	$(x_1 = tr)$	$[x_1 : ap]$	0	∞
α_2	$(x_1 = ap)$	$[x_1 : in]$	10	∞
β	$(x_2 = up \wedge x_3 = \mathtt{L2} \wedge y = ap)$	$[x_2 : down, x_3 : \mathtt{L3}]$	0	u_β
δ	$(x_1 = in \wedge x_2 = down \wedge x_3 = \mathtt{L3})$	$[x_1 : tr, x_2 : up, x_3 : \mathtt{L1}]$	0	∞
m-chan	$(x_3 = \mathtt{L1})$	$[x_3 : \mathtt{L2}, y : x_1]$	0	u_m
skip	$(x_3 = \mathtt{L2} \wedge y \neq ap)$	$[x_3 : \mathtt{L1}]$	0	u_s

Table 1: Transitions for the train-gate example

AS - *Axiom of Succession*

Let $\tau \in \mathcal{T}$ have enabling condition e and transformation function h. Then, for all state-formulas φ not containing any occurrences of $\mathbf{n}$:

$(\mathbf{n} = \tau \wedge \varphi_h^v) \rightarrow (e \wedge \bigcirc\varphi)$

RI -*Rule of Invariance*

Let $\varphi_0 \cdots \varphi_m$ be any state formulas, and let $\varphi = \varphi_0 \vee \cdots \vee \varphi_i \vee \cdots \vee \varphi_m$. Then

(1) $\Theta \rightarrow \varphi_0$
(2) For each i, $\varphi_i \rightarrow \psi$
(3) For each i, $\{\varphi_i\}\mathcal{T}\{\varphi\}$

$\Box\psi$

and *controller* respectively. Then, since

$$\varphi_h^v \stackrel{\text{def}}{=} \varphi_{ap}^{x_1} = (x_2 = up \wedge (x_3 = \mathtt{L1} \vee (x_3 = \mathtt{L2} \wedge y \neq ap)))$$

we have that $\psi_0 \rightarrow \varphi_h^v$, and thus we obtain (1). Similarly, if φ is set to $(\psi_1 \vee \psi_2) \wedge t = T)$, then from AS we obtain

$$(\mathbf{n} = \alpha_1 \wedge \psi_0) \wedge t = T \rightarrow \bigcirc((\psi_1 \vee \psi_2) \wedge t = T) \tag{2}$$

Note that no temporal logic was needed in the derivation of (1). In fact, all the hypotheses in the proof rules (RI, RD and RL) are also deduced using first-order predicate logic; yet the conclusions of the rules are temporal formulas.

In RI, Θ is a state-formula whose satisfying states are exactly the initial states of

RD - *Rule for Delay*

Let $\tau \in \mathcal{T}$ have enabling condition e and lower time bound l. Then, for any φ_0 and φ_1:

(1) $\varphi_0 \rightarrow (\neg e \vee \Theta) \wedge \bigcirc\varphi_1$
(2) $\varphi_1 \rightarrow e$

$(\varphi_0 \vee \varphi_1) \wedge t = T \rightarrow (\mathbf{n} \neq \tau \mathcal{U} t \geq T + l)$

RL - *Rule of real-time Liveness*

Let $\mathcal{T}$ consist of the disjoint union $\{\tau\} \cup \mathcal{T}_s \cup \mathcal{T}_l$, where τ has enabling condition e and finite upper time bound u.
Then, for any φ

(1) $\varphi \rightarrow e$
(2) $\{\varphi\}\mathcal{T}_s\{\varphi\}$

$$(\varphi \wedge t = T) \rightarrow \ (\varphi \mathcal{U}[\varphi \wedge t \leq T + u \wedge (\mathbf{n} = \tau \vee \mathbf{n} \in \mathcal{T}_l)])$$

the plant-controller system **M**. For example, in the train-gate example Θ is given by

$$\Theta \stackrel{\text{def}}{=} (x_1 = travelling \wedge x_2 = up \wedge x_3 = \texttt{L1} \wedge t = 0) \tag{3}$$

To prove specification S1, let ψ in RI be given by

$$\psi \stackrel{\text{def}}{=} \neg(x_1 = ingate \wedge x_2 = up)$$

We then obtain the proof diagram of Figure 3, where the arrows show how a transition leads from one state-formula (*node*) to another. The forked arrow labelled α_1 indicates that (1) must hold true. *Selfloops* (i.e., transitions which lead from a node to itself) are not shown. In other words, the proof diagram only shows those transitions that are enabled and lead to a destination node that is different from the source node. See [21] for heuristics that aid in the construction of proof diagrams.

The proof diagram may be viewed as a "high-level" version of the global transition graph. A single node may represent many states, and it is consequently easier to see "chunked" behaviour patterns. As a simple example, it is clear from the proof diagram that S1 cannot be proved (using only RI), because α_2 leads to a state-formula that does not imply ψ. Intuitively, we must show that the sequence of transitions *skip* .. *m-chan* .. β always occurs before α_2. More formally, if we can demonstrate the $\Sigma_{\mathbf{M}}$-validity of

$$(\mathbf{n} = \alpha_1 \wedge \psi_0) \rightarrow (\mathbf{n} \neq \alpha_2 \mathcal{U} \psi_4) \tag{4}$$

then α_2 cannot leave nodes ψ_1 and ψ_2 (even though α_2 is enabled), and S1 will then be $\Sigma_{\mathbf{M}}$-valid.

To prove (4) we apply the proof rules RD and RL. In RD, let $\varphi_0 = (\mathbf{n} = \alpha_1 \wedge \psi_0)$, let $\tau = \alpha_2$, and let $\varphi_1 = (\psi_1 \vee \psi_2)$. Using (1) we obtain the first hypothesis of RD, and the second hypothesis is trivially true. Thus, we obtain (by weakening the conclusion of RD)

$$(\mathbf{n} = \alpha_1 \wedge \psi_0 \wedge t = T) \rightarrow (\mathbf{n} \neq \alpha_2 \mathcal{U} t \geq T + 10) \tag{5}$$

In RL, let $\tau = skip$, $\varphi = \psi_1$, and let $\mathcal{T}_l = \{\alpha_2\}$. Then the hypotheses (by predicate calculus) hold true, and we thus obtain (using $(w_1 \mathcal{U} w_2 \rightarrow \Diamond w_2)$ to weaken the conclusion of RL)

$$\begin{aligned} \psi_1 \wedge t = T \quad &\rightarrow \quad \Diamond[(\mathbf{n} = skip \vee \mathbf{n} = \alpha_2) \wedge \psi_1 \wedge t \leq T + u_{skip}] \\ &\rightarrow \quad \Diamond(\psi_2 \wedge t \leq T + u_{skip}) \vee \Diamond(\mathbf{n} = \alpha_2 \wedge t \leq T + u_{skip}) \end{aligned} \tag{6}$$

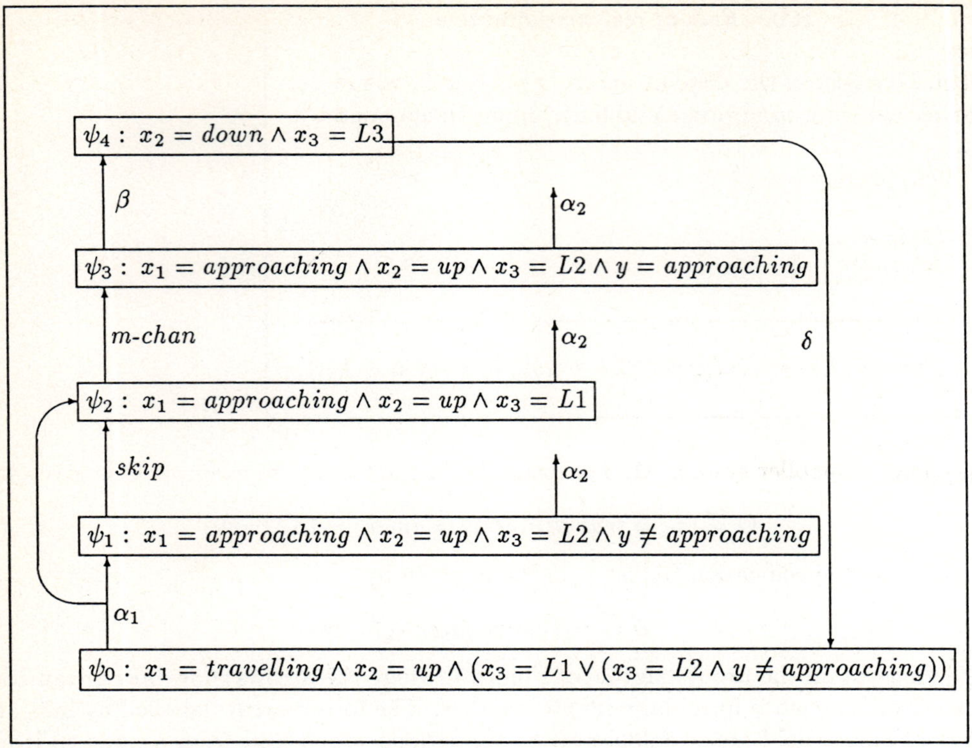

Figure 3: Proof diagram for train-gate example

where AS was used to obtain ($\mathbf{n} = skip \wedge \psi_1 \rightarrow \bigcirc\psi_2$). In RL, $\mathcal{T}_s$ contains those transitions that are selfloops, and $\mathcal{T}_l$ contains those transitions that *leave* the node φ. Using (6), and applying RL again to ψ_2 and *m-chan*, and then again to ψ_3 and β we obtain (using the rules of temporal logic on the time domain)

$$(\psi_1 \vee \psi_2) \wedge t = T \rightarrow \Diamond[(t \leq T + U) \wedge (\psi_4 \vee \mathbf{n} = \alpha_2)] \tag{7}$$

where $U = u_{skip} + u_{m-chan} + u_\beta$.

Finally, by (2),(7) and (5) we obtain

$$\begin{aligned}(\mathbf{n} = \alpha_1 \wedge \psi_0 \wedge t = T) \;\rightarrow\;& (\mathbf{n} \neq \alpha_2 \mathcal{U} t \geq T + 10) \\ & \wedge[\Diamond(\psi_4 \wedge t \leq T + U) \\ & \quad \vee \Diamond(\mathbf{n} = \alpha_2 \wedge t \leq T + U)] \\ \rightarrow\;& (\mathbf{n} \neq \alpha_2 \mathcal{U}(\psi_4 \wedge t \leq T + U))\end{aligned}$$

so long as

$$U = (u_{skip} + u_{m-chan} + u_\beta) < 10 \tag{8}$$

in which case we obtain (4) as required.

The above inequality illustrates the fact that we need not know the exact time bound on each of the individual transitions, so long as the *sequence of transitions* meets the required deadline. In fact, the proof diagram in this case provides the scheduling constraints that must now be imposed on the controlling software.

6 Concluding remarks

ESMs provide clear visual models of DEDS and support the representation of real-time programming constructs. Thus ESMs are a flexible modelling tool for any mix of hardware or software processes in either the plant or controller. RTL is the assertion language for specification and verification. Both qualitative properties (e.g. safety and precedence) as well as quantitative real-time properties (e.g real-time liveness) can be expressed in a fairly "natural" fashion.

Prolog programs are currently under development to partially automate the construction of proof diagrams used to verify RTL properties. So far there is a BUILD program to generate the transition set of a plant-controller system. There is also a VERIFY program that accepts a node specification in a proof diagram, and then computes which transitions exit from the node and the corresponding destination states. The BUILD and VERIFY tools allow for much more rapid construction of proof diagrams, and can also be used to provide insight into the design of controllers.

A major disadvantage of the ESM/RTL framework is that the full plant-controller system must be available before verification can proceed. To beat combinatorial explosion of states in "big" systems a more modular hierarchical approach is needed in which parts of the systems are independently verified without having to refer to the total system. There are two steps that must be taken to achieve this goal. Firstly, when two ESMs are brought together to evolve concurrently, the resulting concurrent object must itself be an ESM that preserves the activity, data and event information of its component ESMs (see [20] for further development of this idea). Secondly, RTL must be adapted to allow for a modular specification of an ESM (which will include assumptions about the environment).

Consider the simple example in which a data variable y in the *plant* ESM is required to satisfy $\Box(y \neq 100)$. How do we express the property that the *plant* never sets the data variable to 100 (this could correspond to the specification that the train never travels at a speed of 100 m.p.h). See [26] for a discussion of this kind of problem in temporal logic for a distributed shared variables model.

Fortunately, we may follow [15] in specifying such a modular property, because event labels in the ESM/RTL framework allow us to distinguish between actions performed by the *plant* and actions performed by the *controller*. Thus we may write:

$$(\mathbf{n} \in \mathcal{T}_{plant}) \wedge (y \neq 100) \rightarrow \bigcirc(y \neq 100)$$

which states that all *plant* transitions preserve the required relationship. Thus all that is left to be done is to show that the *controller* ESM satisfies a similar property. What is not fully resolved at this point is how to deal with those transitions involving synchronized cooperation and communication between plant and controller. If communication is asynchronous then the modular specification problem appears easier to solve.

In conclusion, modularity and automation of design in the ESM/RTL framework should help substantially in the modelling and analysis of larger systems.

7 Acknowledgments

I would like to thank Murray Wonham for supervising my original work on the ESM/RTL framework, and Rick Holt and Eric Hehner for their constructive comments on the real-time aspects of this work.

References

[1] B. Auernheimer and R.A. Kemmerer. RT-ASLAN: a specification language for real-time systems. *IEEE Transactions on Software Engineering*, SE-12(9):879–889, September 1986.

[2] A. Bernstein and P.K. Harter. Proving real-time properties of programs with temporal logic. In *Proceedings of ACM SIGOPS 8th annual ACM Symposium on Operating Systems Principles*, pages 1–11, December 1981.

[3] Challenges to control: a collective view. IEEE Transactions on Automatic Control, April 1987.

[4] N. Francez. *Fairness*. Springer-Verlag, 1986.

[5] D. Harel. Statecharts: a visual formalism for complex systems. *Science of Computer Programming*, 8:231–274, 1987.

[6] D. Harel and A. Pnueli. On the development of reactive systems. In K.R Apt, editor, *Logics and Models of Concurrent Systems*, pages 477–498, Springer-Verlag, 1985.

[7] C. A. R. Hoare. *Communicating Sequential Processes*. Prentice Hall, 1985.

[8] C.A.R. Hoare. Communicating sequential processes. *Communications of the ACM*, 21(8):549–557, 1978.

[9] F. Jahanian and A.K. Mok. Safety analysis of timing properties in real-time systems. *IEEE Transactions on Software Engineering*, SE-12(9):890–904, September 1986.

[10] R. Koymans, J. Bytopil, and W.P. de Roever. Real-time programming and asynchronous message passing. In *Proc. 2nd Annual Symposium on Principles of Distributed Computing*, pages 187–197, Montreal, August 1983.

[11] R. Koymans, R.K. Shyamasundar, W.P. de Roever, R. Gerth, and S. Arun-Kumar. *Compositional Semantics for Real-time Distributed Computing. LNCS 193*, Springer Verlag, June 1985.

[12] J. Kramer and J. Magee. Dynamic configuration for distributed systems. *IEEE Transactions on Software Engineering*, SE-11(4):424–436, April 1985.

[13] J. Kramer, J. Magee, and M. Sloman. A software architecture for distributed computer control systems. *Automatica*, 20(1):93–102, January 1984.

[14] F. Kroger. *Temporal Logics of Programs.* Volume 8 of *EATCS Monographs on Theoretical Computer Science*, Springer-Verlag, 1987.

[15] L. Lamport. Specifying concurrent program modules. *ACM Transactions on Programming Languages and Systems*, 5(2):190–222, April 83.

[16] N.G. Leveson and J.L Stolzy. Safety analysis using petri nets. *IEEE Transactions on Software Engineering*, SE-13(3):386–397, March 1987.

[17] Z. Manna and A. Pnueli. How to cook a temporal proof system for your pet language. *Proceedings of the Symposium on Principles of Programming Languages*, 141–154, January 1983.

[18] R. Milner. *A Calculus of Communicating Systems. LNCS 92*, Springer-Verlag, 1980.

[19] B. Moszkowski. A temporal logic for multilevel reasoning about hardware. *Computer*, 18(2):10–19, February 1985.

[20] J.S. Ostroff. Modularity in the ESM/RTL framework. In *5th IFAC/IFIP Symposium on Software for Computer Control*, April 1988. (to appear).

[21] J.S. Ostroff. *Real-time Computer Control of Discrete Event Systems modelled by Extended State Machines: a Temporal Logic Approach.* Technical Report 8618, Systems Control Group, Dept. of Electrical Engineering, University of Toronto, September 1986. Preliminary draft of Ph.D. thesis - final version presented January 1987.

[22] J.S. Ostroff and W.M. Wonham. Modelling, specifying and verifying real-time embedded computer systems. In *Proceedings of IEEE Computer Society Eighth Real-Time Systems Symposium*, San Jose, December 1987.

[23] J.S. Ostroff and W.M. Wonham. State machines, temporal logic and control: a framework for discrete event systems. In *Procceedings of the 26th IEEE Conference on Decision and Control*, Los Angeles, December 1987.

[24] J.S. Ostroff and W.M. Wonham. A temporal logic approach to real time control. In *Proceedings 24th IEEE Conference on Decision and Control*, Florida, Dec 1985.

[25] J.L. Peterson. *Petri Net Theory and the Modelling of Systems.* Prentice-Hall, Englewood Cliffs, N.J., 1981.

[26] A. Pnueli. In transition from global to modular temporal reasoning about programs. In K.R Apt, editor, *Logics and Models of Concurrent Systems*, pages 123–144, Springer-Verlag, 1985.

[27] A. Pnueli. The temporal logic of programs. In *Proceedings of the 18th Annual Symposium on the Foundations of Computer Science*, pages 46–57, IEEE, Providence, R.I., November 1977.

[28] G. von Bochmann. *Concepts for Distributed System Design.* Springer-Verlag, 1983.

[29] N. Wirth. Towards a discipline of real-time programming. *Communications of the ACM*, 20(8), Aug 1977.

[30] P. Zave. An operational approach to requirements specification for embedded systems. *IEEE Transactions on Software Engineering*, SE-8(3):250–269, May 1982.

AN OVERVIEW OF DISTRIBUTED SYSTEM CONSTRUCTION USING CONIC

Jeff Kramer, Jeff Magee, Morris Sloman

Department of Computing, Imperial College,
180 Queensgate, London SW7 2BZ.

Keywords:

Distributed systems, distributed programming, configuration, host/target environment, dynamic configuration.

Abstract:

Distributed systems offer an attractive implementation architecture for many applications, particularly those environments where the application is itself physically distributed. For the last eight years the Distributed Systems Research Group at Imperial College has conducted research into the development of an environment to support the construction and operation of distributed software. The result has been the Conic Toolkit: a comprehensive set of language and run-time tools for program compilation, building, debugging and execution in a distributed environment. Programs may be run on a set of interconnected host computers running the Unix™ operating system and/or on target machines with no resident operating system.

Two languages are provided, one for **programming** individual task modules (processes) and one for the **configuration** of programs from groups of task modules. In addition the environment supports the reuse of program components and allows the configuration of new components into running systems. This **dynamic configuration** capability is provided by a distributed configuration management tool which is the primary method of creating, controlling and modifying distributed application programs.

This paper describes the main features of the Conic toolkit and illustrates the facilities provided using a simple example.

1. INTRODUCTION

Distributed systems offer an attractive implementation architecture for many applications, particularly those environments where the application is itself physically distributed. Some of the well publicised advantages include the flexibility to incrementally extend or modify a system, fault tolerance by loose coupling of components and the provision of redundancy, and good response by the provision of local processing and concurrency. However, the construction of distributed systems is not an easy task: it requires the integration of many disciplines such as communication networks and associated protocols, operating systems, device drivers, and the application itself. The concepts involved include concurrency, synchronisation and communication, time dependencies and fault handling.

Support for the construction of distributed systems can be by the provision of service routines at the operating system level for communication, multitasking and the like. Besides being difficult to use, this tends to lead to poorly structured systems which are difficult to validate. A better approach is the provision of language based facilities specifically designed for the construction of distributed systems. Providing support for distribution at the language level permits modularity, concurrency, synchronisation and communication facilities to be integrated

™ Unix is a trademark of AT&T Bell Laboratories.

NATO ASI Series, Vol. F47
Advanced Computing Concepts and Techniques
in Control Engineering
Edited by M. J. Denham and A. J. Laub

into a single framework [Strom 85, Hoare 78, Andrews 86, Black 87, Scott 87]. Compile, link and run-time checks can ensure message compatibility between components. Consistent naming, communication and synchronisation can be provided for both local and remote interactions. Thus language environments are generally **simpler** and **safer** to use.

In this paper we describe one such language environment, Conic. All the required facilities for real-time programming, communication and synchronisation are provided by a programming language for tasks (processes). Flexible configuration, modularity and reuse of software components is facilitated by separation of the language for **programming** the individual task modules ("programming in the small") from the language for **configuring** programs from predefined modules ("programming in the large"). The configuration language provides a concise configuration description and hierarchical composition, and is used to specify the configuration of software modules (processes) in terms of instances of components and their logical interconnection.

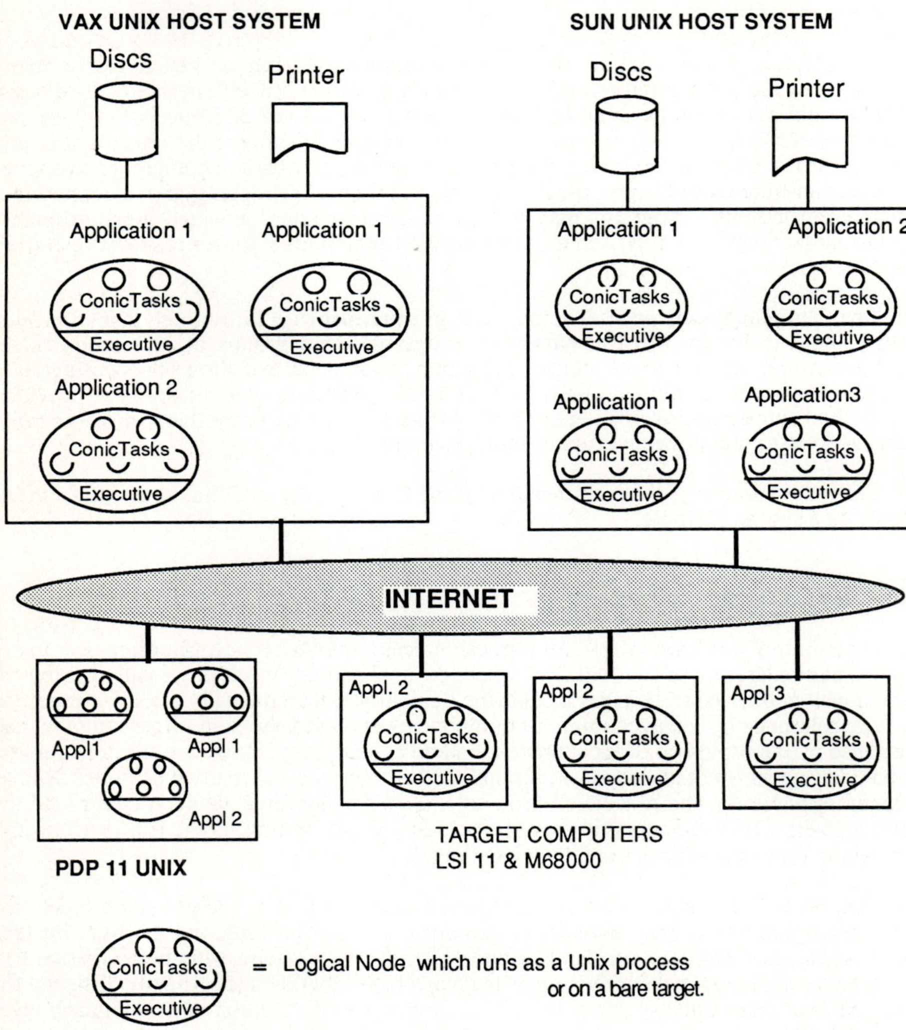

Figure. 1. Distributed Applications in a Conic Environment

Large distributed applications are subject to both **evolutionary** and **operational** changes. Evolutionary changes occur through the need to incorporate new functionality and technology in a manner which is difficult to predict. Operational changes result from the need to redimension to cater for growth and to reorganise to recover from failures. It is impractical and uneconomic to take out of service an entire distributed system simply to modify part of it. Conic caters for these requirements by language and run-time support for **dynamic configuration** [Kramer 85] of logical nodes. This permits on-line modifications to a running Conic system using the configuration language.

Conic was originally designed for support of embedded systems where the simple target computers used for real-time applications lack the facilities for program development. Host computers are used to develop software for subsequent downline-loading into the targets. In practice Conic has been used to construct a wide range of applications, from general distributed algorithms to system support utilities and services, on both targets and host computers. The capability of running in a **mixed** host target environment permits targets to be used for device interaction and real-time response, while the hosts provide access to the file servers, graphics displays and printing services. Fig.1 depicts a typical Conic environment. A **logical node** is the system configuration unit consisting of a set of tasks which execute concurrently within a shared address space on a host as a Unix™ process or directly on a target. Systems are constructed as sets of one or more interconnected logical nodes.

Various versions of the Conic toolkit have been in use for about 8 years at Imperial College, by research groups at other universities and in industry. We have used the environment as the basis for further research, for substantial student research projects and for student exercises on concurrency and communication protocols. The industrial users include British Coal for the implementation of underground monitoring and communication in coal mines; British Petroleum for research into reconfigurable control systems and GEC for the development of an object-oriented support system and front-end security processor. Conic has also been used for a number of years for research on self-tuning adaptive controllers [Gawthrop 84]. It is also being used for research and teaching at universities in Canada, France, Japan, Korea and Sweden.

In this paper we overview the Conic Programming, configuration and dynamic configuration facilities. A simple example is used to illustrate the concepts and languages.

2. PROGRAMMING LANGUAGE

2.1 Task Modules

Modularity is a key property for providing flexibility. The Conic programming language is based on Pascal, with extensions for modularity and message passing [Kramer 84].

The language allows the definition of a **task module type** which is a self-contained, sequential task (process). A task module type is written and compiled independently from the particular configuration in which it will run and so provides **configuration independence** in that all references are to local objects and there is no direct naming of other modules or communication entities. This means there is no configuration information embedded in the programming language and so no recompilation is needed for configuration changes as is the case with other languages such as CSP [Hoare 78] and Ada [USA DOD 80].

At configuration time, **module instances** are created from module types. Module instances exchange messages and perform a particular function in the system such as performing a computation, managing a resource or controlling a device. Multiple instances of a module type can be created on the same or different stations in a distributed system and a station can contain many different modules instances.

An example:

Consider the simplified water supply system illustrated in fig. 2.1. A pump is used to provide water to a reservoir. If the level reaches the high point, the pump is switched off. When the level falls below the low point, the pump is switched on again.

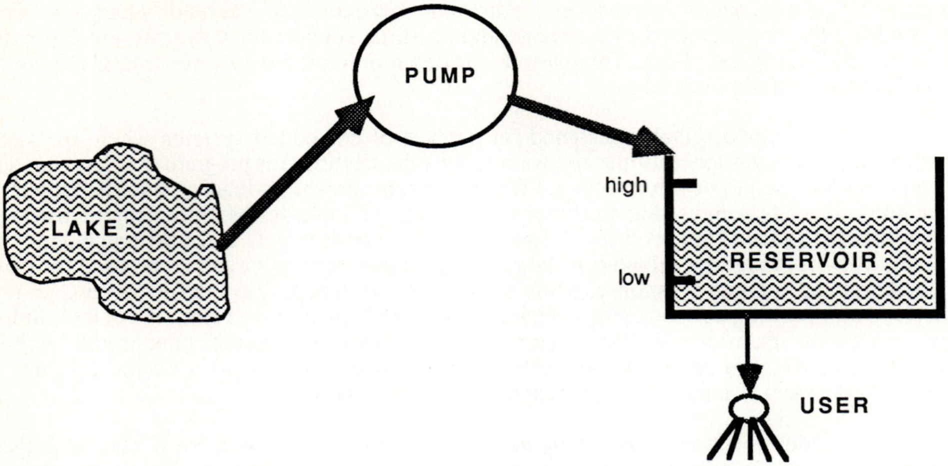

Figure 2.1 The Water Supply System.

This system can be easily modelled in Conic by programming modules to represent each of the physical entities in the system, viz. a lake, a pump, a reservoir, and a user. Fig. 2.2 is a graphic representation and listing of the most complex of these task modules, the reservoir.

CONIC modules have a well defined, strongly typed interface which specifies all the information required to use the module in a system. The interconnections and information exchanged by modules is specified in terms of **ports**. An **exitport** denotes the interface at which message transactions can be initiated and specifies a local name and message type in place of the destination name. In fig. 2.2, signals (messages with no data content used for synchronisation) are sent to the task's exitports *pumpon, pumpoff* to control the action of the pump. At configuration time, the exitports can be linked to any compatible entryport (ie. of type signaltype) of a task which wishes to receive pump control signals. The **entryports** *waterin* and *waterout* denote the interface at which message transactions can be received. Waterin receives some number of units of water for each received message, and waterout receives requests for water a unit at a time. At configuration time, any task with a compatible exitport can be linked to these entryports. The programming language uses local names within the task instead of directly naming the source and destination of messages. The binding of an exitport to an entryport is part of the configuration language and cannot be performed within the programming language. Therefore there is no need to recompile a task module when it is reused in different situations. This provides complete configuration independence for a task module.

At instantiation time, parameters can be passed to a module to tailor a module type for a particular environment. In the example the *capacity* of the reservoir is a parameter, and the high and low levels are set at 90% and 10% respectively of that value.

There are two classes of ports which correspond to the message transaction classes described below. **Request-reply Ports**, such as *waterout* in fig. 2.2, are bidirectional. They specify both a request and reply message type. **Notify Ports,** such as *pumpon, pumpoff* and *waterin* are unidirectional ie. they have no reply part. For convenience, it is possible to define families (arrays) of identical ports.

Ports define all the information required to use a module and so it is very simple to replace a module with a new or different version with the same operational interface.

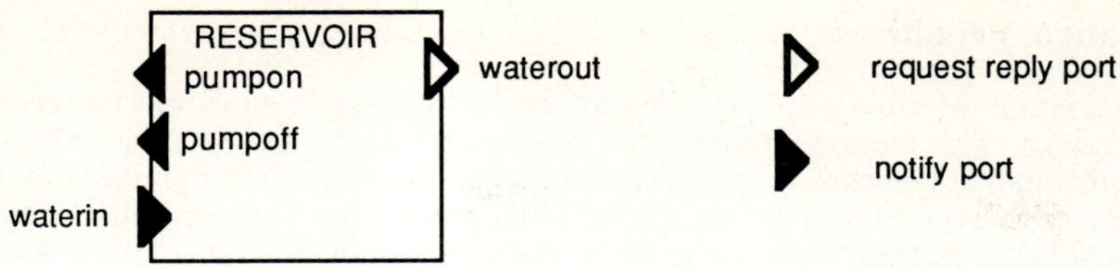


```
task module reservoir (capacity:integer);
     exitport                                          {interface}
          pumpon,
          pumpoff : signaltype;
     entryport
          waterout : signaltype reply integer;
          waterin : integer;

     var                                               {local data}
          level, high, low  : integer;
          pump: (on,off);
          bucket: integer;

procedure initialise;
     begin
          low := capacity/10;  high := capacity -  low;
           level := 0;
     end;

     begin                                                        {task body}
          initialise;
          send signal to pumpon;
          pump := on;
          loop
               select
                    receive bucket from waterin
                         => level := level+bucket;
                              if (level>high) and (pump=on)  then
                              begin
                                   send signal to pumpoff;
                                   pump := off;
                              end;
               or
                    when (level>0)  receive signal from waterout
                         => level := level - 1;
                              if (level<low) and (pump=off) then
                              begin
                                   send signal to pumpon;
                                   pump :=on;
                              end;
                              reply  1  to waterout;
               end;
          end;
end.
```

Fig. 2.2 Reservoir Task Module

2.2 Communication Primitives

Communication primitives are provided to **send** a message to an exitport or **receive** one from an entryport. The message types must correspond to the port types. Although differences in performance between local and remote communication are inevitable due to network delays, the primitives do use the same syntax and provide the same semantics (logical behaviour) for local (intra-station) and remote (inter-station) communication. This is termed **communication transparency** and allows modules to be allocated either to the same or different stations. This property can be particularly useful during the development of embedded systems in that modules can be fully tested together in a large computer with support facilities and then later distributed into target stations to achieve better perofrmance.

There are two classes of message transactions:

a) A **Notify transaction** provides unidirectional, potentially multi-destination message passing (fig. 2.3). The send operation is asynchronous and does not block the sender, although the receiver may block waiting for a message. There is a (dimensionable) fixed size queue of messages associated with each entryport. Messages are held in order of arrival at the entryport. When no more buffers are available the oldest message in the queue is overwritten. The Notify transaction can be used for time critical tasks such as within the communication system, with the queue size corresponding to a flow-control window or for periodic status information, when the latest information is of interest and the entryport specifies a single buffer.

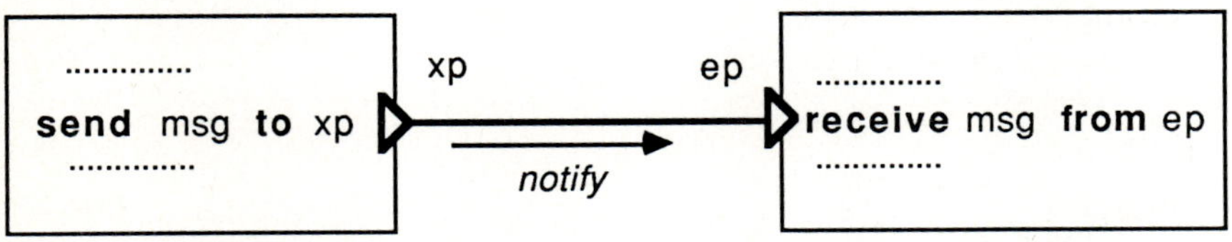

Fig. 2.3 The Notify Transaction

b) A **Request Reply** transaction provides bidirectional synchronous message passing. The sender is blocked until the reply is received from the receiver. A fail clause allows the sender to withdraw from the transaction on expiry of a timeout (*tval* in fig. 2.4) or if the transaction fails. The receiver may block waiting for a request. On receipt of a request, the receiver may perform some processing and return a reply message. In place of a normal reply, the receiver may either **forward** the request to another receiver (thereby allowing third party replies directly to the sender) or it may **abort** the transaction.

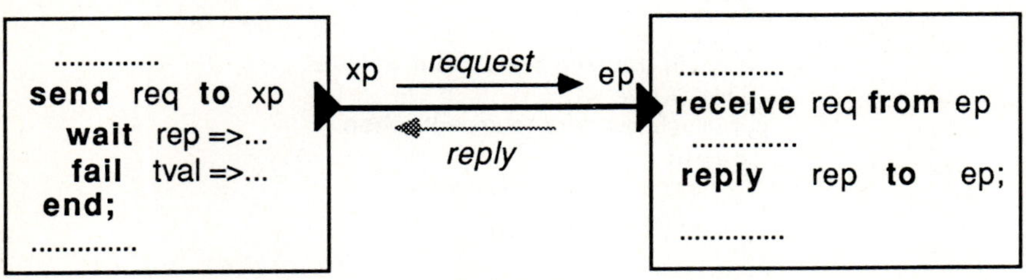

Fig. 2.4 Request-Reply Transaction

Standard functions are provided to determine whether an exitport is linked to an entryport, the number of messages queued at an entryport or the reason for a send-wait failing.

Any of the receive, receive-reply, receive-forward, or receive-abort primitives can be combined in a **select** statement (fig. 2.5). This enables a task to wait on messages from any number of potential entryports. An optional guard can precede each receive to further define conditions upon which messages should be received. A timeout can be used to limit the time spent waiting in the select statement. The order of selection is defined by the textual ordering of the alternatives in the select statement i.e. if there is a message waiting on both *ep1* and *ep2* in fig 2.5, then the the message on *ep1* will be received first.

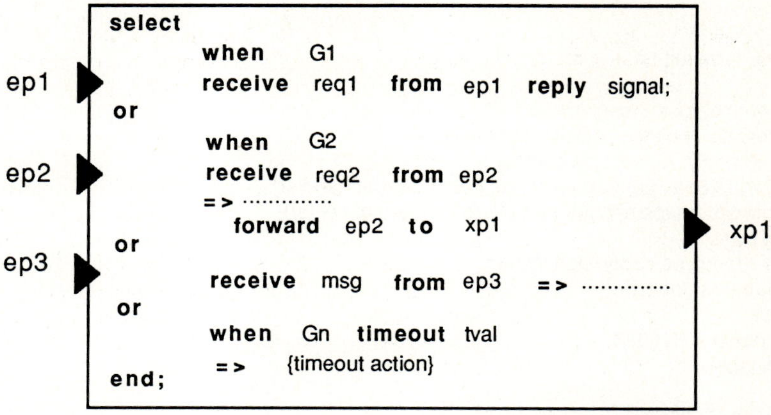

Fig. 2.5 Selective Receive

2.3 Definitions Unit

The module is the basic reuseable software component within a system. However there are many definitions which are common between different modules within a system. Definitions of constants, types, functions and procedures may be defined in separate definitions units. These can be compiled independently and can be imported into a module to define a **context**. This avoids errors introduced by having to redefine message types in communicating modules. For example the definition of message type *valtype* would be imported from a definitions unit called *msgtypes* by means of a declaration such as:

```
use msgtypes : valtype;
```

The definitions unit allows the introduction of language "extensions" without modifying the compiler. For example a set of standard string definitions and manipulation procedures can be made available as a definitions unit as shown in fig. 2.6. This exports 2 functions *strlen* and *strcpy*, and a type *string*.

```
define stringdefs: strlen, strcopy, string;
    const strmax = 128;
    type string  =  record
                        len:integer;
                        ch :array[1..strmax] of char;
                    end;
    function strlen (s:string):integer;
    ......
    procedure strcopy (s1,s2:string);
    ......
end.
```

Fig. 2.6 An Outline Definitions Unit

A definitions unit can encapsulate data, initialisation and access procedures for the data. This is similar to an Abstract Data Type but only a single instance can be declared when it is imported into a task module. However multiple instances of the encapsulating task module can be declared. The encapsulating task can access the data via exported procedures or directly (if the data variables are exported) but other modules must access the data via the encapsulating task's message passing interface.

```
task module transmit (status,vector : natural);
    use
        commstypes : msgtype;
        kercalls : priority, {system, normal etc.}
        setpriority, {to set task priority}
        SendSignal, {special message from interrupt handler}
        intmap; {maps handler procedure to interrupt vector}
    entryport
        tx : msgtype reply signaltype;
        done : signaltype;
    const
        enable = 0100#8;
        disable = 0;
    var
        txstat : ^natural;
        txbuff : ^char;
        msg : msgtype;
        count : integer;

    procedure inthandler;
        begin
            if count <= msg.len
            then begin
                xbuff^ := msg.chars[count];
                count := count + 1;
            end
            else begin
                txstat^ := disable; {completed}
                SendSignal; {to done entryport}
            end;
    end;

begin
    ref (txstat,status); {converts address to pointer type}
    ref (txbuff,status+2);
    setpriority (systempr); {raises task priority}
    intmap (done, vector, inthandler);
    {Kernel sets inthandler to run when interrupt received on vector}
    {done entryport is set up to receive signal from handler}
    loop
        receive msg from tx; {wait for message to transmit}
            count := 1;
            txstat^ := enable; {generates immediate interrupt}
            receive signal from done;
            reply signal to tx;
    end ;
end.
```

Fig. 2.7 Device Driver Task Module

2.4 Input-output

The programming language supports the standard Pascal and C input/output procedures, which can be freely mixed. These are automatically transformed by the compiler into message passing operations on standard, pre-declared task exitports.

In addition, CONIC provides simple primitives to support the programming of device handlers as application tasks. We have experimented with 3 versions of interrupt handling. Initially we used the Modula 1 type of kernel call *waitio (interrupt vector)* [Wirth 77]. This was called by the device handler task whenever it wished to wait for an interrupt. It was not possible to wait for both an interrupt and a message and the *waitio* resulted in a task context switch to the device handler for every interrupt, which slowed down the response time. We then tried the Ada mechanism of the Kernel converting an interrupt to a message [USA DoD 80]. However this was very slow as it resulted in a context switch and a message transfer for each interrupt.

With the current mechanism, a device driver task defines a procedure for each interrupt it handles. Fig.2.7 shows a transmitter driver for a serial port, based on LSI 11 hardware. It makes use of a set of special kernel calls imported from a definitions unit called *kercalls*. The task raises its priority to *system* to ensure that it is not preempted by any other task while transmitting a message. Different device drivers may have different hardware priority levels, allowing nested interrupts. The *intmap* procedure maps a handler procedure to the interrupt generated on the given vector. It also specifies an entryport from which the driver task will receive a signal from the handler procedure. The interrupt procedure runs in the context of the interrupted process, so it cannot use the normal message primitives, but it can make a special kernel call to send a signal when it has completed its function.

The above mechanism is very efficient, yet it means interrupt handlers are not part of the kernel, but are syntactically part of the device driver task. Consequently device drivers can be written and incorporated into a system without modyfing the kernel. This simplifies the writing and configuration of device drivers.

2.5 Discussion

The Request-reply and notify communication primitives have proved to be an excellent choice in that they do cater for most interaction requirements. In an early version of Conic we tried to do without an the asynchronous notify, but this led to deadlocks and a proliferation of tasks which gave immediate replies to make the synchronous send-wait appear asynchronous to the sender. Relying only on a synchronous request-reply primitive definitely complicated the programming of many applications. This is borne out by other systems which previously provided only remote procedure calls and are now introducing asynchronous remote procedure calls [Liskov 87]. There is a need for some form of overwrite strategy in the notify, as otherwise there must be some form of backward flow of information from the receiver to the sender. This leads to less efficient implementation and a send which is not really asynchronous i.e. the sender is delayed or blocked if there are no buffers at the receiver. This can be easily programmed as a flow control protocol, if the overwrite semantics of the notify are inappropriate and blocking the sender on buffer exhaustion is required.

The select statement gives priority on textual ordering. This could result in starvation, but none of the users reported this as a problem. Usually different entryports are used for different types of requests rather than different clients, and requests on a particular port are queued in arrival order. Guarded recieves could be used to overcome starvation problems if required.

The port based indirect addressing for communication primitives has proved very useful. This is one of the most important contributions to Conic's configuration flexibility in that it enables components to be reused in many alternaive configurations.

Although remote procedure calls (RPC) are currently very fashionable, we have no regrets about the provision of message primitives. Our request-reply transaction is similar to an RPC without parameter marshalling. However it is more flexible as it has a clause for handling errors. RPC implementations often provide another synchronisation mechanism (e.g. monitors) as well, whereas the message primitives can be used for both communication and synchronisation.

On the whole, our user experience has been that the minimal extensions we have provided to Pascal have made it sufficiently versatile to be used for programming a wide range of applications. However the Configuration language and dynamic configuration form the most interesting and novel aspects of the Conic toolkit.

3. CONFIGURATION LANGUAGE

One of the key elements in the provision of flexibility is the need to separate the programming of individual software components (task module types) from the building of a system from instances of modules. This has led to the development of the CONIC Configuration Language [Dulay 84] which is used to specify the instances of module types and their interconnection to form a group module. The structure of tasks (and groups) within a group module is described hierarchically.

For example, figure 3.1 describes a group module which represents the *water supplysystem* described earlier. The **use** construct specifies the set of message types necessary to declare a module interface (in this case it is null since the messages are of primitive types) and the set of task and/or group module types used with the group. Instances of task (or group) types are specified by the *create* construct. In the example, one instance of each of the module types *lake*, *pump*, and *reservoir* are created, giving the same name to the instances as their respective type names.

The interface to the water supply system is the following ports, provided to permit possible extension by the connection of further module instances:

lakeout to permit access to the lake to draw water (by a pump or user),
waterin to supply water to the reservoir,
on, off to receive control signals to start, stop delivering water to *waterin*,and
waterout to permit use of water from the reservoir by consumers.

Figure 3.2 indicates a possible extension by the addition of a further pump P1 and users U1 and U2. This is discussed later.

The **link** construct specifies the interconnection of module instances by binding a module exitport to a module entryport e.g. linking the *waterin* exitport on *pump* to the *waterout* entryport on *lake* in fig. 3.1. Both message type and transaction compatibility are checked so an exitport can only be linked to an entryport of the same data and transaction type. Multiple exitports can be linked to a single entryport which is particularly useful for connecting clients to servers (as is used for multiple pumps to the lake, and users to the reservoir). For notify ports, an exitport can be linked to multiple entryports to provide multidestination message transactions (as in the control signals to the pumps). In addition **link** binds group module interface ports (eg. *waterout*) to ports on internal module instances (eg. *reservoir.waterout*). This linking is merely a name mapping and does not entail any run-time overheads, i.e. there is no copying or queuing of messages at interface ports.

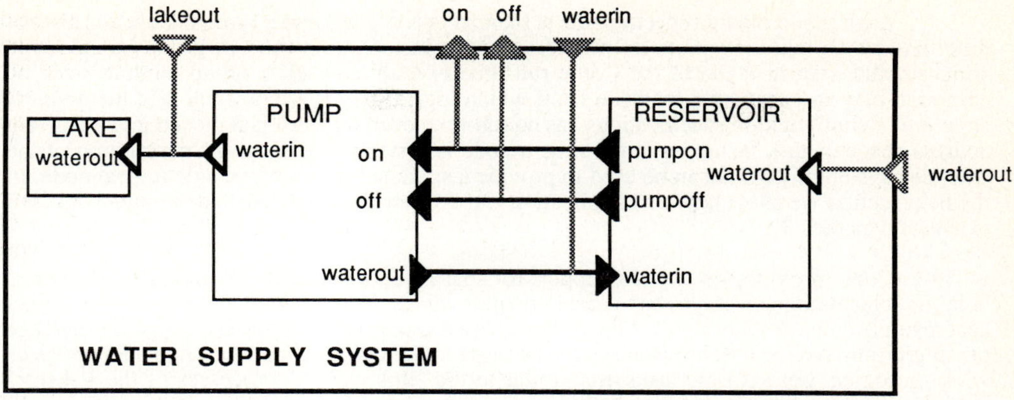


```
group module WaterSupplySystem (capacity : integer = 1000, bucketUnit : integer = 2);
    exitport                                    {interface ports for possible extension}
        on,
        off : signaltype;                       {for addition of extra pumps}
    entryport
        waterin : integer;
        lakeout,
        waterout : signaltype reply integer;    {for addition of users }
    use
        lake, pump, reservoir;
    create
        lake;
        pump (bucketUnit);
        reservoir (capacity);
    link                                        {internal and interface links}
        pump.waterin ,
        lakeout           to  lake.waterout;    {multiple pumps or users of the  lake }
        reservoir.pumpon  to  pump.on ,
                              on;               {multidestination}
        reservoir.pumpoff to  pump.off,
                              off;
        pump.waterout ,
        waterin           to  reservoir.waterin;
        waterout          to  reservoir.waterout;
end.
```

Fig 3.1 Water Supply System Group Module

The Conic Configuration language supports default parameter values. For *capacity*, the default number of instances is 1000 and the default *bucketUnit* is 2. The default value of *capacity* and *bucketUnit* can be overidden by passing a value when creating an instance of a group module.

It should be noted that the interface to a group module is identical to that of a task module. When a group module type has been defined, it may be instantiated and connected in exactly the same way as a task module. Hence complex configurations can be built up by nesting groups and tasks within encapsulating groups to any required level. Figures 3.2 and 3.3 illustrates this by defining an extended water system with two parallel pumps to supply the reservoir which supplies multiple users. We have found the group module abstraction to be a powerful way of structuring the tasks which constitute a **logical node**: the unit of distribution.

Each group module specification is separately compiled into a symbol table and a set of functions which will instantiate its structure at node instantiation time. A group module type which includes an instance of the Conic run-time executive (itself a group module) can be compiled into an executable load file from which logical nodes are created. The hierarchical structure of configuration specifications has no run-time overhead as it is flattened into a uniform address space of task instances at the time a node is instantiated. Thus the description of the extended system in fig.3.2 can be used to provide a static definition of a single logical node. In the next section we show how this application can be executed as a distributed application and extended dynamically.

Conic provides no explicit support for sharing data between task modules. However, within a logical node messages can contain pointer values. Consequently, a task can give direct access to the data it encapsulates. Mutually exclusive access can be enforced using the message passing primitives for synchronisation. In the respect that tasks exist in the same address space within a logical node, Conic tasks are similar to the "lightweight" processes of the V-kernel [Cheriton 84] and Amoeba [Mullender 86].

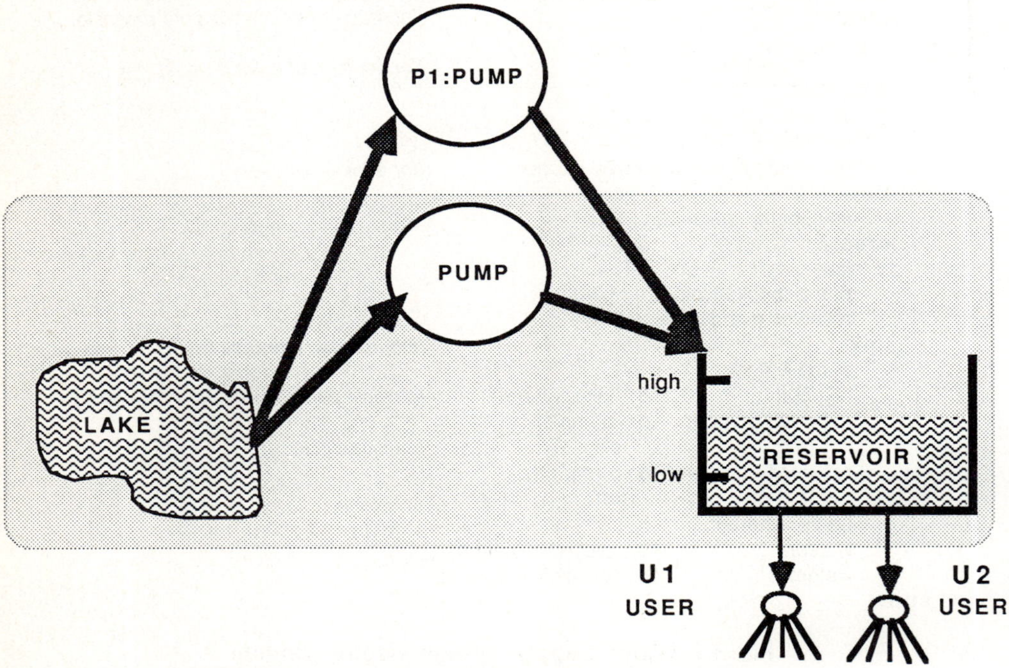

Figure 3.2 Extended Water Supply System

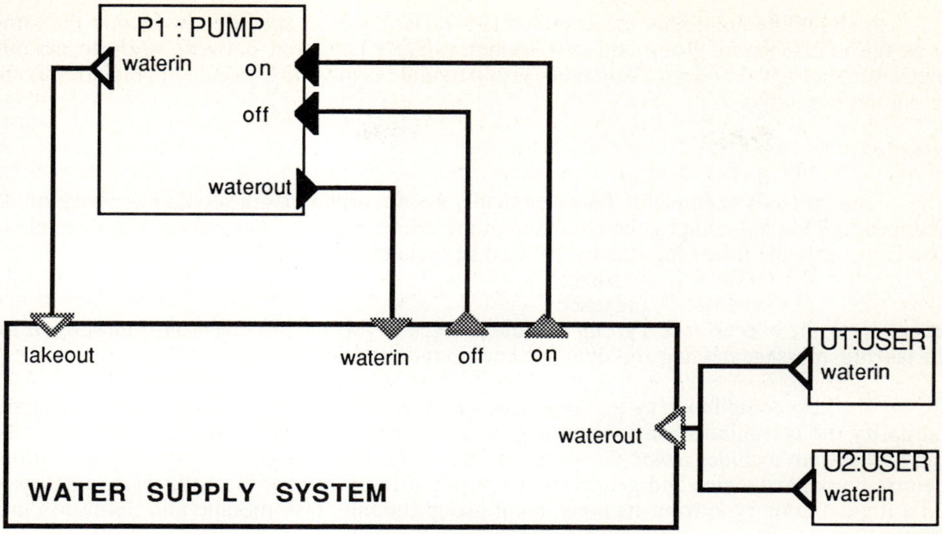


```
group module ExtendedWaterSupplySystem ( bucketUnit : integer = 2);
    use
        WaterSupplySystem, pump, user;
    create
        WSS : WaterSupplySystem (capacity=2000);
        P1 : pump (bucketunit);
        U1,
        U2 : user;
    link
        P1.waterin      to  WSS.lakeout;
        P1.waterout     to  WSS.waterin;
        WSS.on          to  P1.on;
        WSS.off         to  P1.off;
        U1.waterin ,
        U2.waterin      to  WSS.waterout;
end.
```

Fig 3.3 Extended Water Supply System Group Module

4. DYNAMIC CONFIGURATION

4.1 Logical Nodes

Distributed programs in Conic are constructed from sets of pre-compiled logical node types with the aid of the dynamic configuration tools. A logical node may run either as a Unix process if it includes an instance of *unixexec* or on a standalone target if it includes an instance of *targexec*. These run-time executives support multi-tasking, message passing and dynamic configuration operations.

Like group modules, logical nodes are types in the sense that more than one node instance may be created from the code file which represents the node type. Actual parameters substituted at instantiation time can be used to control the numbers of tasks created within nodes and the parameter values passed to those tasks.

It should be noted that the interface to a logical node is specified in exactly the same way as the interfaces of group and task modules. The distinction between a group module implementing a logical node and any other group module is that the logical node includes a run-time support executive.

Example

The previous example of the water supply system group module could have been run as an independent logical node by the inclusion of the relevant type of executive. For example to run on Unix, only the following statements need be included to the definition in fig. 3.1:

```
use      unixexec;
create   unixexec;
```

As mentioned, the executive is a group module [Magee 87] which provides run-time support for multi-tasking, message passing and dynamic configuration facilities.

The host compilation system produces an executable code file for the logical node type. To simplify the compilation and subsequent maintenance of complex logical node types, the Conic host system includes a *makefile generator* tool. This analyses group module specifications to determine dependencies and generates the required input file for the Unix *make* facility to build a logical node type from its constituent group module, task module and definition unit sources.

4.2 Managing an Application Configuration

Conic distributed programs are constructed from pre-compiled logical node types. Each logical node type is contained in an executable code file. In the following, we describe how the water system logical node can be configured with a user node to form a distributed application on the hardware configuration depicted in fig. 4.1. We then show how it can be subsequently extended **dynamically** by the addition of pumps and a user to the targets targ1 and targ2 respectively.

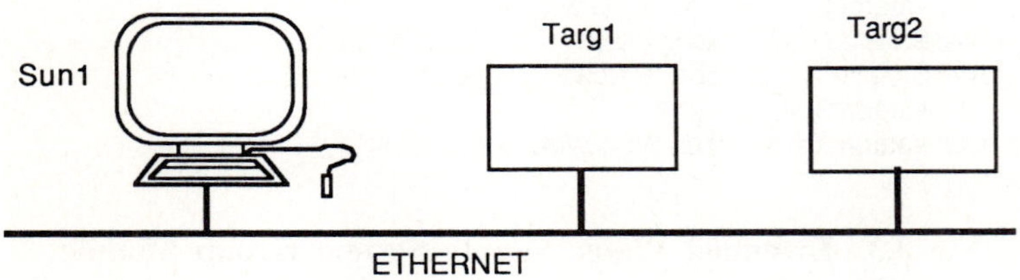

Fig. 4.1 Hardware Configuration

The initial water system of fig. 2.1 is configured by submitting the following set of configuration commands to a configuration manager. The commands may be typed interactively to an invocation of the manager (**iman**) or may be read from a file. The manager may be run in a window on one of the Suns or on a separate machine.

Configuration commands:-

```
manage water

create WSS: WaterSupplySystem(capacity=2000) at sun1
create U1:user at targ2

link U1.waterin to WSS.waterout

start WSS  U1
```

The **manage** command provides a name for the distributed application. A user may thus control one or more distributed applications concurrently. Each time the configuration manager is invoked, the user must specify the application he wishes to control. If omitted this name defaults to the user's Unix login name.

The **create** command creates the specified logical node type **at** a location. In this example *WSS* is created at *sun1* and a user *U1* at *targ2*. The **link** command is used to connect interface ports between logical nodes. The language used to communicate with a configuration manager corresponds with the configuration language used to construct group modules.

Having constructed the initial distributed application structure, it is now possible to modify it to produce the extended system in fig.3.2. Let us assume that executives *targexec* were added to a pump to form the logical node *pumpnode*. Instances of an extra pump and users can be created dynamically at targ1 and targ2 resp., and linked to the running *WSS*.

manage water

create P1:pumpnode **at** targ1
create U2:user **at** targ2

link P1.waterin **to** WSS.lakeout
link P1.waterout **to** WSS.waterin
link WSS.on **to** P1.on
link WSS.off **to** P1.off
link U2.waterin **to** WSS.waterout

start P1 U2

Additional control commands are also available to **stop** logical nodes, **remove** instances and **unlink** ports. As well as providing commands to control a configuration, the manager provides a set of queries to let the user examine the state of his system:

systems lists the set of applications currently running.
nodes lists the set of nodes within a system together with their current state (*started*, *stopped*).
ports <node> lists a node's interface ports and types
links <node> lists the entryports to which a node's exitports are connected.

4.3 Summary and Discussion

This section has provided a user's view of the dynamic configuration facilities of the Conic system. The functionality of an application is implemented by task modules and definition units using the Conic Programming Language. These tasks may be combined into groups to provide extra levels of structuring using the Conic Configuration Language. The set of task and group types is then partitioned into logical node types. These logical node types form the unit of distribution. When defining a logical node type the user must consider the environment in which the node is to execute (host or target) and include the appropriate run-time support executive. Compiling a logical node type results in an executable code file. This compiled node type, although constrained as to whether it may run on a host or target, is unrestricted as to its hardware location and the particular logical configuration in which it will run. Furthermore, the number of task instances contained within a logical node can be specified by parameters at node creation time.

The initial construction and subsequent modification of an application is carried out using a configuration manager which allows the user to create instances of logical nodes at specified locations within his network. These instances are interconnected to form the logical application configuration.

Our initial conception of dynamic configuration management [Kramer 85] involved what was essentially an on-line database which recorded the current configuration state. It was intended that a dynamic configuration manager would use this database to retrieve information on the current application configuration in order to perform changes. The dynamic manager would both change the system and update the configuration database. The database was intended to "mirror" the system providing translations from symbolic names to actual addresses. The database would ensure that only consistent and validated changes could be performed. One motivation for this design was that translation information need not be stored in target nodes which have no backing store and may have limited main store. This translation information would have been significant since we intended to manage systems at all levels down to the level of a task module.

The design outlined above had a number of significant problems, primarily concerned with the implementation of the database. To achieve a distributed and robust management system, it would have required a distributed database implementation with the attendant problems of maintaining replicated data and performing consistent atomic updates. While solutions exist to these problems and a distributed database could have been constructed we felt that this design was overly complex. The database would constrain the speed with which changes could be performed. This speed is particularly important when re-configuration is required as a result of failure. Consequently, we abandoned this design and the current implementation results from two fundamental decisions.

Firstly, it was decided that the user's requirement for dynamic configuration could be satisfied by management at the level of logical nodes. Essentially, the logical node became both the unit of configuration management and the smallest unit of failure. This decision dramatically reduces the quantity of information which must be handled by the management system. In the systems we have constructed to date, the configuration of tasks within a node is more complex than the configuration of nodes which combine to form an application. Nodes typically have 10 to 100 constituent task instances, including the executive.

Secondly, rather than have a separate configuration database, it was decided that a running application would be its own database. Each logical node would contain enough information to describe its own interface and its links to other nodes. The quantity of this information is small enough, as a result of the previous decision, to hold in main memory. A configuration manager obtains information on an application by querying a name server to find the set of logical nodes which constitute the application. Information concerning the node itself is obtained by communicating directly with the node.

5. CONCLUSIONS

Conic has been used at Imperial College, other universities and in industry for implementing communication protocols, operating systems, image processing, adaptive control, distributed discrete event simulation, distributed databases etc. It is gratifying that all our users have found the concepts embodied in Conic, and the facilities provided by its support environment, to be easy to assimilate and use. They are particularly enthusiastic about the use of the configuration language to describe and construct their systems and about **dynamic configuration** using logical nodes. The functionality provided seems to be more than adequate to support the flexibility required in distributed systems.

The separation of programming from configuration has enabled us to maintain the knowledge of the configuration structure and status necessary to make unpredicted configuration changes. It is difficult to envisage how such arbitrary changes can be incorporated in a system where configuration information and control is embedded in the programming language and hence in the program. Planned changes in Conic, such as in response to failures, can be initiated from the programming level by communication with the configuration manager. However, further work is required to investigate the interaction between running programs and the process of dynamic configuration. Identification of the possible points for reconfiguration is related to the notion of module quiescence, where a module is inactive and awaits stimulus before performing further actions. Previous work [Kramer 78] using invariants to characterise module quiescence appears promising.

The selection of **simple** and **efficient** primitives for Conic have provided a sound basis for the implementation of experimental distributed systems. Where functionality was sacrificed for simplicity and/or efficiency, more complex operations can generally be provided at a higher level. For example we have provided atomic transactions by extending the standard facilities provided by the executive [Anido 86] rather than as base primitives as in Argus [Liskov 83]. We have also experimented with the use of passive module redundancy and the reconfiguration facilities to provide fault-tolerance in a transparent manner [Loques 86].

Support for **mixed hosts / targets** has provided an extremely versatile environment. The fact that operational distributed targets can communicate with Conic logical nodes running under Unix has obviated the development of standard facilities such as a file system or printer spooler. It has allowed us to keep targets simple as the complex components of the Conic support environment can run on the host computers. In addition, the ability to test distributed systems on a Unix host prior to down-line loading to a distributed architecture, has speeded up the development process in many cases.

The **uniformity** provided by the use of Conic itself for implementation of the support environment, has proved useful in tailoring the facilities provided. For example the communication system can be configured to include a connection service, routing over interconnected subnets or drivers for different LANs. In addition, the **accessibility** of the system facilities ("open architecture") has even permitted users to adapt and modify the executive to support their requirements. For example, in their development of a run-time environment for an object-oriented system, GEC Research have modified some of the Conic intertask communication primitives and introduced support for manipulating capabilities.

The Conic environment also supports allocation **flexibility** and provides the necessary transformations for a an application to run on a mixed set of non-homogeneous computers. Structuring the executive as Conic modules has meant that the standard Conic configuration tools can be used to build the run-time system for the variety of hosts and targets. It would have been difficult to maintain and support this variety of machines any other way. However, the environment currently supports only a single programming language. This has the advantage that the compiler can check message type compatibility between messages and ports and that port interconnections can be validated for type compatibility at configuration time. Therefore no run time checks are needed. Furthermore, the transformations required for transferring messages between heterogeneous computers are comparatively simple as the compiler generates similar data structure representations in different target computers. Some current work, based on that of Matchmaker[Jones 85] and MLP [Hayes 86] is aimed at supporting additional module programming languages. The Conic configuration facilities will provide the basis of integrating diverse language components with those implemented in Conic.

Work on supporting heterogeneity and dynamic configuration for distributed computer control systems in the manufacturing engineering field is currently being funded by the ACME Directorate of the British SERC (Science and Engineering Research Council). Figure 5. gives a schematic outline of the proposed environment to be developed in the project. The virtual system acts as a 'faster than real time' on-line simulation of the plant to provide control and configuration parameters back to the real system. This work is collaborative with a research group at BP, who are providing the necessary control and automation expertise.

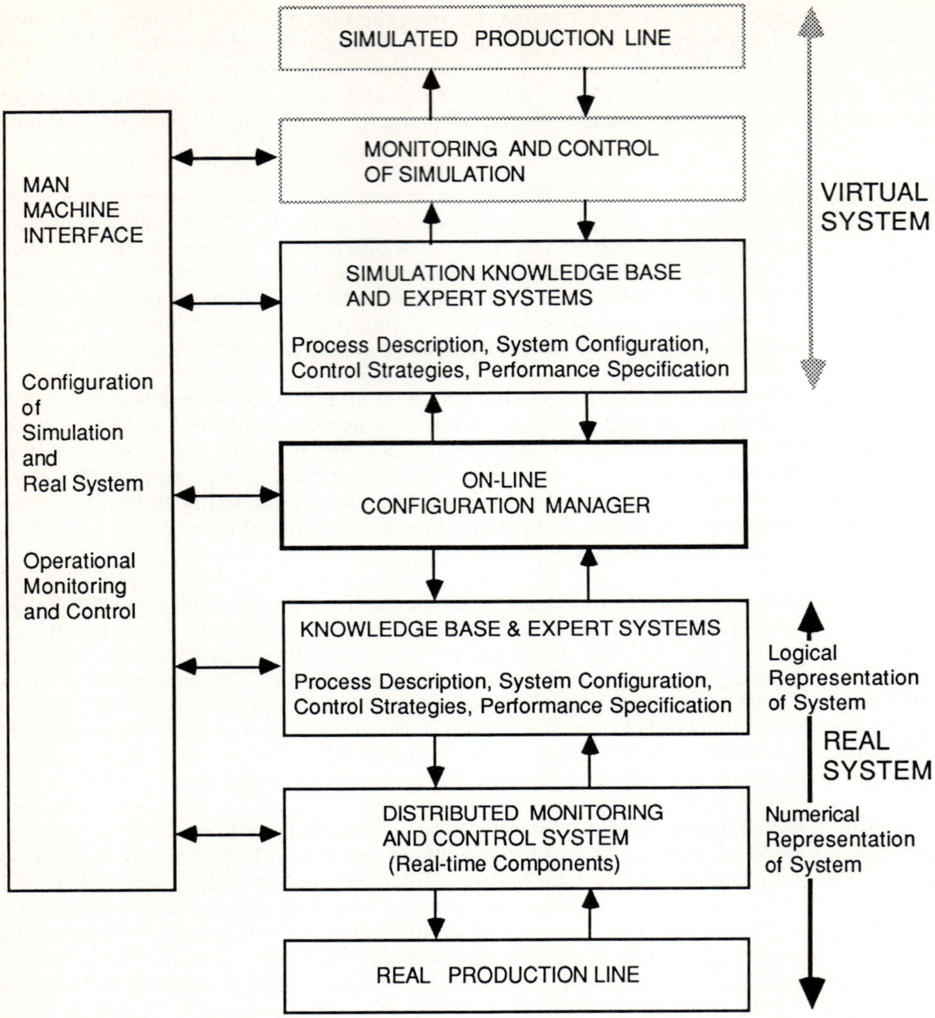

Figure 5. Proposed Environment for Production Line Automation.

We also intend to continue to use Conic as the basis for more general distributed system research such as decentralised algorithms, fault tolerance, design, specification and analysis of distributed applications, and security in management domains.

As can be seen from the above description, Conic provides a flexible and sound environment for the implementation of experimental distributed systems, both to ourselves and our various users. Conic has benefitted from user experience and we intend to continue this fruitful partnership.

Acknowledgements

Acknowledgement is made to British Coal for a grant in aid of the initial work on

Conic, but the views expressed are those of the authors and not necessarily those of British Coal. This work has also been partially funded by the SERC under Grant GR/C/31440. We particularly acknowledge the contribution of our colleagues Naranker Dulay and Kevin Twidle to the concepts described in this paper and to the implementation of the Conic environment itself. Finally we wish to thank Ricardo Anido who provided the first version of the water supply system.

6. REFERENCES

Andrews 86 G. Andrews, R. Olsson, "The evolution of the SR programming language", Distributed Computing, 1, July 1986, pp. 133-149.

Anido 86 R. Anido, J. Kramer, "Synchronised forward & backward recovery", 7th. IFAC DCCS, Germany, Sep. 1986, to be published by Pergamon Press.

Black 87 A. Black, N. Hutchison, E. Jul, H. Levy, L. Carter, "Distribution and abstract types in Emerald", IEEE Trans. on Software Eng. SE-13(1), Jan. 87, pp. 65-76.

Cheriton 84 D.Cheriton, "The V-Kernel a software base for distributed systems", IEEE Software, 1 (2), April 1984, pp. 19-43.

Dulay 84 N.Dulay, J.Kramer, J.Magee, M.Sloman, K.Twidle, "The Conic configuration language, version 1.3", Imperial College Research Report DoC 84/20, Nov. 1984.

Gawthrop 84 P.Gawthrop, "Implementation of distributed self-tuning controllers", EUROCOM 1984, Brighton, Peter Peregrinus, pp384-352.

Hayes 86 R. Hayes, R.D. Schlichting, "Facilitating mixed language programming in distributed systems", TR 85-11a Dept. of Computer Science, University of Arizona, Tucson 85721, March 1986.

Hoare 78 C.A. R. Hoare, "Communicating sequential processes," CACM, 21(8), Aug. 1978, pp. 666-677.

Jones 85 M. Jones, R. Rashid, M. Thomson, "An interface specification language for distributed processing", Proc. 12th ACM SIGACT-SIGPLAN Symposium on Principles of Programming Languages., ACM Jan. 1985.

Kramer 78 J.Kramer,R.J. Cunningham,"Towards a notation for the functional design of distributed processing systems", in IEEE Proc. 1978 Int. Conf. Parallel Processing, Aug. 1978, pp 69-76.

Kramer 84 J.Kramer, J.Magee, M.Sloman, K.Twidle, N.Dulay, "The Conic programming language, version 2.4", Imperial College Research Report DoC 84/19, October 1984.

Kramer 85 J.Kramer, J.Magee, "Dynamic configuration for distributed systems", IEEE Transactions on Software Engineering, SE-11 (4), April 1985, pp. 424-436.

Liskov 83 B.Liskov, R.Sheifler, "Guardians and actions: linguistic support for robust distributed programs", ACM TOPLAS, 5 (3), July 1983, pp. 381-404.

Liskov 87 B. Liskov, L. Shrira, "Promises: Linguistic Support for Efficient Asynchronous Procedure Calls in Distributed Systems", MIT Lab. Computing Science,Cambridge MA 02139, Aug. 1987

Loques 86 O. Loques, J. Kramer, Flexible fault tolerance for distirbuted computer systems" IEE Proc. pt. E, 133(6), Nov. 1986, pp. 319-337.

Magee 87 J.Magee, J.Kramer, M.Sloman, "Constructing Distributed Systems in Conic", to appear in IEEE Trans. in Software Engineering.

Mullender 86 S.J. Mullender, A.S. Tanenbaum, "The Design of a Capability Based Distributed Operating System", Computer Journal, Vol. 29 No.4, Aug 1986, pp. 289-299.

Scott 87 M.L. Scott, "Language support for loosely coupled distributed programs", IEEE Trans. on Software Eng. SE-13(1), Jan. 1987, pp. 77-86.

Strom 85 R. Strom, S. Yemini, "The Nil distributed systems programming language: A status report", ACM SIGPLAN Notices, 20(5), May 1985, pp. 36-44.

USA DOD 80 USA Department of Defense, "Reference manual for the Ada™ programming language", Proposed Standard Document, July 1980.

Wirth 77 N. Wirth, "Modula: a language for modular multiprogramming", Software Practice and Experiences, 12, 1982, pp. 719-753.

Structural Design of Decentralized Control Systems

P. Martin Larsen
Dept. of Electric Power Eng.
Tech. University, Build. 325
DK-2800 Lyngby, Denmark

F. J. Evans
19 Homesdale Road
Orpington, Kent BR5 1JS
England

KEYWORDS/ABSTRACT: structural design of control systems/ Mason's formula / Kalman criteria / decentralized control / APL /

The structural design methods for decentralized control systems determine if the structure of the system to be controlled and the proposed local feedback loops are sufficient to make controllability and observability possible under nearly all conditions.

The structure of the system is represented by a digraph of which the nodes represent the state variables. The structural design methods are based on the Boolean form of the matrices describing the system and therefore they imply only simple Boolean matrix operations. The rank conditions included in the Kalman criteria for controllability and observability are substituted by term rank conditions in the structural case.

The structural design methods for control systems supplement the numerical methods, which determines the specified dynamical system behaviour.

1. Introduction

The design of continuous linear control systems is often based on the concepts of controllability and observability introduced by Kalman. Assuming that the state space equations for the system to be controlled are known, the Kalman criteria imply that the rank of the controllability and the observability matrix respectively is equal to the order n of the system.

In the case of large systems,there can be considerably difficulties in the determination of the rank of the relevant matrices. Therefore it is far more realistic to ask if, in principle, the structure of the system is sufficient to make controllability and observability possible under nearly all conditions.

The structure of the system to be controlled can be represented by a digraph in which the nodes represent the state variables.

NATO ASI Series, Vol. F47
Advanced Computing Concepts and Techniques
in Control Engineering
Edited by M.J. Denham and A.J. Laub

The digraph including the input and the output connections can be drawn directly from the system matrix A, the distribution matrix B, and the output matrix C by inspection.

S.J. Mason (1953 & 1956) established very significant connections between the algebra and the topological structure of the so-called signal-flow graphs, which are digraphs with transmittances associated to each edge. Mason stated that for any signal flow graph the transmittance relating any nodal variable to another is given by the so-called Mason's Formula. The denominator of Mason's Formula is the characteristic polynomial, which determines the eigenvalues of the system. Therefore Mason's Formula becomes the key to detect how the eigenvalues of the system are associated with specific parts of the system topology.

The aim of this paper is to introduce the necessary methology to determine if a dynamic system is potential state controllable and observable, i.e. if the structure is sufficient to make controllability and observability possible. In this case the numerical matrices of the system are substituted by the corresponding Boolean matrices, and the numerical matrix operations are substituted by Boolean operations.

The structural criteria are first derived for centralized control systems, i.e. systems of which the control variables are generated by one control station (multivariable controller) on basis of the available state or output variables, and they are then expanded to the decentralized case, where the control action is distributed among a number of control stations, each control station having access only to a subset of the output variables of the system and influence only on a subset of the input variables of the system.

The structural design methods supplement efficiently the numerical design methods,which finally forms the basis for obtaining a satisfactorily stationary and transient behaviour of the designed control system.

A tool box for numerical and structural design of control systems has been implemented on a Personal Computer programmed first in APL, and later in Nial (Nested Interactive Array Language).

This paper is summarizing the lecture notes on "Structural Design of Control Systems" by F.J. Evans and P. Martin Larsen.

2. Introduction to basic graph theory

A directed graph or a digraph G = (X,U) consists of:

a) a finite set $X = (x_1, x_2, \ldots\ldots\ldots\ldots, x_n)$ the elements of which are called nodes.

b) a subset U of the Cartesian product XxX, the elements of which are called edges. The concept of the Cartesian product intoduces "ordering".

The digraph shown on figure 2.1 includes six nodes:

$X = (x_1, x_2, x_3, x_4, x_5, x_6)$

and eight edges:

$U = ((x_1, x_2), (x_2, x_1), (x_2, x_3), (x_3, x_4), (x_4, x_5),$
$(x_4, x_5), (x_5, x_3), (x_6, x_6))$

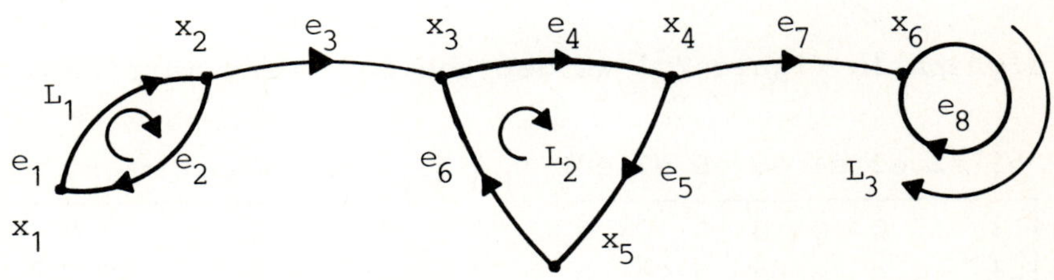

Fig. 2.1. Digraph with six nodes and eight edges.

Digraphs can be represented by matrices in various ways:

The incidence matrix **N**(G) in which the elements n_{ij} are determined in the following way

n_{ij} = 1 if the edge j is incident at node i and is oriented away from node i.

n_{ij} = -1 if the edge j is incident at node i and is oriented towards node i.

n_{ij} = 0 if the edge j is not incident at node i.

The dimension of **N**(G) is n x m when n is the number of nodes and m is the number of edges. It is clear that selfloops, such as e_8 cannot be clearly represented in the incidence matrix.

The adjacency matrix **A**(G) in which the elements a_{ij} are determined in the following way:

a_{ij} = 1 if the edge (i,j) is directed from node i to node j

a_{ij} = 0 if otherwise

The dimension of the **A**(G) is n x n.

The occurrence matrix which is defined as the transposed of A(G)

The circuit matrix **B**(G) in which the elements b_{ij} are determined in the following way:

b_{ij} = 1 if the edge j is in the circuit (loop) i and the orientation of the circuit and the edge coincide

b_{ij} = -1 if the edge j is in the circuit i and the orientations do not coincide

b_{ij} = 0 if the edge j is not in the circuit i.

For the digraph in figure 2.1 we get the following matrices:

N(G) =

	e1	e2	e3	e4	e5	e6	e7	e8
x1	1	-1	0	0	0	0	0	0
x2	-1	1	0	0	0	0	0	0
x3	0	0	-1	1	0	-1	0	0
x4	0	0	0	-1	1	0	1	0
x5	0	0	0	0	-1	1	0	0
x6	0	0	0	0	0	0	0	±1

A(G) = (To nodes: columns; From: rows)

	x1	x2	x3	x4	x5	x6
x1	0	0	0	0	0	0
x2	1	0	0	0	0	0
x3	0	0	0	1	0	0
x4	0	0	0	0	1	1
x5	0	0	1	0	0	0
x6	0	0	0	0	0	1

$\mathbf{B}(G) =$

	e1	e2	e3	e4	e5	e6	e7	e8
L1	1	1	0	0	0	0	0	0
L2	0	0	0	1	1	1	0	0
L3	0	0	0	0	0	0	0	1

$\mathbf{A}(G)^t =$

From nodes:

	x1	x2	x3	x4	x5	x6	
x1	0	1	0	0	0	0	To
x2	1	0	0	0	0	0	n
x3	0	1	0	0	1	0	o
x4	0	0	1	0	0	0	d
x5	0	0	0	1	0	0	e
x6	0	0	0	1	0	1	s

The rank of the incidence matrix **N**(G) is equal to the number of tree branches in the digraph, provided no self-loops exist.

The rank of the circuit matrix **B**(G) is equal to the number of links or chords, termed the nullity of the graph.

Hence the rank of **N**(G) plus the rank of **B**(G) equals the number of edges of the graph.

Certain graph theoretical terms are commonly used in the following sections. Therefore, the following definitions are given here:

A directed walk in a digraph is an alternating sequence of nodes and edges following the orientation of the latter.

The topological length of a directed walk is determined by the number of edges in it.

If there is a path from node x_i to node x_j, then x_j is said to be reachable from the node x_i.

A closed walk has the same first and last node; a closed walk whose nodes are distinct (except the first and last node) is called a cycle; a spanning closed walk contains all the nodes.

A diagraph is said to be strongly connected or simply strong if and only if it has a spanning closed walk. A strong component of a digraph is a subgraph which is strong.

The <u>term rank</u> of a digraph is the maximal number of edges contained in a uni-directional path (not necessarily connected) which spans the maximum number of nodes, such that the number of edges going into and the number of edges going out from any node is not greater than unity. The term rank can be shown to be the number of elements contained in the maximal permutation matrix to be found in the occurence matrix.

Returning to the digraph in figure 2.1 it is obvious that the uni-directional path containing the edges (e1, e2, e6, e4, e7) spans all the nodes. The maximal permutation matrix to be found in the occurence matrix is shown by the ringed elements below:

$$
\begin{array}{ccc|cccccc}
 & & & \multicolumn{6}{c}{\text{From nodes}} \\
 & \text{To} & & x1 & x2 & x3 & x4 & x5 & x6 \\
\cline{4-9}
 & & x1 & 0 & \textcircled{1} & 0 & 0 & 0 & 0 \\
 & n & x2 & \textcircled{1} & 0 & 0 & 0 & 0 & 0 \\
\mathbf{A}(G)_t = & o & x3 & 0 & 1 & 0 & 0 & \textcircled{1} & 0 \\
 & d & x4 & 0 & 0 & \textcircled{1} & 0 & 0 & 0 \\
 & e & x5 & 0 & 0 & 0 & \textcircled{1} & 0 & 0 \\
 & s & x6 & 0 & 0 & 0 & 1 & 0 & \textcircled{1}
\end{array}
$$

Finally the <u>Reachability matrix</u> is defined as a matrix $\mathbf{R}(G)$ in which the elements r_{ij} are determined in the following way:

r_{ij} = 1 if node x_i is reachable from node x_j.

r_{ij} = 0 if otherwise.

3. From state space equations to digraphs

Multivariable time in variant linear dynamic systems (processes) are conventionally represented by state space matrix equations:

$$
\begin{aligned}
\underline{\dot{x}} &= \mathbf{A}\,\underline{x} + \mathbf{B}\,\underline{u} \\
\underline{y} &= \mathbf{C}\,\underline{x} + \mathbf{D}\,\underline{u}
\end{aligned}
$$

in which $\underline{x}$ is the state vector, $\underline{u}$ is the input or control vector, and $\underline{y}$ is the output vector.

The number n of the state variables defines the order of the system. The state vector will trace a state or phase trajectory in an n-dimentional state space with passing time.

The system matrix **A**, the distribution matrix **B**, the output matrix **C**, and the feedforward matrix **D** contain constant numerical elements which not only define the internal structure, but also the magnitude of the coupling which exists between the states, the inputs, and the outputs. It is from this description that the whole of the structural approach proceeds.

Let us consider the following simple 3-order system:

$$\begin{bmatrix} \dot{x}_1 \\ \dot{x}_2 \\ \dot{x}_3 \end{bmatrix} = \begin{bmatrix} 5 & 1 & 3 \\ 2 & 0 & 0 \\ 0 & 4 & 0 \end{bmatrix} \begin{bmatrix} x_1 \\ x_2 \\ x_3 \end{bmatrix} + \begin{bmatrix} 1 \\ 0 \\ 0 \end{bmatrix} u$$

$$y = [\ 0 \quad 0 \quad 6\ ]\ [\ x_1 \quad x_2 \quad x_3\ ]^t$$

By assuming that the initial conditions are $\underline{x}(0) = 0$, the Laplace transformed statespace equations become:

$$x_1 = s^{-1} \quad (5\ x_1 + 1\ x_2 + 3\ x_3 + 1\ u)$$
$$x_2 = s^{-1} \quad (2\ x_1 + 0\ x_2 + 0\ x_3 + 0\ u)$$
$$x_3 = s^{-1} \quad (0\ x_1 + 4\ x_2 + 0\ x_3 + 0\ u)$$

$$y = \quad (0\ x_1 + 0\ x_2 + 6\ x_3)$$

On the basis of these transformed equations the corresponding signal flow graph or digraph can be drawn as shown in figure 3.1.

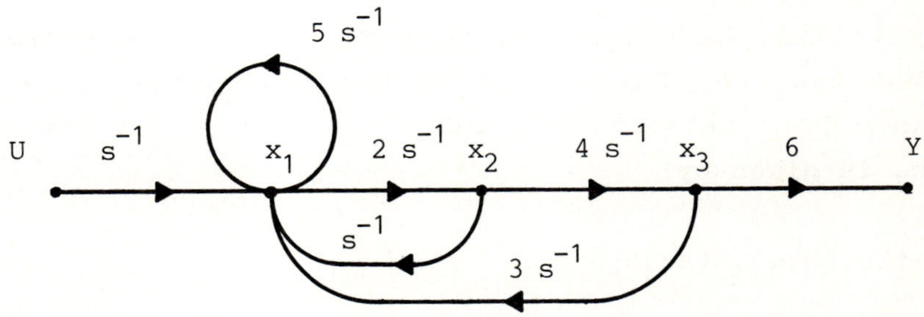

Fig. 3.1. Signal flow graph or digraph for a 3-order system.

The digraph consists of three nodes corresponding to the three states, and one node corresponding to the single input variable u.

Now from a structural point of view we will temporarily ignore the transmittances attached to each edge, and consider only the structure of the digraph as:

The occurrence matrix for the digraph in figure 3.1 can derived by inspection.

$$\mathbf{A}(G)^t = \quad \text{To} \quad \begin{array}{c|ccc} & \multicolumn{3}{c}{\text{From nodes}} \\ & x1 & x2 & x3 \\ \hline x1 & 1 & 1 & 1 \\ x2 & 1 & 0 & 0 \\ x3 & 0 & 1 & 0 \end{array}$$

However, the occurrence matrix could have been derived simply from the state space equation above by obtaining the Boolean form of the system matrix A. The Boolean system matrix $\mathbf{A}_B$ is obtained from the A-matrix by assigning every non-zero element the value 1, and all zero elements the value 0.

Similarly, the Boolean form of the distribution matrix **B** determines the connections between the input variables and the states, and the Boolean forms of the output matrix **C** and the feed-forward matrix **D** describe the connections to the output from the state variables and the input variables respectively.

S.J. Mason (1953) gave an ingenious and elegant connection between the topology of the signal flow graph and the algebra associated with the characteristic polynomial for the system. He stated that for any flow graph of the type shown above, the transmittance T_{ij} relating the nodal variable x_i to the nodal variable x_j is given by:

$$T_{ij} = (\Sigma_k \, (P_{ij})_k \, (\Delta_{ij})_k) \; / \; \Delta$$

in which $(P_{ij})_k$ is the transmittance of the k-th path from nodal variable x_i to nodal variable x_j.

Δ is the determinant of the system expressed by the transmittances L_i of the circuits or loops of the digraph:

$\Delta = 1 - \Sigma_i L_i$

$+ \Sigma L_i L_j$ (non-touching loops taken 2 at a time)

$- \Sigma L_i L_j L_k$ (non-touching loops taken 3 at a time)

$+ \ldots\ldots\ldots\ldots\ldots\ldots\ldots$ etc.

$(\Delta_{ij})_k$ is the co-factor of the path $(P_{ij})_k$ and is calculated as the value of the determinant Δ for each forward path making the loop transmittance for all touching loops equal to zero.

Returning to the digraph in figure 3.1, we can identify 3 loops with the loop gains:

$L1 = 5\ s^{-1}$ $\quad L2 = 2\ s^{-2}$ $\quad L3 = 24\ s^{-3}$

Hence the determinant is (all loops are touching node 1):

$\Delta = 1 - 5\ s^{-1} - 2\ s^{-2} - 24\ s^{-3}$

The transmittance of the direct path from x_1 to x_3 is:

$P_{13} = 8\ s^{-2}$

Including the input path $P_u = s^{-1}$ and the output path $P_y = 6$ the transmittance or transfer function from input u to output y becomes:

$G(s) = Y(s)/U(s) = (P_u\ P_{13}\ P_y) / \Delta$

$= 48\ s^{-3} / (1 - 5\ s^{-1} - 2\ s^{-2} - 24\ s^{-3})$

$= 48 / (s^3 - 5\ s^2 - 2\ s^1 - 24)$

Although its importance has always been recognised, Mason's work has not been fully exploited by control engineers, because of the difficulties associated with manipulating the formula for any system other than the simpliest. Computer aided techniques now available enable these difficulties to be overcome, and Mason's formula is now included as a useful part of the tool-kit for the designer of complex control systems.

It is clear from the work of Mason that the identification of the loops in any digraph is central to the successful exploitation of the strucutre in control system design. Clearly, a loop in graph theoretic terminology is a cyclic component. Although we now have CAD tools, which enable us to perform Mason-type analysis on a graph, it is necessary to be able to relate, if possible, specific roots of the characteristic polynomial to specific components of the graph. This may not always be possible, and if that is the case, then it is useful to know that this situation exists.

Fortunately, there is a fundamental theorem, first stated by Harary (1955), which allows us to relate specific roots to specific components of the graph. A modified form of this theorem is as follows:

> *The set of eigenvalues for any matrix is the union (including multiplicity) of the sets of eigenvalues of the submatrices corresponding to the strong components of the digraph of the matrix, plus the number of nodes, which belongs to any acyclic parts of the digraph.*

To identify all cyclic components from the digraph itself can be very difficult, except in the simplest of cases. Therefore, we need to develop ways in which any digraph can be decomposed into cyclic components analytically.

Having derived a particular set of state space equations which have some physical relevance, i.e. the chosen set of state variables have some relationship to measurable inputs and outputs of the system, we should aim to preserve the structural properties of this mathematical model. Hence, if the structure is to be preserved we can only resort to a <u>reordering</u> of the nodes in order to make the components more obvious.

In the case of the canonical state transformation, the eigenvalues of the system is exposed along the diagonal of the transformed system matrix, and all the numerical values of the elements of the system matrix are changed.

In the case of reordering the nodes, i.e. changing the numbering of the state variables, the aim is to expose the cyclic components along the diagonal of the Boolean system matrix. Any reordering of the rows and columns of a matrix can be associated with a permutation matrix. Therefore, a permutation matrix P should be generated from the system matrix **A** such that the reordered system matrix $\mathbf{P}^t\mathbf{AP}$ has a lower block diagonal form. Each diagonal block represents a cyclic component. Diagonal blocks which are themselves of lower diagonal form with zero along the main diagonal, form the acyclic components of the system. Any entries below the diagonal blocks represent the structural connections between parts of the system.

To understand the APL functions which are used in the decomposition procedure, the important digraph property of Reachability must be introduced.In the reachability matrix the element r_{ij} is equal to unity if it is possible to reach the node x_i from the node x_j along a uni-directional path.

It is well known, and easily verified, that the powers of the Boolean system matrix $\mathbf{A_B}$ indicate the existence of directed paths of increasing topological length. By definition, the Boolean matrix $\mathbf{A_B}^0$ is the unit matrix **I**, which indicates that each node of the digraph can be reached from itself through a path of topological length 0.

We shall find it significant to form a Boolean tensor:

$$\mathbf{Q_B} = \mathbf{A_B}^0, \mathbf{A_B}^1, \mathbf{A_B}^2, \ldots\ldots\ldots\ldots \mathbf{A_B}^{n-1}$$

for any digraph, where n is the number of nodes. The tensor Q_B is 3-dimensional, and can be visulized as the n powers of $\mathbf{A_B}$ placed behind each other along the third axis.

If we now use the APL function v/ (or compress) along axis [1] we obtain the reachability matrix:

```
R←v/[1] QB
```

The permutation matrix P is generated on the basis of the reachability matrix R in the following way:

Form the transposed matrix of R:	`RT ← ⍉R`
Form the negated matrix of RT:	`RTN ← ~⍉R`
Use an and operation between R and RTN:	`RRTN ← R ∧ RTH`
Form a vector by add compress:	`J ← +/[1] RRTN`
Downgrade J	`DJ ← ⍒J`

The J vector represents the required permutation matrix, which shows that the "old" numbers listed in J should be changed to the "new" numbers 1,2,3, ... n.

Example 3.1

A fourth-order linear system has the following system matrix:

$$\mathbf{A} = \begin{bmatrix} 0 & 0 & 0 & 8 \\ 6 & -4 & 0 & 0 \\ 0 & 0 & 0 & 3 \\ 0 & 0 & 2 & 0 \end{bmatrix}$$

The Boolean form of the system matrix, and the corresponding digraph is shown in figure 3.2.

$$\mathbf{A}_B = \begin{bmatrix} 0 & 0 & 0 & 1 \\ 1 & 1 & 0 & 0 \\ 0 & 0 & 0 & 1 \\ 0 & 0 & 1 & 0 \end{bmatrix}$$

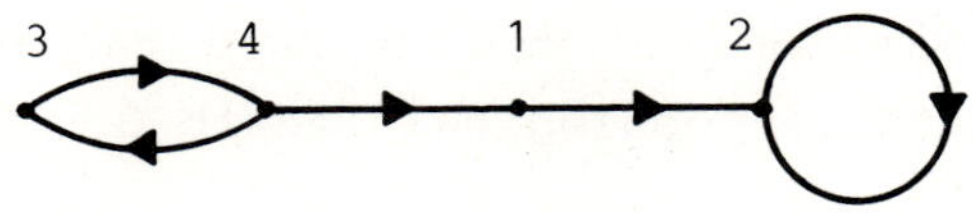

Figur 3.2. Digraph for 4-order system.

The reachability matrix is derived directly from the digraph by inspection, and the matrices RT, RTN and RRTN are found:

$$R = \begin{bmatrix} 1 & 0 & 1 & 1 \\ 1 & 1 & 1 & 1 \\ 0 & 0 & 1 & 1 \\ 0 & 0 & 1 & 1 \end{bmatrix} \qquad RT = \begin{bmatrix} 1 & 1 & 0 & 0 \\ 0 & 1 & 0 & 0 \\ 1 & 1 & 1 & 1 \\ 1 & 1 & 1 & 1 \end{bmatrix}$$

$$RTN = \begin{bmatrix} 0 & 0 & 1 & 1 \\ 1 & 0 & 1 & 1 \\ 0 & 0 & 0 & 0 \\ 0 & 0 & 0 & 0 \end{bmatrix} \qquad RRTN = \begin{bmatrix} 0 & 0 & 1 & 1 \\ 1 & 0 & 1 & 1 \\ 0 & 0 & 0 & 0 \\ 0 & 0 & 0 & 0 \end{bmatrix}$$

The J vector is found as: J← +/[1] RRTN ← [1 0 2 2]

The "downgrade" of J is: DJ ← [3 4 1 2]

The DJ vector represents the required permutation matrix P, which shows that the "old" numbers 3,4,1,2 should be changed to the "new" numbers 1,2,3,4.

The permutation matrix is then:

$$\mathbf{P} = \begin{bmatrix} 0 & 0 & 1 & 0 \\ 0 & 0 & 0 & 1 \\ 1 & 0 & 0 & 0 \\ 0 & 1 & 0 & 0 \end{bmatrix}$$

The decomposed system matrices (numerical and Boolean) are determined by changing the numbers according to the permutation matrix:

$$\mathbf{A} = \left[\begin{array}{cc|c|c} 0 & 3 & 0 & 0 \\ 2 & 0 & 0 & 0 \\ \hline 0 & 8 & 0 & 0 \\ \hline 0 & 0 & 6 & -4 \end{array}\right] \qquad \mathbf{A_B} = \mathbf{P^t A_B P} = \left[\begin{array}{cc|c|c} 0 & 1 & 0 & 0 \\ 1 & 0 & 0 & 0 \\ \hline 0 & 1 & 0 & 0 \\ \hline 0 & 0 & 1 & 1 \end{array}\right]$$

The new system matrix has been reordered into a lower block diagonal form, which clearly indicates the fact that nodes 1 and 2 form a small cyclic component, that node 3 is acyclic, and that there is a cyclic loop at node 4. From the new A-matrix with the numerical elements it appears that the system has the following eigenvalues exposed along the diagonal: $\lambda = \pm \sqrt{6}$, $\lambda = 0$, $\lambda = -4$. These values are found according to Harary's theorem as the union of the eigenvalues for the three subsystems.

From Harary's theorem it also appears that the elements 8 and 6 below the diagonal blocks have no influence on the eigenvalues of the system; these numbers indicate (feedforward) connections between the subsystems. Introduction of elements above the diagonal blocks will represent feedback connections in the total system, and will effect the eigenvalues. This principle can be utilized in the design of more complex control systems, when introduction of extra feedback connections is considered.

4. Centralized multivariable control

The concepts of controllability and observability were first introduced into linear control theory by Kalman (1960).

A system is controllable if it is possible to find a control vector $\underline{u}(t)$ which, in a specified time t_f, will transfer the system from an arbitrary finite initial state $\underline{x}(0)$ to an arbitrary finite state $\underline{x}(t_f)$.

Kalman showed that a system is completely state controllable if and only if the rank of a matrix Q_C is equal to n, where n is the number of states,and the controllability matrix is given by:

$$Q_C = (B,\ AB,\ A^2B,\ \ldots\ldots\ A^{n-1})$$

A system is observable if there exists a time $0<t_f<\infty$ such that given the vector $\underline{y}(t)$ of the outputs in the time interval $(0, t_f)$, it is possible to deduce the initial state vector $\underline{x}(0)$.

Kalman showed that a system is completely state observable if, and only if, the rank of the matrix Q_0 is equal to n, where n is the number of states and the observability matrix is given by:

$$Q_0 = (C,\ CA,\ CA^2,\ \ldots\ldots\ CA^{n-1})^t$$

In case the system is completely state controllable and observable it is possible to control the system by means of one control or input variable derived as a linear combination of all the states of the system.

There can be considerable difficulties in the determination of the rank of very large matrices. Furthermore, in real life problems the accuracy of the numerical data given for a particular system is probably no more than ± 10%. Hence, this type of test is not only difficult to perform in many cases, but also can be unreliable as a guide to the actual controllability and observability. It would be far more realistic to ask if, in principle, adequate structure exists in the system to make controllability and observability possible under nearly all conditions -or perhaps conversely,is the structure inadequate so that controllability and observability are impossible under all conditions (i.e.whatever the actual values of the non-zero elements of the matrices).

In the structural controllability test to be introduced now, the determination of a term rank must be expected as an alternative to the Kalman's rank test.

The fact that in both Kalman criteria we find the power series of the A-matrix: A^0, A^1, A^2 ...A^{n-1} suggests that from a structural point of view the reachability is important, because the Boolean tensor Q_B includes the power series of the Boolean A-matrix.

The equivalent structural criteria for structural (or potential) controllability and observability is summarized in the following table, and amount to the fact that the system must satisfy both a term rank and a reachability criterion simultaneously if it is to have potential state controllability and observability.

The reachability tests simply ensure that every state is reachable from at least one input, and that each state is able to reach at least one output.

Potential state **Controllability**	Potential state **Observability**
Form the Reachability Matrix $R \leftarrow \vee/[1]A_B{}^0,\ A_B{}^1, \ldots\ldots\ A_B{}^{n-1}$	
$KB \leftarrow R \vee.\wedge\ B_B$	$MB \leftarrow C_B \vee.\wedge\ R$
$KC \leftarrow \vee/KB,\quad K \leftarrow \wedge/KC$	$MC \leftarrow \vee/[1]\ MB\quad M \leftarrow \wedge/MC$
If K = 1, then the system might be potential state controllable. If K = 0, then the system is uncontrollable.	If M = 1, then the system might be potential state observable. If M = 0, then the system is nonobservable.
If K = 1, and if **Term rank** $TR\ [A_B\ B_B] = n$ **Then the system is potentially state controllable.**	**If M = 1, and if** **Term rank** $TR\begin{bmatrix}A_B\\C_B\end{bmatrix} = n$ **Then the system is potentially state observable.**

Let us consider a system with the Boolean matrices:

$$\mathbf{A} = \begin{bmatrix} 0 & 1 & 0 & 0 \\ 1 & 1 & 1 & 0 \\ 0 & 1 & 0 & 1 \\ 0 & 0 & 0 & 1 \end{bmatrix} \quad \mathbf{BB} = \begin{bmatrix} 1 & 0 \\ 1 & 0 \\ 0 & 1 \\ 0 & 0 \end{bmatrix} \quad \mathbf{CB} = \begin{bmatrix} 0 & 1 & 1 & 0 \\ 0 & 0 & 0 & 1 \end{bmatrix}$$

The corresponding digraph is shown in figure 4.1.

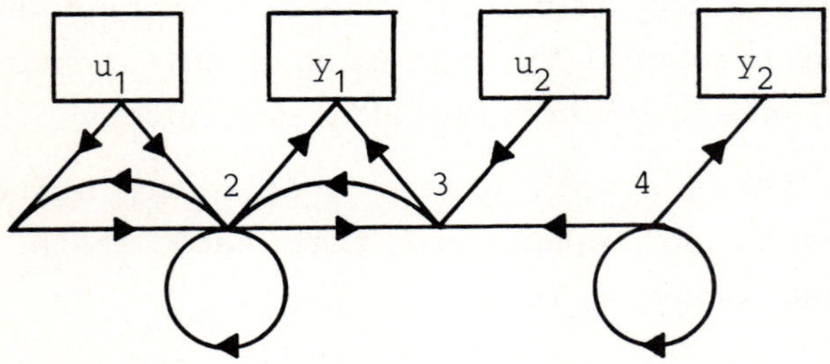

Fig. 4.1. Digraph for 4th order system with 2 inputs and 2 outputs.

The reachability matrix is obtained by inspection

$$\mathbf{R} = \begin{bmatrix} 1 & 1 & 1 & 1 \\ 1 & 1 & 1 & 1 \\ 1 & 1 & 1 & 1 \\ 0 & 0 & 0 & 1 \end{bmatrix}$$

The potential state controllability test show that the system is uncontrollable:

$$\mathbf{KB} \leftarrow \mathbf{R} \vee.\wedge \ \mathbf{BB} = \begin{bmatrix} 1 & 1 \\ 1 & 1 \\ 1 & 1 \\ 0 & 0 \end{bmatrix} \quad \mathbf{KC} \leftarrow \vee/\mathbf{KB} = \begin{bmatrix} 1 \\ 1 \\ 1 \\ 0 \end{bmatrix} \quad \mathbf{K} \leftarrow \wedge/\mathbf{KC} = 0$$

The potential state observability test is satisfied:

$$MB \leftarrow CB \vee.\wedge R = \begin{bmatrix} 1 & 1 & 1 & 1 \\ 0 & 0 & 0 & 1 \end{bmatrix} \quad MC \leftarrow \vee/[1]MB = [1 \; 1 \; 1 \; 1] \qquad M = 1$$

and the term rank of the augmented matrix:

$$TR \begin{bmatrix} \mathbf{AB} \\ \mathbf{CB} \end{bmatrix} = \begin{bmatrix} 0 & \textcircled{1} & 0 & 0 \\ \textcircled{1} & 1 & 1 & 0 \\ 0 & 1 & 0 & \textcircled{1} \\ 0 & 1 & 0 & 1 \\ 0 & 1 & \textcircled{1} & 0 \\ 0 & 0 & 0 & 1 \end{bmatrix} = \mathbf{4}$$

which is equal to the order of the system. The term rank is equal to the number of elements contained in the maximal permutation matrix to be found, shown by the ringed elements.

In the design of a centralized multivariable control system, the feedback can be generated either from the states or from the outputs of the system.

The state space equations:

$$\dot{x} = \mathbf{A}\,\underline{x} + \mathbf{B}\,\underline{u}$$

$$\underline{y} = \mathbf{C}\,\underline{x}$$

represent the <u>open-loop</u> system, and the control vector is either generated form the states $\underline{x}$ or the $\underline{y}$:

$$\underline{u} = \mathbf{F}\,\underline{x} = \mathbf{F}\,\mathbf{I}\,\underline{x} \quad \text{or} \quad \underline{u} = \mathbf{F}\,\underline{y} = (\mathbf{F}\,\mathbf{C}\,\underline{x})$$

In the case of state feedback we can use the same formula for $\underline{u}$ as in the case of output feedback by setting **C** equal to **I**.

For the <u>closed-loop</u> system the state space equations become:

$$\underline{x} = (\mathbf{A} + \mathbf{B}\,\mathbf{F}\,\mathbf{C})\,\underline{x}$$

It should be noted that the state space equations for the closed-loop system determine the dynamic properties of the system considered to be autonomous.

Hence, in the most general terms, the control problem can be stated as follows:

> Given the constant matrices **A, B, and C** of the open-loop system, a suitable feedback matrix **F** should be found in order to assign the eigenvalues of the closed-loop system matrix (**A + B F C**) to some acceptable values.

The introduction of feedback changes the elements in the A-matrix to the elements in (**A + B F C**).Therefore, the determination of the elements in the F-matrix can be supported by studying the sensitivity matrices for the system. In the sensitivity matrix for the eigenvalue λ the element s_{ij} indicates the change in λ for a unit change in the corresponding element a_{ij} of the A-matrix, i.e. $s_{ij} = \delta\lambda/\delta a_{ij}$. In this way a desired set of eigenvalues for the closed loop system results in desired values of the elements in the matrix (**A + B F C**). However, these values can only be obtained if the feedback matrix includes n^2 elements, i.e. the input vector $\underline{u}$ must have n elements.

In the case of complex higher order systems the utilization of the sensitivity matrices becomes impossible, but it should be recalled that any controllable and observable system is completely state controllable by means of <u>one</u> control variable under full state feedback, i.e. by means of n feedback elements an arbitrary set of closed loop eigenvalues can be obtained.

Clearly, in the closed-loop system, assignment of eigenvalues may not be the sole criteria for closed loop response, and questions such as steady state error may also have to be considered.

5. Decentralized multivariable control

In multivariable control system design it becomes desirable to be able to decide if in principle it is possible to achieve the closed loop performance desired (i.e. if it is possible to assign the closed loop poles to the desired values), and if it is, what is the minimum complexity of the controller which will achieve this, and if it is not possible, what is the best we can do?

In a multivariable control system one could initially assume that all the states were available to generate feedback signals (the state feedback principle), and the controller should be designed within this very loose constraint. This relative freedom could itself produce a controller of unnecessary complexity. In most systems, however, not all state variables are available for measurement, and there is some functional dependency between the states and the so-called outputs of the system. Therefore, most controllers will have to be designed under this added constraint.

The availability of cheap micro processors for controllers has also brought the ability to introduce a far greater variety of controller structure than before, and in large plants to provide complex control laws in a local environment.

The closed-loop performance of a linear system can freely be adjusted or assigned with centralized state feedback control, provided the system is both controllable and observable. However, with the added constraints of decentralization or output feedback, controllability and observability are no longer sufficient, and it becomes increasingly important to know under what conditions there exists a set of appropriate decentralized control laws with which to stabilize the system.

In figure 5.1 is shown the basic structures of centralized and decentralized control systems. In the decentralized system, each of the control stations (local controllers) has access to only a sub-set of the system outputs and is able to effect changes to only a sub-set of the system inputs.

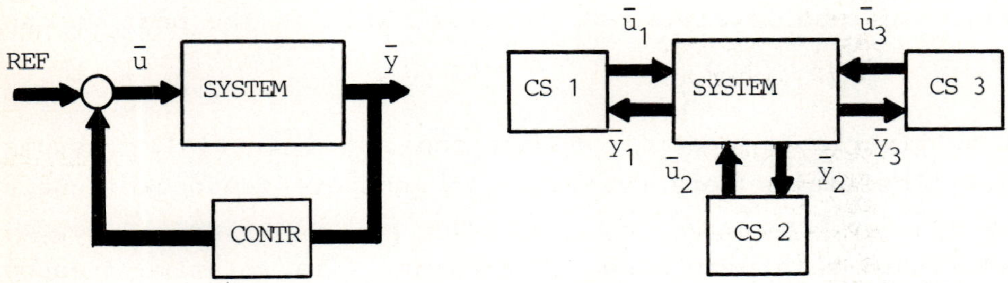

Fig. 5.1. Centralized control versus decentralized control.

Therefore, it has to be decided if within such a given structure the control objectives can be achieved, and if not, what is the minimum communication required between the control stations.

In decentralized control the structural design procedure will include the following steps:

1. Determine for each control station the subspace of states which can be reached from at least one of the available inputs (reachable subspaces).

2. Determine for each control station the subspace of states which potentially can be controlled from that control station (controlable subspaces).

3. Determine if the control stations in combination potentially can control the complete state space of the system.

4. Determine for each control station the subspace of states which can reach at least one of the available outputs.

5. Determine if the control stations in combination potentially can observe the complete state space for the system (observable subspaces).

6. Provided the control stations in combination are able potentially to control and observe the complete state space, determine if the introduction of local feedbacks in each control station makes it possible from each control station potentially to control and observe the complete state space.

The procedure will be developed step by step and illustrated by means of the decentralized control system with two control stations shown in figure 5.2. The system has two input vectors $\underline{u}_1$ and $\underline{u}_2$, one from each control station, and has two output vectors $\underline{y}_1$ and $\underline{y}_2$, one for each control station.

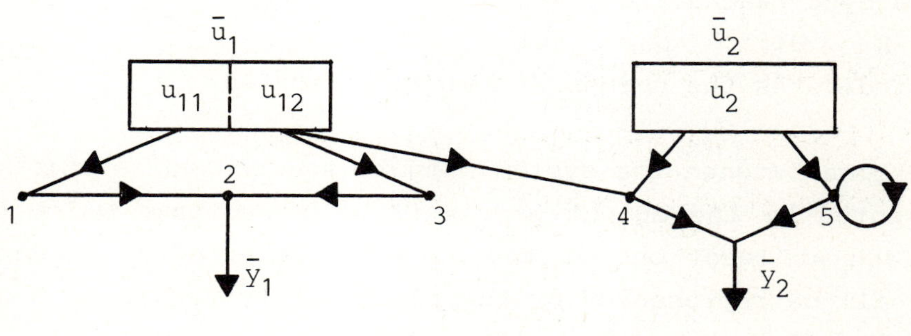

Fig. 5.2. Decentralized control system with two control stations.

The state space equations on Boolean form are:

$$\dot{\underline{x}} = A\,\underline{x} + B1\,\underline{u}_1 + B2\,\underline{u}_2$$

$$\underline{y}_1 = C1\,\underline{x} \qquad \underline{y}_2 = C2\,\underline{x}$$

$$A = \begin{bmatrix} 0 & 0 & 0 & 0 & 0 \\ 1 & 0 & 1 & 0 & 1 \\ 0 & 0 & 0 & 0 & 0 \\ 0 & 0 & 0 & 0 & 0 \\ 0 & 0 & 0 & 0 & 1 \end{bmatrix} \quad B1 = \begin{bmatrix} 1 & 0 \\ 0 & 0 \\ 0 & 1 \\ 0 & 1 \\ 0 & 0 \end{bmatrix} \quad B2 = \begin{bmatrix} 0 \\ 0 \\ 0 \\ 1 \\ 1 \end{bmatrix}$$

$$C1 = \begin{bmatrix} 0 & 1 & 0 & 0 & 0 \end{bmatrix} \quad C2 = \begin{bmatrix} 0 & 0 & 0 & 1 & 1 \end{bmatrix}$$

Reachable subspaces

In decentralized control every control station will have associated with it a subspace of state variables which it is possible to reach from at least one of the inputs available form that control station.

The connections between control station I and the inputs are represented by the input matrix **BI**, and the connections between the inputs and the states are represented by the reachability matrix **R**. The subspace reachable from control station I will be represented by a reachability vector:

RUI ← v/R v.∧ **BI**; rui = +/[1] **RUI**

where rui indicates the number of reachable states.

The connections between the system outputs and control station I are represented by the output matrix **CI**. The subspace which is able to reach at least one of the outputs connected to control station I will be represented by the reachability vector:

RYI ← v/[1] **CI** v.∧ **R**; ryi = +/**RYI**

where ryi indicates the number of states, which can reach control station I through the outputs.

From these definitions, reduced forms of the system matrix **A** can be obtained. For every zero element RUI_j and RYI_k in the reachability vectors, all elements in the j-th row and column and the k-th row and column respectively of the A matrix are put equal to zero. The corresponding subsystem matrices are called **AUI and AYI.**

Returning to the decentralized control system in figure 5.2, the reachability matrix **R** (obtained by inspection) and the reachability vectors are:

$$\mathbf{R} = \begin{array}{c} \begin{matrix} x1 & x2 & x3 & x4 & x5 \end{matrix} \\ \begin{bmatrix} 1 & 0 & 0 & 0 & 0 \\ 1 & 1 & 1 & 0 & 0 \\ 0 & 0 & 1 & 0 & 0 \\ 0 & 0 & 0 & 1 & 0 \\ 0 & 0 & 0 & 0 & 1 \end{bmatrix} \end{array} \qquad \mathbf{R} \vee . \wedge \mathbf{B1} = \begin{array}{c} \begin{matrix} u11 & u12 \end{matrix} \\ \begin{bmatrix} 1 & 0 \\ 1 & 1 \\ 0 & 1 \\ 0 & 1 \\ 0 & 0 \end{bmatrix} \end{array} \qquad \mathbf{RU1} = \begin{array}{c} u1 \\ \begin{bmatrix} 1 \\ 1 \\ 1 \\ 1 \\ 0 \end{bmatrix} \end{array} \quad ru1 = 4$$

$$\mathbf{RY1} = \begin{bmatrix} 1 & 1 & 1 & 0 & 0 \end{bmatrix}$$

$$\mathbf{ry1} \leftarrow +/[1]\mathbf{RY1} = 3$$

$$\mathbf{RU2} = \begin{array}{c} u2 \\ \begin{bmatrix} 0 \\ 0 \\ 0 \\ 1 \\ 1 \end{bmatrix} \end{array} \quad ru2 = 2$$

$$\mathbf{RY2} = \begin{bmatrix} 0 & 0 & 0 & 1 & 1 \end{bmatrix}$$

$$\mathbf{ry2} = +/[1]\mathbf{RY2} = 2$$

The states 1,2,3, and 4 can be reached from control station 1. The states 4 and 5 can be reached from control station 2. The states 1,2, and 3 can reach control station 1, and the states 4 and 5 can reach control station 2.

The corresponding subsystem matrices are:

$$\mathbf{AU1} = \begin{bmatrix} 0&0&0&0&0 \\ 1&0&1&0&0 \\ 0&0&0&0&0 \\ 0&0&0&0&0 \\ 0&0&0&0&0 \end{bmatrix} \mathbf{AU2} = \begin{bmatrix} 0&0&0&0&0 \\ 0&0&0&0&0 \\ 0&0&0&0&0 \\ 0&0&0&0&0 \\ 0&0&0&0&1 \end{bmatrix} \mathbf{AY1} = \begin{bmatrix} 0&0&0&0&0 \\ 1&0&1&0&0 \\ 0&0&0&0&0 \\ 0&0&0&0&0 \\ 0&0&0&0&0 \end{bmatrix} \mathbf{AY2} = \begin{bmatrix} 0&0&0&0&0 \\ 0&0&0&0&0 \\ 0&0&0&0&0 \\ 0&0&0&0&0 \\ 0&0&0&0&1 \end{bmatrix}$$

Controllable and observable subspaces

In this brief introduction to structural design of decentralized control systems we shall restrict ourselves to the case of two control stations. A generalisation of the results for more than two control stations is possible but becomes increasingly complex.

For each control station we must examine if its inputs can potentially control its reachable subspace. It can be shown that the reachable subspace **AUI** is controllable if and only if the number of states in the reachability vector RUI is equal to the term rank of the augmented matrix

[AUI, BI]

If the term rank is less than the number of reachable states rui, the controllable subspace can be chosen in several ways, including a number of states equal to the term rank of the augmented matrix. The controllable subspaces can be represented by a number of column vectors representing the controllable states:

1KI, 2KI, ...

These vectors are determined by the maximum permutation matrices, which can be identified in the augmented matrix.

The total system is completely potentially controllable from the two control stations in combination if it is possible to chose a pair of controllable subspaces, i.e.:

iK1 v **jK2** = unit vector

For each control station we must also examine if the states which can reach the outputs connected to that control station also can be observed. It can be shown that the subspace **AYI** which can reach control station I is potentially observable if the number of states in the reachability vector RYI is equal to the term rank of the augmented matrix:

$$\begin{bmatrix} \mathbf{AYI} \\ \mathbf{CI} \end{bmatrix}$$

If the term rank of the augmented matrix is less than the number of states ryi in the reachability vector RYI, the observable subspace can be chosen in several ways. The observable subspaces, which have a number of states equal to the term rank of the augmented matrix, can be represented by the row vectors:

1MI, 2MI, ...

These vectors can be determined by the maximum permutation matrices which can be identified in the augmented matrix.

The total system is completely potentially observable from the two control stations in combination if a pair of compatible subspaces can be chosen such that:

iM1 v jM2 = unit vector

Returning again to the decentralized control system shown in fig. 5.1, the controllable subspaces for control station 1 is determined by the augmented matrix:

$$[\mathbf{AU1},\ \mathbf{B1}] = \begin{bmatrix} 0&0&0&0&0&\textcircled{1}&0 \\ 1&0&\textcircled{1}&0&0&0&0 \\ 0&0&0&0&0&0&\textcircled{1} \\ 0&0&0&0&0&0&1 \\ 0&0&0&0&0&0&0 \end{bmatrix} = \begin{bmatrix} 0&0&0&0&0&\textcircled{1}&0 \\ \textcircled{1}&0&1&0&0&0&0 \\ 0&0&0&0&0&0&1 \\ 0&0&0&0&0&0&\textcircled{1} \\ 0&0&0&0&0&0&0 \end{bmatrix}$$

The term rank of the augmented matrix is 3, and less than ru1 = 4. The control station 1 cannot potentially control its reachable subspace. The maximum permutation matrix can be chosen in two different ways as indicated by the two sets of ringed elements:

$$1K1 = [1\ \ 1\ \ 1\ \ 0\ \ 1]^t \quad \text{or} \quad 2K1 = [1\ \ 1\ \ 0\ \ 1\ \ 0]^t$$

Only three state variables can be controlled, and we can either choose the state variables 1,2, and 3, or 1,2, and 4.

The controllable subspaces for control station 2 are determined by the augmented matrix:

$$[\mathbf{AU2},\ \mathbf{B2}] = \begin{bmatrix} 0&0&0&0&0&0&0 \\ 0&0&0&0&0&0&0 \\ 0&0&0&0&0&0&0 \\ 0&0&0&0&0&0&\textcircled{1} \\ 0&0&0&0&0&\textcircled{1}&1 \end{bmatrix} \qquad 1K2 = \begin{bmatrix} 0 \\ 0 \\ 0 \\ 1 \\ 1 \end{bmatrix}$$

The term rank of the augmented matrix is 2 and equal to ru2. Therefore the control station 2 potentially can control its reachable subspace.

The total system is completely potentially controllable from the two control stations in combination, because it is possible to choose a pair of compatible subspaces, namely 1K1 and 1K2.

The observable subspaces for control station 1 is determined by the augmented matrix:

$$\begin{bmatrix} \mathbf{AY1} \\ \mathbf{C1} \end{bmatrix} = \begin{bmatrix} 0 & 0 & 0 & 0 & 0 \\ \textcircled{1} & 0 & 1 & 0 & 0 \\ 0 & 0 & 0 & 0 & 0 \\ 0 & 0 & 0 & 0 & 0 \\ 0 & 0 & 0 & 0 & 0 \\ 0 & \textcircled{1} & 0 & 0 & 0 \end{bmatrix} = \begin{bmatrix} 0 & 0 & 0 & 0 & 0 \\ 1 & 0 & \textcircled{1} & 0 & 0 \\ 0 & 0 & 0 & 0 & 0 \\ 0 & 0 & 0 & 0 & 0 \\ 0 & 0 & 0 & 0 & 0 \\ 0 & \textcircled{1} & 0 & 0 & 0 \end{bmatrix}$$

The term rank of the augmented matrix is 2, and less than the number of states ry1 in RY1. Therefore, the observable subspace for control station 1 can be chosen in two ways, as indicated by the two sets of ringed elements in the augmented matrix:

1M1 = [1 1 0 0 0] or 2M1 = [0 1 1 0 0]

Only two state variables can be observed, either state 1 and 2, or state 2 and 3.

The observable subspaces for control station 2 is determined by the augmented matrix:

$$\begin{bmatrix} \mathbf{AY2} \\ \mathbf{C2} \end{bmatrix} = \begin{bmatrix} 0 & 0 & 0 & 0 & 0 \\ 0 & 0 & 0 & 0 & 0 \\ 0 & 0 & 0 & 0 & 0 \\ 0 & 0 & 0 & 0 & 0 \\ 0 & 0 & 0 & 0 & \textcircled{1} \\ 0 & 0 & 0 & \textcircled{1} & 1 \end{bmatrix} \qquad 1M2 = [0\ 0\ 0\ 1\ 1]$$

The term rank is 2 and equal to the number of states ry2 in RY2. Therefore control station 2 potentially can observe the subspace, which reaches the output 2.

The total system is not potentially observable from the two control stations in combination, because it is not possible to choose a pair of compatible subspaces.

Decentralized controllability and observability

Up to this point we have considered the two control stations as independant of each other. Henceforth it is assumed that the two control stations in combination are able to reach the state space of the total system through the inputs of the system, and that the total state space is able to reach the two stations in combination through the outputs of the system.

In each control station is now established a local feedback, i.e. the output $\underline{y}_1$ is connected to $\underline{u}_1$ and the output $\underline{y}_2$ is connected to the input $\underline{u}_2$.

The first question now is whether the introduction of local feedbacks will extend the reachable subspaces of each control station to the total state space of the system. Assuming that the two-control-stations in combination can reach the whole state space, the two control station system is completely reachable from con- trol station 1 if and only if

$$RU1 \wedge RY2^t \neq 0$$

i.e. the two reachable subspaces should at least have one common state.

Similarly, in any two-station-system the outputs connected to control station 1 can be reached from the whole state space, if and only if:

$$RY1^t \wedge RU2 \neq 0$$

i.e. the two subspaces should have at least one common state.

The second question is whether the introduction of local feedbacks in a two-station-system will extend the controllable subspaces of each control station to the total state space of the system. Assuming that the reachability condition above is fulfilled, the two station system is completely potentially controllable from control station 1, if and only if the term rank of the combined matrix:

$$TR \begin{bmatrix} A & B1 \\ C2 & 0 \end{bmatrix} \quad \text{is} \geqslant n \quad \text{(the order of the total system)}$$

The third question is whether the introduction of local feedbacks in a two-station-system will extend the observable subspaces of each control station to the total state space of the system. Assuming that the reachability condition above is fulfilled, the two-station-system is completely potentially observable from control station 1, if and only if the term rank of the combined matrix:

$$TR \begin{bmatrix} A & B2 \\ C1 & 0 \end{bmatrix} \quad \text{is} \geqslant n \quad \text{(the order of the total system)}$$

The conditions for complete potential controllability and observability from control station 2 can easily be derived from the conditions above by interchanging the indices 1 and 2.

The condition for controllability from station 2 is identical with the condition for observability from control station 1.

The condition for observability from station 2 is identical with the condition for controllability from station 1.

This is example of the "duality", that pervades the whole subject of controllability and observability.

In the two station system shown on figure 5.2, the control station 1 can control the whole system with decentralized feedback control, because:

$$RU1 \wedge RY2_t = [0 \quad 0 \quad 0 \quad 1 \quad 0] \neq 0$$

and the term rank of the combined matrix is equal to the order of the system:

$$TR \begin{bmatrix} A & B1 \\ C2 & 0 \end{bmatrix} = TR \begin{bmatrix} 0 & 0 & 0 & 0 & 0 & \textcircled{1} & 0 \\ \textcircled{1} & 0 & 1 & 0 & 0 & 0 & 0 \\ 0 & 0 & 0 & 0 & 0 & 0 & \textcircled{1} \\ 0 & 0 & 0 & 0 & 0 & 0 & 1 \\ 0 & 0 & 0 & 0 & \textcircled{1} & 0 & 0 \\ 0 & 0 & 0 & \textcircled{1} & 1 & 0 & 0 \end{bmatrix} = 5 = n$$

Due to the duality mentioned above the control station 2 is potentially able to observe the total state space.

However control station 2 cannot control the whole system with decentralized feedback control, because its controllable subspace is not enlarged by the local feedback in control station 1; the condition:

$$RU2 \wedge RY1^{t} = [0 \;\; 0 \;\; 0 \;\; 0 \;\; 0] \neq 0$$

is not fulfilled because the subspaces RU2 and RY1 have no states in common.

Consequently, the control station 1 cannot potentially observe the total state space.

The example in figure 5.2 is rather simple, but it illustrates the use of the structural design rules without overly complicated calculations. From the conclusions above, it is clear that the structure of the system should be changed, e.g. by connecting the control variable u_2 to one of the states 1,2, or 3.

6. Conclusions

The structural design methods for decentralized control systems determine if the structure of the system to be controlled and the proposed local feedback loops are sufficient to make controllability and observability possible under nearly all conditions, and in this way they supplement efficiently the numerical methods, which determine the specified dynamical properties of the system.

References

Mason, S.J. (1953) Feedback Theory - Some Proporties of Signal Flow Graphs. Proc. IRE 41, pp 1144-56.

Mason, S.J. (1956) Feedback Theory - Further Proporties of Signal Flow Graphs. Proc. IRE, p 920.

Harary, F. (1959) A Graph Theoretic Method for the Complete Reduction of a Matrix with a view towards Finding its Eigenvalues. J. Math. and Phys. 38. pp 104-111.

Kalman, R.E. (1960) On the General Theory of Control Systems. Proc. 1st IFAC Congress, Moscow 1960. Vol 1, pp 481-92. Butterworth, London.

Evans, F.J. & Larsen, P. Martin (1983) Structural Design of Control Systems. Lecture Notes, Electric Power Eng. Dept. Tech.Univ. DK

III. ALGORITHMS FOR ADVANCED ARCHITECTURES

Matrix Computations on Shared–Memory Multiprocessors[1]

K. A. Gallivan and A. H. Sameh

Center for Supercomputing Research and Development

and

Department of Computer Science

University of Illinois

Urbana, Illinois 61801

1. Introduction

The main driving force for higher supercomputer performance is the fact that some important applications in engineering and science currently consume excessive amounts of time or are infeasible to attempt at all on available vector processors. To describe physical phenomena, one must resort to simulation of complex models on the computer. The closer the model is to a physical phenomenon, the more extensive are the required computational resources. Important uses of parallel processors are in the simulation of gauge theory and elementary particle physics, multidimensional semiconductor devices, electronic circuits, weather circulation, and oil reservoirs, as well as studies in chemical quantum dynamics and molecular scattering, seismic imaging and dynamic structural analysis.

On the basis of experiments with automatic program restructuring for multiprocessing, it has been found that potential bottlenecks in taking advantage of parallelism in these applications are most obvious from analyzing a few important building blocks. One such building block includes the various numerical linear algebra algorithms. This brief survey reflects our efforts in designing such algorithms for a multiprocessor with hierarchical shared memory.

[1]This work was supported in part by the **National Science Foundation** under Grants No. US NSF DCR84-10110 and US NSF DCR85-09970, the US **Department of Energy** under Grant No. US DOE DE-FG02-85ER25001, the **Air Force Office of Scientific Research** under Grant No. AFOSR–85–0211, and the IBM Donation.

NATO ASI Series, Vol. F47
Advanced Computing Concepts and Techniques
in Control Engineering
Edited by M. J. Denham and A. J. Laub

We present algorithms for dense matrix computations contained in the Cedar scientific library (which has the functionality of the well-known mathematical packages Linpack and Eispack) and the strategies used to design and analyze them. These algorithms include block LU and block orthogonal factorization schemes as well as algorithms for handling the symmetric dense eigenvalue problem and the singular value decomposition. These algorithms depend in a significant way on some modules for handling basic matrix operations which are designed and implemented on the Alliant FX/8 (a single Cedar cluster) so as to manage the hierarchical memory system as efficiently as possible. We hasten to add, however, that the algorithms presented here may have to be redesigned for maximum efficiency on different architecture. The use of such basic modules, which can be finely tuned for various machines, is not new. The concept originated with the well-known BLAS in Linpack which dealt only with vector operations. Later, these were extended to matrix-vector operations (BLAS2) for better register management on register-to-register vector machines. We call our matrix modules BLAS3. In fact, an effort is currently underway to establish standards for such modules [DDDH87].

Our survey also deals with the important problem of solving large narrow-banded, or block tridiagonal, systems. We present an algorithm that could make use of the efficient dense system solvers mentioned earlier. Finally, we devote a section to elliptic problem solvers. First, we introduce a domain decomposition-based rapid solver that relies on FFTs, and a scheme that modifies the well-known block cyclic reduction algorithm so as to enhance parallelism. Two iterative schemes are also outlined for handling elliptic difference equations: a preconditioned conjugate gradient algorithm for self-adjoint problems and a block Kaczmarz scheme for nonself-adjoint problems. This latter scheme proves to be more robust and more amenable to parallel processing than other preconditioned CG-like algorithms which are available in the literature.

For some of these algorithms we present generalizations suitable for implementation using multiple Cedar clusters.

2. Architectures and Models for Performance Analysis

In this section, descriptions of the target architecture and the methodology used to analyze the behavior of the primitives and block algorithms are presented.

2.1. Target Architecture

The architectures of interest in this paper are multivector processors with a hierarchical shared memory system. Such an architecture typically has p vector processors which share a small fast cache (whose size in words is denoted CS below) and a larger slower global memory. An analysis for a system with private caches can be derived using arguments contained in this paper. The p vector processors are assumed to be connected to the shared cache in such a way that a processor can read or write any location in the cache. It is also assumed that the processors are provided with a synchronization mechanism which allows MIMD execution.

An example of such an architecture which is commercially available is the ALLIANT FX/8. This machine is being used as the basis for a single cluster of the Cedar machine, [KDLS86], [Yew86], see Figure 1. Eight pipelined computational elements (CE's) allow processing at a rate of 68 Mflops including vector startup. Each CE contains eight 64–bit vector registers of length thirty–two. A concurrency control bus connects the eight CE's and acts as a synchronization facility. The CE's share a 128K–byte write–back cache which allows up to eight simultaneous 64–bit word accesses per 170ns cycle. The cache is connected to memory by a memory bus which is capable of delivering up to four 64–bit words per cycle, i.e. half of the cache–CE bandwidth, but, in practice, the rate achieved is less mainly due to memory–cache bus traffic and cache–miss management.

2.2. Methodology

In order to make the investigation of the effect of data locality, concurrency and vectorization on the performance of an algorithm tractable, a decoupling methodology is used. In this methodology, two time components, whose sum is the total time for the algorithm, are analyzed separately. A region in the parameter space, i.e. the space of possible blocksize choices, which provides near–optimal behavior is produced for each time component. The intersection of these two regions yields a set of blocksizes which should give near–optimal performance for the time function as a whole. These regions and their intersection are characterized in terms of certain architectural parameters in such a way as to allow conclusions to be drawn concerning the behavior of the algorithm across the class of multivector processors with a shared hierarchical memory system.

The first component considered is called the *arithmetic time* and is denoted T_a. This time represents the raw computational speed of the algorithm and is derived by ignoring the hierarchical nature of the memory system: it is the time required by the algorithm given that the cache is infinitely large and contains all of the initial data. Given a schedule of the computations and instruction timings for a machine it is a straightforward albeit tedious task to write T_a in terms of the blocksizes used and architectural parameters of the processors. One could then, theoretically, attempt to determine the optimal choice of blocksizes. In some cases, such as the matrix multiplication primitive, this task is reasonable; in others, such as the block LU decomposition, deriving a detailed function is not practical and one must settle for general observations which yield near–optimal values of T_a. A simplified form of T_a is used later in this section. This form is derived from the fact that regardless of the complexity of the scheduling the arithmetic time must essentially be proportional to the number of arithmetic operations performed. Therefore, the arithmetic time component can be written

$$T_a = n_a \tau_a, \tag{2.1}$$

where n_a is the number of operations and τ_a is the average time for a single operation.

The second component of the time function considered is the degradation of the raw computational speed of the algorithm due to the use of a cache of size CS and a slower main memory. This component is called the *data loading overhead* and is denoted Δ_l. The component Δ_l is proportional to the number of data transfers, from memory to cache, required by the algorithm; therefore, the total time for the algorithm is

$$T = T_a + \Delta_l = n_a \tau_a + n_l \tau_l, \tag{2.2}$$

where n_l is the number of data transfers and τ_l is the proportionality constant or the 'average' time for a data load. *Note that no assumptions have been made concerning the overlap of computation and the loading of data in order to write T as a sum of these two terms.* The effect of such overlapping is seen through a reduction in τ_l. This overlap effect can cause τ_l to vary from 0 for machines which have a perfect prefetch capability from memory to cache to t_l, where t_l is the amount of time it takes to transfer a single data element, for machines which must fetch data sequentially from memory to cache.

Clearly, the function Δ_l is, due to τ_l, even more complicated than T_a so there is little hope making general statements concerning Δ_l directly. But the more modest goal of producing a region in the parameter space where the relative cost of the data loading $\Delta_l \ / \ T_a$ is small, can be achieved independent of the complexity of τ_a and τ_l. This analysis is accomplished by expressing $\Delta_l \ / \ T_a$ in terms of two ratios: a *cache-miss* ratio and a *cost* ratio. Specifically,

$$\frac{\Delta_l}{T_a} = \lambda \, \mu \tag{2.3}$$

where $\mu = n_l / n_a$ is the cache–miss ratio and $\lambda = \tau_l / \tau_a$ is the cost ratio. The product in (2.3) shows the complementary nature of prefetch mechanisms and caches. For algorithms which have predictable data fetch patterns, implementation on a machine with an appropriate prefetch mechanism can reduce the factor λ so that the cache–miss ratio becomes less important. Similarly, for algorithms which do not lend themselves to extensive prefetching or for machines with limited prefetch capability the cache–miss ratio is the term of most interest.

For most machines in the class of multivector processors with hierarchical shared memory, λ can be bounded based on worst case hardware considerations. The analysis of the impact of the hierarchical memory, therefore, reduces to the consideration of the behavior of the cache–miss ratio μ in terms of the blocksizes used to implement the algorithm. It is assumed throughout the analysis of μ that the cache is controlled by an optimal replacement policy or is fully user controlled.

The utility of the results of the decoupling form of analysis depends upon the fact that the intersection of the near–optimal regions for each term is not empty or at least that the arithmetic time does not become unacceptably large when using parameter values in the region where small relative costs for data loading are achieved. For some algorithms, as is shown below, this is not true; reducing the arithmetic time may directly conflict with reducing the relative cost of data loading. As observed above, the minimization of T_a is highly machine dependent. Therefore, there is no hope of developing a strategy, applicable for a large number of machines, of altering the algorithm so that the near–optimal region for T_a moves in the parameter space to create an intersection with the near–optimal region for Δ_l. The minimization of Δ_l / T_a is, however, fairly generic across machines with a shared memory hierarchy and a technique, called multilevel blocking, is presented in this paper which resolves the conflicts between the arithmetic time and the data loading time by altering the algorithm to shift the near–optimal region for Δ_l to create an intersection with the near–optimal region for T_a.

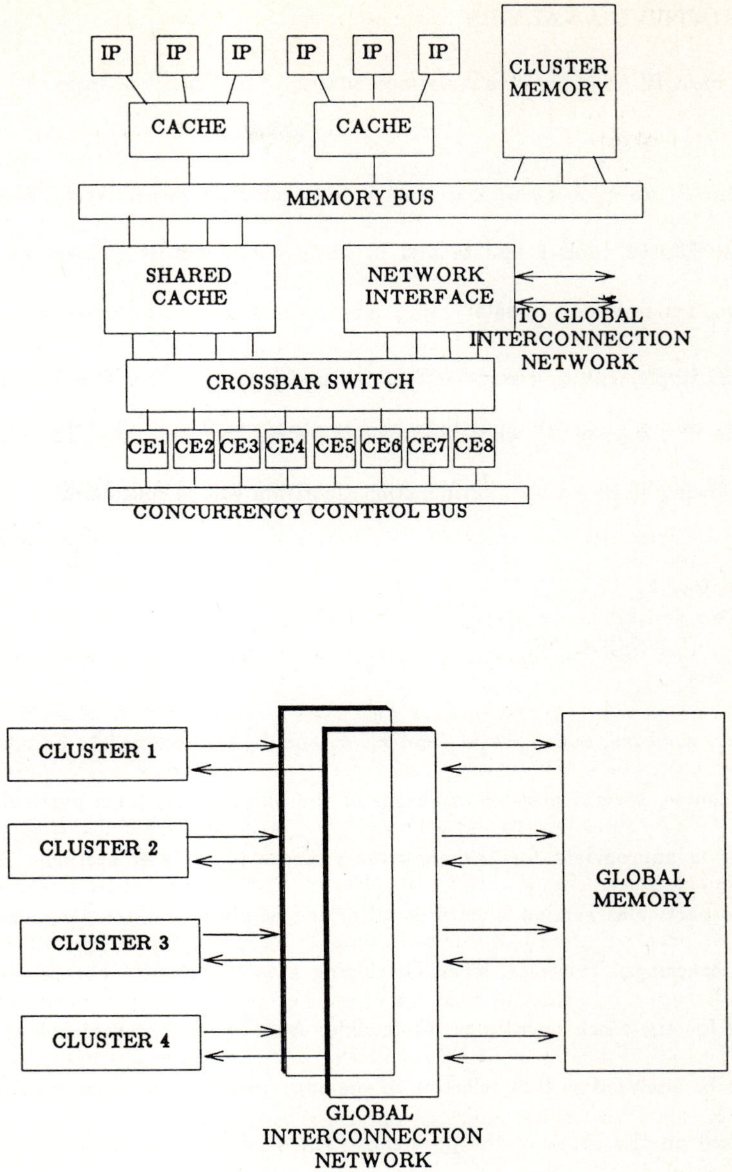

Figure 1. A Four Cluster CEDAR

3. BLAS3 PRIMITIVE ANALYSIS

The most basic BLAS3 primitive is a simple matrix operation of the form

$$C \leftarrow C + AB,$$

where C, A and B are $n_1 \times n_3$, $n_1 \times n_2$ and $n_2 \times n_3$ matrices respectively. When $n_2 = 1$ the primitive is the BLAS2 rank–1 update and in block algorithms it is most often used with $n_2 = \omega \ll n_1, n_3$, i.e. a rank–$\omega$ update.

The BLAS3 implementation assumes that the matrices C, A and B have been partitioned into submatrices C_{ij}, A_{ik} and B_{kj} whose dimensions are $m_1 \times m_3$, $m_1 \times m_2$ and $m_2 \times m_3$ respectively. The following block matrix multiplication algorithm will be considered

```
do i=1,k1
  do k=1,k2
    do j=1,k3                                   (3.1)
      Cij=Cij+Aik*Bkj
end do
```

where $n_1 = k_1 m_1$, $n_2 = k_2 m_2$ and $n_3 = k_3 m_3$ and k_1, k_2 and k_3 are assumed to be positive integers. There are, of course, several possible orderings of the block loops. This particular ordering is chosen since it is appropriate for the algorithms discussed in later sections. Moreover, the analysis of this particular version illustrates all of the points which are pertinent to investigations of other orderings. However, when developing a robust BLAS3 library, as was done at CSRD, kernels for the block operations which differ from those discussed below and alternate orderings must be analyzed so that selection of the appropriate form of the routine can be done at runtime based on the *shape* of the problem. (The $i-j-k$ ordering of the block loops, for example, produces distinctly different blocksizes and shapes.) This is especially important for cases with extreme shapes, e.g. guaranteeing smooth performance characteristics as the shapes become BLAS2–like.

The block operations $C_{i,j}=C_{i,j}+A_{i,k}*B_{k,j}$ possess a large amount of concurrent and vectorizable computations, so the algorithm proceeds by first partitioning the matrices and then dedicating the full resources of the p vector processors to each of the block operations in turn.

The following sections contain summaries of the analyses of the arithmetic time and the data loading overhead according to the model presented in the previous section [GJMS87]. The objective of these analyses is to determine values for the blocking parameters m_1, m_2 and m_3 which yield high–performance on a multivector processor with a hierarchical shared memory.

3.1. Arithmetic time optimization

The basic block operation kernel uses the algorithm

```
do r=1,m3
   do s=1,m1
      do t=1,m2
         c(s,r)=c(s,r)+a(s,t)b(t,r)
end do
```

(3.2

where c_{sr}, a_{st} and b_{tr} denote the elements of C_{ij}, A_{ik} respectively B_{kj}.

Parallelism is introduced by distributing the m_3 iterations of the r–loop across the p processors. Within each processor the s–loop is vectorized by performing an m_2–adic operation ($v_0 \leftarrow \sum_1^{m_2} \alpha_i v_i$); the product of the submatrix A_{ik} with a column of B_{kj}. The use of an m_2–adic operation allows the efficient use of the vector registers and memory bandwidth of each processor by accumulating the results of m_2 vector operations in registers. It is also possible with this operation to exploit the chaining of the multiplier, adder and data fetch available on many multivector processors.

Values of m_1,m_2, and m_3 which yield near–optimal values of the arithmetic time for the kernel are highly dependent on the particular machine under consideration, but the following discussion for register–to–register architectures illustrates the reasoning required. The value of m_1 is the length of the vector operations used in the kernel algorithm. Therefore, if m_1 is small it should be taken to be a multiple of the length of the vector registers. For larger values of m_1, performance is not as sensitive to nonmultiples of register length and any convenient value may be used. Since m_2 is the order of the m_2-adic operation performed by each processor, it should be chosen large enough to achieve near–peak performance of the operation. Since concurrency is used on the columns of the submatrices of B, the value of m_3, if small, should be a multiple of the number of processors available or large enough so that the processors are all kept busy for a significant fraction of the time.

The ALLIANT FX/8 is a multivector processor with a shared hierarchical memory upon which the Cedar cluster is based [Yew86]. For this machine the values of m_1, m_2 and m_3 chosen according to the preceding reasoning are: $m_1 = 32k$ or $m_1 > 160$; $m_2 \geq 32$; and $m_3 = 8k$ or is large.

3.2. Data Load Overhead Optimization

Since the submatrices A_{ik} are associated with the inner loop, it is assumed that each A_{ik} is loaded once and kept in cache for the duration of the j loop. Similarly, it is assumed that each of the C_{ij} and B_{kj} are loaded into cache repeatedly. It is easily seen then that the number of loads required is

$$L = k_1 k_2 [m_1 m_2 + k_3 (m_1 m_3 + m_2 m_3)] \qquad (3.3)$$
$$= n_1 n_2 + n_1 n_2 n_3 \, \rho(m_1, m_2),$$

where $\rho(m_1, m_2) = m_1^{-1} + m_2^{-1}$.

There are constraints associated with this load function. These constraints are generated by determining what amount of data must fit into cache at any given point in time and requiring that this quantity be bounded by the cache size CS.

Due to the assumptions above concerning A_{ik} the constraint must contain the term $m_1 m_2$. There is also a term due to the m_2-adic operation. If $m_1 > rv_l$, where r is the number of vector registers used to accumulate the m_2-adic operation and v_l is the length of a single vector register then the column of B_{kj} associated with the m_2-adic operation must be kept in cache since it is used more than once. Whether or not m_1 is large enough to cause the inclusion of this extra term in the constraint is, of course, implementation dependent, but in the sequel the conservative approach is taken and the term is included in the constraint. The above considerations, as well as the fact that the submatrices cannot be larger than the matrices being multiplied, imply that the minimization of the number of loads performed by the BLAS3 primitive is equivalent to the solution of the minimization problem

$$\min \rho(m_1, m_2) \tag{3.4}$$

subject to

$$m_2(m_1 + p) \leq CS$$

$$1 \leq m_1 \leq n_1$$

$$1 \leq m_2 \leq n_2,$$

where CS is the cache size (in floating point word size) and p is the number of processors. The constraints trace a rectangle and an hyperbola in the (m_1, m_2)-plane.

The resulting optimal blocksizes for the BLAS3 matrix multiplication primitive can be summarized as:

(i) The value of m_3 is arbitrary.

(ii) If $n_2(n_1 + p) \leq CS$ (Regime 1)

$$m_1 = n_1 \text{ and } m_2 = n_2.$$

(iii) If $n_2(n_1 + p) > CS$ and $n_2 \leq CS(\sqrt{CS} + p)^{-1}$ (Regime 2)

$$m_1 = \frac{CS}{n_2} - p \quad \text{and} \quad m_2 = n_2.$$

(iv) If $n_2(n_1 + p) > CS$ and $n_1 \leq \sqrt{CS}$ (Regime 3)

$$m_1 = n_1 \quad \text{and} \quad m_2 = \frac{CS}{n_1 + p}.$$

(v) If $n_2(n_1 + p) > CS$, $n_1 > \sqrt{CS}$ and $n_2 > CS(\sqrt{CS} + p)^{-1}$ (Regime 4)

$$m_1 = \sqrt{CS} \quad \text{and} \quad m_2 = \frac{CS}{\sqrt{CS} + p}.$$

These four regimes are shown plotted in the (n_1,n_2)–plane in Figure 2. The number of loads L and the cache–miss ratio μ are given by

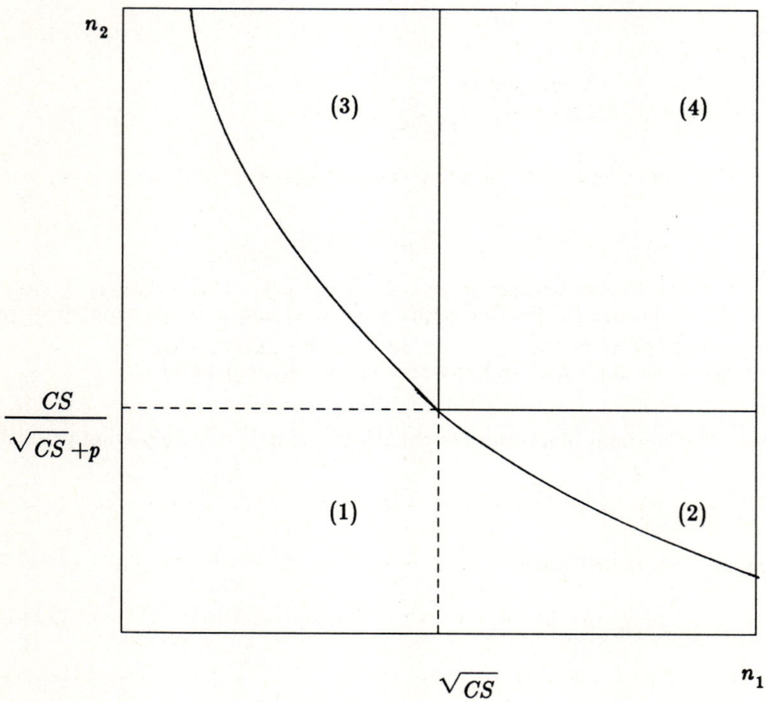

Figure 2. The Four Regimes

$$L = n_1 n_2 + n_1 n_2 n_3 (\frac{1}{m_1} + \frac{1}{m_2}) \tag{3.5a}$$

and

$$\mu = \frac{1}{2m_1} + \frac{1}{2m_2} + \frac{1}{2n_3}. \tag{3.5b}$$

Note that since the near–optimal region for the arithmetic time component was unbounded in the positive direction, there is a nontrivial intersection between it and the near–optimal region for the data loading component. This implies that, except for some boundary cases where n_1, n_2, and/or n_3 become small, the decoupling methodology does yield a strategy which can be used to choose near–optimal blocksizes for BLAS3 primitives. (The troublesome boundary cases can be handled by altering the block–loop ordering or choosing a different form of the block multiplication kernel.)

3.3. Observations on the Results of the Analysis

The minimum number of loads possible for this BLAS3 primitive is

$$L_{\min} = n_1 n_2 + n_2 n_3 + n_1 n_3$$

which occurs when each data element is loaded exactly once. The number of operations required is $n_a = 2n_1 n_2 n_3$ and therefore the minimum cache–miss ratio possible is

$$\mu_{\min} = \frac{1}{2n_1} + \frac{1}{2n_2} + \frac{1}{2n_3}.$$

Note that the value of $\mu_{\min}$ is greater than or equal to 1/2 if any of the three dimensions is taken to be 1, i.e. as in a BLAS2 primitive. Therefore, since the theoretical lower bound on the number of loads is independent of the algorithm used to implement the primitive it is clear that BLAS2 primitives do not have the necessary data locality to deliver high–performance on systems with a hierarchical shared memory. Indeed, the same can be said about BLAS3 primitives in some parts of regime (1). In this regime, the intrinsic data locality of the primitive is too small and achieving the minimum cache–miss ratio does not guarantee high performance.

The key observation with respect to the behavior of μ for BLAS3 primitives is that it is a decreasing function over the regimes, as n_1 and n_2 increase, which achieves its global minimum in regime (4) where it is independent of n_1 and n_2,

$$\mu = \frac{1}{\sqrt{CS}} + \frac{p}{2CS} + \frac{1}{2n_3}. \tag{3.6}$$

Therefore, assuming that n_3 is much larger than $\sqrt{CS}$, the minimum cache–miss ratio achievable for the loop ordering (3.1) is approximately $1/\sqrt{CS}$. This limit on the cache–miss ratio reduction due to blocking is consistent with the bound derived in [HoKu81]. Note also that for the often–used rank–ω update μ is proportional to $1/\omega$; therefore, for these updates and for block algorithms whose load behavior is dominated by them the return on blocking should be diminishing.

Exactly how large m_1 and m_2 must be in order to reduce the data loading overhead to an acceptable amount depends on the cost ratio λ of the machine under consideration, since the relative amount of time spent loading data is of the form $\lambda\mu$. The existence of a lower bound on the cache–miss ratio achievable by blocking does, however, have implications with respect to the blocksizes used in block versions of linear algebra algorithms. Blocksizes larger than $\sqrt{CS}$ will not improve the cache–miss ratio of the algorithm beyond the limit of $1/\sqrt{CS}$. The only reason for increasing the blocksize beyond this point and move farther into regime (4) is to improve the performance with respect to the arithmetic time component of the algorithm.

3.4. Experimental Results

In this section, the results of experiments run with single–cluster implementations of the rank–k update and dense matrix–matrix multiplication are presented.

As noted in Section 2, a single Cedar cluster is a slightly modified ALLIANT FX/8. This machine is a multivector processor with a 16K–word (64–bit word) cache shared by eight CE's. The ALLIANT conforms to the architectural model assumed in all ways except that the cache is hardware controlled using a direct–mapped strategy. Despite this fact, it is possible to demonstrate the trends predicted by the analyses of the previous sections.

The experiments were performed executing the particular kernel many times and averaging to arrive at an estimate of the time spent in a single instance of the kernel. This technique was used to minimize the experimental error present on the ALLIANT when measuring a piece of code of short duration. As a consequence of this technique, the curves have two distinct parts. The first is characterized by a peak of high performance. This is the region where the kernel operates on a problem which fits in cache. The performance rate in this region gives some idea of the arithmetic component of the time function. It is interesting to compare this peak to the rest of the curve which corresponds to the kernel operating on a problem whose data is initially in main memory. When the asymptotic performance in the second region is close to the peak in cache the number of loads are being managed effectively.

Figure 3 illustrates the effect of blocksize on the performance of the BLAS3 primitive $C \leftarrow C - AB$ where all three matrices are square and of order n. The blocksizes used for each curve are from low to high performance : $m_1 = 32$, $m_2 = 32$ and $m_3 = 32$; $m_1 = 64$, $m_2 = 64$ and $m_3 = 64$; and $m_1 = 128$, $m_2 = 96$ and $m_3 = n$. It is clear from the asymptotic performance of the top curve that a significant portion of peak performance can be achieved by choosing the correct blocksizes. In this case an asymptotic rate of just below 52 Mflops is achieved on a machine with a peak rate, including vector startup, of 68 Mflops.

Figures 4 and 5 show the performance of various rank–k updates on the ALLIANT. The parameters m_2 and m_3 are taken as k and n as recommended by the analysis of the BLAS3 primitive. The parameter m_1 is taken to be 96 and 128 in the two figures respectively. This parameter is kept constant for each figure to allow a fair comparison between the performances of the various kernels. Further, the BLAS3 analysis recommends $m_1 = (CS/k) - p$. In fact, for the values of k considered here, if $m_1 \geq 96$ then the term in the expression for the number of loads for the rank-k kernel which involves m_1 is not significant compared to the term involving m_2.

These curves clearly show that increasing k yields increased performance. Also note that the diminishing return on performance as the order of the update increases is clearly shown. Indeed, the $k = 96$ curve was not included in Figure 4 since it delivers performance virtually identical to the $k = 64$ kernel.

It is instructive to compare the performance of the rank-k kernel to typical BLAS and BLAS2 kernels. The BLAS kernels $\alpha = x^T y$ and $z = y \pm \alpha x$ both have cache-miss ratios of approximately 1 and achieve 11 Mflops and 7 Mflops respectively with their arguments in main memory. The BLAS2 kernels for a rank-1 update and a matrix-vector product have cache-miss ratios of approximately 1/2 and achieve 7 Mflops and 18 Mflops respectively. The superiority of the BLAS3 kernels on a single cluster of the Cedar machine is clear.

The astute reader will have noticed that the primitives of each of the two sets above have similar cache-miss ratios yet drastically different performances. This is due to the fact that all of the primitives are capable of saturating the main memory bandwidth and the poorer performing primitives must dedicate a significant portion of the bandwidth to writing results back to memory. The model from which the cache-miss ratio is derived only counts data fetches. The problem of write-back is only significant for kernels which have high cache-miss ratios and does not significantly affect the analysis of the BLAS3 primitives.

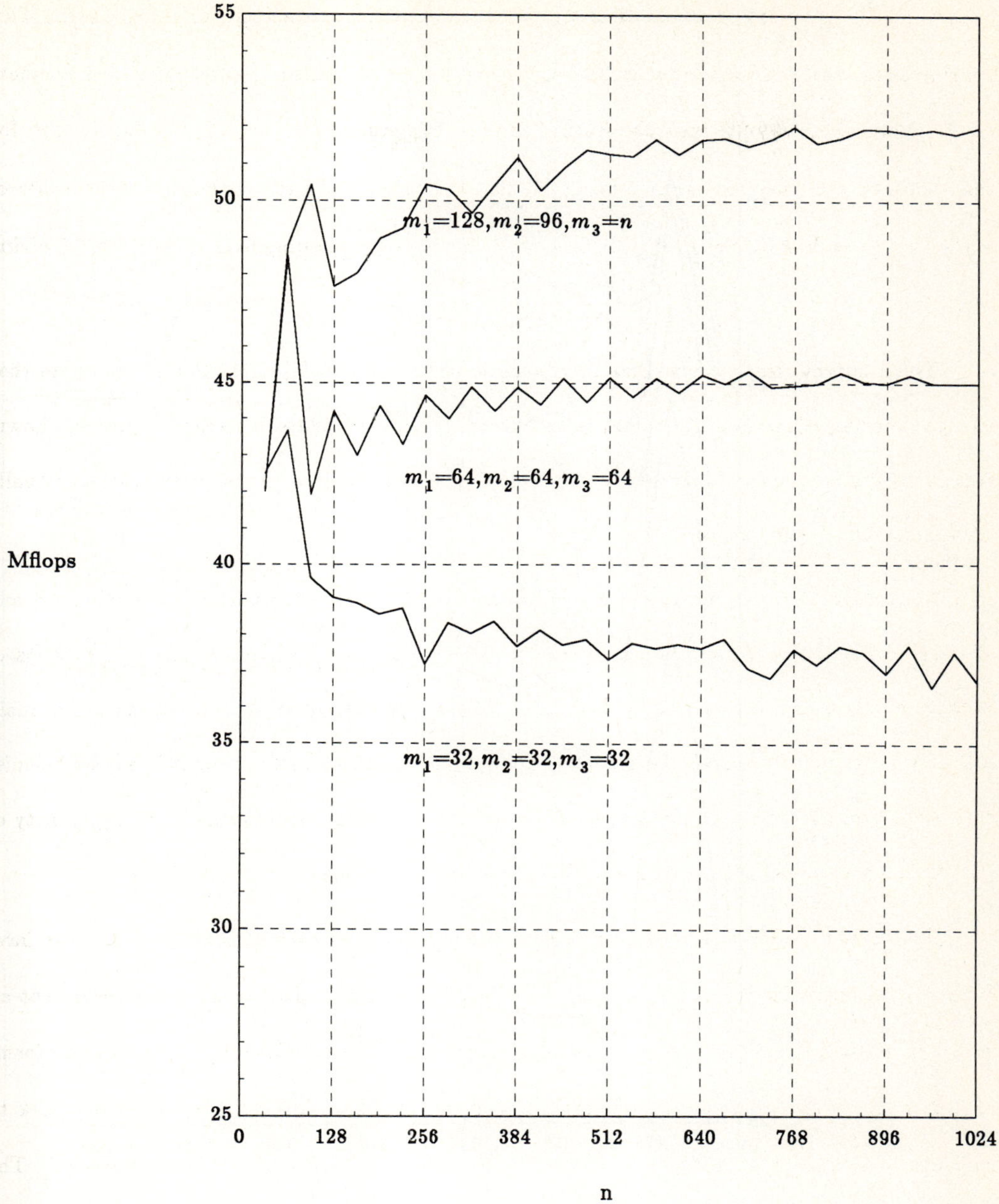

Figure 3. Dense Matrix Primitive Performance

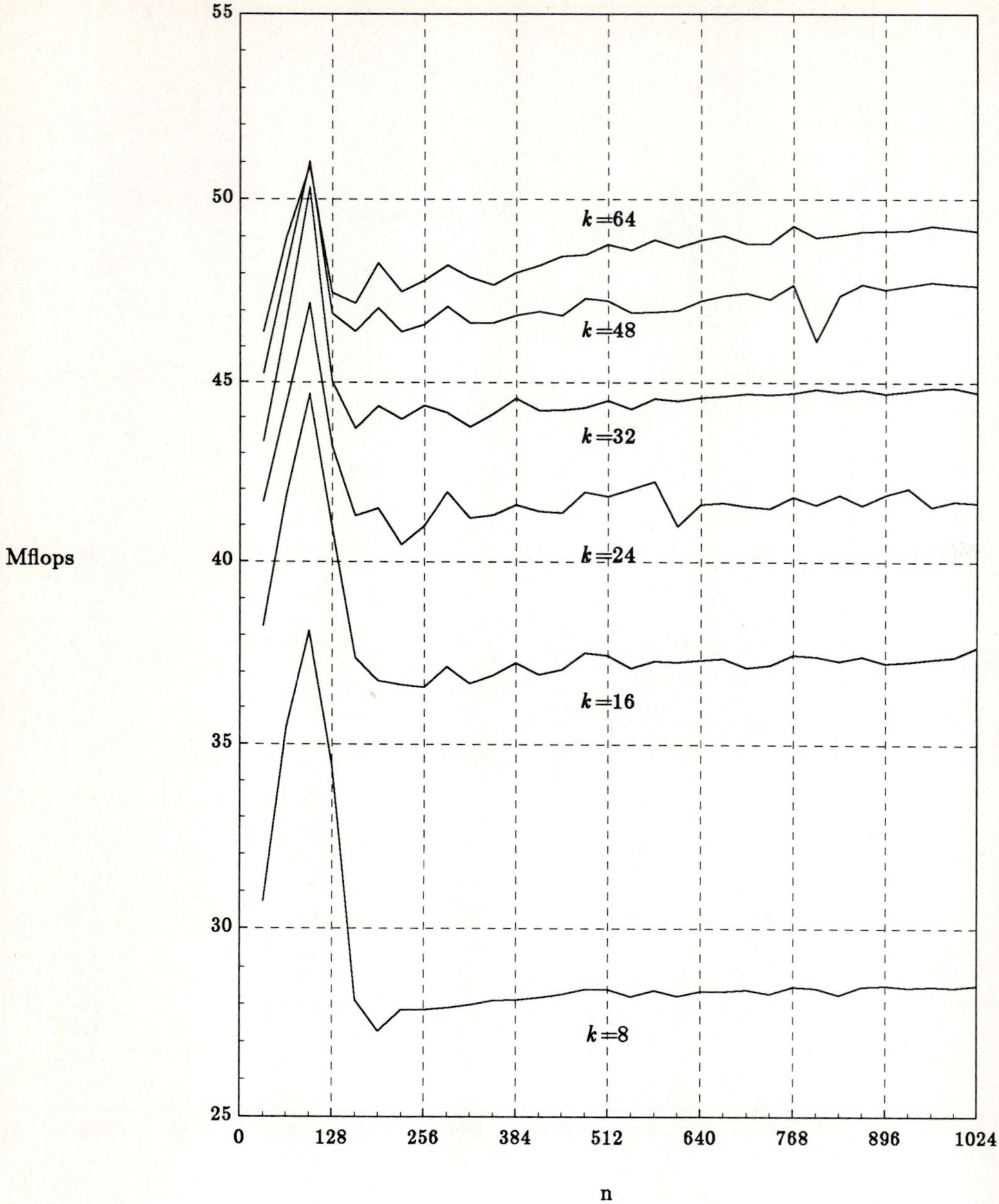

Figure 4. Rank–k Update, m1 = 96, m2 = k, m3 = n

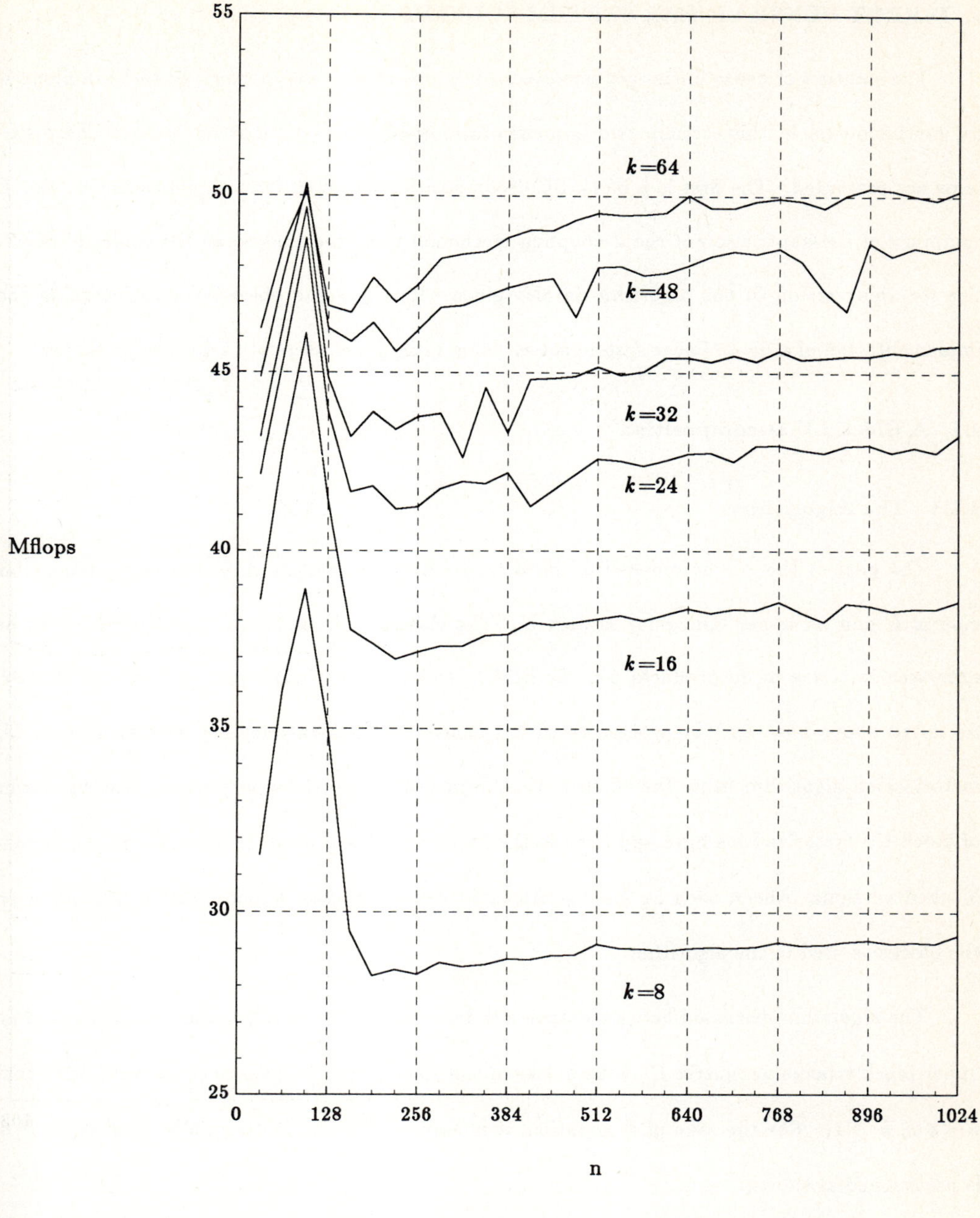

Figure 5. Rank–k Update, m1 = 128, m2 = k, m3 = n

4. DIRECT DENSE LINEAR SYSTEM SOLVERS

The solution of dense linear systems is certainly one of the most performed tasks in numerical mathematics. In this section, two implementations of direct methods for solving linear systems are presented. The first is a block LU decomposition designed for a single Cedar cluster. A summary of the application of the decoupling methodology to the analysis of the choice of blocksize for this version of the algorithm is also given. Finally some observations concerning the implementation of a dense linear system solver using multiple clusters of Cedar are presented.

4.1. A Block LU Decomposition

4.1.1. The Algorithm

The goal of the LU decomposition is to factor an $n \times n$–matrix A into a lower triangular matrix L and an upper triangular matrix U. The classical LU factorization, [GoVa83], can be expressed in terms of dotproducts, i.e. the BLAS. or in terms of rank one updates, i.e. BLAS2. As noted above however, for processors with a shared hierarchical memory system a block LU factorization algorithm using the efficient BLAS3 primitives must be performed. Various forms of block LU factorizations have appeared in the literature. Most produce the same result as non-blocked versions; others, such as the one discussed below, produce a factorization dependent on the blocksize used in the algorithm.

The algorithm discussed here decomposes A into a lower block triangular matrix L_ω and an upper block triangular matrix U_ω with blocks of the size ω by ω (it is assumed for simplicity that $n = k\,\omega$, $k > 1$). For the sake of illustration, it is assumed that A is diagonally dominant. If A is partitioned as shown

$$A = \begin{pmatrix} A_{11} & A_{12} \\ A_{21} & A_{22} \end{pmatrix} = \begin{pmatrix} I & 0 \\ L_{21} & I \end{pmatrix} \begin{pmatrix} U_{11} & U_{12} \\ 0 & B \end{pmatrix},$$

where A_{11} is square and of order ω, then the block LU algorithm is given by:

(i) $A_{11} \leftarrow A_{11}^{-1}$

(ii) $A_{21} \leftarrow L_{21} = A_{21} A_{11}$

(iii) $A_{22} \leftarrow B = A_{22} - L_{21} A_{12}$

(iv) Proceed recursively on the matrix B.

Note that the BLAS2 version of the classical LU decomposition is a special case of this algorithm ($\omega=1$). If A is positive definite the resulting decomposition is a block LDL^T. Of course, the algorithm can be altered to exploit the symmetry of A and the results of the analysis of this section can be applied directly since the number of loads and the number of operations scale appropriately.

Statements (i) and (ii) can be implemented in several ways depending upon assumptions made concerning the matrix A. Since A is assumed to be diagonally dominant, explicit inversion of the diagonal blocks can be done via the Gauss–Jordan algorithm [PeWi75] or an LU decomposition without pivoting. If the Gauss–Jordan kernel is used, as is assumed below, the block LU algorithm is more expensive by a factor of approximately $(1+2/k^2)$ than the classical LU factorization which requires about $2n^3/3$ operations. Note also that if A is not diagonally dominant partial pivoting is easily introduced in a manner similar to that used in the block algorithms which produce the standard LU factorization, e.g. [Cala86]. This increases the number of loads by a factor of $O(n^2)$ which is not significant to the analysis below. In general, the analysis of this paper holds for all forms of the algorithm, and, the simplifying assumption that A is diagonally dominant does not detract from the generality of the analysis.

Another organizational difference between the various forms of the algorithm that appear in the literature concerns the amount of information updated on each block step. Some versions update only the elements of the matrix necessary to compute the next block transformation. For the algorithm above, one could only update the first ω rows and columns of A_{22} rather than performing the entire update. Essentially such an approach trades a reduction in the amount of information written at each step for an increase in the amount read, which can be a valuable tradeoff on some machines.

For the remainder of the discussion of the block LU algorithm, it is assumed that $n = k\omega$, and $n^2 \geq CS$ where CS is the cache size and k is a positive integer.

4.1.2. Minimization of Arithmetic Time

There are two general aspects of the block LU decomposition through which ω influences the arithmetic time: the number of redundant operations; and the relationship, as a function of ω, between the performance of each of the primitives and the distribution of work among the primitives. These are considered in turn in this section.

As noted above, the number of operations in the block LU algorithm using the Gauss–Jordan primitive increases by a factor of $1 + 2/k^2$, where $k = n/\omega$, compared to the standard algorithm. Clearly, for small values of k a significant amount of redundant computations must be performed. Redundant operations in block methods are usually very costly in performance since they typically originate in low–performance kernels; in this case the Gauss–Jordan primitive. Of course, if the size of the system, n, is very large it is unlikely that ω will be large enough to make k small; but, for moderately sized systems, $n^2 \leq 10CS$, a small value of k is very likely as ω nears $\sqrt{CS}$. (It is shown in the next section that data loading considerations favor ω near $\sqrt{CS}$.)

The relationship, as a function of ω, between the performance of each of the primitives and the distribution of work among the primitives is usually the dominant concern in minimizing the arithmetic time component.

The number of operations performed by the Gauss–Jordan primitive is a quadratically increasing function in ω. The operations in the matrix multiplication primitive increase quadratically, achieving a maximum at $\omega \approx n/2$, and then decrease. The rank–ω primitive operation count is quadratically decreasing on the interval of interest. If θ is that value of ω for which all three primitives have effectively reached their peak performance then increasing ω beyond that value can cause a decrease in algorithm performance since, as ω increases operations from the two high–performance BLAS3 primitives are redistributed to the Gauss–Jordan primitive which is the worst performing of the three. The significance of the decrease depends upon the size of the performance difference between the Gauss–Jordan primitive and the BLAS3 primitives and the fraction of the number of operations required which are performed by the Gauss–Jordan primitive. This fraction is approximately $3/(k^2+2)$ which, as in the redundant computation discussion, is most significant when k is small.

Of course, the actual optimal blocksize for the arithmetic time component may not be θ but it is clear from this discussion that it can not be greater than θ. Furthermore, since ω dictates the length of vector operations in the Gauss–Jordan primitive and the degree of the m_2–adic operation in the BLAS3 primitives, θ should be significantly smaller than $\sqrt{CS}$ on most machines. Therefore, it is reasonable to conclude that the arithmetic time component decreases as ω increases until reaching some point which is usually much less than $\sqrt{CS}$. The significance of the increase in the arithmetic time component which results from increasing ω beyond this optimal blocksize is considerable for smaller values of k and inconsequential for large systems of equations.

4.1.3. Single Level Load Analysis

The cache–miss ratio of the single–level algorithm can be written as a weighted average of the cache–miss ratios of the various instances of each primitive. The weights are the ratio of the number of operations in the particular instance of the primitive to the total number of operations required. Specifically,

$$\mu = \gamma_{GJ}\,\mu_{GJ} + \sum_{i=1}^{k-1} \gamma_i^M \mu_i^M + \sum_{i=1}^{k-1} \gamma_i^R \mu_i^R,$$

where $n = k\omega$; γ_{GJ}, γ_i^M and γ_i^R are the operation count ratios for the k instances of the Gauss–Jordan primitive and the i–th instances of the matrix multiplication and rank–ω update primitives respectively; and μ_{GJ}, μ_i^M and μ_i^R are the corresponding cache–miss ratios. In practice, some of the cache–miss ratios are zero or very small due to the interaction between the instances of the primitives. This occurs when the remaining part of the matrix to be decomposed approaches the size of the cache and latter instances of primitives find an increasing proportion of their data left in cache by earlier instances. The formulae presented in this section are derived using the conservative assumption of no interaction between instances of primitives. For more details concerning the derivation of the results in this section the reader is directed to [GJMS87].

It can be shown that μ is a decreasing function of ω on the interval $1 \le \omega \le \sqrt{CS}$. For small values of ω, i.e. $\omega \le 16$, the cache–miss ratio is of the form:

$$\mu \approx \frac{1}{2\omega} \gamma_R + \eta_1,$$

where η_1 is proportional to $1/n$ and γ_R is the sum of the γ_i^R. This result is expected since the computations are dominated by the rank–ω update which achieves a similar cache–miss ratio. In particular, it is clear that the data locality of a BLAS2 version, $\omega = 1$, is very poor. In the middle of the interval of interest the cache–miss ratio is of the form:

$$\mu \approx \frac{1}{\omega} \gamma_R + \eta_2,$$

where η_2 is proportional to $1/n$. Finally, when ω approaches $\sqrt{CS}$, the cache–miss ratio is

$$\mu \approx \frac{1}{\sqrt{CS}} \gamma_R + \eta_3,$$

where η_3 is proportional to $1/n$.

When ω increases beyond $\sqrt{CS}$, μ remains decreasing for some (typically small) interval and then becomes a rapidly increasing function with the value of approximately 1/4 At the point $\omega = n$. The exact point where this transition occurs is dependent on the implementation of the Gauss–Jordan primitive, but the decrease in μ between $\omega = \sqrt{CS}$ and the transition point is insignificant unless $n/\sqrt{CS}$ is small.

In summary, the cache–miss ratio of the block LU algorithm decreases more or less hyperbolically on the interval $1 \leq \omega \leq \sqrt{CS}$. For larger blocksizes, the cache–miss ratio remains essentially constant until it begins to increase rapidly reaching a value of approximately 1/4 at $\omega = n$.

4.1.4. Double–level Blocking

It is clear that under certain circumstances the reduction of the data loading overhead can directly conflict with reducing the arithmetic time component. In this section, a technique for mitigating this conflict called double–level blocking is presented.

There are two basic approaches to double–level blocking for the block LU algorithm each of which corresponds to a different view of the source of the conflict. Both approaches work with a pair of blocksizes (θ,ω) where θ is the blocksize preferred for reducing the arithmetic time component and ω that which reduces data loading overhead to an acceptable level.

The first approach, called the *outer–to–inner* approach, alters the single–level algorithm with blocksize ω to reduce the performance bottleneck caused by inverting systems of order ω via the Gauss–Jordan primitive by replacing the Gauss–Jordan primitive and the matrix multiplication primitive with, respectively, a block LU decomposition with blocksize θ and block forward and back solves of the resulting triangular systems. For the interval $1 \leq \omega \leq \sqrt{CS}$, this double–level algorithm has the same number of loads as the single–level algorithm with blocksize ω. The arithmetic time component clearly improves due to replacement of the Gauss–Jordan primitive.

The second approach, appropriately called the *inner–to–outer* approach, alters the single–level algorithm with blocksize θ to reduce the data loading overhead to acceptable levels while maintaining its arithmetic time component. The basic idea is to rearrange the computations performed in the block LU decomposition with block size θ so that several rank–θ updates are grouped together into a single more efficient rank–ω update. This technique reduces the number of loads considerably while still maintaining the arithmetic time benefit of working with a block size of θ. In order to accomplish this rearrangement, the multi–level form of the algorithm performs m iterations of the decomposition with block size θ, altered so that only certain portions of the rank–θ update are computed, followed by a rank–ω update, where $\omega = m\theta$. Since only a small portion of the rank–θ update is computed on each of the m iterations preceding the rank–ω update, the number of loads are essentially the same as if the block size were ω.

To illustrate this double–level blocking form of the algorithm assume that $\omega = 2\theta$ and partition the matrix A as shown in Figure 6, where the square submatrices A and C are of order θ, the square matrix G is of order $n - \omega$ and the others are dimensioned accordingly. The first θ—iteration performs

(i) $A \leftarrow A^{-1}$
(ii) $D \leftarrow DA$
(iii) $E \leftarrow EA$
(iv) $C \leftarrow C - DB$
(v) $F \leftarrow F - EB$
(vi) $H \leftarrow H - DJ$

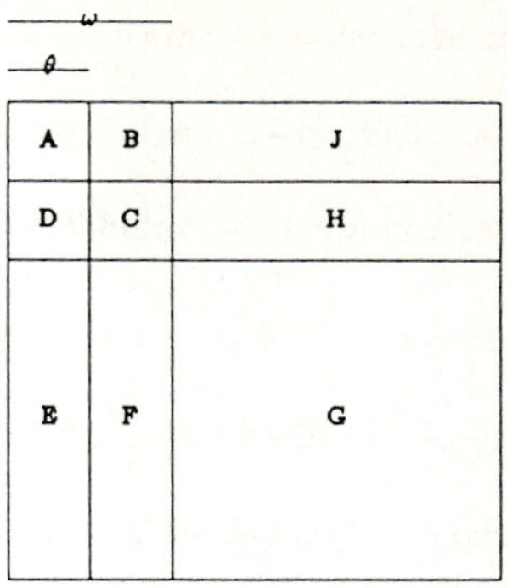

Figure 6. Inner–to–outer double–level partitioning

Statements (ii) and (iii) are, of course, performed as a single matrix multiplication. Note that G is not touched. This is due to the fact that the computations correspond to performing only the part of the rank–θ update needed to compute the next rank–θ update.

The next θ–iteration of the algorithm repeats the computations of steps (i) and (iii) with the submatrices C and F. A rank–ω update is then performed on the submatrix G. This rank–ω update accomplishes the effect of the two rank–θ updates on G. In general, this update is performed after m such θ–iterations.

At this point it is clear how this double–level blocking procedure reduces the number of loads. Only a single pass through the submatrix G is required. If rank–θ updates were used the submatrix G would have been loaded m times. Of course, the number of loads required before the rank–ω update is performed is increased compared to the number performed in a single–level block LU decomposition with block size ω, but this increase is $O(n^2)$, which does not significantly affect the expression for the number of loads presented in the last section.

4.1.5. Experimental Results

Figures 7 and 8 illustrate the relative performances of single–level and inner–to–outer double–level versions of the block LU algorithm. Note that the performance is measured in *pseudo–megaflops:* time is scaled by $(4n^3 - 3n^2)/6$ yielding pseudo–megaflops. This is necessary in order to ensure a fair comparison between methods due to the fact that the number of operations are a function of n, θ, and ω. The actual megaflop rate is higher for all points on the curves since the scale factor is a lower bound on the number of operations.

The curves for the single–level version in Figure 7 conform to the predictions made by the preceding analysis. The importance of the distribution of operations among the primitives for systems of small and moderate size is demonstrated by the relationship of the curves for $\omega = 32, 64, 96$. For the smaller systems $\omega = 32$ achieves the best performance of the three, even though it has the largest number of loads, because it does not require performing as many operations in the scalar Gauss–Jordan primitive as the $\omega = 64$ and $\omega = 96$ versions. As the size of the system increases, however, the number of data loads becomes as important as operation distribution. This is demonstrated by the crossing of the $\omega = 32$ and $\omega = 64$ curves. For larger systems, the importance of operation distribution becomes negligible while the number of data loads assumes dominance. This trend is clearly shown in the curves. Note that for the larger systems the $\omega = 32$ curve has leveled off at its asymptotic level while the $\omega = 96$ implementation, which performed dismally for small and moderately sized systems, has reached a level comparable to the $\omega = 64$ implementation and both are still headed upward at $n = 1024$.

The curve for $\omega = 8$ demonstrates that the tradeoff between number of loads and operation distribution is only dominant when ω is large enough. If ω is too small, two negative trends determine the behavior of the algorithm. First, as expected, the number of data loads is far too large to allow high performance. Second, if ω is small none of the computational primitives will achieve reasonable performance. It is clear from the figure that with $\omega = 8$ the influence of these factors causes low performance.

For all of these blocksizes, however, the performance is vastly superior to the performance achieved by a BLAS2 implementation of the algorithm (approximately 10 to 11 Mflops, see [GaJM87]).

The shifting of the preferred blocksize in the single–level version is somewhat bothersome when developing library routines which must select blocksizes when the user is not able to decide. The motivation for a double–level version is to remove the shift by mitigating the conflict between arithmetic time and data loading overhead minimization. Figure 8 demonstrates that this is indeed achievable in practice. The double–level implementation illustrated used an outer blocksize of 64. The curves show the performance with an inner blocksize of 16 and 64 (a single–level code included from Figure 7 for the purpose of comparison). Note that that double–level version yields performance higher than all of the single–level implementations of Figure 7 over the entire interval.

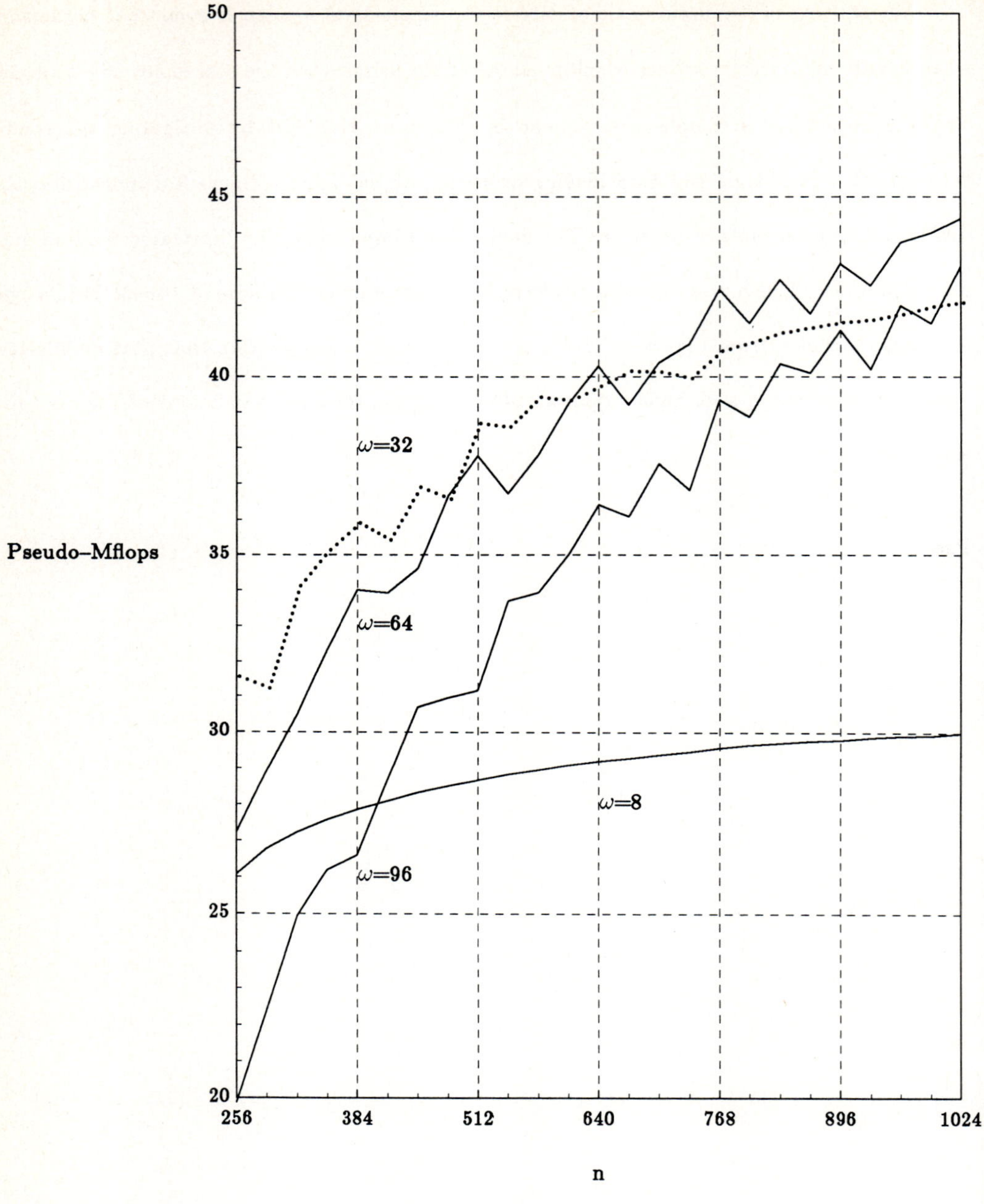

Figure 7. Block LU single–level, (n,n)–matrix

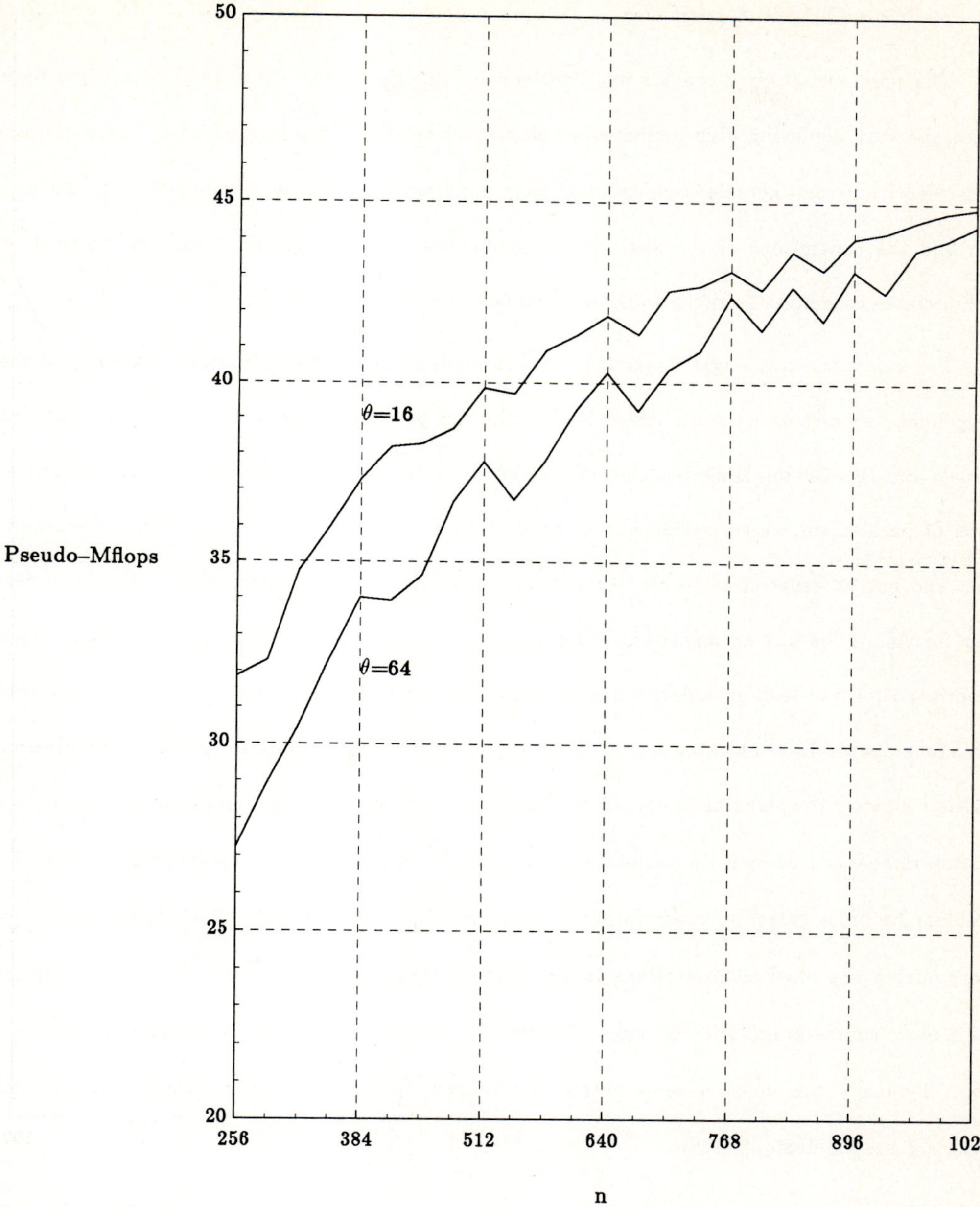

Figure 8. Block LU double–level version, outer blocksize = 64

4.2. Multiple Cluster Algorithms

The previous sections concerning the block LU algorithm and the BLAS3 primitives were concerned with achieving high performance on an architecture like a single Cedar cluster. While these algorithms and kernels form an invaluable building block for algorithms on the Cedar system and the conclusions of the analysis are applicable over a fairly wide range of multivector architectures care must be taken to not generalize these conclusions too far.

For example, on a single Cedar cluster (and similar architectures) routines for many of the basic linear algebra tasks encountered in practice can be designed as a series of calls to BLAS3 kernels and BLAS2–implemented algorithms thereby masking all of the architectural considerations of parallelism, vectorization and communication. This method of algorithm design, however, can not be generalized to all hierarchical shared memory machines. One of the main reasons for this is the fact an algorithm designed via this method may have problems with an inappropriate choice of task granularity and the resulting communication requirements. The need to introduce double–level blocking forms of the algorithm indicated the onset of such a problem on a Cedar cluster: the attempt to spread the BLAS2–implemented kernel across the processors in a cluster introduced serious limitations on the performance of the block algorithm. When this problem becomes extreme, other forms of the algorithm must be used which typically involve reorganizing the block computations to more efficiently map the algorithm to the architecture via a more course granularity of tasks with more attention focusing on the required communication. Typically this involves some notion of *pipelining* (possibly multidimensional) at the block level, e.g. see [Same85], [Bisc87].

An example of such a situation is the solution of a dense linear system using more than one cluster of Cedar (probably a subset of the total number available). In this case the algorithm design must take into account that intercluster communication is rather costly. There are several possible designs for such an algorithm. One of the most straightforward is based on the outer–to–inner double–level block form presented above. The block computations can be pipelined across clusters using the necessary Cedar synchronization primitives. A second possibility uses the control structure of the pipelined Given's factorization on a ring of processors described in [Same82]. A block of rows rather than a single row is communicated between processors and the row rotation is replaced with a block Gaussian elimination procedure. The remainder of this section discusses another algorithm for solving dense linear systems on the four cluster Cedar shown in Figure 1 which requires a relatively small amount of intercluster communication.

Let A, a nonsingular matrix of order n, be partitioned as

$$A^T = (A_1^T \ , \ A_2^T \ , \ A_3^T \ , \ A_4^T)$$

where A_i resides in the i–th cluster memory. The algorithm consists of two major stages. In the first stage, using a block–LU scheme with partial pivoting, each A_i is factored into the form

$$P_i A_i = L_i U_i \qquad i = 1,2,3,4$$

where P_i is a permutation, L_i is unit lower triangular and U_i is upper trapezoidal.

Assuming, without loss of generality, that each U_i has a nonsingular upper triangular part, the factorization of A may be completed in the second stage which consists of $3n/4$ computational waves pipelined across the four clusters. These computational waves comprise three groups of $n/4$ waves. During the k–th group the latest values for the rows of U_k are used by clusters $k+1$ to 4 in a pipelined fashion to further reduced their segments of the decomposition. It should be noted that cluster k is idle during the k–th group of waves and the remainder of the algorithm since the other clusters will update the rows of U_k that it has produced and placed in global memory. (For example, cluster 1 only performs the initial reduction of A_1 and is then released for other tasks within the application code of which solving the system is a part or the tasks of other users since Cedar is a multiuser system.) The first group of $n/4$ computational waves which use the rows of U_1 produced by cluster 1 is described below. The pattern of the remaining two groups follows trivially.

Wave 1

Let $U_k \equiv [\mu_{i,j}^k]$. The first row of U_1 is transmitted via the global memory to cluster 2 where it is used, with pairwise pivoting, annihilate the first element of the (possibly new due to pairwise pivoting) first row of U_2, $\mu_{1,1}^2$. The updated first row of U_1 is then transmitted to cluster 3 so as to annihilate $\mu_{1,1}^3$ and then to cluster 4 where $\mu_{1,1}^4$ is eliminated and the final version of the first row of U_1 resides in global memory.

As soon as $\mu_{1,1}^k$ is annihilated in cluster k, $k = 2,3,4$, the nonzero portion of U_k is a $n/4 \times (n-1)$ upper Hessenberg matrix, e.g. for n=24 it is of the form

$$\begin{bmatrix} x & x & x & x & x & x & . & . & x \\ x & x & x & x & x & x & . & . & x \\ & x & x & x & x & x & . & . & x \\ & & x & x & x & x & . & . & x \\ & & & x & x & x & . & . & x \\ & & & & x & x & . & . & x \end{bmatrix}$$

The cluster then proceeds to reduce U_k to upper trapezoidal form through a pipelined Gaussian elimination process using pairwise pivoting.

Wave $2 \leq j \leq n/4$

Similar to the first wave, the j-th row of U_1 is transmitted to clusters 2, 3, and 4 to annihilate $\mu_{1,j}^2$, $\mu_{1,j}^3$ and $\mu_{1,j}^4$, respectively. After these annihilations occur, each cluster reduces U_k, which at this point is upper Hessenberg, to upper trapezoidal form.

Note that after this first group of computational waves U_1 is in its final form in global memory. The matrix U_2 is in its penultimate form since it will only change due to the pairwise pivoting done by clusters 3 and 4 in the second group of computational waves. This implies that cluster 2 is now available for other work. The second and third computational groups proceed in the same way as the first did with each cluster fetching the appropriate row from the source matrix, U_2 followed by U_3, transforming U_k to upper Hessenberg form and then reducing it back to an upper trapezoidal matrix.

For the purposes of illustration, a pairwise pivoting form of the algorithm has been presented. In practice on Cedar one of the many possible block forms of this algorithm would be used.

5. BANDED LINEAR SYSTEM SOLVERS

In this section, multicluster algorithms for the solution of narrow-banded diagonally dominant linear systems are considered, see [SaKu78], [LaSa83] and [DoSa84]; for symmetric positive definite systems see [DoJo87].

Let the linear system under consideration be denoted by $Ax = f$, where A is a banded diagonally dominant matrix of order n. It is assumed that the number of superdiagonals $m << n$ is equal to the number of sub-diagonals and that, for simplicity of presentation, $n = pq$. On a sequential machine such a system would be solved via Gaussian elimination, see [DoBM79] for example. The algorithm described below (and informally called the *reduced-system* method) assumes a Cedar system with p clusters. Here, for the sake of illustration, p is assumed to be four, but the method is obviously applicable to any shared memory multiprocessor with p processing units.

Let the matrix A be partitioned into the block-tridiagonal form with block row $[C_i, A_i, B_i]$ and conformally x and f, e.g.

$$\begin{bmatrix} A_1 & B_1 & 0 & 0 \\ C_2 & A_2 & B_2 & 0 \\ 0 & C_3 & A_3 & B_3 \\ 0 & 0 & C_4 & A_4 \end{bmatrix} \begin{bmatrix} x_1 \\ x_2 \\ x_3 \\ x_4 \end{bmatrix} = \begin{bmatrix} f_1 \\ f_2 \\ f_3 \\ f_4 \end{bmatrix}$$

where, in general, $A_i, 1 \leq i \leq p$, is a banded matrix of order $q = n/p$ and bandwidth $2m+1$ (same as A),

$$B_i = \begin{bmatrix} 0 & 0 \\ \hat{B}_i & 0 \end{bmatrix}$$

and

$$C_{i+1} = \begin{bmatrix} 0 & \hat{C}_{i+1} \\ 0 & 0 \end{bmatrix}$$

$1 \leq i \leq p-1$, in which $\hat{B}_i$ and $\hat{C}_{i+1}$ are lower and upper triangular matrices, respectively, each of order m.

The algorithm consists of three stages.

Stage 1

If both sides of $Ax = f$ are premultiplied by $diag(A_1^{-1}, A_2^{-1}, \ldots, A_p^{-1})$ a system is obtained with the form

$$\begin{bmatrix} I_q & E_1 & 0 & 0 \\ F_2 & I_q & E_2 & 0 \\ 0 & F_3 & I_q & E_3 \\ 0 & 0 & F_4 & I_q \end{bmatrix} \begin{bmatrix} x_1 \\ x_2 \\ x_3 \\ x_4 \end{bmatrix} = \begin{bmatrix} g_1 \\ g_2 \\ g_3 \\ g_4 \end{bmatrix}$$

where

$$E_i = (\hat{E}_i, 0),\ F_i = (0, \hat{F}_i),$$

in which $\hat{E}_i$ and $\hat{F}_i$ are matrices of m columns given by

$$\hat{E}_i = A_i^{-1} \begin{bmatrix} 0 \\ \hat{B}_i \end{bmatrix}$$

and

$$\hat{F}_i = A_i^{-1} \begin{bmatrix} \hat{C}_i \\ 0 \end{bmatrix}$$

and will, in general, be full.

In Stage 1, $\hat{E}_i, \hat{F}_i$ and g_i are obtained by solving the associated linear systems. In each cluster $2 \leq k \leq p$ one solves $2m + 1$ linear systems of the form $A_k \nu = r$, while clusters 1 and p each solves $m+1$ linear systems of the same form. It can be shown that the A_k are diagonally dominant. Note that no intercluster communication is needed.

The method of solution for these p systems with multiple right–hand sides, of course, varies with the architecture. On Cedar, where the full resources of a cluster is dedicated to solving the independent systems, the method depends on the size of m. Berry and Sameh have shown, [BeSa87a], that for small m (less than eight or so) block cyclic reduction or another application of the reduced–system method should be used. For $8 \leq m \leq 16$ (approximately), block cyclic reduction is the most effective and for larger m a block Gaussian elimination is recommended. For architectures where the cluster of Cedar is replaced by a single scalar or vector processor simple Gaussian elimination without pivoting is most likely adequate.

Stage 2

Let $\hat{E}_i$ and $\hat{F}_i$ be partitioned, in turn, as follows

$$\hat{F}_i = \begin{bmatrix} P_i \\ M_i \\ Q_i \end{bmatrix}, \quad \hat{E}_i = \begin{bmatrix} S_i \\ N_i \\ T_i \end{bmatrix}$$

where P_i , Q_i , S_i , and $T_i \epsilon R^{mxm}$. Also, let g_i and x_i be conformally partitioned:

$$g_i = \begin{bmatrix} h_{2i-2} \\ w_i \\ h_{2i-1} \end{bmatrix}, \quad x_i = \begin{bmatrix} y_{2i-2} \\ z_i \\ y_{2i-1} \end{bmatrix}.$$

The structure of the resulting partitioned system is such that the unknown vectors y_j, $1 \leq j \leq 2p - 2$, (each of order m) are disjoint from the rest of the unknowns. In other words, the m equations above and the m equations below each of the $p - 1$ partitioning lines form an independent system of order $2m(p-1)$, which is referred to as the *reduced system* $Ky = h$, which, for $p = 4$, is of the form

$$\begin{bmatrix} I_m & T_1 & - & - & - & - \\ P_2 & I_m & - & S_2 & - & - \\ Q_2 & - & I_m & T_2 & - & - \\ - & - & P_3 & I_m & - & S_3 \\ - & - & Q_3 & - & I_m & T_3 \\ - & - & - & - & P_4 & I_m \end{bmatrix} \begin{bmatrix} y_1 \\ y_2 \\ y_3 \\ y_4 \\ y_5 \\ y_6 \end{bmatrix} = \begin{bmatrix} h_1 \\ h_2 \\ h_3 \\ h_4 \\ h_5 \\ h_6 \end{bmatrix}$$

Since A is diagonally dominant, it can be shown that the reduced system is also diagonally dominant and hence there are a number of options available for solving it. Typically, it is small enough to be sent to a single cluster and solved with the appropriate algorithm based on the reasoning in [BeSa87a].

When it is large enough to warrant a multicluster approach the reduced–system approach could be applied again. Note, however, that the bandwidth of the system has doubled compared to the original system. Block–column permutations can reduce the bandwidth back to its original value but this destroys diagonal dominance and extensive pivoting will usually be required to solve the permuted reduced system. It is also possible to use all of the clusters to solve the reduced system via an iterative technique such as Orthomin(k), [DoSa84].

Finally, if the original linear system is *sufficiently* diagonally dominant, one can ignore the matrices Q_i and S_i as $||S_i||_\infty$ and $||Q_i||_\infty$ are much smaller than $||T_i||_\infty$ and $||P_i||_\infty$ respectively. This results in a block–diagonal reduced system in which each block is of the form

$$\begin{bmatrix} I_m & T_k \\ P_{k+1} & I_m \end{bmatrix}$$

for $1 \leq k \leq p-1$.

Stage 3

Once the y_i's are obtained, the rest of the components of the solution vector of the original system may be retrieved as follows:

$$z_k = w_k - M_k\, y_{2k-3} - N_k\, y_{2k},$$

for $1 \leq k \leq p$,

$$y_0 = h_0 - S_1 y_2,$$

and

$$y_{2p-1} = h_{2p-1} - Q_p\, y_{2p-3}$$

This stage requires no intercluster communication with respect to blocks of the matrix. (The y_i, of course, must be fetched from global memory.)

Note that throughout the algorithm the BLAS3 play an important role in achieving high performance on each cluster and the entire Cedar multiprocessor.

There are several modifications and reorganizations possible on the reduced–system method for solving banded systems discussed above. These can be used to alter the form of the algorithm to more efficiently map to a variety of shared memory architectures. For one such alternative see [Same84].

6. LINEAR LEAST SQUARES PROBLEMS

In solving the linear least squares problem:

$$\min \| f - Ax \|_2,$$

where A is an $m \times n$ matrix of rank n, $(m > n)$, one needs to obtain the factorization,

$$QA = \begin{bmatrix} R \\ 0 \end{bmatrix},$$

in which Q is an orthogonal matrix and R is a nonsingular upper triangular matrix of order n. Such a factorization may be realized on multiprocessors via plane rotations, see [DoSS84], [Same82] and [SaKu78], elementary reflectors, see [Same82] or the Modified Gram–Schmidt algorithm, see [BGHJ86]. (Although the latter algorithm is more commonly associated with the calculation of an orthogonal basis of the range of A.) In this section block versions of the reflector–based algorithm and the Modified Gram–Schmidt algorithm are presented.

6.1. A Block Householder Reduction

If $A \equiv A_1 = [a_1^{(1)}, a_2^{(1)}, \cdots, a_n^{(1)}]$, then it is possible to generate elementary reflectors $P_k = I - \alpha_k u_k u_k^T$, $k = 1, \cdots, n$, such that forming $P_k A_k$ produces the k–th row of R and the $(m-k)\times(n-k)$ matrix $A_{k+1} = [a_{k+1}^{(k+1)}, \cdots, a_n^{(k+1)}]$ by annihilating all but the first element in $a_k^{(k)}$. The two basic tasks in such a procedure are [Stew73]: (i) generation of the reflector P_k such that $P_k a_k^{(k)} = (\rho_{kk}, 0, \cdots, 0)^T$, $k = 1, 2, \cdots, n$; and (ii) updating the remaining $(n-k)$ columns, $P_k a_j^{(k)} = (\rho_{kj}, a_j^{(k+1)T})^T$, $j = k+1, \cdots, n$. Note that on a parallel processor these two tasks may take place simultaneously for successive reflectors. The communication requirements are such that the algorithm maps readily onto multiprocessors with a shared memory or a simple topology such as a ring, [Same82].

The above parallel organization based on overlapping the execution of the individual reflectors needed for the orthogonal reduction does not offer the data locality needed for high performance on an architecture such as that of a single Cedar cluster (an Alliant FX/8). A block scheme proposed by [BiVa87], see also the related scheme in [ScPa87], offers such data locality. This scheme depends on the fact that the product of k elementary reflectors $Q_k=(P_k, \cdots, P_2, P_1)$, where $P_i = I_m - w_i w_i^T$, can be expressed as a rank–k update of the identity of order m, i.e.

$$Q_k = I_m - V_k U_k^T$$

where $V_1 = U_1 = w_1$, $V_j = (P_j V_{j-1}, w_j)$ and $U_j = (U_{j-1}, u_j)$, for $j = 2,\dots,k$.

The block algorithm may be described as follows. Let the $m \times n$ matrix $(m \geq n)$ whose orthogonal factorization is desired be given by

$$A = [A_1, B]$$

where A is of rank n, and A_1 consists of the first k columns of A. Next, proceed with the usual Householder reduction scheme by generating the k elementary reflectors P_1 through P_k such that

$$(P_k \cdots P_2 P_1) A_1 = \begin{pmatrix} R_1 \\ 0 \end{pmatrix}$$

where R_1 is upper triangular of order k without modifying the matrix B. If one accumulates the product $Q_k = P_k \cdots P_1 = I - V_k U_k^T$ as each P_i is generated, the matrix B is updated via

$$B \leftarrow (I - V_k U_k^T) B$$

which relies on the high efficiently of one of the most important kernels in BLAS3. The process is then repeated on the modified B with another *well–chosen* block size, and so on until the factorization is completed. One may also wish to accumulate the various Q_k's, one per block, to obtain the orthogonal matrix that triangularizes A.

It was shown in [BiVa87] that this block algorithm is as numerically stable as the classical Householder scheme. The block scheme, however, requires roughly $(1 + 2/p)$ times the arithmetic operations needed by the classical sequential scheme, where $p = n/k$ is the number of blocks (assuming a uniform block size throughout the factorization). An example, of the performance achieved by a BLAS3 implementation of the block Householder algorithm (PQRDC) compared to a BLAS2 version (DQRDC) on Alliant FX/8 [Harr87] is shown in Figure 9.

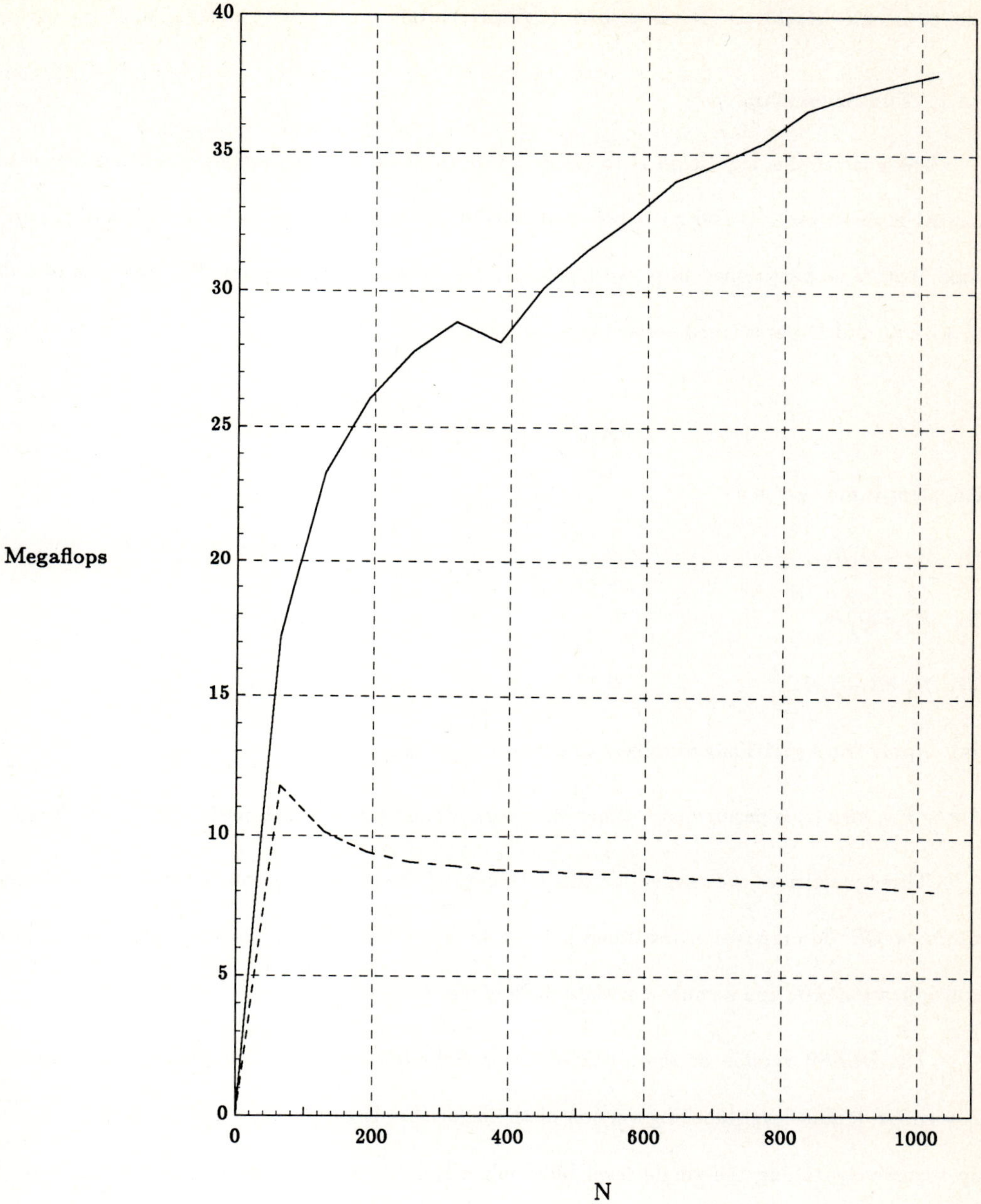

Figure 9. PQRDC and DQRDC on 1024 x N Matrices

6.2. A Block–Modified Gram–Schmidt Algorithm

6.2.1. The Algorithm

The goal of this algorithm is to factor an $m \times n$–matrix A of maximal rank into an orthonormal $m \times n$–matrix Q and an upper triangular R of order n where $m > n$ and A is of maximal rank. Let A be partitioned into two blocks A_1 and B where A_1 consists of ω columns of order m, with Q and R partitioned accordingly:

$$\left(A_1, B\right) = \left(Q_1, P\right) \begin{pmatrix} R_{11} & R_{12} \\ 0 & R_{22} \end{pmatrix}.$$

The algorithm is given by:

(i) $A_1 = Q_1 R_{11}$,

(ii) $R_{12} = Q_1^T B$,

(iii) $B_1 = B - Q_1 R_{12}$.

(iv) Apply the algorithm recursively to produce $B_1 = PR_{22}$.

If $n = k\omega$, step (i) is performed k times and steps (ii) and (iii) are each performed $k - 1$ times.

Three primitives are needed for the j–th step of the single–level block Gram–Schmidt algorithm: a QR decomposition (assumed here to be a modified Gram–Schmidt routine); a matrix multiplication AB; and a rank–ω update of the form $C \leftarrow C - AB$.

The BLAS2 version of the modified Gram–Schmidt algorithm is obtained when $\omega = 1$ or $\omega = n$. A double–level blocking version of the algorithm is derived in a straightforward manner by recursively calling the single–level block algorithm to perform the QR factorization of the $m \times \omega$ matrix A_1.

In the following sections, a summary of a blocksize analysis similar to that presented in detail for the block LU algorithm is presented along with experimental results on a single Cedar cluster for the single and double–level versions of the algorithm.

6.2.2. Blocksize Analysis

An analysis of the blocksize for the block Modified Gram–Schmidt algorithm is presented in [GJMS87]. The analysis is more complex than that of the block LU algorithm discussed above, but the conclusions are similar. The arithmetic time component analysis shows that double–level blocking is probably required for some systems due to the BLAS2–based MGS primitive. Indeed, the need for double–level blocking is greater for this algorithm than for the block LU since the fraction of operations performed by the BLAS2 routines is much higher.

The behavior of the algorithm with respect to the number of data loads can be discussed most effectively by considering approximations of the cache–miss ratios. For the interval $1 \leq \omega \leq l \approx CS/m$ the cache–miss ratio is

$$\mu \approx \frac{1}{2\omega} + \eta_1,$$

where η_1 is proportional to $1 / n$, which achieves its minimum value $m/(2CS)$ at $\omega = l$. Under certain conditions the cache–miss ratio continues to decrease on the interval $l \leq \omega \leq n$ where it has the form

$$\mu \approx \frac{1}{2\omega}\left(1 - \frac{\gamma}{n}\right) + \frac{\omega}{2}\left(\frac{1}{n} + \frac{1}{CS}\right) + \eta_2,$$

where η_2 is proportional to $1 / n$, which reaches its minimum at a point less than $\sqrt{CS}$ and increases thereafter, as expected. (See [GJMS87] for details.) When $\omega = n$ the cache–miss ratio for the second interval is 1/2 corresponding to the degeneration from a BLAS3 method to a BLAS2 method. The composite cache–miss ratio function over both intervals behaves like a hyperbola before reaching its minimum; therefore the cache–miss ratio does not decline as rapidly in latter parts of the interval as it does near the beginning.

A load analysis of the double–level algorithm shows that double–level blocking either reduces or preserves the cache–miss ratio of the single–level version while improving the performance with respect to the arithmetic component of time. (This is the added degree of complexity compared to the analysis of the block LU algorithm referred to above.)

6.2.3. Experimental Results

Figures 10 and 11 illustrate, respectively, the results of experiments, run on an Alliant FX/8, using single–level and double–level versions of the algorithm to orthogonalize square matrices. The cache size on this particular Alliant is $16K$ double precision words.

For the range of n, the order of the matrix, shown in Figure 10, the single–level optimal blocksize due to the data loading analysis starts at $\omega = 64$ decreases to $\omega = 21$ for $n = 768$ and then increases to $\omega = 28$ at $n = 1024$. Analysis of the arithmetic time component recommends the use of a blocksize between $\omega = 16$ and $\omega = 32$. Therefore, due to the hyperbolic nature of μ and the arithmetic time component analysis it is expected that the performance of the algorithm should increase until $\omega \approx 32$. The degradation in performance as ω increases beyond this point to, say $\omega = 64$ or 96, should be fairly significant for small and moderately sized systems due to the rather large portion of the operations performed by the BLAS2 MGS primitive.

The results of the experiments confirm the trends predicted by the theory. The version using $\omega = 32$ is clearly superior. The performance for $\omega = 8$ is uniformly dismal across the entire interval since the blocksize is too small for both data loading overhead and arithmetic time considerations. Note that as n increases the gap in performance between the $\omega = 32$ version and the larger blocksize versions narrows. This is due to both arithmetic time considerations as well as data loading. As noted above, for small systems, the distribution of operations reduces the performance of the larger blocksize version; but, as n increases, this effect decreases in importance. (Note that this narrowing trend is much slower than that observed for the block LU algorithm.

This is due to the fact that the fraction of the total operations performed in the slow primitive is ω/n for the block Gram–Schmidt algorithm and only ω^2/n^2 for the block LU.) Further, for larger systems, the optimal blocksize for data loading is an increasing function of n; therefore, the difference in performance between the three larger blocksize must decrease.

Figure 11 shows the increase in performance which results from double–level blocking. Since the blocksize indicated by arithmetic time component considerations is between 16 and 32 these two values were used as the inner blocksize θ. For $\theta = 16$ the predicted outer blocksize ranges from $\omega = 64$ up to $\omega = 128$; for $\theta = 32$ the range is $\omega = 90$ to $\omega = 181$. (Recall, that the double–level outer blocksize is influenced by the cache size only by virtue of the fact that $\sqrt{CS}$ is used as a maximum cutoff point.) For these experiments the outer blocksize of $\omega = 96$ was used for two reasons. First, it is a reasonable compromise for the preferred outer blocksize given the two values of θ. Second, the corresponding single–level version of the algorithm, i.e. $(\theta,\omega) = (96,96)$, did not yield high–performance and a large improvement due to altering θ would illustrate the power of double–level blocking. (To emphasize this point the curve with $(\theta,\omega) = (96,96)$ is included.)

The curves clearly demonstrate that double–level blocking can improve the performance of the algorithm significantly. Over most of the interval, an improvement of around 10 Mflops compared to the $(\theta,\omega) = (96,96)$ single–level version is achieved and up to 5 Mflops over the best performing single–level version with $\omega = 32$. As expected, the performance of the (16,96) and (32,96) versions are similar. Notice that for small systems, however, the double–level algorithm does not improve on the performance of the single–level version with $\omega = 32$. Two factors contribute to this situation. For the double–level version with $\theta = 16$ the preferred value of ω is close to 64 and therefore the data loading overhead does not improve when choosing $\omega = 96$. (For small systems a double–level version with $(\theta,\omega) = (16,64)$ did indeed, achieve higher performance than the $(\theta,\omega) = (32,32)$ version, but not by the impressive margin shown here for larger systems.) The more important factor, however, is the distribution of operations. Recall that the distribution of operations in the double–level algorithm is the same as in the single–level algorithm with blocksize ω, i.e. the BLAS2 MGS primitive in the single–level algorithm performs the same fraction of the total operations as the block GS primitive with blocksize θ in the double–level algorithm; and this distribution is determined roughly by the *width* of the system: the fraction of operations performed by the MGS primitive is approximately ω / n. Hence, if the system is not wide enough relative to ω then the improvement in performance of the block GS primitive with blocksize θ over the BLAS2 MGS primitive may not be enough to achieve performance significantly higher than some other high–performance single–level version (in this case $(\theta,\omega) = (32,32)$). This can be especially troublesome in the case of tall and narrow systems.

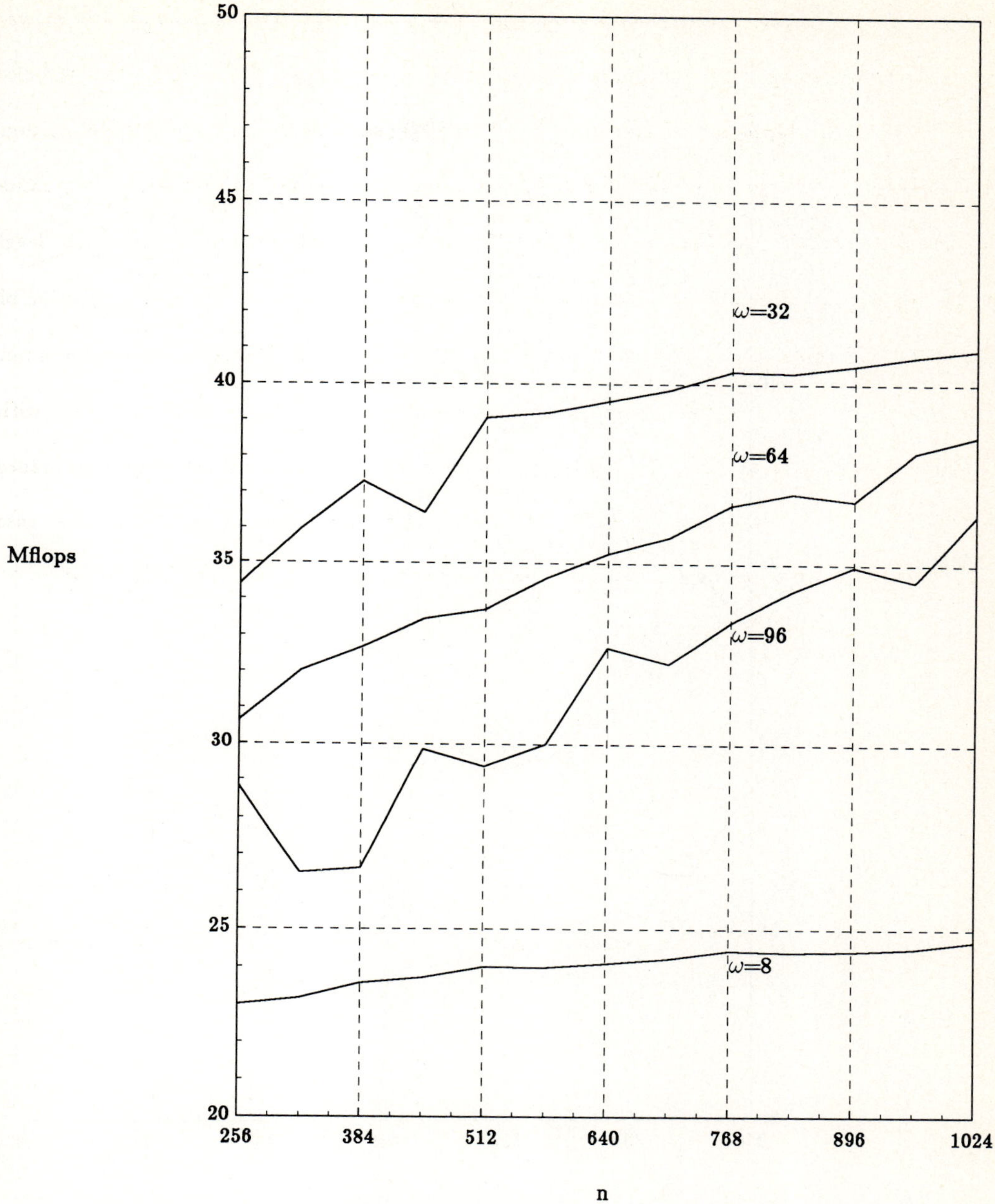

Figure 6.1. 1–level Block GS for a (n,n)–matrix

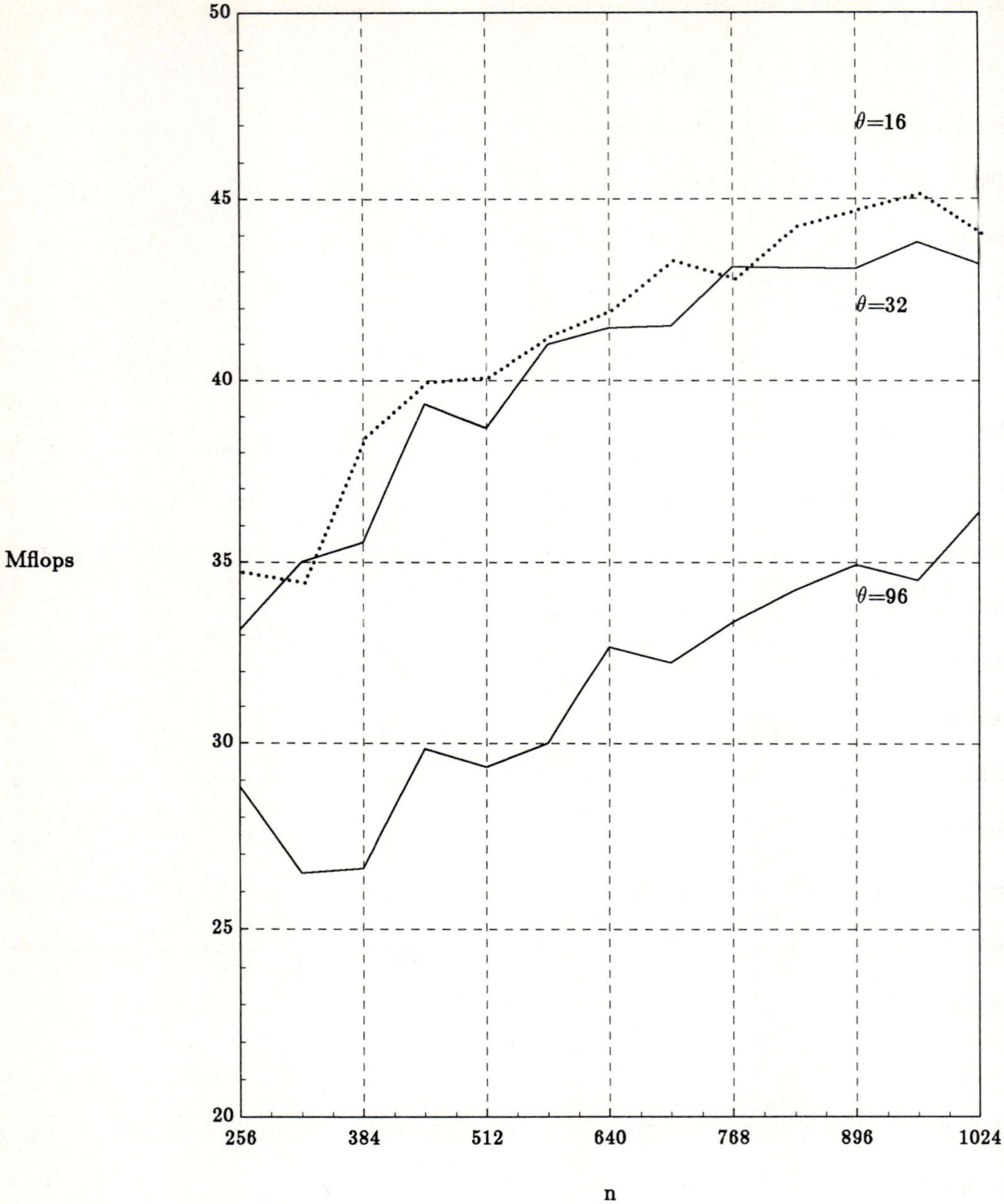

Figure 6.2. 2–level Block GS for a (n,n)–matrix, outer block size: 96

7. ELLIPTIC PROBLEM SOLVERS

In this section, three topics concerning the solution of elliptic problems are considered. First, a brief survey of rapid solvers for the Poisson equation on the unit square is presented. The final two sections consider the application of two iterative techniques, the preconditioned conjugate gradient scheme and the Block–Kaczmarz method, to the systems of equations derived from finite difference discretizations of self–adjoint and nonself–adjoint elliptic equations respectively.

7.1. Fast Poisson Solvers on the Cedar

Efficient direct methods for solving the finite–difference approximation of the Poisson equation on the unit square have been developed by Buneman [Bune69], Hockney [Hock65],[Hock70], and Golub [BuGN70]. The most effective sequential algorithm combines the block cyclic reduction and Fourier analysis schemes. This is Hockney's $FACR(l)$ algorithm, [Hock70]. Excellent reviews of these methods on sequential machines have been given by Swarztrauber [Swarz77] and Temperton [Temp79],[Temp80]. In [Swarz77] it is shown that the asymptotic operation count for $FACR(l)$ on an $n \times n$ grid is $O(n^2 \log_2 \log_2 n)$, and is achieved when the number l of the block cyclic reduction steps preceding Fourier analysis is taken approximately as $(\log_2 \log_2 n)$. Using only cyclic reduction, or Fourier analysis, to solve the problem on a sequential machine would require $O(n^2 \log_2 n)$ arithmetic operations.

Buzbee [Buzb73] observed that Fourier analysis, or the matrix decomposition Poisson solver (MD–Poisson solver), is ideally suited for parallel computation. It consists of performing a set of independent sine transforms, and solving a set of independent tridiagonal systems. On a parallel computer consisting of n^2 processors, with an arbitrarily powerful interconnection network, the MD–Poisson solver for the two–dimensional case requires $O(\log_2 n)$ parallel arithmetic steps [SaCK76]. It can be shown, [Peas68] and [Ston71], that a perfect shuffle interconnection network is sufficient to keep the communication cost to a minimum. Ericksen [Eric72] considered the implementation of $FACR(l)$, [Hock70], and $CORF$, [BuGN70], on the ILLIAC IV; and Hockney [Hock82] compared the performance of $FACR(l)$ on the Cray–1, Cyber–205, and the ICL–DAP.

In this survey we consider only the MD–algorithm and a modified block cyclic reduction scheme for solving the 5–point finite difference approximation of the Poisson equation on the unit square with a uniform $n \times n$ grid.

For the sake of illustration, we consider only Dirichlet boundary conditions. The multiprocessor algorithms presented below can be readily modified to accommodate Neumann and periodic boundary conditions.

Using natural ordering of the grid points, we obtain the well–known linear system of order n^2

$$\begin{bmatrix} T & -I & & & & \\ -I & T & -I & & & \\ & \cdot & \cdot & \cdot & & \\ & & \cdot & \cdot & \cdot & \\ & & & -I & T & -I \\ & & & & -I & T \end{bmatrix} \begin{bmatrix} u_1 \\ u_2 \\ \cdot \\ \cdot \\ u_{n-1} \\ u_n \end{bmatrix} = \begin{bmatrix} f_1 \\ f_2 \\ \cdot \\ \cdot \\ f_{n-1} \\ f_n \end{bmatrix}$$

where $T = [-1,4,-1]$ is a tridiagonal matrix of order n.

7.1.1. A domain decomposition MD–scheme

The MD–scheme consists of 3 stages:

Stage 1:

(1) Each cluster j, $1 \le j \le 4$, (as before a four cluster Cedar is assumed) forms the subvectors $f_{(j-1)q+1}, f_{(j-1)q+2}, \cdots, f_{jq}$ of the right–hand side, where $q = n/4$.

(2) Next, each cluster j obtains $\hat{g}_j^T = (g_{(j-1)q+1}^T, \cdots, g_{jq}^T)$, where $g_k = Qf_k$, in which $Q = [(2/[n+1])^{1/2}\sin(lm\pi/[n+1])]$, $l,m = 1, 2, \ldots, n$, is the eigenvector matrix of T. This amounts to performing in each cluster, q sine transforms each of length n.

(3) Now we have the system

$$\begin{bmatrix} M & E & & \\ E^T & M & E & \\ & E^T & M & E \\ & & E^T & M \end{bmatrix} \begin{bmatrix} \hat{v}_1 \\ \hat{v}_2 \\ \hat{v}_3 \\ \hat{v}_4 \end{bmatrix} = \begin{bmatrix} \hat{g}_1 \\ \hat{g}_2 \\ \hat{g}_3 \\ \hat{g}_4 \end{bmatrix}$$

where each cluster memory contains one block row. Here, $\hat{v}_j^T = (v_{(j-1)q+1}^T, \cdots, v_{jq}^T)$ with $v_k = Qu_k$, $M = [-I, \Lambda, -I]$ is a block tridiagonal matrix of order qn, and

$$E = \begin{bmatrix} & 0 \\ -I_n & \end{bmatrix}.$$

This system may be reduced to,

$$\begin{bmatrix} I & F & & \\ G & I & F & \\ & G & I & F \\ & & G & I \end{bmatrix} \begin{bmatrix} \hat{v}_1 \\ \hat{v}_2 \\ \hat{v}_3 \\ \hat{v}_4 \end{bmatrix} = \begin{bmatrix} \hat{h}_1 \\ \hat{h}_2 \\ \hat{h}_3 \\ \hat{h}_4 \end{bmatrix} \tag{7.1}$$

where $I \equiv I_{qn}$, $\hat{h}_j^T = (h_{(j-1)q+1}^T, \cdots, h_{jq}^T)$, F and G are given by: $M\hat{h}_j = \hat{g}_j$, $1 \le j \le 4$, $MF = E$, and $MG = E^T$. Observing that M consists of n independent tridiagonal matrices $T_k = [-1, \lambda_k, -1$ each of order q, where $\lambda_k = 4 - 2\cos(k\pi/[n+1])$, $k = 1, 2, \cdots, n$, the right–hand side of the above system is obtained by solving in each cluster j the n independent systems

$$T_k r_k = s_k,$$

for $k = 1, \cdots, n$, where $\hat{e}_i^T s_k = e_k^T g_{(j-1)q+i}$, and $\hat{e}_i^T r_k = e_k^T h_{(j-1)q+i}$, for $i = 1, 2, \cdots, q$, and $1 \le j \le 4$. Here, $\hat{e}_i$ and e_i are the i–th columns of I_q and I_n, respectively.

The matrices F and G can be similarly obtained by solving, in each cluster j, the independent systems $T_k c_k = \hat{e}_1$, and $T_k d_k = \hat{e}_q$, for $k = 1,2, \cdots ,n$. Since T_k is a Toeplitz matrix, however, we have $c_k = J d_k$, where $J = [\hat{e}_q, \cdots ,\hat{e}_1]$, see [KaVm78] for example. As a result, in order to obtain F and G we need only solve in each cluster the n systems $T_k d_k = \hat{e}_q$, $k = 1,2, \cdots ,n$. Hence, F and G are of the form,

$$F = \begin{bmatrix} \Gamma_q & \\ . & \\ . & 0 \\ . & \\ \Gamma_1 & \end{bmatrix}$$

and

$$G = \begin{bmatrix} & \Gamma_1 \\ & . \\ 0 & . \\ & . \\ & \Gamma_q \end{bmatrix},$$

where $\Gamma_i = -diag(\gamma_i^{(1)}, \cdots ,\gamma_i^{(n)})$, in which $\gamma_i^{(k)} = \hat{e}_i^T c_k$, for $i = 1, \cdots ,q$, and $k = 1, \cdots ,n$.

Stage 2:

From the structure of (7.1) it is seen that the three pairs of n equations above and below each partition are completely decoupled from the rest of the n^2 equations [SaKu78]. This reduced system, of order 6n, consists of interlocking blocks of the form:

$$\begin{bmatrix} I & \Gamma_1 & 0 & 0 & & \\ \Gamma_1 & I & 0 & \Gamma_q & & \\ \Gamma_q & 0 & I & \Gamma_1 & 0 & 0 \\ & & \Gamma_1 & I & 0 & \Gamma_q \\ & & \Gamma_q & 0 & I & \Gamma_1 \\ & & & & \Gamma_1 & I \end{bmatrix}$$

This system, in turn, comprises n independent pentadiagonal systems each of order six, which can be solved in a very short time.

Stage 3:

(1) Now, that the subvectors v_{kq}, v_{kq+1}, $k = 1, \cdots, 3$, are available, each cluster j obtains

$$v_{(j-1)q+i} = h_{(j-1)q+i} - (\Gamma_i v_{(j-1)q} + \Gamma_{q-i+1} v_{jq+1})$$

for $i = 2, \cdots, q-1$, where $v_0 = v_{4q+1} = 0$.

(2) Finally, each cluster j retrieves the q subvectors $u_{(j-1)q+i} = Qv_{(j-1)q+i}$, for $i = 1, \cdots, q$, of the solution via q sine transforms, each of length n.

Note that one of the key computational kernels in this algorithm is the calculation of multiple sine transformations. In order to design an efficient version of this kernel it is necessary to perform an analysis of the influence of the memory hierarchy similar to that presented above for the block LU algorithm. Such an analysis is contained in [GaJa87].

7.1.2. A modified block cyclic reduction algorithm

Recently, work has been done which improves the performance of block cyclic reduction methods for solving Poisson's equation [GaSa87]. (Similar work has also been done by Sweet [Swee87].) This work is concerned with removing a computational bottleneck in Buneman's form of the algorithm.

The most time–consuming part of Buneman's algorithm involves the solution of the matrix equation $A^{(r)}X = Y$ where the matrix $A^{(r)}$ is a known polynomial in A. Specifically,

$$A^{(r)} = 2T_{2^r}(\frac{A}{2}) \equiv \prod_{i=1}^{2^r}(A-\lambda_i I)$$

where T_k denotes the Chebyshev polynomial of the first kind of degree k and therefore

$$\lambda_i = 2\cos\frac{(2i-1)\pi}{2^{r+1}},\ i = 1, 2, ..., 2^r.$$

The problem which must be solved at every step of the algorithm thus has the form

$$\prod_{i=1}^{i=k}(A-\lambda_i I)X=Y.$$

When implemented on a serial computer, the linear systems corresponding to each of the factors and each of the right–hand–sides are solved in succession. Moreover, Gaussian elimination is used to solve these tridiagonal systems. On vector and parallel machines, these facts result in less efficient utilization of the available computational resources. An obvious but partial remedy is to still use Gaussian elimination but solve for all the right–hand–sides simultaneously. Unfortunately such an approach achieves high performance only on particular steps of the algorithm due to the fact that the number of right–hand sides vary considerably over the execution of the algorithm.

Another way of introducing parallelism is to use (scalar) cyclic reduction for the tridiagonal systems. However, the cost of cyclic reduction in terms of operation count is high and its use may lead to disappointing speed–ups. This was in particular pointed out in [Saad87] in the context of Alternating Direction Implicit (ADI) methods.

The modification of Gallopoulos and Saad exploits the following lemma:

Lemma

Every rational function

$$f(t) = \frac{u_l(t)}{g_k(t)} = \frac{u_l(t)}{\prod_{i=1}^{k}(t-\lambda_i)},$$

where the degree l of u_l is less than k and where the poles λ_i are all distinct, can be expanded in terms of elementary fractions as follows

$$f(t)=\sum_{i=1}^{k} \frac{\alpha_i}{t-\lambda_i}$$

where the scalars α_i are given by

$$\alpha_i=\frac{u_l(\lambda_i)}{g_k'(\lambda_i)}$$

This lemma is applied to the matrix problem to be solved resulting in the following decomposition of the solution

$$X=\sum_{i=1}^{k}\alpha_i(A-\lambda_i I)^{-1}Y$$

where α_i are constants given in [GaSa87]. The clear advantage of this form over the standard approach is that the tridiagonal systems involving the matrix $(A-\lambda_i I)$ are now decoupled and can be solved in parallel. In essence, a problem involving sequential solutions of tridiagonal linear systems has been replaced by one which involves the simultaneous solution of independent tridiagonal linear systems and the linear combination of the partial results.

For results concerning the stability of the altered algorithm and performance on a single Cedar cluster the reader is referred to [GaSa87].

7.2. The Preconditioned Conjugate Gradient Algorithm

The preconditioned conjugate gradient scheme (e.g., see [Vors82]) has become one of the most effective algorithms for handling those linear systems that arise from the 5–point finite difference discretization of second–order self–adjoint problems. Consider, for example, the problem

$$\frac{-\partial}{\partial x}(a(x,y)\frac{\partial x}{\partial y}) - \frac{\partial}{\partial y}(b(x,y)\frac{\partial x}{\partial y}) + c(x,y)u = g(x,y) \tag{7.2}$$

on the unit square with Dirichlet boundary conditions, where $a,b \geq 0$, and $c \geq 0$. Using a uniform mesh of width $h = 1/(n+1)$, n–even, with point red–black ordering, we obtain (after some diagonal scaling) a linear system of the form.

$$\begin{bmatrix} I & H \\ H^T & I \end{bmatrix} \begin{bmatrix} x_R \\ x_B \end{bmatrix} = \begin{bmatrix} f_R \\ f_B \end{bmatrix}, \tag{7.3}$$

where $H = [\Gamma_i, H_i, \Omega_i]$ is block–tridiagonal of order $n^2/2$ in which Γ_i and Ω_i are diagonal matrices, and H_i is upper bidiagonal for i–even and lower bidiagonal for i–odd. We consider now solving the reduced system,

$$(I - H^T H)x_B = (f_B - H^T f_R), \tag{7.4}$$

using the conjugate gradient method with the diagonal of $(I - H^T H)$ as the preconditioner. This is mathematically equivalent to solving (7.3) via C.G. with incomplete Cholesky factorization with zero fill–in, i.e., ICCG(0). In any case, the basic computational tasks in a preconditioned C.G. iteration for solving (7.4) are:

(i) inner–products: $\alpha = x^T y$,

(ii) scalar * vector + vector: $z = \alpha x + y$,

(iii) matrix * vector: $q = (I - H^T H)p$,

(iv) solving a system of the form: $Ms = r$, where M is the preconditioner.

Forming the matrix $(I - H^T H)$ explicitly as a part of the preprocessing stage, and multiplying it by a vector using the effective scheme of multiplication by diagonals (it contains 11 diagonals), step (iii) consumes the highest percentage of time per iteration for grids with $n \geq 32$. Taking M as the diagonal of $(I - H^T H)$, the performance of this PCG scheme is superior to other preconditioning schemes such as point [Vors82] or block [CoGM85] incomplete factorization, or polynomial preconditioners, [Meur84] and [Saad85], on systems resulting from natural ordering. This reduced system approach achieves a performance on a single Cedar cluster which reaches a peak of 18 Mflops for n around 32 but drops sharply to roughly 8 Mflops for larger values of n. (See [JaMS87] for complete details concerning the implementation of this algorithm on a single Cedar cluster.) This is much inferior to performance achieved for dense matrix calculations using basic primitives and algorithms that enhance date locality. In any case, the above reduced system approach with a simple preconditioner indicates that high performance depends on choosing a preconditioner that is powerful enough to reduce the number of iterations without sacrificing a high degree of vectorization and parallelism in solving the systems $Mz = r$.

On multiple clusters of Cedar, the above reduced system PCG scheme can be implemented by performing those vector operations in steps (i), (ii), and (iv) using the global memory. In such an implementation all vectors are stored in the global memory, thus bypassing the cluster caches for those vector operations that exhibit poor data locality. The only operation that utilizes the cluster memories is that of step (iii), multiplying the matrix $(I - H^T H)$ by a vector. This may be outlined as follows. Let $(I - H^T H) = [D_{i-2}, J_{i-1}^T, T_i, J_i, D_i]$, where the matrices D are diagonal and J, T are tridiagonal, be partitioned as $[E_{i-1}^T, S_i, E_i]$ for $i \leq i \leq 4$. Here, each block row is stored in the local memory of cluster i, and cluster i requires the subvectors x_{i-1}, x_i and x_{i+1} from the global memory to obtain the subvector y_i of $y = (I - H^T H)x$.

7.3. The Block-Kaczmarz Algorithm

We have developed an iterative scheme for solving large block-tridiagonal nonsymmetric linear systems of equations such as those arising from the finite element or finite difference discretization of elliptic partial differential equations [KaSa86]. This method, namely, an accelerated block-SSOR scheme, is ideally suited for multiprocessors and can be used when the eigenvalue distribution of the matrix of coefficients is not known, or when the eigenvalues are known to lie on both sides of the imaginary axis. This, hopefully, addresses a gap in the literature since most iterative methods which have been proposed thus far for solving nonsymmetric systems require some restriction on the eigenvalue distribution of the matrix or its symmetric part.

The block–SSOR scheme is a projection method that is obtained by applying the SSOR method to the normal equations, with either block row or block column partitioning of the matrix. By suitable permutation of the block rows of our block–tridiagonal matrix, we partition the system as

$$\begin{bmatrix} B_1^T \\ B_2^T \end{bmatrix} x = \begin{bmatrix} f_1 \\ f_2 \end{bmatrix}$$

then the k–th iteration of the block–row SSOR method is given by

$$z_1 = x_k$$

$$z_2 = z_1 + \omega\,(g_1 - P_1 z_1)$$

$$z_3 = z_2 + \omega\,(g_2 - P_2 z_2)$$

$$z_4 = z_3 + \omega\,(g_2 - P_2 z_3)$$

$$z_5 = z_4 + \omega\,(g_1 - P_1 z_4)$$

$$x_{k+1} = z_5$$

where $g_i = B_i(B_i^T B_i)^{-1} f_i = B_i^{+} f_i$, i=1,2 and P_i, i=1,2 is the projector onto the space spanned by the columns of B_i. Here, the permutation is chosen such that each B_i is a block–diagonal matrix. Thus, the linear least squares problem in the iteration above actually consists of several smaller independent linear least squares problems. This enables us to implement the algorithm on a multiprocessor such as the Cray X–MP/48 with high efficiency. We show that the block SSOR method converges for $0 < \omega < 2$, irrespective of the eigenvalue distribution of the matrix. Further, our numerical experiments show that the optimal value of ω for the block–SSOR method is 1.0. This observation simplifies, considerably, the iterative scheme and makes the method independent of any input parameters.

Our iterative scheme is then equivalent to solving the linear system

$$(I - Q) = Rf$$

where

$$Q = Q_1 Q_2 Q_1 ,$$

$$Q_i = (I - P_i) \quad i = 1,2$$

and

$$R = \quad [(I + Q_1 Q_2)(B_1^+)^T \quad , \quad Q_1(Q_2 + I)(B_2^+)^T \quad]$$

via the iteration

$$x_{k+1} = Q\ x_k + Rf$$

The matrix $(I - Q)$ is symmetric positive definite and thus ideally suited for acceleration by the conjugate gradient method. The matrix–vector product

$$(I - Q)u = v$$

required by the the C.G. acceleration involves the solution of linear least squares problems, making the CG accelerated block–SSOR method well suited for implementation on a multiprocessor.

We have also investigated various methods for the solution of the linear least squares problems and various permutations that could be used to split the larger linear least squares problems into smaller independent ones. For the two–dimensional elliptic P.D.E's. we consider, direct factorization methods for solving the linear least squares problems are the most efficient. We believe, however, that for three–dimensional problems (our eventual target) one must resort to iterative schemes for solving these linear least squares problems. Comparisons with the best methods available namely those of Yale's PCGPAK (i.e. Orthomin(k), GCR(k) and GMRES(k)) reveal that, when PCGPAK schemes are applicable, and when used in combination with good preconditioners, such as MILU, they are the fastest available on one CPU of the Cray X–MP. However, there is no guarantee that these methods would lead to convergence in the case of a general problem. In addition, these preconditioners may become unstable, leading to early stagnation of the residuals. PCGPAK also requires the user to be familiar with both the problem and the various CG–like methods it contains, as well as provide a substantial amount of input. In comparison, the block–SSOR method is reliable, equally accurate, and does not require any user input.

Finally, the suitability of the block–SSOR method for multiprocessors is demonstrated by the fact that we have achieved a speedup of 3.7 on the four–processor Cray X–MP/48, with a computational rate of 210 Mflops.

Recently, we have investigated the effect of partitioning the matrix into three independent blocks, rather than two blocks. This led to an increase in the number of iterations required for convergence. The amount of work required per iteration, however, is decreased leading to an overall reduction in the time for scalar calculations. Performance of this scheme on one CPU of the Cray X–MP is inferior to the original 2–block organization. It should be noted that the 3–block scheme requires less storage than the original algorithm, hence it may be more suitable for multiprocessors with modest local storage such as the Alliant FX/8 (cache size is currently 16K 64–bit words) ,or the Cray–2.

8. THE ALGEBRAIC EIGENVALUE PROBLEM

Solving the algebraic eigenvalue problem, either standard $Ax = \lambda Bx$, or generalized $Ax = \lambda Bx$, is an important and potentially time–consuming task in numerous applications. In this brief review, only the dense symmetric case is considered. First, we dispense with the generalized eigenvalue problem. The first step is reduction to the standard form; this is achieved by the block–Cholesky factorization of the symmetric positive–definite matrix, $B = LL^T$, and forming the matrix $L^{-1}AL^{-T}$ explicitly. Such a step can be realized with high efficiency on the Alliant FX/8 using the above–mentioned BLAS3. Throughout this review we assume that the dense eigenvalue problem involves only matrices of moderate size, and hence these problems are relegated only to a single cluster of Cedar, i.e., the Alliant FX/8. After solving the standard problem the eigenvectors x_i, if desired for the generalized problem, can be recovered from the eigenvectors y_i of the standard problem via the transformation $x = L^{-T}y$.

The most common method for handling the standard dense symmetric eigenvalue problem consists of first reducing the symmetric matrix to tridiagonal form via elementary reflectors followed by handling the tridiagonal eigenvalue problem by a QR–based method if all the eigenvalues are required or a bisection–inverse iteration combination if only selected eigenpairs are desired, e.g., see the routines in [WiRe71]. For parallel processors with hierarchical memories, high performance in the tridiagonalization stage depends on reorganizing Householder's reduction to produce a block version of this tridiagonalization scheme, and on accumulating the elementary reflectors for each block so as to make use of the powerful matrix rank–k update primitive for each step of the block algorithm. An Alliant FX/8 implementation of such a scheme [Harr87], achieves higher performance over the best vectorized version of the classical Householder's tridiagonalization procedure that depends on matrix–vector modules (BLAS2). This block scheme proves to be faster than BLAS2–based TRED2 for matrices of order 400 or higher reaching a factor of 2 for matrices of order 1,000. Such superiority over the classical scheme is realized even though the block algorithm requires roughly $(1+2/p)$ times more arithmetic operations, where p is the number of blocks used in the procedure, assuming a uniform block–size.

Once the tridiagonal matrix is obtained, two algorithms may be used on the Alliant FX/8 so as to exploit its two levels of parallelism. The first algorithm depends on a multisectioning strategy [LoPS87], which is a generalization of the well-known bisection scheme that uses the nonlinear recurrence form for the Sturm sequence. This multisectioning algorithm has three main stages: (i) isolation of the eigenvalues, one per interval using repeated multisectioning on the eight computational elements (CE's) of the FX/8, (ii) extraction of each eigenvalue using either bisection or the ZEROIN procedure, e.g., see [FoMM77], and (iii) inverse iteration to obtain the corresponding eigenvectors together with orthogonalization for those vectors corresponding to *poorly* separated eigenvalues. This scheme proved to be the most effective on the Alliant FX/8 for obtaining all or few of the eigenvalues. Compared to its execution time on one CE, it achieves a speedup of 7.9 on eight CE's, and is more than four times faster than TQL1 [WiRe71], for the tridiagonal matrix [-1,2,-1] of order 500 with the same achievable accuracy for the eigenvalues. Even if all the eigenpairs of the above tridiagonal matrix are required, this multisectioning scheme is more than thirteen times faster than the best BLAS2-based version of TQL2, twenty-seven times faster than Eispack's pair Bisect and Tinvit, and five times faster than its nearest competitor [DoSo87] with the same accuracy in the computed eigenpairs. For matrices with clusters of poorly separated eigenvalues, however, the multisectioning algorithm may not be competitive if all the eigenpairs are required with high accuracy. For example, for the well-known Wilkinson matrices $W^{\dagger}_{2n+1}$, e.g., see [Wilk65], which have pairs of very close eigenvalues, the multisectioning method requires roughly twice the time required by the scheme in [DoSo87] in order to achieve the same accuracy for all the eigenpairs. Note that the algorithm in [DoSo87] currently requires the computation of all the eigenvalues and eigenvectors.

The second algorithm for handling the tridiagonal eigenvalue problem on the Alliant FX/8 is naturally that of [DoSo87]. It is a robust scheme which is based on Cuppen's algorithm [Cupp81], which in turn depends on the construction of the eigenpairs of a tridiagonal matrix of order 2m given the eigenpairs of its block diagonal, $diag(T_1, T_2)$, where each T_i is of order m. This parallel implementation of Cuppen's algorithm is faster than the best BLAS2-version of

TQL2 on eight CE's by a factor of roughly 2.4 for the tridiagonal matrix [–1,2,–1] of order 500, achieving the same accuracy as that of TQL2.

An alternative to the above approach which depends on a tridiagonalization procedure, is that of using one of the Jacobi schemes for obtaining all the eigenvalues or all the eigenvalues and eigenvectors. Numerous experiments on the Alliant FX/8 indicate that the one–sided Jacobi scheme for orthogonalizing the columns of a symmetric nonsingular matrix is the most effective Jacobi variation for this hierarchical memory system. Note that if A is not known to be nonsingular, one treats the eigenvalue problem $\tilde{A}x = (\lambda + \alpha)x$, where $\tilde{A} = A + \alpha I$, with α being chosen such that $\tilde{A}$ is diagonally dominant with positive elements. In the one–sided Jacobi scheme one determines independent plane rotations that orthogonalize up to $floor(n/2)$ independent plane rotations After several sweeps, in which a sweep consists of $(n-1)$ transformations with each consisting of $floor(n/2)$ independent plane rotations, the resulting matrix has all its columns orthogonal within a given tolerance. From such a matrix one extracts the eigenvalues of $\tilde{A}$, and if the plane rotations are accumulated one obtains the eigenvectors as well. If $\alpha \neq 0$, it is from these eigenvectors that one estimates the eigenvalues of the original matrix A. This Jacobi scheme is superior to algorithms that depend on tridiagonalization, on the Alliant FX/8, for matrices of size less than 150 or for matrices that have few clusters of eigenvalues.

The same one–sided Jacobi scheme has also proved to be most effective, on the hierarchical memory system of the Alliant FX/8, for obtaining the singular–value decomposition of $m \times n$ matrices where $m \gg n$. Here, the Jacobi scheme is preceded by a block–Householder reduction to the upper–triangular form. The one–sided Jacobi algorithm is applied to this $n \times n$ triangular matrix. Such a procedure resulted in a performance that is superior to the best vectorized version of Eispack's or Linpack routines which are based on the algorithm in [GoRe70]. Experiments showed that the block–Householder reduction and the one–sided Jacobi scheme combination is up to five times faster, on the Alliant FX/8, than the best BLAS2–version of Linpack's routine for matrices of order 16,000×128 , [BeSa87b].

REFERENCES

[BeSa87a] M. Berry and A. Sameh, *Multiprocessor Schemes for Solving Block Tridiagonal Systems,* CSRD Report, CSRD University of Illinois at Urbana–Champaign, 1987.

[BeSa87b] M. Berry and A. Sameh, *Multiprocessor Algorithms for the Singular Value Decomposition,* CSRD Report, CSRD University of Illinois at Urbana–Champaign, 1987.

[BGHJ86] M. Berry, K. Gallivan, W. Harrod, W. Jalby, S. Lo, U. Meier, B. Phillipe and A. Sameh, *Parallel Numerical Algorithms on the CEDAR System,* In: **CONPAR 86, Lecture Notes in Computer Science** , W. Handler et. al., Eds., Springer–Verlag, Aachen, F.R. Germany, 1986.

[Bisc87] C. Bischof, *A Pipelined Block QR Algorithm for a Ring of Vector Processors,* To appear in the Proc. Third SIAM Conf. on Par. Proc. for Sci. Comp., Los Angeles, December 1987.

[BiVa87] C. Bischof and C. Van Loan, *The WY Representation for Products of Householder Matrices,* SIAM J. Sci. Stat. Comput., Vol. 8, No. 2, March 1987.

[BuGN70] B. Buzbee, G. Golub and C. Nielson, *On Direct Methods for Solving Poisson's Equation,* SIAM J. Numer. Analysis 7, pp. 627–656, 1970.

[Bune69] O. Buneman, *A Compact Non–iterative Poisson Solver,* Report 294, Stanford University Institute for Plasma Research, Stanford, California, 1969.

[Buzb73] B. Buzbee, *A Fast Poisson solver Amenable to Parallel Computation,* IEEE Trans. Comput. C–22, pp. 793–796, 1973.

[Cala86] D. Calahan, *Block–Oriented , Local–Memory–Based Linear Equation Solution on the*

CRAY-2: Uniprocessor Algorithms, Proceedings of ICPP 1986, IEEE Computer Society Press, Washington D.C., August 1986.

[CoGM85] P. Concus, G. Golub and G. Meurant, *Block Preconditioning for the Conjugate Gradient Method,* SIAM J. Sci. Stat. Comput., Vol. 6, 1985, pp. 220–252.

[Cupp81] J. Cuppen, *A Divide and Conquer Method for the Symmetric Tridiagonal Eigenproblem,* Numer. Math., Vol. 36, 1981, pp. 177–195.

[DDHH87] J. Dongarra, J. Du Croz, I. Duff and S. Hammarling, *A Proposal for a Set of Level 3 Basic Linear Algebra Subprograms,* Technical Memo No. 88, MCS Division, Argonne National Laboratory, Argonne, Illinois, 1987.

[DoBM79] J. Dongarra, J. Bunch, C. Moler and G. W. Stewart, **Linpack User's Guide** , SIAM, 1979.

[DoJo87] J. Dongarra and L. Johnsson, *Solving Banded Systems on a Parallel Processor,* Parallel Computing, Vol. 5, 1987, pp. 219–246.

[DoSa84] J. Dongarra and A. Sameh, *On Some Parallel Banded System Solvers,* Parallel Computing, Vol. 1, 1984, pp. 223–235.

[DoSo87] J. Dongarra and D. Sorensen, *A Fully Parallel Algorithm for the Symmetric Eigenvalue Problem,* SIAM J. Sci. Stat. Comput., Vol. 8, 1987, pp. s139–s154.

[DoSS84] J. Dongarra, A. Sameh and D. Sorensen, *Implementation of Some Concurrent Algorithms for Matrix Factorization,* Mathematics and Computer Science Division Report, Argonne National Laboratory, Argonne, Illinois, 1984.

[Eric72] J. Ericksen, *Iterative and Direct Methods for Solving Poisson's Equation and Their Adaptability to Illiac IV,* CAC Document No. 60, University of Illinois at Urbana–Champaign, December 1972.

[FoMM77] G. Forsythe, M. Malcolm and C. Moler, **Computer Methods for Mathematical Computations** , Prentice–Hall, 1977.

[GaJa87] D. Gannon and W. Jalby, *The Influence of Memory Hierarchy on Algorithm Organization: Programming FFTs on a Vector Multiprocessor.* in The Characteristics of Parallel Algorithms, L. Jamieson, D. Gannon and R. Douglass, Eds., MIT Press, Cambridge, 1987.

[GaJM87] K. Gallivan, W. Jalby and U. Meier, *The Use of BLAS3 in Linear Algebra on a Parallel Processor with a Hierarchical Memory,* SIAM J. Sci. Stat. Comput., Vol. 8, No. 6, November 1987.

[GaSa87] S. Gallopoulos and Y. Saad, *A Parallel Block Cyclic Reduction Algorithm for the Fast solution of Elliptoc Equations,* CSRD Report, CSRD University of Illinois at Urbana–Champaign, 1987.

[GJMS87] K. Gallivan, W. Jalby, U. Meier and A. Sameh, *The Impact of Hierarchical Memory Systems on Linear Algebra Algorithm Design,* CSRD Report, CSRD University of Illinois at Urbana–Champaign, 1987. To appear in Inter. Jour. of Supercomputing Applications, Spring, 1988.

[GoRe70] G. Golub and C. Reinsch, *Singular Value Decomposition and Least Squares Solutions,* Numer. Math., Vol. 14, 1970, pp. 403–420.

[GoVa83] G. Golub and C. Van Loan, **Matrix Computations** ,The Johns Hopkins University Press, 1983.

[Harr87] W. Harrod, *Programming with the BLAS,* in **The Characteristics of Parallel Algorithms** ,L. Jamieson, D. Gannon and R. Douglass, Eds., MIT Press, Cambridge, 1987.

[Hock65] R. Hockney, *A Fast direct solution of Poisson's Equation Using Fourier Analysis,* JACM 12, pp. 95–113, 1965.

[Hock70] R. Hockney, *The Potential Calculation and Some Applications,* in Methods of Computational Physics, Alder, Fernback and Rotenberg, Eds., Vol. 9, pp. 135–211, Academic Press, 1970.

[Hock82] R. Hockney, *Optimizing the FACR(l) Poisson Solver on Parallel Computers,* Proc. ICPP 1982, IEEE Computer society Press, 1982.

[HoKu81] J. W. Hong and H. T. Kung, *I/O Complexity: The Red-Blue Pebble Game,* Proc. of the 13th Ann. Symp. on Theory of Computing, October 1981, pp. 326–333.

[JaMS87] W. Jalby, U. Meier and A. Sameh, *The Behavior of Conjugate Based Algorithms on a Multi-vector Processor with Memory Hierarchy,* CSRD Report, CSRD University of Illinois at Urbana–Champaign, 1987.

[KaSa86] C. Kamath and A. Sameh, *A Projection Method for Solving Nonsymmetric Linear Systems on Multiprocessors,* CSRD Report, CSRD University of Illinois at Urbana–Champaign, 1986.

[KaVM78] T. Kailath, A. Vieira and M. Morf, *Inversion of Toeplitz Operators, Innovations and Orthogonal Polynomials,* SIAM Review 20, pp. 106–119, 1978.

[KDLS86] D. Kuck, E. Davidson, D. Lawrie and A. Sameh, *Parallel Supercomputing Today and the Cedar Approach,* Science, Vol. 231, 1986, pp. 967–974.

[LaSa83] D. Lawrie and A. Sameh, *The Computations and Communication Complexity of a Parallel Banded System Solver,* ACM TOMS, Vol. 10, 1984, pp. 185–195.

[LoPS87] S. Lo, B. Philippe and A. Sameh, *A Multiprocessor Algorithm for the Symmetric Tridiagonal Eigenvalue Problem,* SIAM J. Sci. Stat. Comput., Vol. 8, 1987, pp. s155–s165.

[Meur84] G. Meurant, *The Block Preconditioned Conjugate Gradient Method on Vector Computers,* BIT, Vol. 24, 1984, pp. 623–633.

[Peas68] M. Pease, *An Adaption of the Fast Fourier Transform for Parallel Processing,* JACM 15, pp. 252–264, 1968.

[PeWi75] G. Peters and J. Wilkinson, *On the Stability of Gauss–Jordan Elimination with Pivoting,* CACM 18, pp. 20–24, January 1975.

[Saad85] Y. Saad, *Practical Use of Polynomial Preconditioning for the Conjugate Gradient*

Method, SIAM J. Sci. Stat. Comput., Vol. 6, 1985, pp. 865–881.

[Saad87] Y. Saad, *On the Design of Parallel Numerical Methods in Message Passing and Shared–Memory Environments,* Proc. Inter. Seminar on Scientific Supercomputers, Paris, France, February 1987.

[SaCK78] A. Sameh, S. Chen and D. Kuck, *Parallel Poisson and Biharmonic Solvers,* Computing 17, pp. 219–230, 1976.

[SaKu78] A. Sameh and D. Kuck, *On Stable Parallel Linear System Solvers,* JACM 25 (1978) pp. 81–91.

[Same82] A. Sameh, Purdue Workshop on Algorithmically–Specialized Computer Organizations, Academic Press, New York, 1982.

[Same84] A. Sameh, *On Two Numerical Algorithms for Multiprocessors,* in **High Speed Computing** , J. S. Kowalik, Ed., Series F: Computer and Systems Sciences, Vol. 7, 1984.

[Same85] A. Sameh, *On Some Parallel Algorithms on a Ring of Processors,* Computer Physics Communications, Vol. 37, 1985, pp. 159–166.

[ScPa87] R. Schreiber and B. Parlett, *Block Reflectors: Theory and Computation,* Dept. Comp. Sci. Tech. Report 87–11, Rensselaer Polytechnic Institute, Troy, New York, March 1987.

[Stew73] G. W. Stewart, **Introduction to Matrix Computations** ,Academic Press, New York, 1973.

[Ston71] H. Stone, *Parallel Processing with the Perfect Shuffle,* IEEE Trans. Comput. C–20, pp. 153–161, 1971.

[Swarz77] P. Swarztrauber, *The Methods of Cyclic Reduction, Fourier Analysis and the FACR Algorithm for the Discrete Solution of Poisson's Equation on a Rectangle,* SIAM Review 19, pp. 490–501, 1977.

[Swee87] R. Sweet, *A Parallel and Vector Variant of the Cyclic Reduction Algorithm,* to appear in

SIAM J. Sci. Stat. Comput.

[Temp79] C. Temperton, *Direct Methods for the Solution of the Discrete Poisson Equation: some Comparisons,* J. Comput. Phys. 31, pp. 1–20, 1979.

[Temp80] C. Temperton, *On the FACR(l) Algorithm for the Discrete Poisson Equation,* J. Comput. Phys. 34, pp. 314–329, 1980.

[Vors82] H. van der Vorst, *A Vectorizable Variant of Some ICCG Methods,* SIAM J. Sci. Stat. Comput., Vol. 3, 1982, pp. 350–356.

[Wilk65] J. Wilkinson, **The Algebraic Eigenvalue Problem** , Oxford, 1965.

[WiRe71] J. Wilkinson and C. Reinsch, **Handbook for Automatic Computation** , Vol. 2, Linear Algebra, Springer–Verlag, 1971.

[Yew86] P.C. Yew, *Architecture of the CEDAR Parallel Supercomputer,* CSRD Report, CSRD University of Illinois at Urbana–Champaign, 1986.

Hypercube Implementation of Some Parallel Algorithms in Control †

Alan J. Laub
Judith D. Gardiner

Department of Electrical and Computer Engineering
University of California
Santa Barbara, CA 93106
USA

KEYWORDS / ABSTRACT : parallel algorithms / control theory / control engineering / hypercube multiprocessors / numerical analysis / second-order systems / frequency response / Riccati equations

A tutorial introduction is given to some parallel algorithms for certain computational problems arising in state-space-based modeling in control engineering. The need for parallel algorithms is motivated by considering computations derived from state-space models involving high-dimensional matrices. Such models occur frequently in, for example, finite element modeling and are typically in so-called second-order form. Specific problems addressed in this paper and used for illustrative purposes include the solution of linear systems of equations, calculation of multivariable frequency response matrices from state-space data, and the solution of algebraic Riccati equations. Parallel algorithm implementations are described for a 32 processor Intel iPSC Hypercube.

† This research was supported by the National Science Foundation (and AFOSR) under Grant No. ECS84-06152 and the Office of Naval Research under Contract No. N00014-85-K-0553.

NATO ASI Series, Vol. F47
Advanced Computing Concepts and Techniques
in Control Engineering
Edited by M. J. Denham and A. J. Laub

1. INTRODUCTION

Computation has long been recognized as an indispensable component of control. It is needed both for off-line functions such as simulation, analysis, and synthesis of control systems as well as for on-line functions associated with control system implementations in embedded processors. As system models grow larger and more complex, computational requirements are pushing the limits of conventional computers and algorithms.

For off-line functions, major contributions have been made over the last decade in providing efficient, reliable, and portable tools (numerical algorithms and software) for many of the common control design procedures. These include things like linear algebra and matrix algebra support routines, Lyapunov and Riccati equation solvers, frequency response calculators, and so forth, which now work routinely and reliably on problems with state dimensions "in the hundreds." (This is true for general problems without special structure such as bandedness or symmetry. For the latter, of course, much larger problems are also routine.) Future design problems, however, involve models with dimensions "in the thousands" or even larger. These arise in complex distributed systems such as large space structures or interconnected networks of power systems whose dynamics extend over wide frequency ranges or spatial domains and cannot readily be reduced to simpler low-dimensional form. Traditional design algorithms must therefore be extended in dimensional capability by at least an order of magnitude, with little or no sacrifice in reliability. Technological advances in microelectronics have made it possible to design and build economically computer systems capable of meeting the challenge of solving enormously complex and large-order numerical and nonnumerical problems at extremely high rates of speed. Parallelism in both processors themselves and communications among processors and memories holds the promise of major breakthroughs in computation for control.

There is intense interest throughout the scientific and engineering communities in the design of novel computer architectures and algorithms for coping with present and future computational problems. Many existing control algorithms can be restructured to exploit the power of parallel machines but completely new algorithms will need to be invented to solve problems considered heretofore impractical or impossible. Since very large state space models often have sparse or otherwise structured matrices, the new algorithms must take advantage of structure better than most current algorithms do.

Research into control algorithms that exploit advanced specialized architectures is only now in its infancy. We are beginning to see some research into control algorithms (principally for off-line calculations) on parallel and vector machines. There is also research into implementation of control algorithms on specialized VLSI chips and systolic arrays. It does appear certain that research into algorithms implemented on specialized architectures offers genuine opportunity and hope for efficient solution of many classes of presently intractable high-order problems as well as many classes of real-time calculations.

This tutorial-level paper will focus on a particular class of parallel computers called hypercube multiprocessors, and some algorithms and programming techniques that have been developed for them. The computational problems described are of rather generic interest so that the algorithms and software discussed have widespread application to aerospace systems, mechanical systems, power systems, and so forth. Briefly, the rest of the paper is organized as follows. In Section 2 some background material is provided on second-order models. These ubiquitous linear models motivate the need to deal with very high-dimensional matrices in control and system theory. A number of "open" numerical problems are discussed. In Section 3 a brief overview is given of hypercube computers, including their architecture and programming. Two simple examples of parallel algorithms are described in Section 4: a simplified frequency response problem and the solution of a system of linear equations. In both cases, a specific Fortran implementation on an Intel iPSC/d5 Hypercube computer is given. Actual code is listed and used to illustrate clearly many basic programming ideas for parallel message-passing computers. Finally, detailed hypercube implementations of two more "sophisticated" algorithms of interest in control are given in Section 5. These algorithms are the computation of a multivariable frequency response matrix from state space data and the solution of an algebraic Riccati equation.

2. SECOND-ORDER MODELS

Recent activity in the control of large flexible space structures has led to an increasing algorithmic interest in the second-order linear models by which they are often approximated. The matrices comprising such models are frequently of very high order and may also be sparse. They generally have structure associated with them that can be exploited to advantage in numerical calculations. Sparse matrices of the order of many hundreds or thousands are not at all unusual in these models, so techniques intended for state space calculations with unstructured (dense) matrices may not necessarily be appropriate. Large, dense matrices are also encountered and parallel algorithms appear to offer a promising vehicle for handling them efficiently.

In this section we shall first outline some of the salient features of large space structure modeling. We then define a generic second-order model which has inspired much of our research into parallel algorithms for control engineering and describe some of its properties. It should be noted that our work deals only with second-order models and not with large space structures *per se* or other applications to which our results may equally well apply.

2.1. Large Space Structures

Some of the defining characteristics of large space structures (LSS) are given in Balas [1] and Mackay [2]. The structures are physically quite large. Because of mass limitations they are very flexible, often having many closely spaced resonances at low frequencies. Stringent requirements for pointing accuracy, vibration suppression, and maintenance of shape may call for active control with a high controller bandwidth, leading to dynamic interactions between the controller and the structural vibration modes.

Large space structures are usually described in terms of partial differential equation (PDE) models which, in theory, implies an infinite-dimensional state. Although the state is infinite-dimensional, the control and observation vectors are usually finite-dimensional. That is, the sensors and actuators are considered point devices, making the controller finite-dimensional.

The PDE model may be discretized using the finite element method (FEM) to produce a large finite-dimensional model for use in control design computations. Alternatively, in some cases the design may be carried out on the infinite-dimensional model and the result discretized [2]. The reasoning behind the latter approach is that discretization introduces new errors and the PDE model leads to a better design. Hughes [3] argues that the since the PDE model itself is an approximation, with poor accuracy at high frequencies, the finite-dimensional approach is actually preferable.

Since most control design is done with finite-dimensional models, the question arises of how well the closed-loop stability properties of the finite-dimensional model approximate those of the actual system or (not the same thing) the infinite-dimensional model. Greene and Stein [4] discuss the importance of damping models to this problem. Junkins and Rew [5] attempt to solve it by means of robust eigenstructure assignment methods. Finally, a controller that has been designed must be implemented. Hanks [6] has studied the problem of approximating the performance of an optimal controller through structural changes.

2.2. Second-Order Models

The generic form of what we shall call a *second-order model* or *Rayleigh model* is a system of differential equations

$$M\ddot{q} + C\dot{q} + Kq = Du \tag{2.1}$$

with, possibly, an associated output equation

$$y = Pq + S\dot{q} + D_y u \tag{2.2}$$

where $q \in \mathbb{R}^n$ and where, in many situations of interest, the model matrices are assumed to have the following structure:

$$M = M^T > 0$$

$$K = K^T \geq 0$$

$$C = C_1 + C_2 \;\; ; \;\; C_1^T = C_1 \geq 0 \;\; , \;\; C_2^T = -C_2 \;\; .$$

In this model, M is the mass matrix, C is the Rayleigh matrix with C_1, the dissipation matrix, representing structural damping forces and C_2, the gyroscopic matrix, representing gyroscopic forces, and K is the Hooke matrix or stiffness matrix. A general K matrix could be considered which can then be decomposed into a nonnegative definite conservative response matrix and a skew-symmetric circulatory response matrix, but for most applications circulatory forces are ignored and only conservative force fields are studied. Various additional assumptions (such as proportional damping with $C_2 = 0$, $C_1 = \alpha M + \beta K$ for constant scalars α and β) are often made to try to force some sort of analytic tractability. However, it is of considerable research interest to try to carry through as much analysis as possible for the case of arbitrary damping C. For further discussion of the significance and limitations of this model, see Arnold [7], Balas [1], and Meirovitch [8].

The second-order model (2.1) arises naturally throughout the sciences and engineering in at least three ways: (i) as a model of the system itself (e.g., "$F = ma$"), (ii) from a finite-dimensional Lagrangian linearized around a constant or periodic solution, or (iii) from a finite element method applied to a dynamic continuum problem. In the latter case, finite element codes such as NASTRAN are used for modeling large structures and produce mass and stiffness matrices for the model (2.1). These matrices can range in size from small to very large.

Various first-order realizations are immediately derived for these equations and this is what is usually done to precede any control calculations. The special structure of the resulting first-order model is typically ignored in most standard computational algorithms. There are several ways to express (2.1) as an equivalent first-order model. Most introduce the state vector $x \in \mathbb{R}^{2n}$ where $x = \begin{bmatrix} q \\ \dot{q} \end{bmatrix}$. The standard realization then takes the form

$$\dot{x} = \begin{bmatrix} 0 & I \\ -M^{-1}K & -M^{-1}C \end{bmatrix} x + \begin{bmatrix} 0 \\ M^{-1}D \end{bmatrix} u \tag{2.3}$$

$$y = (P\ ,S)x + D_y u \tag{2.4}$$

For numerical reasons, it is often desirable to avoid the need for M^{-1} on the right-hand-side of (2.3) so the generalized state space realization

$$\begin{bmatrix} I & 0 \\ 0 & M \end{bmatrix} \dot{x} = \begin{bmatrix} 0 & I \\ -K & -C \end{bmatrix} x + \begin{bmatrix} 0 \\ D \end{bmatrix} u \tag{2.5}$$

can be used to replace (2.3).

Other first-order realizations of (2.1) are possible; see, for example, [9]. These realizations may exploit, for example, symmetry or skew-symmetry depending on special structure in the damping matrix C. In each case, the first-order realization is of the "standard" $\dot{x} = Ax + Bu$ or $E\dot{x} = Ax + Bu$ form. However, except in particular cases, much of the special structure of the second-order model matrices is lost. One attempt to retain symmetric structure can be described as follows [10]. Consider a general transformation

$$w = E_2\dot{q} + F_2 q \quad .$$

Setting

$$M\ddot{q} + C\dot{q} + Kq = E_1\dot{w} + F_1 w + Gq$$

we find the following three equations that must be solved

$$M = E_1 E_2$$

$$C = E_1 F_2 + F_1 E_2$$

$$K = F_1 F_2 + G \quad .$$

A classical solution to these equations is $E_2 = M$, $F_2 = 0$, $E_1 = I$, $F_1 = CM^{-1}$, and $G = K$. This gives essentially the realization (2.3) or (2.5). However, a "symmetric" solution is

$$E_1 = E_2 = M^{1/2}$$

$$F_1 = F_2 = F$$

where F solves the Lyapunov equation

$$M^{1/2}F + FM^{1/2} = C$$

and

$$G = K - F^2 \quad .$$

This gives rise to the realization

$$\begin{bmatrix} M^{1/2} & 0 \\ 0 & M^{1/2} \end{bmatrix} \begin{bmatrix} \dot{q} \\ \dot{w} \end{bmatrix} = \begin{bmatrix} -F & I \\ F^2 - K & -F \end{bmatrix} \begin{bmatrix} q \\ w \end{bmatrix} + \begin{bmatrix} 0 \\ D \end{bmatrix} u \tag{2.6}$$

It is of interest to try to develop some control and system theory which applies directly to models of the form (2.1), (2.2) in the hopes of exploiting the structure therein rather than going to one of the first-order realizations above in which it is considerably more difficult to take advantage of symmetry and so forth. First-order realizations do, however, have the advantage of the availability of a large number of canonical forms which are of great utility in numerical calculation. Numerically useful and reliable canonical forms for systems of the form (2.1), (2.2) do not presently exist.

One fairly simple example in which remaining in second-order form pays dividends over first-order form is in determining controllability (or, dually, observability) of (2.1). When posed in terms of the "A" matrix of (2.3) or the "(E, A)" pair of (2.5) or (2.6) standard controllability tests are in terms of $2n \times 2n$ matrices and some of these tests can turn out to be quite complicated. However, for the controllability problem, it is possible to apply the Hautus-type pencil test directly whereby it can be shown [9], [11] that (2.1) is controllable if and only if

$$rank\,[\lambda^2 M + \lambda C + K \;, D] = n \tag{2.7}$$

for all $\lambda \in W$ where W, the set of modes of the model (2.1), is the set of generalized eigenvalues of the pencil

$$\begin{bmatrix} 0 & I \\ -K & -C \end{bmatrix} - \lambda \begin{bmatrix} I & 0 \\ 0 & M \end{bmatrix} \tag{2.8}$$

Clearly, controllability of particular modes can be checked selectively and tests for other "abilities" such as observability, stabilizability, etc. follow similarly. However, even in (2.7) the full structure of the second-order model is not being exploited. For example, the symmetry or definiteness of M or K is not being used explicitly.

Some interesting open problems can be posed by examining the structure of Lyapunov and Riccati equations derived from state-space models in second-order form.

Consider the Lyapunov equation

$$\mathcal{A}\mathcal{X} + \mathcal{X}\mathcal{A}^T + \mathcal{B}\mathcal{B}^T = 0 \tag{2.8}$$

where

$$\mathcal{A} = \begin{bmatrix} 0 & I \\ A_1 & A_2 \end{bmatrix} \tag{2.9}$$

with $A_i \in \mathbb{R}^{n \times n}$ and

$$\mathcal{B} = \begin{bmatrix} 0 \\ B \end{bmatrix} \tag{2.10}$$

with $B \in \mathbb{R}^{n \times m}$. There are many interesting sets of assumptions possible on the coefficient matrices of (2.8) to guarantee various properties of the solution $\mathcal{X}$. For convenience in this discussion let us make the common control-theoretic assumptions that $\mathcal{A}$ is asymptotically stable and the pair $(\mathcal{B}^T, \mathcal{A})$ is detectable. In that case there is a unique nonnegative definite solution of (2.8) which we shall denote in compatibly partitioned form by

$$\mathcal{X} = \begin{bmatrix} X & Y \\ Y^T & Z \end{bmatrix} = \mathcal{X}^T \tag{2.11}$$

Then the $2n \times 2n$ Lyapunov equation (2.8) is easily seen to be equivalent to the following three coupled $n \times n$ equations:

$$Y + Y^T = 0 \tag{2.12}$$

$$Z + A_1 X + A_2 Y^T = 0 \tag{2.13}$$

$$A_2 Z + Z A_2^T + B B^T + A_1 Y + Y^T A_1^T = 0 \;. \tag{2.14}$$

Once Y has been chosen, (2.14) is a standard Lyapunov equation for Z. The interesting question is which skew-symmetric matrix Y should be chosen so that, for example, Z is nonnegative definite or at least symmetric. Moreover, X then needs to be found from (2.13) and the choice of Y clearly has an effect on whether X is nonnegative definite or at least symmetric. Finally, X, Y, and Z must be chosen so that $\mathcal{X} \geq 0$.

The discrete-time analog of (2.8) with $\mathcal{A}$ and $\mathcal{B}$ as in (2.9), (2.10) can also be investigated, along with the appropriate analog of the coupled matrix equations (2.12)-(2.14).

The Lyapunov equation (2.8) is the linear analog of another important class of matrix equations in control and system theory, namely Riccati equations. There are close connections between the two and progress in the former will likely be necessary for the latter. Consider the following simple linear-quadratic optimal control problem:

$$Min \ \tfrac{1}{2} \int_0^{+\infty} [y^T Qy + u^T Ru]dt \tag{2.15}$$

$$s.t. \quad M\ddot{x} + C\dot{x} + Kx = Du \tag{2.16}$$

$$y = Px + S\dot{x} \tag{2.17}$$

where $x(t) \in \mathbb{R}^n$, $y(t) \in \mathbb{R}^r$, and $u(t) \in \mathbb{R}^m$. This problem is of the "standard form" except that the linear constraint is in second-order form. Now, assuming the necessary stabilizability and detectability conditions to guarantee a unique nonnegative definite solution, $\mathcal{X} = \mathcal{X}^T \in \mathbb{R}^{2n \times 2n}$, of the Riccati equation associated with this problem, write this solution in the form

$$\mathcal{X} = \begin{bmatrix} X & Y \\ Y^T & Z \end{bmatrix} \tag{2.18}$$

where $X, Y, Z \in \mathbb{R}^{n \times n}$. Then substituting a first-order realization of this problem (in, say, generalized state space form (2.5)) into a generalized algebraic Riccati equation [12] yields the following three matrix equations for X, Y, and Z:

$$YK + KY^T + YDR^{-1}D^T Y^T - P^T QP = 0 \tag{2.19}$$

$$MZC + C^T ZM + MZDR^{-1}D^T ZM - S^T QS - MY^T - YM = 0 \tag{2.20}$$

$$X - YC - KZM - YDR^{-1}D^T ZM + P^T QS = 0 \tag{2.21}$$

The optimal feedback control takes the form

$$u(t) = -R^{-1}D^T(Y^T x(t) + ZM\dot{x}(t)) \ . \tag{2.22}$$

Now observe that (2.19) is a Riccati-like equation which can be solved for Y. This solution for Y can then be substituted in (2.20) which becomes a standard generalized Riccati equation for Z. The solutions Y and Z can then be substituted into (2.21) to get X directly. Note that X is not required in the feedback gain matrix in (2.22).

Many open questions are immediate and obvious. The most difficult is, of course, the nature of the solutions to (2.19). This equation presumably has many solutions and, of these, which one(s) must be chosen so that when substituted in (2.20) and (2.21) gives nonnegative definite X and Z and nonnegative definite $\mathcal{X}$. Equation (2.19) is Riccati-like, in the sense of being "quadratic" in Y, but it is not a standard Riccati equation.

Many special cases are of interest such as the case of zero damping ($C = 0$). In particular, consider a model with no structural damping ($C = -C^T$), rate feedback only ($P = 0$), and collocated actuators and sensors ($DR^{-1}D^T = S^T QS$). For this problem it is easily verified that the Riccati equation solution is given (in terms of 2.18) by $X = K$, $Z = M^{-1}$, and $Y = 0$. The discrete-time analog of the above can also be investigated.

It is hoped that the material in this section will give the reader some indication of the kinds of control-related and numerical problems that arise in so-called second-order models. As mentioned above, these models typically involve matrices of very high order thereby providing a fertile source of problems requiring parallel algorithms and computers for their solution. In the next section we shall give an overview of one particular class of parallel computers which may be used to solve such problems.

3. HYPERCUBE "MINI-SUPERCOMPUTERS"

One of the most popular classes of parallel computers is the hypercube multiprocessor. In this paper we shall be interested specifically in so-called distributed-memory hypercubes in which the processors execute programs independently and communicate only by passing messages. The word "hypercube" actually refers only to the topology of the communication network connecting the processors, but much of what follows applies to distributed-memory multiprocessors based on other topologies as well.

This section describes the hypercube architecture, provides some simple programming examples, and discusses some algorithms and software that have been developed for the hypercube. We shall consider only scientific computation, although there is a considerable amount of work being done on artificial intelligence applications as well. For more information the reader is referred to tutorials by Wiley [13] and Karp [14] and to the proceedings of two hypercube conferences [15], [16].

Hypercube promoters often call their machines "mini-supercomputers" or "personal supercomputers." Hypercubes are much less expensive than vector processing supercomputers, such as the Cray, yet they are intended for similar applications. With the right algorithms and software, hypercubes can theoretically achieve computing speeds comparable to supercomputer speeds.

Although they were introduced only a short time ago, in 1983, hypercubes are already making the transition from experimental to practical machines. There are at least four commercial manufacturers of hypercubes, as well as several research-level projects. Table 3.1 summarizes the features of these computers. The software base, while still limited, is growing as researchers convert old algorithms to parallel form and develop new algorithms. Enough support software exists to make the hypercube a useful research tool.

Before describing the hypercube architecture in detail, it will be beneficial to see where hypercubes fit into the world of parallel processors and supercomputers. While a complete comparison of the hypercube with other architectures is far beyond the scope of this paper, a few comments will provide some perspective. Many high performance scientific computers utilize more than one form of parallelism, so this is a discussion of architectural concepts rather than specific machines.

At the high end of the scale in terms of performance, ease of programming, and, of course, price are the vector processing supercomputers. Vectorizing compilers allow old algorithms and code to be run on a vector processor with only minor modifications, although for maximum performance more work is required. Hypercubes typically cost much less, their theoretical maximum performance is less (at least for now), and efficient programs are more difficult to develop. Parallelizing compilers are still very much in the early research stages, so parallelism must be explicitly identified and exploited by the programmer.

Another popular parallel architecture is the shared-memory multiprocessor, in which several processors share a common memory. A hypercube of the kind that we are discussing has no shared memory. Each processor operates only on its own data or its own copy of common data, and any sharing of computational results is handled through message-passing. For a shared-memory

Machine	Developer	Year	Topology	Maximum number of nodes	Maximum memory per node, Kbytes	Node CPU	Estimated node performance, megaflops
Waterloop/64	University of Waterloo Ontario, Canada	1983	Loop	64	128	8086/87	0.025
Cosmic Cube	Calif. Inst. of Technology Pasadena, California	1983	Hypercube	64	128	8086/87	0.025
Mark II	Jet Propulsion Lab(Caltech) Pasadena, California	1985	Hypercube	64	256	80286/287	0.035
iPSC	Intel Scientific Computers Beaverton, Oregon	1985	Hypercube	128	512	80286/287	0.035
System 14	AMETEK, Inc. Arcadia, California	1985	Hypercube	256	256	80286/287	0.035
NCube/ten	NCube Beaverton, Oregon	1986	Hypercube	1024	128	Special	0.3-0.5
Computing Surface	Meiko Kanagawa, Japan	1986	2-dimensional mesh	84	128	Transputer	-
iPSC-VX	Intel Scientific Computers Beaverton, Oregon	1986	Hypercube	64	1.5K	Vector	6-20
T-Series	Floating Point Systems Beaverton, Oregon	1986	Modified Hypercube	16384	1.0K	Vector	16-20
Connection Machine *	Thinking Machines Corp. Cambridge, Massachusetts	1986	Hypercube	65536	0.5	Special (1 bit serial)	-
Butterfly *	Bolt, Beranek, & Newman Cambridge, Massachusetts	1986	Banyan Switch	256	1.0K	68020/881	0.1
Mark III	Jet Propulsion Lab(Caltech) Pasadena, California	1986	Hypercube	1024	4.0K	Vector	20

* synchronized execution

Table 3.1. Characteristics of Some Distributed Memory Multiprocessors
Adapted from Wiley [13]

computer, the programmer has to allocate processing tasks among the processors, but all the data is available to any processor. Thus the problems of distributing the data and trying to localize access are not as critical as they are on the hypercube, although they do arise when hierarchical memory systems are used. The disadvantage of the shared-memory model is that the number of processors is generally limited to a small number, while a distributed-memory machine may have thousands of processors.

Systolic arrays, in which a large number of processors execute in lock-step, may have even more potential processing power than the hypercube, but they are correspondingly more difficult to work with. Computers of this type are not yet practical for general purpose scientific computing.

3.1. Architecture

The hypercube topology is a particular pattern of interconnections among the processors of a computer. A hypercube can be viewed as a cube of some dimension d (Figure 3.1). Each vertex of the cube is a processor, often called a node, and the edges are communication links. The number of processors, p, is always a power of 2, $p=2^d$. Each node can communicate directly only with its nearest neighbors; messages destined for a more distant node must be forwarded through one or more intermediate nodes.

Other interconnection schemes have been used in some-distributed memory multiprocessors, but the hypercube has been the most successful. The maximum distance between any two nodes is the dimension d. The total number of connections is $(p/2)\log_2 p$, or $(p/2)d$. By contrast, a fully

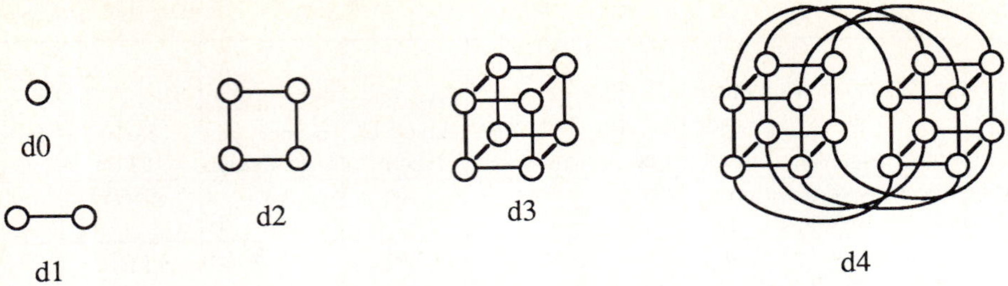

Figure 3.1. Hypercube topology, dimensions 0 through 4

connected system lets any processor communicate directly with any other, but the number of communication links is proportional to p^2, which is prohibitively expensive with a large number of processors. A ring, which is one of the simplest topologies, requires only p links but has a maximum internode distance of $p/2$, leading to slow communications for many tasks. The hypercube architecture is a good compromise between these two extremes.

Another advantage of the hypercube topology is its flexibility in simulating other connection schemes. By using only a subset of the available communications paths, a programmer can view the hypercube as a ring, a 2-dimensional or 3-dimensional mesh, or a tree. The optimal use of the topology depends heavily on the application.

Current commercial hypercubes range in size from 16 nodes to about 16000 nodes. The individual processing nodes can be as simple or as powerful as one desires, from microprocessors to vector processors. Each node is an independent computer in the sense that it has its own processor and memory, a copy of the operating system, and copies of any applications programs that are being run. The nodes may run different programs or they may have copies of the same program operating on different data.

In addition to the processing nodes of the cube, there is usually another processor called a programming host or cube manager which handles program development and controls access to the cube itself. Input and output, as well as interface with the user, are handled by the host. While the host is sometimes used for computation or to coordinate computation in the nodes, this is not a recommended practice since it creates a bottleneck and usually slows a program down. Communication between a node and the host may be much slower than between two nodes.

3.2. Programming

Programming on a hypercube is usually done in a high level language, such as Fortran or C for scientific applications, with the aid of special subroutines for message-passing. A program running on the host computer takes care of administrative functions like reading input data, displaying results, and loading and starting the node programs. All the real work is done by the node programs. Most scientific applications use a single node program, which is duplicated in all the nodes. A node can determine its identity (node number from 0 to $p-1$), so it can specialize its processing if necessary. For example, if the topology is to be viewed as a mesh, each node knows its coordinates in the mesh; in matrix computations, it knows which columns of the matrix it is working on.

Unless an application is perfectly parallel, the node programs must communicate with each other in the course of solving a problem. All the hypercube manufacturers provide a node operating system and utility subroutines to handle the details of internode communications. The basic functions required for passing messages are "send" and "receive". To send a message, a process (node program) calls the send subroutine, passing it the node number of the destination node, an array of data containing the message, and possibly a message type. To receive a message, the process calls the receive subroutine with an array into which the message is to be stored.

Communication between nodes is typically slower than computation. High startup times on messages make it desirable to send a few long messages rather than many short messages. In matrix computations, a message is often a vector or a column of a matrix. For efficiency, communications should be between adjacent processors (nearest neighbors) whenever possible. Each node is adjacent to d other nodes, where d is the dimension of the cube. Most hypercube operating systems will forward messages automatically if a program wants to send a message to a nonadjacent node.

There are several ways to decompose a problem for parallel execution. Some problems have an explicitly parallel structure. The problem domain may be a physical rod or plate and each processor handles a particular region. In this case, the parallelism is obvious and communication is between nearest neighbors only. For numerical integration the range of integration may be split among the processors; the only communication required is a summation of the partial results at the end of the computation. Most matrix computations are split by data parallelism. Each node is assigned certain columns of the matrix which it operates on. Communications in this case are not just between nearest neighbors, but they are quite regular. Usually broadcasts and tree-oriented operations are used.

3.3. Status of hypercube algorithms and software

The versatility of the hypercube architecture has attracted researchers from many computational areas. Some problems are naturally well suited for parallel implementation while others can be quite difficult to parallelize. Here we shall briefly review the status of hypercube algorithms and software. Only algorithms that have been implemented and made to work are described; "paper studies" have been excluded. Most of the software is still in experimental form, i.e., not user-friendly. The main areas of activity have been in linear systems, symmetric eigenvalue problems, and partial differential equations.

An amazing number of different algorithms have been proposed for the parallel solution of large, dense linear systems. For the most part, the algorithms are based on the standard sequential algorithms of Gaussian elimination or QR factorization followed by forward and backward substitution. The differences in the parallel algorithms are in the way the matrix elements are distributed among the processors, the order in which certain operations are performed, and the communications pattern. One simple yet effective algorithm is described in detail in the next section.

The symmetric eigenvalue problem has been solved on the hypercube by at least three different algorithms: multisection, Cuppen's method, and the Jacobi method. All require an initial reduction to tridiagonal form, which can be accomplished efficiently with the use of Householder transformations in the form of rank one updates. For computing the eigenvalues of a symmetric tridiagonal matrix, the most straightforward method is multisection, a generalization of bisection. Cuppen's method is a fairly new algorithm that was developed with parallel computation in mind. It has proved to be quite efficient for both sequential and parallel architectures. The Jacobi method, which is not particularly impressive as a sequential algorithm, has been implemented on several parallel architectures, including the hypercube. Interestingly enough, the most popular sequential algorithm, the QR algorithm, seems poorly suited to parallel computation.

Partial differential equations (PDE's) have probably received more attention than any other class of problems in the area of parallel processing. In addition to the hypercube conference proceedings already mentioned, reference [17] gives a good overview of this subject.

Other algorithms have also been moved to the hypercube, with varying degrees of success. Of particular interest to control engineers are the unsymmetric eigenvalue problem, frequency response calculation, and solution of algebraic Riccati and Lyapunov equations. Algorithms for the latter two problems are described in detail in the next section. The unsymmetric eigenvalue problem is an example of a problem that is very difficult to parallelize. The shifted QR algorithm has been implemented on the hypercube, but high communication requirements keep it from being as efficient as one would like.

4. PARALLEL ALGORITHMS: EXAMPLES

The most effective way to gain an understanding of parallel programming concepts is through actual program examples. This section presents two simplified programs for the hypercube followed by two detailed algorithms. All are genuine programs that have been run on an Intel iPSC/d5 hypercube (32 processors). The simplified examples serve to illustrate several important parallel programming techniques. The Fortran code is included. The selected problems are two that arise regularly in control engineering, namely computation of the peak of a frequency response and the solution of a system of linear equations. Detailed algorithms are then given in Section 5 for a general multivariable frequency response problem and the solution of algebraic Riccati equations. The detailed algorithms are based largely on the authors' work [18].

4.1. Simplified Example: Frequency Response

Calculation of frequency response for a multivariable state space model requires a large amount of computing time, although it is conceptually a simple process. A matrix function of the form $C(sI-A)^{-1}B$ must be evaluated repeatedly, where A, B, and C are matrices and s is a complex-valued scalar (usually on the imaginary axis or unit circle). For an efficient sequential algorithm see [19]. This problem is almost perfectly parallel, since the response at each value of the parameter s can be computed independently. For this example it will be assumed that the model is small enough that the response at a single frequency can be computed on a single processor. This constraint allows us to create a parallel program from a sequential one with almost no modification.

We define our specific problem as follows: Compute the maximum principal gain, or peak frequency response, over a specified frequency range. As input the program reads the model matrices A, B, and C, minimum and maximum frequencies ω_{min} and ω_{max}, and the number *NFREQS* of data points to compute. The output is the value of the maximum principal gain with its corresponding frequency. For simplicity of the example program we require that *NFREQS* be a multiple of the number of processors p. This requirement is easily relaxed, but it complicates the code somewhat. The model size is limited to 64 states, 32 inputs, and 32 outputs. The procedure is to compute $G(j\omega) = C(j\omega I-A)^{-1}B$ for logarithmically spaced ω between ω_{min} and ω_{max}. ($G(j\omega)$ is a complex-valued matrix.) For each ω we compute the singular value decomposition (SVD) of $G(j\omega)$. (The maximum principal gain is the maximum singular value.) If the maximum principal gain for this frequency is larger than the previously computed maximum, it is saved.

This problem can easily be solved on a single processor, but given its natural parallelism, it can be solved much faster on 32 processors. All of the computation is done with standard, published code which required no modification. Most of the communication is accomplished with routines supplied by Intel, so very little new code had to be written. Next we shall examine the host and node programs in detail. Please remember that this software is meant to illustrate certain parallel programming principles. It does not necessarily represent the most efficient way to solve the problem, and it certainly is not robust.

The host program is shown in Figure 4.1. It begins by setting up a communications area using the system call COPEN. Next it initializes the cube by loading the node program into all the processors. The call to LOAD requires the name of the node program (unimaginatively called "NODE" for this example), the identities of the processors to be loaded (-1 for all of them), and the "process id" (PID). Since we are running only one program on each node, we always use PID 0.

Communications with the user are always handled by the host program. Our example program prompts the user for the size of the matrices, the frequency range, and the number of points to compute. These parameters are copied into a message buffer and sent to all the node processors using SENDMSG. The model matrices A, B, and C are read from disk files and sent as three separate messages to the nodes. The call to RECVMSG causes the host program to wait until a message arrives from one of the nodes (actually node 0, but the host does not need to know that). The received message consists of two values, the peak gain and its corresponding frequency, which are then displayed for the user.

The node program (Figure 4.2) executes simultaneously in all the node processors and handles all the calculations. An important feature, and one shared by most hypercube programs, is that the code is independent of the dimension of the cube. This particular code has been run on cubes of dimension 0 (d0, a single processor) through d5 (32 processors).

The node program first performs some initialization tasks. It opens a communications area using COPEN, determines the id of the node it is executing on with MYNODE, and reads the dimension of the cube with CUBEDIM. The number of nodes P is computed from the cube dimension. The node id, ID, is a number from 0 to P-1.

Next the program reads the messages sent by the host. Each call to RECVW causes the program to wait for a message to arrive if there is no message already there. The first message has to be decoded, since it contains several different parameters. The next three messages are read directly into the arrays for the A, B, and C matrices. In this case, all the nodes receive copies of the same data.

For this example we decided to split the work among the processors by assigning each node a contiguous portion of the frequency range. Each node computes the frequency response for NFREQS/P different frequencies. The starting frequency (actually the logarithm of it) is calculated using the node id, so that each node starts at a different point in the overall range. This is the key computational point in the whole example. The variable ID parameterizes the calculations being performed in parallel by processors 0 through P-1.

The next section of code is identical to the corresponding code on a sequential computer. The subroutine SFRMG is an unmodified copy of published code [20] for calculating frequency response. CSVDC is the Linpack [21] routine for finding the singular values of a complex matrix. Each node saves the highest gain in its subrange in the variable PEAK and the frequency in FPEAK.

Finally, after each node has found its local maximum, the global maximum must be found. This step requires communication among the nodes. The subroutine GMAX (global maximum) is a modified version of the protocode subroutine GOP (global operation) provided by Intel. The code for GMAX is shown in Figure 4.3. It uses the spanning tree view of the hypercube shown in Figure 4.4. Values are passed up the tree until the root node (in this case node 0) has the maximum value. The root node passes the result to the host, which displays it for the user.

```
      PROGRAM HOST
C
C HOST PROGRAM  --  COMPUTE PEAK FREQUENCY RESPONSE GAIN (SIMPLIFIED)
C
      INTEGER CID,TYPE,CNT,NID,PID,RNID,RPID
      INTEGER COPEN
      INTEGER N,M,R,NFREQS
      REAL FMIN,FMAX
      REAL RMSG(6),TMP(64*64)
C
C OPEN COMMUNICATIONS AREA
C
      PID = 0
      CID = COPEN(PID)
C
C LOAD NODE PROGRAMS
C
      CALL LOAD ('NODE', -1, PID)
C
C READ PROBLEM SIZE, FREQUENCY RANGE, AND NUMBER OF POINTS FROM THE USER
C
      WRITE (*,*) ' ENTER NUMBER OF STATES, INPUTS, AND OUTPUTS: '
      READ (*,*) N, M, R
      WRITE (*,*) ' ENTER FMIN, FMAX, NUMBER OF POINTS TO COMPUTE: '
      READ (*,*) FMIN, FMAX, NFREQS
C
C SEND MESSAGE TO ALL NODES
C
      RMSG(1) = N
      RMSG(2) = M
      RMSG(3) = R
      RMSG(4) = NFREQS
      RMSG(5) = FMIN
      RMSG(6) = FMAX
      CALL SENDMSG (CID,1,RMSG,24,-1,PID)
C
C READ MATRICES AND SEND TO ALL NODES
C
      READ (2,*) (TMP(I),I=1,N*N)
      CALL SENDMSG (CID,2,TMP,4*N*N,-1,PID)
      READ (3,*) (TMP(I),I=1,N*M)
      CALL SENDMSG (CID,3,TMP,4*N*M,-1,PID)
      READ (4,*) (TMP(I),I=1,R*N)
      CALL SENDMSG (CID,4,TMP,4*R*N,-1,PID)
C
C RECEIVE THE RESULT FROM ONE OF THE NODES
C
      CALL RECVMSG (CID,TYPE,RMSG,8,CNT,RNID,RPID)
      WRITE (*,30) RMSG(1), RMSG(2)
   30 FORMAT (' PEAK GAIN IS ',F10.2,
     &        ' AT FREQUENCY ',F10.2,' RADIANS/SEC')

      END
```

Figure 4.1. Host program for frequency response example

```
      PROGRAM NODE
C
C NODE PROGRAM  --  COMPUTE PEAK FREQUENCY RESPONSE (SIMPLIFIED)
C
      INTEGER PID,CID,ID,DIM,CNT,RNID,RPID,HOSTID
      INTEGER COPEN, CUBEDIM, MYNODE
      INTEGER P,N,M,R,NFREQS,IWRK(64)
      REAL A(64*64), B(64*64), C(64*64)
      REAL TMP(2), RMSG(6), WRK1(64), WRK2(64), WRK3(64)
      COMPLEX FREQ, G(64*64), S(64)
      COMPLEX CWRK1(64), CWRK2(64*64), CWRK3(64*64)
      LOGICAL FIRST
C
      PARAMETER (HOSTID=-32768)
C
C OPEN COMMUNICATIONS AREA
C FIND NODE NUMBER, CUBE DIMENSION, NUMBER OF PROCESSORS
C
      PID = MYPID()
      CID = COPEN(PID)
      ID = MYNODE()
      DIM = CUBEDIM()
      P = 2**DIM
C
C GET PROBLEM SIZE, FREQUENCY RANGE, AND MODEL MATRICES FROM THE HOST
C
      CALL RECVW (CID,1,RMSG,24,CNT,RNID,RPID)
      N = RMSG(1)
      M = RMSG(2)
      R = RMSG(3)
      NFREQS = RMSG(4)
      FMIN = RMSG(5)
      FMAX = RMSG(6)
C
      CALL RECVW (CID,2,A,4*N*N,CNT,RNID,RPID)
      CALL RECVW (CID,3,B,4*N*M,CNT,RNID,RPID)
      CALL RECVW (CID,4,C,4*R*N,CNT,RNID,RPID)
C
C COMPUTE FREQUENCY RANGE ASSIGNED TO THIS NODE, INITIALIZE
C
      FMINLOG = LOG10(FMIN)
      FMAXLOG = LOG10(FMAX)
C
      NFLOC = NFREQS / P
      FSTEP = (FMAXLOG-FMINLOG) / (NFREQS-1)
      FSTART = ID*NFLOC*FSTEP + FMINLOG
```

```
C
C COMPUTE FREQUENCY RESPONSE AND PRINCIPAL GAINS OVER THIS NODE'S
C RANGE, SAVING PEAK VALUE
C
      FIRST = .TRUE.
      IERR = 0
      INFO = 0
      PEAK = 0.0
C
      DO 20 I = 1,NFLOC
C
         W = 10.0 ** (FSTART + (I-1)*FSTEP)
         FREQ = CMPLX(0.0,W)
C
         CALL SFRMG (FIRST,N,N,R,R,M,N,A,B,C,FREQ,CWRK2,G,
     &               CWRK3,IWRK,CWRK1,WRK1,WRK2,WRK3,IERR,INFO,RCOND)
C
         CALL CSVDC (G,64,R,M,S,CWRK1,G,R,G,R,CWRK2,0,INFO)
C
         S1 = REAL(S(1))
         IF (S1 .GT. PEAK) THEN
            PEAK = S1
            FPEAK = W
         ENDIF
C
   20 CONTINUE
C
C FIND GLOBAL MAXIMUM AND SEND RESULT TO HOST
C
         TMP(1) = PEAK
         TMP(2) = FPEAK
         CALL GMAX (CID,5,TMP,2,HOSTID,DIM,WRK1)
C
      STOP
      END
```

Figure 4.2. Node program for frequency response example

4.2. Simplified Example: Linear Systems

The next example illustrates the use of the hypercube to solve a problem that may be too big to fit into the memory of a single processor. We want to solve the linear equation

$$Ax = b$$

where A is a real $n \times n$ matrix and b is a vector of length n. The standard sequential solution, implemented in Linpack [21] as subroutines SGEFA and SGESL, is to factor A using Gaussian elimination with partial pivoting, then solve the resulting equations using forward and back substitution. That is, A is factored as

$$PA = LU \quad ,$$

where P is a permutation matrix, L is unit lower triangular, and U is upper triangular. The systems that must be solved are then

```
      SUBROUTINE GMAX (CI, TYPE, X, N, ROOT, DIM, WORK)
      INTEGER CI, TYPE, N, ROOT, DIM
      REAL X(N), WORK(N)
c
c  Global maximum using spanning tree.  Maximum is taken with respect
c  to first element in vector X.  Other elements are carried along.
c
c  Adapted from Intel's protocode routine GOP.
c
      INTEGER BIT, BYTES, CNT, DIFF, RSIZE, I, IGNORE, ME, MYNODE,
     &        MYPID, P, PARENT, PID, TROOT, XOR, HOST
      PARAMETER (RSIZE = 4, HOST = -32768)
c
      ME = MYNODE()
      P = 2**DIM
c
c     Find temporary root (either the real root, or the lowest
c     numbered node in the active subcube--found by zeroing the
c     DIM lowest bits in mynode).
c
      TROOT = ROOT
      IF (IABS(ROOT) .EQ. IABS(HOST)) TROOT = (ME/P)*P
c
      PID = MYPID()
      DIFF = XOR(ME,TROOT)
c
c     Accumulate contributions from children, if any
c
      BIT = P/2
      BYTES = RSIZE*N
    5 IF (BIT .LE. DIFF) GO TO 20
         CALL RECVW(CI,TYPE,WORK,BYTES,CNT,IGNORE,PID)
         IF (WORK(1) .GT. X(1)) THEN
            DO 10 I=1,N
               X(I) = WORK(I)
   10       CONTINUE
         ENDIF
         BIT = BIT/2
      GO TO 5
c
c     Pass result back to parent
c
   20 CONTINUE
      IF (BIT .NE. 0) THEN
         PARENT = XOR(ME, BIT)
         CALL SENDW(CI,TYPE,X,BYTES,PARENT,PID)
      ELSE
         IF (IABS(ROOT) .EQ. IABS(HOST))
     &      CALL SENDW(CI,TYPE,X,BYTES,HOST,PID)
      ENDIF
      RETURN
      END
```

Figure 4.3. Subroutine for finding global maximum over all nodes

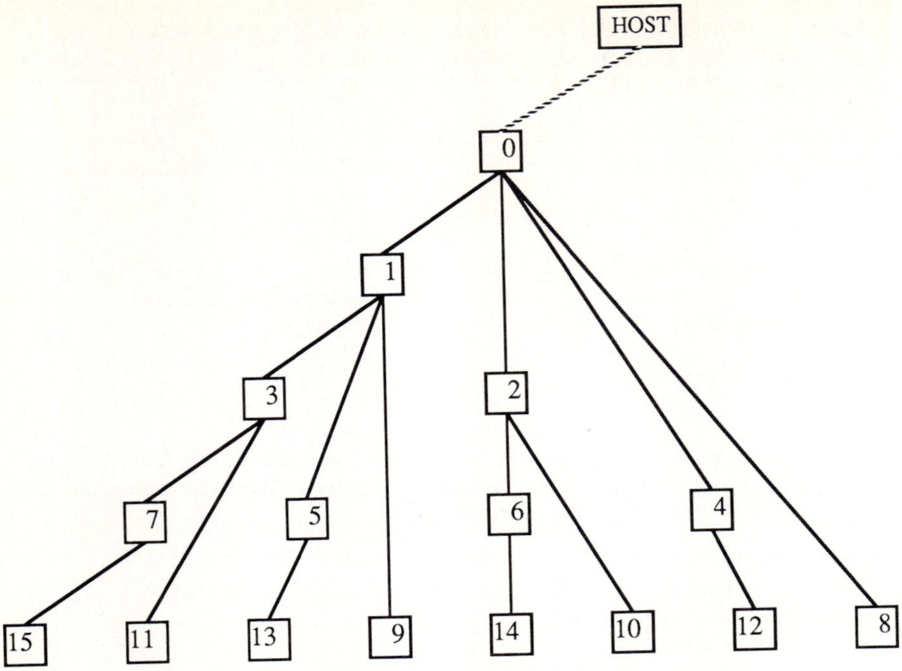

Figure 4.4. Spanning tree for a d4 hypercube

$$Ly = Pb$$

and

$$Ux = y$$

The parallel solution described here was developed by Cleve Moler, formerly with Intel Scientific Computers. His philosophy is to use known algorithms with known numerical properties but to develop new data structures to fit the parallel computing environment. Thus the algorithm is the one used in Linpack, but the matrix A is distributed among all the nodes. To retain the column orientation of the original algorithm and also keep the computational load balanced among the processors, a column-wrapped distribution is used, as illustrated in Figure 4.5. Each node "owns" approximately n/p columns of A, i.e., it holds these columns in its memory and performs any operations that must be done on them.

As in the frequency response example, all the processors run identical code. Each node determines which columns it owns and tailors its operations accordingly. The code takes the form of two subroutines PGEFA and PGESL, corresponding to Linpack's DGEFA and DGESL (double precision versions of SGEFA and SGESL). PGEFA computes the LU factorization and PGESL solves the two triangular systems.

For large n most of the work is in the LU factorization, which requires $O(n^3)$ flops. The parallel version, shown in Figure 4.6, is straightforward and efficient. The basic algorithm zeros out the subdiagonal elements of each column of A in turn by means of elementary transformations. The transformations must be applied to all columns of A, which is where the parallelism comes in. The node that owns the current column computes the pivot and the transformation and broadcasts this information to all the other nodes using a spanning tree (subroutine GSEND). All the nodes then apply the transformation to their columns, and control passes to the owner of the next column.

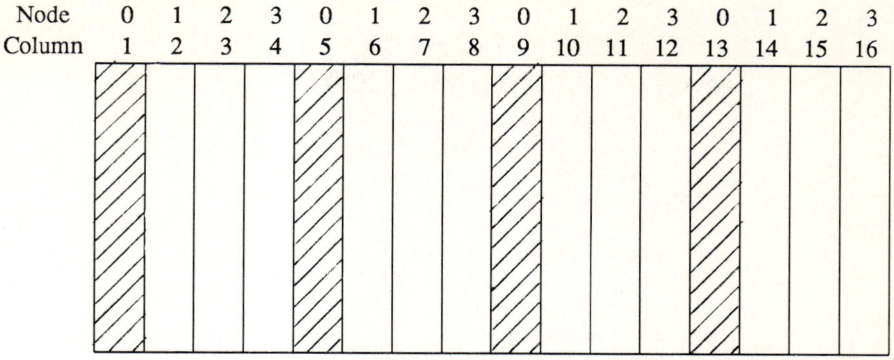

Figure 4.5. Column-wrapped distribution of a matrix in a 4 node hypercube. Column j is in node $(j-1) \bmod p$.

The solve algorithm for a single right-hand side is less straightforward to parallelize. Since it requires only $O(n^2)$ operations and since for multiple right-hand sides it possesses a natural parallelism, we present here only a very simple algorithm. The code is shown in Figure 4.7. The vector b is passed from node to node in a ring, first in one direction for the forward solve ($Ly = Pb$) and then in the other direction for the back substitution ($Ux=y$). The variables PRED and SUCC are the node's predecessor and successor, respectively, within the ring. Each node updates b with its portion of the transformation before passing it on. The solution x ends up in node 0.

```
      SUBROUTINE PGEFA(A,LDA,N,M,P,CID,ID,IPVT,BUF)
      INTEGER LDA,N,M,P,CID,ID,IPVT(M)
      DOUBLE PRECISION A(LDA,M),BUF(N)
c
c     Parallel version of LINPACK DGEFA
c     LU matrix factorization by Gaussian elimination.
c
      DOUBLE PRECISION T
      INTEGER J,K,KP,L,R,IDAMAX
c
c     Gaussian elimination with partial pivoting
c     L = starting index of active matrix
c
      L = 1
      DO 60 K = 1, N
c
c        Process R (for Root) owns the K-th column.
c
         R = MOD(K-1,P)
         IF ( R .EQ. ID ) THEN
c
c           Find KP = pivot index.  KP = 0 signals zero column.
c
            KP = IDAMAX(N-K+1,A(K,L),1) + K - 1
            IF (A(KP,L) .EQ. 0.0D0) KP = 0
            IPVT(L) = KP
```

```
c
c           Interchange if necessary
c
            IF (KP .NE. K) THEN
               T = A(KP,L)
               A(KP,L) = A(K,L)
               A(K,L) = T
            ENDIF
c
c           Compute multipliers
c
            IF (KP .NE. 0) THEN
               T = -1.0D0/A(K,L)
               CALL DSCAL(N-K,T,A(K+1,L),1)
            ENDIF
c
c           Prepare message for broadcast
c
            CALL DCOPY(N-K,A(K+1,L),1,BUF,1)
            BUF(N-K+1) = KP
            L = L + 1
c
         ENDIF
c
c        Broadcast elimination information.
c
         CALL GSEND (ID,R,K,P,CID,BUF,N-K+1)
         KP = BUF(N-K+1)
c
c        Row elimination with column indexing
c
         IF (KP .NE. 0) THEN
            DO 30 J = L, M
               T = A(KP,J)
               IF (KP .NE. K) THEN
                  A(KP,J) = A(K,J)
                  A(K,J) = T
               ENDIF
               CALL DAXPY(N-K,T,BUF,1,A(K+1,J),1)
   30       CONTINUE
         ENDIF
   60 CONTINUE
      RETURN
      END
```

Figure 4.6. Subroutine for parallel LU factorization

```
      SUBROUTINE PGESL(A,LDA,N,M,P,CID,ID,IPVT,B)
      INTEGER LDA,N,M,P,CID,ID,IPVT(M)
      DOUBLE PRECISION A(LDA,M),B(N)
c
      DOUBLE PRECISION T
      INTEGER K,L,KP,PRED,SUCC,TYPE,BYTES,PID
      DATA TYPE /4/
c
      PRED = MOD(ID-1+P,P)
      SUCC = MOD(ID+1+P,P)
      BYTES = 8*N
      PID = 0
c
c     First solve  L*y = b
c
      K = ID + 1
      DO 20 L = 1, M
         IF (K.GT.1 .AND. PRED.NE.ID)
     &      CALL RECVW(CID,TYPE,B,BYTES,BYTES,PRED,PID)
         KP = IPVT(L)
         T = B(KP)
         IF (KP .NE. K) THEN
            B(KP) = B(K)
            B(K) = T
         ENDIF
         CALL DAXPY(N-K,T,A(K+1,L),1,B(K+1),1)
         IF (K.LT.N .AND. SUCC.NE.ID)
     &      CALL SEND(CID,TYPE,B,BYTES,SUCC,PID)
         K = K + P
   20 CONTINUE
c
c     Now solve  U*x = y
c
      DO 40 L = M, 1, -1
         K = K - P
         IF (K.LT.N .AND. SUCC.NE.ID)
     &      CALL RECVW(CID,TYPE,B,BYTES,BYTES,SUCC,PID)
         B(K) = B(K)/A(K,L)
         T = -B(K)
         CALL DAXPY(K-1,T,A(1,L),1,B(1),1)
         IF (K.GT.1 .AND. PRED.NE.ID)
     &      CALL SEND(CID,TYPE,B,BYTES,PRED,PID)
   40 CONTINUE
      RETURN
      END
```

Figure 4.7. Subroutine for forward and back substitution

5. DETAILED IMPLEMENTATIONS OF TWO ALGORITHMS ON A HYPERCUBE

In this section we present complete algorithms, without code, for two key computational problems from control engineering. The first problem is frequency response calculation for large or small models; the second is the solution of algebraic Riccati equations. Riccati equations are particularly ubiquitous and play a central role in optimal estimation and filtering, as well as in optimal control. These particular problems were chosen for parallel implementation for three reasons: 1) They are important in control systems engineering, 2) they are computationally expensive, and 3) straightforward (i.e., easily parallelized) algorithms were already available for them. Reference [22] gives further background in control design computations.

Both problems are based on the linear, time-invariant, multivariable state space model

$$\dot{x} = Ax + Bu \qquad (5.1)$$

$$y = Cx$$

where

$$x \in \mathbb{R}^n \ , \quad u \in \mathbb{R}^m \ , \quad y \in \mathbb{R}^r \ , \quad A \in \mathbb{R}^{nxn} \ , \quad B \in \mathbb{R}^{nxm} \ , \quad C \in \mathbb{R}^{rxn} \ .$$

The coefficient matrices A, B, and C range in size from rather modest, say on the order of 10-100, through quite large, say on the order of 500-1000. We shall further assume these matrices to be dense, with no exploitable structure. Such a model might represent, for example, a large space structure (LSS) as approximated by the finite element method (FEM). If the matrices possess useful structure (e.g., tridiagonal) or have exploitable sparsity then other more specialized algorithms should be used.

5.1. Frequency Response

Given the linear system (5.1) we want to compute the $r \times m$ complex-valued frequency response matrix

$$G(j\omega) = C(j\omega I - A)^{-1}B$$

for N values of the scalar parameter ω, where typically $N \gg n$. We assume that the combined memory of the hypercube nodes is large enough to store the data matrices plus some workspace, but the data need not fit into a single node. Thus we cannot use the unmodified sequential algorithm, as we did in the small example of section 4.1.

An efficient and reliable algorithm for frequency response computation is:

1. Transform A to upper Hessenberg form using orthogonal similarity transformations, also transforming B and C appropriately.
2. For each value of ω, "solve $G(j\omega)=C(j\omega I-A)^{-1}B$ ", where (A, B, C) is the system resulting from step 1. This step will be described further below. For additional details, see [19], [20]. Under the assumption that $N \gg n$, most of the computation time is spent on this step.

It should also be noted, as in [22], that the above method applies equally well to the case of evaluating a frequency response matrix of the form

$$G(j\omega) = C(j\omega E - A)^{-1}B$$

arising from an implicit or generalized state space model of the form

$$E\dot{x} = Ax + Bu \ .$$

The principal difference is that Step 1 above uses instead the "front-end" of the QZ algorithm, i.e., simultaneous reduction of E and A to upper Hessenberg and upper triangular form, respectively, with appropriate modification of B and C. In either case, the key observation is that the transformed $(j\omega I - A)$ or $(j\omega E - A)$ coefficient matrix remains in upper Hessenberg form as ω varies.

As noted in Sec. 2.2, large second-order models can be put conveniently in the above implicit form (cf. (2.5)). We also note an interesting open problem. With reference to the notation of (2.4),(2.5), it would be useful to find an analog of the above "Hessenberg trick" which works directly on the frequency response matrix

$$(P + j\omega S)(-\omega^2 M + j\omega C + K)^{-1} D \ .$$

That is, one would like to find an equivalence which simultaneously compresses the three matrices M, C, and K. Of course, if special assumptions are made (for example, $C = 0$, $M = M^T > 0$, $K = K^T \geq 0$), highly efficient simultaneous diagonalization methods are available.

We shall now describe a hypercube implementation of the Hessenberg algorithm for the "standard (A,B,C)" model. An analysis of this algorithm reveals a considerable amount of parallelism on several levels. After the Hessenberg reduction has been done, the computations at the N different frequencies ω are independent. Assuming that the coefficient matrices are small enough, the processing nodes can work independently on different frequencies, making the problem almost perfectly parallel. If the matrices must be split among multiple nodes, then they are allocated by columns (column-wrapped distribution). Each processor receives approximately n/p columns of A, m/p columns of B, etc., and produces m/p columns of G, where p is the number of processors.

To allow for any size problem we combine these two approaches. Let d be the dimension of the smallest subcube capable of computing $G(j\omega)$ for a single value of ω, and let $d_{\max}$ be the dimension of the entire cube. We treat the system as $2^{d_{\max}-d}$ independent subcubes of dimension d. The subcubes receive identical matrices but different frequency ranges. The columns are distributed among the nodes of each subcube. With this implementation the Hessenberg reduction step is performed redundantly by all the subcubes, but it is an insignificant part of the total computation.

The Hessenberg reduction is accomplished by Householder similarity transformations:

$$A \longleftarrow P_{n-2} \cdots P_2 P_1 A \, P_1 P_2 P_{n-2}$$

$$B \longleftarrow P_{n-2} \cdots P_2 P_1 B$$

$$C \longleftarrow C \, P_1 P_2 \cdots P_{n-2}$$

where

$$P_j = I - \sigma_j u_j u_j^T, \quad \sigma_j = \frac{u_j^T u_j}{2} \ .$$

Each u_j is selected to zero out the elements of column j below the subdiagonal, without affecting the previously processed columns.

The key step is the computation of PAP, which can be written (dropping all subscripts) as

$$PAP = (I - \sigma u u^T) A \ (I - \sigma u u^T)$$

$$= A - \sigma u v^T - \sigma w u^T$$

where

$$v = A^T u$$

$$w = \tilde{A} u$$

$$\tilde{A} = A - \sigma u v^T$$

Note that

$$w = \tilde{A}u = [\tilde{a}_1 \ \cdots \ \tilde{a}_n] \begin{bmatrix} \mu_1 \\ \cdot \\ \cdot \\ \mu_n \end{bmatrix} = \sum_i \mu_i \tilde{a}_i$$

(We are following Householder's notational conventions. Thus $\tilde{a}_i$ is a column of the matrix $\tilde{A}$, μ_i is an element of the vector u, and ν_i is an element of v.) The parallel algorithm is:

For $j = 1, \ldots, n-2$

1) Owner of a_j: Compute u and σ from a_j as for sequential algorithm
Others: wait *

2) Broadcast u, σ *

3) All: Compute $\nu_i = a_i^T u$ for each local a_i †

4) All: Compute $a_i \longleftarrow a_i - \sigma \nu_i u$ for local a_i (Now have $A - \sigma u v^T$) †

5) All: Compute w (partial sum) := $\sum \mu_i a_i$ for local a_i †

6) Global add: Add up pieces of w from all nodes *

7) Broadcast w *

8) All: Compute $a_i \longleftarrow a_i - \sigma \mu_i w$ for local a_i †

Updating B is similar to steps 3 and 4; updating C resembles steps 5 through 8.

The steps marked with * involve idle time for some nodes, because of either communication delays or a sequential portion of the algorithm. The steps marked with † possess another level of parallelism that could be exploited by vector processors in the nodes.

Next consider the computation of the frequency response matrix

$$G = C(j\omega I - A)^{-1}B$$

for a single value of ω. Defining

$$H = j\omega I - A \ ,$$

we have

$$\begin{aligned} G &= CH^{-1}B \\ &= C(LU)^{-1}B \\ &= (CU^{-1})(L^{-1}B) \\ &= YX \end{aligned}$$

where

$$\begin{aligned} H &= LU \\ YU &= C \\ LX &= B \end{aligned}$$

The factorization of the complex Hessenberg matrix H is done by Gaussian elimination with partial pivoting. Note that QR factorization could also have been used and is, in fact, the factorization employed in a systolic (highly parallel) implementation of the Hessenberg frequency response algorithm described in [23].

Examining the problem for high-level functional parallelism, we immediately see that Y and X can be computed in parallel. Next we consider the factorization of the upper Hessenberg matrix $H =: [\eta_{ij}]$, which has the form

For j=1, . . . , n–1

1) Compute a transformation L_j to eliminate $\eta_{j+1,j}$ (based on η_{jj} and $\eta_{j+1,j}$)

2) Apply L_j to the $(n+1-j) \times (n+1-j)$ trailing submatrix of H (affects only rows j and j+1)

3) L_j and the jth row of U are in their final form at this point

$\mu_{nn} := \eta_{nn}$ (the last row of U) is available at this point

Each factor of L can be applied to B as soon as it is produced, affecting rows j and j+1 of B and yielding the jth row of X; there is no need to store L or its factors.

The triangular system $YU = C$ can be written in the form

$$y_i = \frac{1}{\mu_{ii}} [c_i - \sum_{j=1}^{i-1} \mu_{ji} y_j] .$$

So if we define

$$y_i^{(0)} = c_i \ , \ i=1, \ldots, n$$

then at step j we can compute

$$y_j = \frac{1}{\mu_{jj}} y_j^{(j-1)}$$

$$y_i^{(j)} = y_i^{(j-1)} - \mu_{ji} y_j \ , \ i=j+1, \ldots, n \ .$$

The rows of U are used as they are produced and so need not be stored.

Finally, since Y emerges by columns and X emerges by rows, G is naturally computed as an outer product. Storage is required for Y because its columns are computed recursively, but not for X.

Figure 5.1 shows a data flow diagram for the entire procedure. This structure can be regarded as data pipelining at a high level. Ideally each box should execute whenever it has input available, without regard to what the other boxes are doing internally. As implemented, the processing within each node is synchronous, but the overlap is still important for minimizing wasted time.

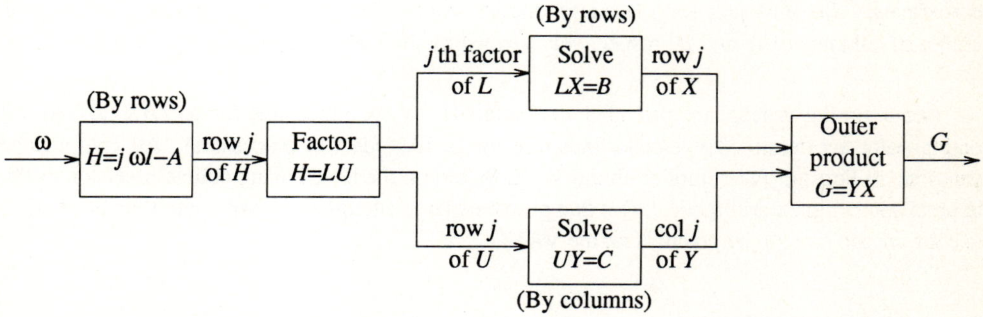

Figure 5.1. Data Flow Diagram for Frequency Response Calculation

The parallel algorithm, given $H = j\omega I - A$, is:

For j=1,...,n−1

1) Owner of h_j, c_j: Compute transformation L_j; compute y_j †
   Others: wait *

2) Broadcast L_j and y_j *

3) All: Apply L_j to local h_i, i=j+1, . . . , n and local b_i, i=1, . . . , m †

4) All: Update local y_i, i=j+1, . . . , n †

5) All: Update local $g_i \longleftarrow g_i + x_{ji} y_j$ (outer product) †

Again, * indicates idle time and † indicates additional parallelism.

The actual code consists of a host program and a node program. The node program runs on as many nodes of the cube as are to be used, normally all of them but always a power of 2.

The host program reads the input matrices and determines the size of the smallest subcube, $p=2^d$, capable of computing the frequency response at a single frequency. The node programs are then loaded and configured to operate as $2^{d_{max}-d}$ independent subcubes of dimension d.

The A, B, and C matrices are sent to all the subcubes and distributed by columns to the nodes within the subcube. Each subcube transforms the system to Hessenberg form. If there are multiple subcubes this results in duplicated work, but it is insignificant if a large number of frequencies are to be used. The duplication of work could be avoided by having all the nodes work together on the Hessenberg reduction, then redistributing the matrices for the appropriate subcube configuration. This would increase both the complexity of the program and the amount of communication between nodes.

The host gets the desired frequency range and number of points from the user and divides the range among the subcubes. Each subcube computes the frequency response at logarithmically spaced frequencies within its range and returns the results to the host. Some sort of optimal choice of the frequencies is still an open problem.

Communication within the cube is limited to broadcasts and "global operations", the dual of broadcasts. Both are accomplished by means of a spanning tree. Each subcube has its own spanning tree, so its communication is independent of any other subcube.

By pipelining data through the program and not estimating condition numbers, we avoid storing the complex matrix $H := j\omega I - A$ and its factors L and U. No storage is required for X since a column of Y and a row of X are used directly to form G. The storage requirement per node is approximately $(n+3r+4)(n_{loc}+m_{loc})+4(n+r)$ double words. The terms n_{loc} and m_{loc} refer to the number of columns of A and B, respectively, owned by the node. Recall that n_{loc} is approximately n/p.

Some timing results are provided in Table 5.1. VAX times are for a VAX 11/780 with floating-point accelerator. Hypercube times are for an iPSC/d5 (32 processors). All computations were done in Fortran (f77 compiler on the VAX, ftn286 on the iPSC) using double precision or double precision complex arithmetic. Note that problems large enough to require more than one node on the cube are too large to be practical on the VAX.

5.2. Riccati Equations

The continuous-time generalized algebraic Riccati equation (GARE), which arises in linear-quadratic optimal control and filtering problems, is

$$A^T XE + E^T XA - E^T XGXE + Q = 0 . \tag{5.2}$$

					Hessenberg reduction			Freq. resp. (per freq)		
n	m	r	subcube dim.	# of subcubes	iPSC time	VAX time	speedup	iPSC time	VAX time	speedup
10	3	2	0	32	<1	.1	-	.0146	.05	3.4
50	10	10	0	32	16	8.2	.5	.367	2.7	7.4
100	15	15	0	32	118	56	.5	1.82	13	7.1
180	20	20	1	16	334	*	*	7.13	*	*
300	32	32	3	4	444	*	*	63	*	*
520	32	32	4	2	1175	*	*	91	*	*

* (too large for VAX)
Note: Times are in seconds.

Table 5.1. Timing Results for Frequency Response Calculation

The coefficients A, E, G, and Q and the solution X are $n \times n$ real matrices; G and Q are symmetric nonnegative definite; E is nonsingular. Under certain technical conditions, (5.2) has a unique symmetric nonnegative definite solution. For details, see [12].

A well-studied and reliable sequential algorithm for solving the GARE is the Schur vector method described in [12], which is based on the QZ algorithm. Since this method does not lend itself readily to parallel implementation, we have chosen to use an alternative algorithm based on the matrix sign function. For details, see [24] and references contained therein.

The sign function algorithm is an iterative algorithm related to the Hamiltonian pencil associated with (5.2). With

$$Z_0 = \begin{bmatrix} -Q & -A^T \\ -A & G \end{bmatrix}$$

and

$$F = \begin{bmatrix} 0 & E^T \\ -E & 0 \end{bmatrix}$$

the iteration is

$$\begin{aligned} Z_{k+1} &= \frac{1}{2}\left(\frac{1}{c} Z_k - cF^T Z_k^{-1} F\right) \\ &= \frac{1}{2}\left(\frac{1}{c} Z_k - cF^T U^{-T} D^{-1} U^{-1} F\right) \\ &= \frac{1}{2}\left(\frac{1}{c} Z_k - cYD^{-1} Y^T\right) \end{aligned}$$

where

$$c = \left(\frac{|\det Z_k|}{|\det F|}\right)^{\frac{1}{2n}}$$

$$Z_k = UDU^T$$

$$UY^T = F$$

Note that Z_k is symmetric at each step, and that F is skew-symmetric. Under the technical conditions already alluded to, it can be shown that the sequence $\{Z_k\}$ converges. Let the final value of Z_k be

$$\bar{Z} = \begin{bmatrix} \bar{Z}_{11} & \bar{Z}_{12} \\ \bar{Z}_{12}^T & \bar{Z}_{22} \end{bmatrix}$$

The solution to (5.2) is then defined by the linear equation

$$(\bar{Z}_{12} + E^T)XE = -\bar{Z}_{11} \tag{5.3}$$

The only operations involved are very basic ones: solution of linear systems, matrix multiplication, scaling, and addition. We have again chosen a column-oriented approach for our implementation and column-wrapped distribution for the matrices Z, F, and Y.

For the decomposition $Z_k = UDU^T$ we parallelized the Linpack [21] algorithm for factoring symmetric indefinite matrices (DSIFA). Here, U is unit upper triangular and D is block diagonal with 1×1 and 2×2 blocks. Unfortunately, the symmetric matrix factorization involves a lot more communication among nodes than the unsymmetric LU factorization. Pivoting requires column exchanges (exchanges between nodes) as well as row exchanges (within a single node).

Figure 5.2 shows the data flow diagram for computing Z_{k+1} from Z_k. The matrix U is generated in factored form, and the factors can be applied to F as they emerge. The Y matrix emerges by columns. The outer product $\sum \delta_j^{-1} y_j y_j^T$ is accumulated to give $YD^{-1}Y^T$. All the nodes compute the determinant of Z_k from D and then compute the scale factor c.

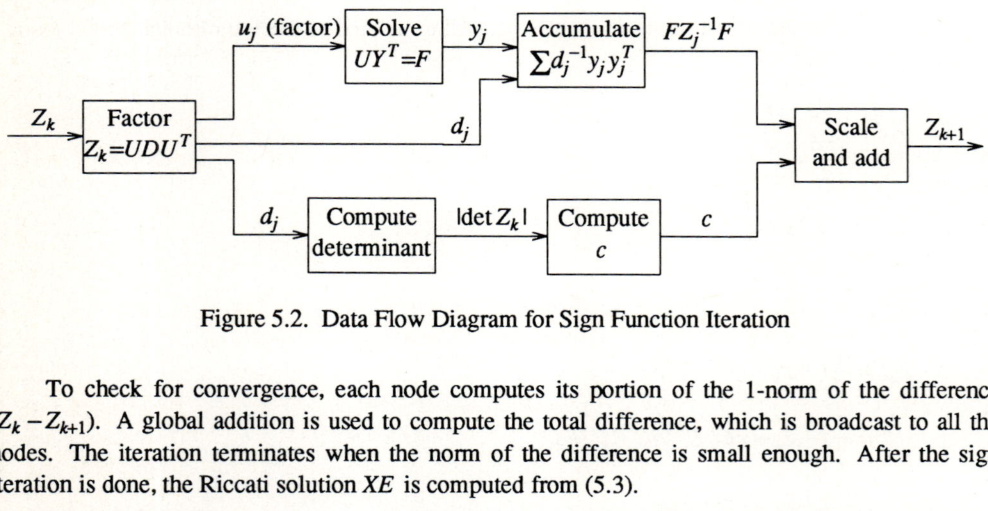

Figure 5.2. Data Flow Diagram for Sign Function Iteration

To check for convergence, each node computes its portion of the 1-norm of the difference $(Z_k - Z_{k+1})$. A global addition is used to compute the total difference, which is broadcast to all the nodes. The iteration terminates when the norm of the difference is small enough. After the sign iteration is done, the Riccati solution XE is computed from (5.3).

Some timing results are given in Table 5.2. VAX times are for a VAX 11/780 with floating-point accelerator using Laub's Schur vector method for standard algebraic Riccati equations. Hypercube times are with 32 processors. All computations were done in Fortran (f77 compiler on the VAX, ftn286 on the iPSC) in double precision arithmetic.

It should be emphasized here that the algorithms compared for the "speedup" measured in the above table are different. That is, the parallel algorithm is based on a matrix sign function iteration while the sequential algorithm is based on an ordered generalized eigenvalue problem. In general, the latter is faster on a conventional sequential computer than a matrix sign iteration.

n	Hypercube time (sec.)	VAX time (sec.)	Speedup
20	47	32	0.68
40	84	210	2.5
80	253	1547	6.1
100	383	too long	*

Table 5.2. Timing Results for Riccati Equation Solution

References

[1] Balas, M.J., "Trends in Large Space Structure Control Theory: Fondest Hopes, Wildest Dreams," *IEEE Trans. Aut. Contr.*, vol. AC-27, pp. 522-535, 1982.

[2] Mackay, M.K., *Active Control of Large Flexible Space Structures,* Doctoral Dissertation, Mechanical Engineering, UCLA , 1983.

[3] Hughes, P.C., "Space Structure Vibration Modes: How Many Exist? Which Ones Are Important?," *IEEE Control Sys. Mag.*, vol. 7, pp. 22-28, Feb. 1987.

[4] Greene, C.S. and G. Stein, "Inherent Damping, Solvability Conditions, and Solutions for Structural Vibration Control," *Proc. 18th Conf. Decision and Control*, pp. 230-232, Ft. Lauderdale, FL, 1979.

[5] Junkins, J.L. and D.W. Rew, "Unified Optimization of Structures and Controllers," in *Large Space Structures: Dynamics and Control*, ed. S.N. Atluri and A.K. Amos, Springer-Verlag, New York, 1987.

[6] Hanks, B.R., "Structural Approximation of Control Performance," in *Optimization Issues in the Design and Control of Large Space Structures*, ed. M.P. Kamat, pp. 16-30, 1985.

[7] Arnold, W.F., *Numerical Solution of Algebraic Matrix Riccati Equations,* Ph.D. Dissertation, Dept. EE-Systems, USC, Feb. 1984.

[8] Meirovitch, L., *Computational Methods in Structural Dynamics,* Sijthoff & Noordhoff, Alphen aan den Rijn, The Netherlands, 1980.

[9] Laub, A.J. and W.F. Arnold, "Controllability and Observability Criteria for Multivariable Linear Second-Order Models," *IEEE Trans. Aut. Contr.*, vol. AC-29, pp. 163-165, 1984.

[10] Leipnik, R.B., Personal communication, 1985.

[11] Bender, D.J. and A.J. Laub, "Controllability and Observability at Infinity of Multivariable Linear Second-Order Models," *IEEE Trans. Aut. Contr.*, vol. AC-30, pp. 1234-1237, 1985.

[12] Arnold, W.F. and A.J. Laub, "Generalized Eigenproblem Algorithms and Software for Algebraic Riccati Equations," *Proc. IEEE*, vol. 72, pp. 1746-1754, 1984.

[13] Wiley, P., "A Parallel Architecture Comes of Age at Last," *IEEE Spectrum*, vol. 24, pp. 46-50, 1987.

[14] Karp, A.H., "Programming for Parallelism," *Computer*, vol. 20, pp. 43-56, 1987.

[15] Heath, M.T. (Ed.), *Hypercube Multiprocessors 1986,* SIAM, Philadelphia, 1986.

[16] Heath, M.T. (Ed.), *Hypercube Multiprocessors 1987,* SIAM, Philadelphia, 1987.

[17] McBryan, O.A. and E.F. Van de Velde, "Hypercube Algorithms and Implementations," *SIAM J. Sci. Stat. Comput.*, vol. 8, pp. 227-287, 1987.

[18] Gardiner, J.D. and A.J. Laub, "Implementation of Two Control System Design Algorithms on a Message-Passing Hypercube," in *Hypercube Multiprocessors 1987*, ed. M.T. Heath, pp. 512-519, SIAM, Philadelphia, 1987.

[19] Laub, A.J., ''Efficient Multivariable Frequency Response Computations,'' *IEEE Trans. Aut. Contr.*, vol. AC-26, pp. 407-408, 1981.

[20] Laub, A.J., ''Algorithm 640: Efficient Calculation of Frequency Response Matrices from State Space Models,'' *ACM Trans. Math. Software*, vol. 12, pp. 26-33, 1986.

[21] Dongarra, J.J., J.R. Bunch, C.B. Moler, and G.W. Stewart, *LINPACK Users' Guide,* Society for Industrial and Applied Mathematics, Philadelphia, 1979.

[22] Laub, A.J., ''Numerical Linear Algebra Aspects of Control Design Computations,'' *IEEE Trans. Aut. Contr.*, vol. AC-30, pp. 97-108, 1985.

[23] Cappello, P.R. and A.J. Laub, ''Systolic Computation of Multivariable Frequency Response,'' *IEEE Trans. Aut. Contr.*, 1988 (to appear).

[24] Gardiner, J.D. and A.J. Laub, ''A Generalization of the Matrix-Sign-Function Solution for Algebraic Riccati Equations,'' *Int. J. Control*, vol. 44, pp. 823-832, 1986.

Load Balancing and Partitioning for Parallel Signal Processing and Control Algorithms

George Cybenko
Department of Computer Science
Tufts University
Medford, MA 02155
and
ALPHATECH, Inc.
111 Middlesex Turnpike
Burlington, MA 01803

ABSTRACT

Large scale signal processing and control applications are now being implemented using parallel computers. Some applications such as tracking and data fusion require novel partitioning and load balancing schemes in order to make parallel processing most effective. In this paper, we survey recent work on these topics. We present diffusion schemes for load balancing and Multidimensional Binary Partitions for spatial partitioning appropriate for hypercube computations.

1. Introduction

Parallel computing technology is having a profound impact on how algorithms for control and signal processing are designed and implemented. The technology promises higher performance computation and therefore expands the domain of what is feasible in an engineered system. This is especially true in real time and embedded systems where processing limitations have traditionally constrained capabilities. Commercial parallel computers have now been available for over three years and to some extent their use in large scale scientific applications (such as the solution of partial differential equations) is becoming routine. One of the ways in which the computational requirements of real time systems in control and signal processing differ from most large scale scientific computations is the way in which data is input. Loosely speaking, partial differential equations that govern physical systems describe the evolution of the system from some initial state and so only the initial state description is typically needed. The data input is thus often part of the program initialization and is done once and for all in a computation. By

This research was partially supported by National Science Foundation under grant DCR-8619103, the Office of Naval Research under contracts N000-86-G-0202 and N00014-87-K-01821 and the Rome Air Development Center under contract F30602-85-C-0303 at ALPHATECH, Inc. while the author was on sabbatical leave during 1986-1987.

NATO ASI Series, Vol. F47
Advanced Computing Concepts and Techniques
in Control Engineering
Edited by M.J. Denham and A.J. Laub

information in the form of new measurements. A system is evolving externally to the computation and the point of the computation is to identify or control the state evolution using such measurements. This turns out to be a nontrivial problem - that is delivering data to the appropriate processors in a parallel system. This is especially true in the case of computations that evolve dynamically and work is distributed according to a rule that balances the computational load. A consequence of this is that the location of specific tasks is not known centrally and the data distribution must be done without benefit of this centralized information.

The applications where we have encountered such problems are multiobject tracking and data fusion. These problems have a simple high level description. Multiple data sets are received that include measurements of the spatial positions of many different objects. These data sets may be obtained from a single sensor but at different times (single sensor multiobject tracking) or from many sensors at roughly the same time (multispectral data fusion). If it is known which measurements are referring to the same objects, then one can use standard filtering techniques and the filtering equations are quite straightforward to implement. The hard part of this application lies in matching the various measurements either from time to time or from field of view to field of view. A discussion of these applications together with algorithms and issues for parallel computation can be found in [1, 2]. In this paper we discuss two combinatorial aspects of parallel computation that arise quite naturally when trying to solve large scale tracking and data fusion problems as outlined in [1, 2]. The first concerns load balancing using decentralized information and general interconnection networks of parallel machines. We present a simple diffusion scheme that uses local information for sharing work locally in a multiprocessor but has globally convergent properties.

This scheme is analyzed and its convergence properties for all networks is derived in a general theory. This general theory does not take into account the specific spatial basis for tasks in a tracking or fusion application and so we present some more specialized results about problems that have a spatial foundation. The technique for load balancing that we introduce is called Multidimensional Binary Partitioning and actually leads to an elegant solution to the measurements distribution problem as well. Multidimensional Binary Partitions are distributed data structures that are well suited to hypercube parallel machines and can be computed in parallel easily. They implement a binary search tree for distributing new measurements according to the spatial location of a measurement in the ambient state space.

The organization of the paper is as follows. In Section 2 we introduce the problem of load balancing from the point of view of parallel tracking and fusion algorithms. Section 3 presents a general diffusion approach to solving such problems and an analysis of the method is outlined. Section 4 addresses the special case of hypercube parallel machines and Section 5 shows that the notion of Multidimensional Binary Partitions is intrinsically related to spatial partitioning of work and allows for extremely efficient measurement distribution.

2. The Need for Load Balancing

In parallel implementations of tracking and data fusion algorithms, each processor manages the information and subsequent estimation relevant to a collection of objects. The position of these objects change in the ambient state space, they are created and they are removed as time progresses. Therefore, the number of objects that a processor is currently managing can change dramatically thereby leading to severe load imbalances if different processors end up being responsible for widely varying numbers of objects. This load imbalance results in processor underutilization and performance degradation in the parallel algorithm since the underutilized processors are idle while the overutilized processors are completing their allocated work loads.

The goal of load balancing is for each processor to perform an equitable share of the total work load. In many applications, such as dense linear systems solving, it is possible to make a priori estimates of work distribution so that a programmer can build load balancing right into a specific applications program. Such an off-line a priori determination can naturally be called *static* load balancing. By contrast, our interest here is in situations where no a priori estimates of load distribution are possible. It is only during actual program execution that it becomes apparent how much work is being assigned to individual processors. This can occur in applications such as partial differential equations solvers using adaptively generated grids [3, 4], and data fusion problems [2, 1]. In such problems, as the computation evolves different processors end up being responsible for differing amounts of work. We shall refer to any strategy for balancing work during an execution as *dynamic*.

There has been considerable interest in load balancing in recent years, particularly since the introduction of large scale commercial multiprocessors. Unfortunately the simplest form of load balancing, in which tasks requiring differing times for completion are to be as equally distributed as possible between two processors, is clearly equivalent to the partition problem. A variety of other multiprocessor optimal scheduling problems are NP-Complete as well [5]. As a result of this fact, research on dynamic load balancing has focused on suboptimal procedures that use local information in a distributed memory architecture. Generally speaking, these procedures describe rules for migrating tasks on overutilized processors to underutilized processors in the network. Tradeoffs exist between achieving the goal of completely balancing the load and the communications costs associated with migrating tasks. Approaches that use such a paradigm include [6, 7, 8].

We quantify work in terms of *tasks* and these tasks are directly related to the number of objects being managed since work is proportional to the number of objects. All tasks require an equal amount of computational work (that is, time) to be completed and they are indecomposable into subtasks - a single task must be performed by a single processor only. We assume that all tasks are independent and therefore the order in which they are done is irrelevant. Another aspect of this independence is that it does not matter which processor in the network executes which task. (In subsequent sections, we see that if the parallel computing network has a regular structure and spatial partitioning is a reasonable basis for work partitioning, different approaches become possible and have many attractive properties.) Furthermore, tasks are self-contained and can be viewed as segments of data in memory. In order to relocate a task from one processor to another, it is only necessary to communicate the data associated with that task. Of course, in order to make dynamic load balancing an issue, we must consider problems in which the cost of moving a task (as measured in terms of communication delay and time) is significantly smaller than the ultimate time cost of executing the task in place. Recall that in data fusion and tracking applications a task is associated with a data record - each data record is repeatedly updated using new information and these data records are independent of one another [2, 1]. Data records are constantly being added and removed and the distribution of data records is essentially equivalent to the distribution of work.

In spite of the existence of applications where such a situation does arise, this model has some limitations for general applicability. For example, in dynamic grid generation problems, it is important to maintain some locality of tasks (which might correspond to grid points in an irregular grid) to one another. Discussions of load balancing strategies for spatially oriented tasks on regular multiprocessor networks such as hypercubes and grids can be found in [9, 4, 10]. Nonetheless, we believe that the model of this paper is suitable for at least a preliminary study of the general problem.

The diffusion approach to load balancing is of interest for a number of reasons. First of all, the load balancing protocols described for general distributed memory networks are extremely simple in comparison with previously studied methods. Secondly, the general method we study is completely analyzable with respect to its performance and convergence properties. This second fact is noteworthy because most other load balancing schemes have not been analytically studied and in spite of being based on convincing heuristics, their behavior under even simple load evolution has not been completely understood.

3. A General Dynamic Load Balancing Scheme

Assume that we have a distributed memory multiprocessor network with n processors labeled 1 through n. At this point we make no assumption about the topology of the network except that it is connected - that is, there is a path between any two processors in the network. In terms of the node adjacency matrix of the graph representing the network, this means that the matrix is irreducible (note that the matrix is symmetric since we assume that communications channels are bidirectional). We will use $N = (V,E)$ to denote the network in standard graph notation - V is the set of processors and E is the edge set (edges are the communications channels between processors).

We quantify the work distribution at time t by an n vector, $w^{(t)}$, where $w_i^{(t)}$ is the number of tasks to be done by processor i at time t. Although the number of tasks to be done by processor i is clearly a nonnegative integer, we will treat them as real quantities. In applications with fine task granularity this is a reasonable approximation.

Assume that at time t we have a work distribution of $w^{(t)}$. Our diffusion model for dynamic load balancing has the following form:

$$w_i^{(t+1)} = w_i^{(t)} + \sum_j \alpha_{ij}(w_j^{(t)} - w_i^{(t)}) + \eta^{(t+1)}, - c \tag{1}$$

where α_{ij} are non-negative constants. The summation is over all processors j in the network with the convention that $\alpha_{ij} = 0$ if i and j are not connected in the network. The interpretation of this formula hinges on the summation terms: processors i and j compare work loads at time t and the amount

$$\alpha_{ij}(w_j^{(t)} - w_i^{(t)})$$

is exchanged. If this quantity is positive, then precisely this amount of work is transferred from processor j to processor i. Should the quantity be negative, the transference is in the other direction. The term $\eta_j^{(t+1)}$ accounts for new work generated at time t for processor j and we will model this as a stochastic phenomenon shortly. The term c is a constant amount of work that any processor can accomplish between times t and times $t+1$.

This approach to load balancing was first described in [11] and we review the main results here.

It is appealing to think of work diffusing in a natural way through the multiprocessor network. Another interpretation of this approach involves analogies with finite Markov chain models. Work distribution can be considered to be an initial probability distribution and the diffusion of work is mathematically identical to the evolution of state occupation probabilities. Accordingly, it is not surprising that most of the mathematical tools and ideas are in fact derived from both Markov chain theory and the numerical analysis of matrix iterative schemes.

In order to analyze the behavior of (1), we must first study a simpler static model and then use linearity of the basic equations to get results for the general model. Our static model is given by the simplified equations

$$w_i^{(t+1)} = w_i^{(t)} + \sum_j \alpha_{ij}(w_j^{(t)} - w_i^{(t)}) \tag{1a}$$

which are obtained from (1a) by dropping the inhomogeneous terms. Rewriting (1a), we have

$$w_i^{(t+1)} = (1 - \sum_j \alpha_{ij})w_i^{(t)} + \sum_j \alpha_{ij} w_j^{(t)} \tag{2}$$

which clearly illustrates the need for two fundamental constraints on the constants α_{ij}:

a. $\alpha_{ij} \geq 0$ for each i and j;

b. $1 - \sum_j \alpha_{ij} \geq 0$ for every i.

Figure 1 illustrates the behavior of such a scheme in a simple five processor network. The circles are processors and the numbers in the circles are the current work loads. The numbers beside the arcs are the α_{ij} coefficients for the diffusion scheme. In the example, we exclude the introduction of new work and the completion of existing work, demonstrating the behavior on a static problem. The two graphs illustrate the evolution of the system over two time intervals.

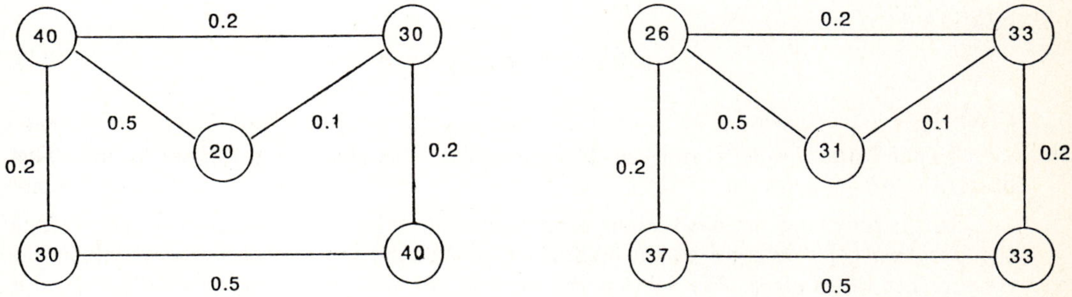

Figure 1

Note that the relationship given by (2) for the dynamic updating of work distribution is linear and we can write a simple vector equation for the update. Specifically, we have that

$$w^{(t+1)} = M w^{(t)} \tag{3}$$

where $M = (m_{ij})$ is given by

$$m_{ij} = \begin{cases} \alpha_{ij} & \text{if } i \neq j \\ 1 - \sum_k \alpha_{ik} & \text{if } i = j \end{cases} \tag{4}$$

Note that

$$\sum_i m_{ij} = 1 \tag{5}$$

and that M is symmetric. In particular, M is a doubly stochastic matrix and virtually all of our analysis depends on this fact. We shall call a matrix M obtained from a dynamic load balancing rule given by the α_{ij} as in (4) a *diffusion* matrix. A final bit of terminology concerns the notion of the *average* work distribution. For a given work distribution w, we let $\overline{w}$ be the vector whose every entry is exactly

$$\sum_i w_i / n = w^* u / n$$

where u is the n-vector all of whose components are 1 and w^* denotes matrix transposition. This uniform distribution allocates the same amount of work to every processor while keeping the total amount of work in the network constant. Convergence of an iterative dynamic load balancing strategy is thus meant to mean convergence of the network work distribution to the uniform one. The distance from one distribution to another is given by the Euclidean distance between distributions considered as n-vectors.

The matrix M leaves the average work load invariant as a consequence of property (5) above. For completeness we have:

LEMMA 1

$$\overline{w}^{(t)} = \overline{w}^{(0)} \text{ for all } t \geq 0.$$

PROOF

Expand $Mw^{(0)} = w^{(1)}$, and use (5) to simplify. The result then follows by induction on t.//

In the following, we use various properties of M to establish conditions under which the iteration (3) converges to the uniform load distribution and the rate at which the convergence takes place. We will use the powerful Perron-Frobenius theory of irreducible non-negative matrices and we refer the reader to [12] for an exposition and derivation of the results we invoke. In the terminology of [12], observe that M is a symmetric irreducible non-negative matrix. The Perron-Frobenius theory asserts a number of facts about such a matrix m.

Lemma 1 actually demonstrates that an eigenvector for 1 is the uniform distribution u and that within scaling this must therefore be the only eigenvector for 1. Now since the eigenvalues of M are bounded by 1, standard arguments [12] show that the iteration (3) converges to the uniform distribution if and only if -1 is not an eigenvalue as well.

Figures 2 and 3 show the possible behaviors of diffusion schemes for simple networks. In Figure 2, the loads cycle between two values while in Figure 3 there is clearly convergence if the process is allowed to continue. Readers familiar with Markov chain theory might recognize the two periodic behavior of the example in Figure 2. In fact, the convergence results for such schemes depend precisely on the type of analysis that arises in Markov chain theory.

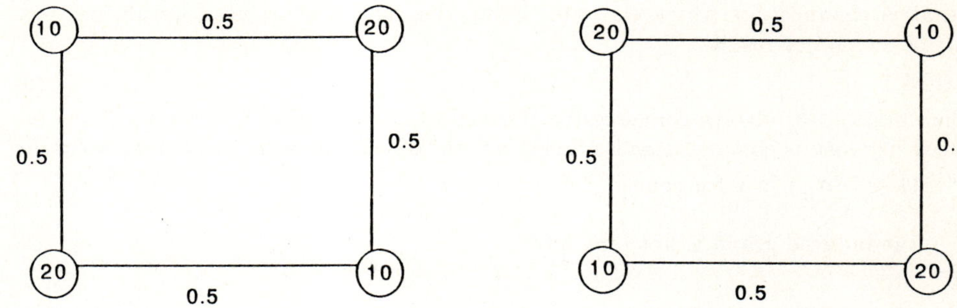

Figure 2

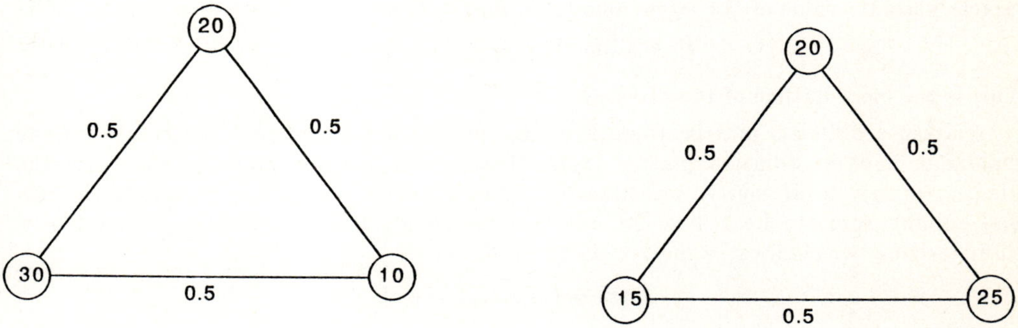

Figure 3

Let

$$\lambda_n \leq \lambda_{n-1} \leq \cdots \leq \lambda_2 < \lambda_1 = 1$$

be the ordered eigenvalues of M with $\lambda_1 = 1$. Note that the last inequality is strict since 1 is a simple eigenvalue. Let

$$\gamma(M) = \max_{i > 1} |\lambda_i|$$

Our goal is to determine situations where $\gamma < 1$. We first need an auxiliary idea.

DEFINITION

Given a network N and a diffusion matrix M as above, the *induced* network or graph is given by $N_M = (V, E_M)$ where for $i,j \in V$ we have $(i,j) \in E_M$ if and only if $\alpha_{ij} > 0$.

The induced network is essentially the original network with possibly some channels deleted - the channels removed are precisely the ones on which no exchange of work is made, that is channels for which $\alpha_{ij} = 0$. Using this notion of induced graph, we now state one of our major results.

THEOREM 1

The iteration (3) always converges to the uniform distribution if and only if the induced network is connected and either (or both) of the following conditions hold:

i. $(1 - \sum_i \alpha_{ij}) > 0$ for some j;

ii. the induced graph is not bipartite.

PROOF

We only sketch the proof here, pointing the reader to [11] for complete details. The diffusion matrix M determined by the diffusion rule is a doubly stochastic matrix and the scheme converges, as noted above, if and only if the smallest eigenvalue of M is greater than -1. By the Perron-Frobenius Theorem, -1 is an eigenvalue of M if and only if M can be permuted into two periodic form and this is possible if and only if the conditions of the theorem do not hold. If -1 is not an eigenvalue of M then standard arguments about matrix iteration show that the scheme given by (1a) converge. Letting γ be the largest absolute value of the eigenvalues excluding 1, it is easily shown that

$$|| Mw - \overline{m} || \leq \gamma || w - \overline{w} ||$$

This is the basic outline of the proof.//

Given a diffusion matrix, there is a class of simple scalings that can transform the matrix to another diffusion matrix. Note that adding a scalar multiple, say κ, of the identity matrix to M merely translates the eigenvalues of M by κ and changes the row and column sums to be $1+\kappa$. M can then be renormalized by dividing it by $1+\kappa$. Summarizing, we shall call a matrix of the form

$$M(\kappa) = (M + \kappa I_n)/(1+\kappa)$$

a *linear* modification of M. Notice that not all linear modifications lead to physically realistic diffusion matrices. Clearly $\kappa > 0$ never presents a problem but $\kappa < 0$ can if some diagonal entries of M are zero. In general, note that

$$\kappa \geq -\min_j m_{jj}$$

leads to the only legal linear modifications since we cannot have a diffusion rule that distributes away more work than exists at a node.

There is a simple rule for determining linear modifications that give optimal rates of convergence for the resulting matrices, subject to the feasibility constraint above.

LEMMA 2

Suppose that M is a diffusion matrix. Then the optimal feasible linear modification is obtained by setting κ equal to the larger of

$$-(\lambda_2+\lambda_n)/2 \quad , \quad -\min_j m_{jj}$$

where as before λ_2 is the second largest eigenvalue of M and λ_n is the smallest eigenvalue.

PROOF

Again complete details are in [11] and we refer readers to that reference. The proof essentially depends on an analysis of the spectrum of $M(\kappa)$ as κ varies over values that do not violate the feasibility constraints.//

Theorem 1 and Lemma 2 give a complete analysis of general diffusion schemes for general topologies with a static work distribution. Our assumption in this static model was that the work load was initially determined by $w^{(0)}$ and that no work was created nor completed between iterations of the diffusion scheme. Let us now return to the case where new work is created and that a constant amount of work is done by every processor between iterations. Our original model from (1) is recalled here:

$$w^{(j+1)} = Mw^{(j)} - cu + \eta^{(j+1)} \tag{12}$$

with $w^{(0)} = \eta^{(0)}$. The reader is reminded that the interpretation of the two additional terms on the right of (12) is :

- each processor completes exactly c tasks between iterations of the dynamic scheme thereby reducing the amount of work by c units;
- $\eta^{(j+1)}$ is a random vector, each element of which is drawn independently from a probability distribution on the positive reals whose mean is μ and variance is σ^2.

The three cases where $\mu > c$, $\mu = c$, $\mu < c$ correspond to three distinct situations. In the first case, the expected uniform load distribution grows unboundedly as t grows indefinitely while the case $\mu < c$ is the case where the expected work distribution reaches zero eventually. The case of real practical interest is when $\mu = c$ since this means that the expected work distribution is a finite constant distribution throughout time. Our analysis of this model however does not make any explicit assumptions about the size of μ relative to c since we state our results in terms of the expected deviation from the uniform distribution regardless of what it is or how it changes in time.

Note that (12) does not guarantee that every entry of $w^{(j)}$ is positive because of the randomness assumption and the fact that we decrement by a constant amount of work independently of how much work there is pending at a processor. If we modify (12) by allowing only positive entries, the system equation (12) becomes nonlinear and impossible to analyze using these techniques. To deal with this problem we consider the following interpretations:

a. assume that $w^{(0)} = Cu$ where C is a large positive constant so that with very high probability, the entries of $w^{(j)}$ remain positive throughout the iteration ;

b. use (12) with the negative entries as is and treat it as an approximation to the realistic situation (we can actually think of carrying negative work loads as the ability of a processor to store up computational power if it is not totally used during one iteration - this may correspond to performing some excess of system overhead that frees future computation for the application being balanced);

c. let $c = 0$ in cases where the work w quantifies the number of objects that have to be *managed* by the various processors in the network (such as distributed searching or data updating).

In any case, we can proceed with an analysis of the convergence of (12) without specifying what the actual situation is. The statement of our main result for this general model is in terms of variances and expectations. We let E denote expectation with respect to the probability distribution of the entries of $\eta^{(j)}$.

THEOREM 2

Suppose that M is a diffusion matrix with $\gamma(M) < 1$. Assume (12) and that

$$E(\eta_i^{(j)}) = \mu \quad \text{for} \quad j > 0$$

$$E(\eta_i^{(0)}) = C$$

$$E(\,|\eta_i^{(j)} - \mu\,|^2) = \sigma^2 \quad \text{for } j > 0$$

$$E(\,|\eta_i^{(0)} - C\,|^2) = \sigma_0^2$$

Then $E(w^{(j+1)})$ is a uniform work distribution and

$$E(\,||w^{(j+1)} - \overline{w}^{(j+1)})\,||^2) \leq (n-1)\sigma\frac{(1-\gamma^{2j+2})}{(1-\gamma^2)} + (n-1)\sigma_0^2\gamma^{2j+2}$$

PROOF

We use the results of Theorem 1 together with linearity of the basic scheme. Complete details are in [11]. //

Letting j grow, meaning that we let the computation run indefinitely, we see that the limiting variance is

$$(n-1)\sigma^2/(1-\gamma^2)$$

which can be viewed as the expected steady state deviation from a uniform distribution. By contrast, consider the variance of the load distribution without balancing - the load at a node of the network is simply the sum of the contributions locally over time. So at time t, this sum includes $t+1$ independent random variables with variance σ^2 each and the variance of the sum is the sum of the variances. Thus the variance of the unbalanced distribution is $(t+1)\sigma^2$ at time t in spite of the fact that the expected value of the distribution is uniform. Put another way, although the unbalanced load has an expected value that is uniform, its expected deviation from the uniform distribution is unbounded as time grows indefinitely. This shows that while the variance of an unbalanced distribution would become arbitrary large, the dynamic load balancing scheme controls variance growth and keeps it bounded. Note that as expected, the convergence factor γ is bounded by 1 and the variance is largest when the convergence is slowest.

4. Spatial Partitioning

In the introduction, we discussed load balancing for problems that involve some spatial basis for the work distribution and we indicated there that more could be done. In this section, we focus on the issues surrounding this type of load balancing strategy which in fact is shared with numerical methods in partial differential equations. To set the framework for the discussion, recall that in tracking and data fusion problems, objects are the basic unit of work and objects occupy a location in space. The density or distribution of objects in space then determines the work distribution in the same ambient state space so that partitioning space so that the subregions have the same number of points is equivalent to load balancing the computation for most purposes. fortunately, as with the general dynamic load balancing problem discussed above, we can abstract the details so that the problems does reduce to one involving points in space.

Suppose we are given a set of points, S, in k-dimensional Euclidean space, E^k. We describe an easily computable and maintainable spatially based partition

$P_1, P_2, \ldots, P_N$ of S that satisfies the following loosely stated criteria:

a. each element of the partition should have almost the same number of points from S;

b. if different sets in the partition are allocated to different processing nodes in a distributed memory multiprocessor (specifically a hypercube), then we require that points from S that are close together in the Euclidean metric should be in elements of the partition that get assigned to processors that are close to one another in the multiprocessor network. (Closeness in a distributed memory multiprocessor network refers to the notion of distance given by shortest communications paths in the network.)

Our approach to satisfying both criteria uses a dynamically defined distributed data structure determined by the actual point distribution. Moreover, the allocation condition b. above is somewhat artificial because future distributed memory machines are likely to be less sensitive to process allocation and the subsequent communication patterns. Nonetheless, it turns out that by addressing the locality condition b. we obtain data structures that are easily maintained and easily generated.

It is only recently that researchers have focused attention on applications where work must be balanced with special care being taken with regard to the relative position of the target processor within a multiprocessor system.

Such geometrically sensitive issues in load balancing arise naturally in a variety of applications. The partitioning problem occurs in computational geometry [13] where one uses the partition as a data structure that permits efficient searching strategies. A properly constructed partition turns out to be easy to distribute in a hypercube parallel computer thereby allowing very efficient parallel algorithms for simultaneously searching. Bentley first developed the notion of a multidimensional binary tree that is so fundamental to this work [14, 15]. Using Bentley's ideas, Berger and Bokhari [4] made the first thorough study of how multidimensional binary trees can be mapped to standard multiprocessor network topologies. Their work generates bounds and estimates for communications costs when embedding binary partitions into grids, trees and hypercubes of processors. Their motivation for studying the partitioning problem arises primarily from numerical solutions of elliptic and hyperbolic partial differential equations using domain decomposition techniques. A general discussion of that application can be found in [3]. Recently, Baden used binary partitions for dynamic load balancing in implementing vortex methods for some fluid mechanics problems [9]. Boris and Lambrakos have developed a multidimensional ordering of points in high dimensional space that turns out to be easily derivable from the dual graph or alternately a hypercube embedding of an RBP. The multidimensional ordering has been coined a Monotonic Logical Grid [16]. Other applications include molecular dynamics [16].

In this section we develop a framework for studying a large class of spatial partitioning problems and identify algorithms for dynamically maintaining and adapting such partitions during program execution on binary hypercube structures. The algorithms we present for constructing some partitions are quite similar to the parallel quicksort algorithm recently studied in [17]. Our interest in dynamic properties and algorithms for these partitions arises from a class of applications involving particle evolution. Some other important applications involving spatial partitioning include vortex methods for fluid dynamics problems and n-body problems [18, 9]. A common aspect of these applications is that the particle or point distribution within sets of the partition is not static and can vary over time. A further complication is that new points can be added and existing points can be deleted during program execution. This insertion/deletion issue certainly arises in the area of database management using hypercube computers.

5. Multidimensional Binary Partitions

5.1. A generic algorithm

Suppose that S is a set of N points in Euclidean k space, E^k. We will describe a generic partitioning algorithm based on the simple notion of recursively splitting sets into disjoints pairs of subsets.

Let us formally introduce a splitting function:

$$(A_1, A_2) = split(A, depth)$$

where A_1 , A_2 , A are sets. The parameter *depth* is included to allow for different splitting mechanisms to be used at different depths of the recursion. A generic recursive partitioning scheme is now given by the procedure and illustrated in Figure 4 below:

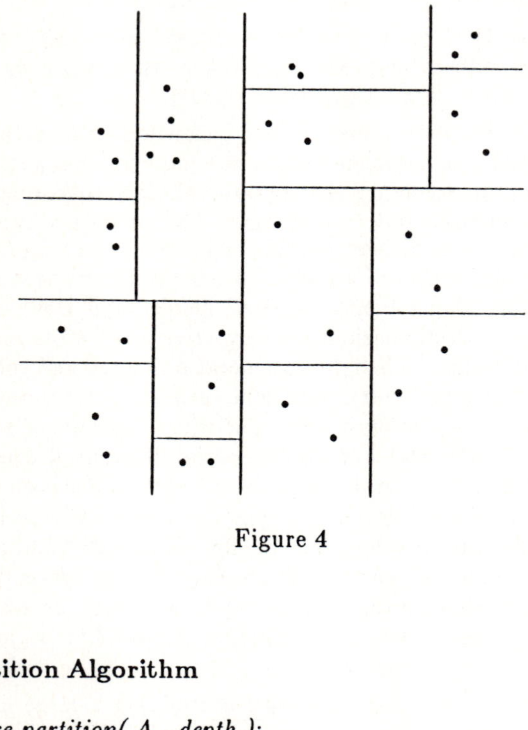

Figure 4

Generic Partition Algorithm

```
procedure partition( A , depth );
begin
        if not (termination) then
        begin
                ( A1 , A2 ) = split( A , depth );
                partition( A1 , depth + 1 );
                partition( A2 , depth + 1 );
        end;
end;
```

In order to define a Multidimensional Binary Partition, we need to specify two ingredients: the termination criterion and the splitting function. The termination criterion is

$$depth > d$$

while the rule for *split* is given by:

> compute the median $x_{depth \bmod 2+1}$ coordinate from *A.points* and cut *A.rect* using that value.

Then A_1 and A_2 are the resulting subrectangles together with the points contained therein. By median, we take to mean a value ensuring that $|A_1.points|$ and $|A_2.points|$ are as close to each other as possible. If points are right on this median line, it makes no difference into which subset they get assigned so long as the cardinality is balanced.

It is easy to see that any number of different partitioning schemes can be imagined by using different estimates for a median. In fact, our distributed algorithm for computing balanced recursive binary partitions uses such an estimate of the median (resulting in an imbalanced but geometrically correct partition) and then repartitions the imbalanced partition into a balanced one.

5.2. Monotonic logical grids

It should be pointed out that MBP's include certain types of monotonic logical grids (MLG), also known as monotonic Lagrangian grids, introduced by Boris, Lambrakos and Picone [16]. A monotonic logical grid is an assignment of integer k-tuple labels to points from E^k that preserves monotonicity in each coordinate direction. Thus if we assign tuples of the form

$$(i_1 , i_2 , \ldots , i_k) \qquad 1 \leq i_j \leq N_j$$

to $N = N_1 N_2 \cdots N_k$ points in E^k and denote by

$$p(i_1, \ldots , i_k)(j)$$

the jth coordinate of the point labeled $(i_1, \ldots , i_k)$ then for $1 \leq m \leq k$, we get

$$p(i_1, \ldots , i_m, \ldots , i_k)(m) \leq p(i_1, \ldots , i_m+1, \ldots , i_k)(m)$$

for every $i_1, \ldots , i_k$ and m. In [16], such grids are generated by a nested sorting algorithm. The diagram below illustrates the close relationship between multidimensional binary partitions and monotonic logical grids. Since there are natural embeddings of grids into hypercubes, these labelings trivially inherit the monotonic logical grid property.

5.3. Parallel Algorithms for Generating MBP's

Th reader is referred to the original reference [10] for complete details and performance characteristics of the partition generation algorithm on an NCUBE hypercube parallel computer. In this subsection, we will briefly describe the basic ideas behind hypercube generation of Multidimensional Binary Partitions.

From the Generic Partitioning Algorithm, we see that a major step in the process is the identification of the median coordinate of a set of points which will initially be distributed over the hypercube. We have used iterative bisection to find this median - a

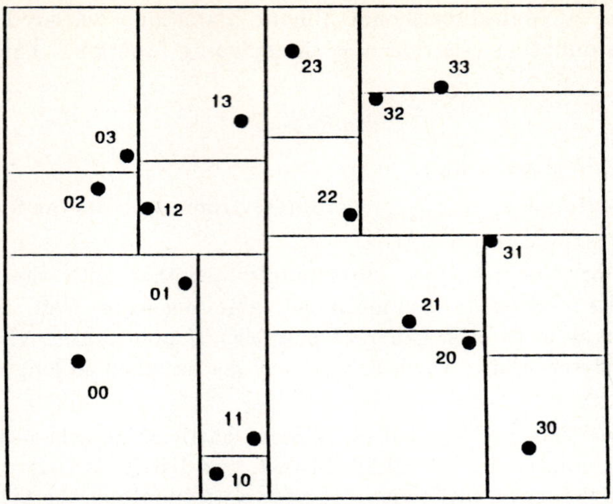

Figure 5.

candidate median is selected according to some rule and all processors report the number of points they have on either side of the candidate median. This count is collected and shared and an adjustment is made to the candidate median according to which side of the line had more points on it. When a balance is reached, all processors exchange points which are on the wrong side of the median that they locally have. The protocol for which processors receive which points is simple but left to the reader to either calculate himself. Reference can be made to the paper [10] as well.

This describes one step of the algorithm - a single median is decided and a set is split into two subsets that reside on two disjoint subcubes of the hypercube. Now the process can be repeated recursively in the other coordinate directions, alternating directions at each level of recursion. Figure 6 illustrates the process for a three dimensional cube and a partition of the plane. Note that the basic scheme does not depend on the underlying dimension of either the hypercube or the ambient Euclidean space.

Performance of the partitioning algorithm on a 64 node hypercube built by NCUBE Corporation is shown in Figure 7.

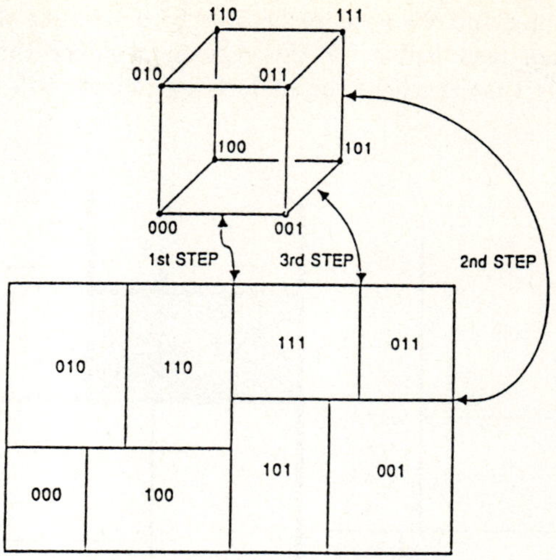

Figure 6

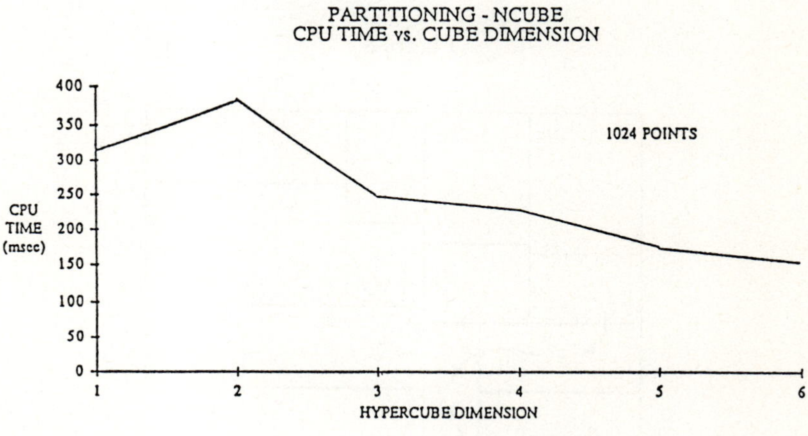

Figure 7

6. Measurement Distribution in a Multidimensional Binary Partition

One of the distinguishing features of signal processing and control problems is that data is continually being received and must be distributed to the processors in a parallel network. In tracking and fusion problems, the measurements are observations of objects that are typically moving and we need to be able to distribute the appropriate measurements to the processors responsible for the regions of space that contain the measurements since presumably that is where the underlying objects will be processed.

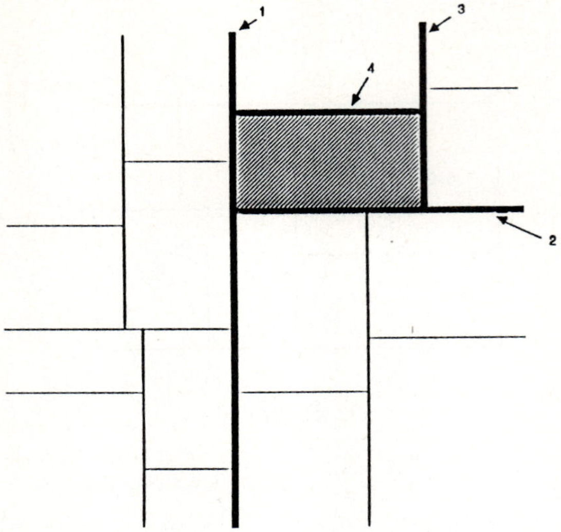

Figure 8

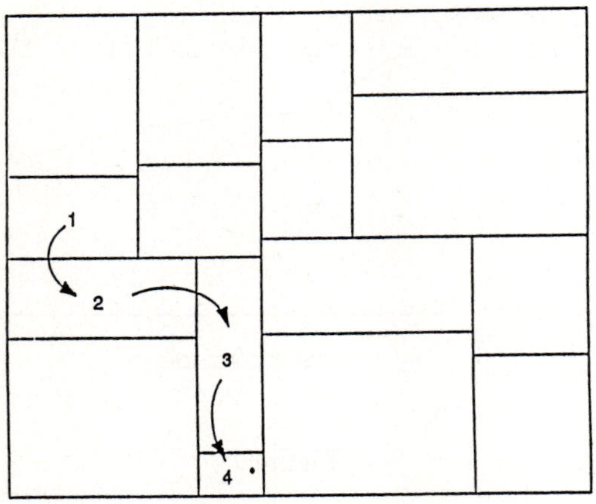

Figure 9

It turns out that Multidimensional Binary Partitions are suited for not only load balancing but also for implementing binary search strategies for making the measurements distribution problem quite simply and elegantly solvable. Recall that an MBP is computed by a sequence of partitions or halvings of sets into two subsets. This sets up a

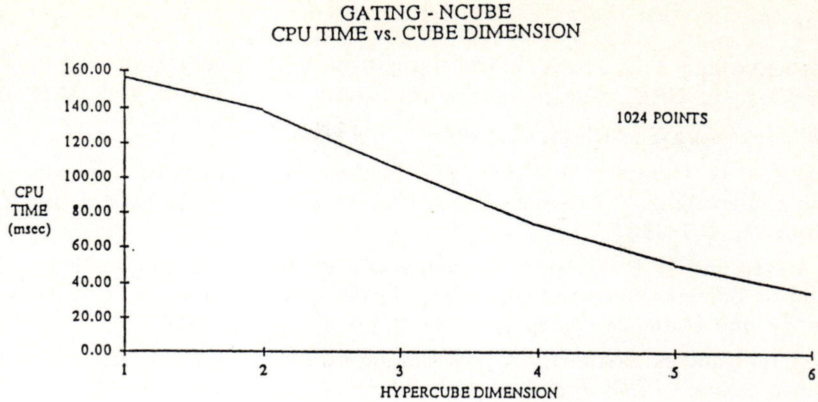

Figure 10

collection of dividing lines that determine each subset of the partition and the collection of dividing lines is a distributed data structure that implements the MBP. Figure 8 shows the information known to the shaded processor in the MBP of that figure - the solid lines are the only lines that the processor needs to have local information about.

We initially send measurements to any processor and using a logarithmic number of comparisons and communications (logarithmic in the number of sets in the partition) the measurement can be sent to the correct processor in the hypercube. Figure 9 shows such a sequence of compares and communications - the point starts out on the processor labeled 1 and the sequence of moves made by the point are shown by the arrows. Figure 10 is a graph showing the performance of the measurement distribution scheme using an MBP. This was computed on an NCUBE parallel hypercube as well. As before, we refer the reader to [10] for details about the algorithm and more performance discussions.

7. Summary

In this paper, we have shown that load balancing is not only an issue of concern for implementing signal processing and control problems on parallel machines but that problems in load balancing are quite simply addressable using novel mathematical and geometric tools. This paper surveyed techniques for general dynamic load balancing in arbitrary interconnection networks and spatial partitioning techniques for hypercubes.

Acknowledgements - The work on spatial partitioning and binary partitions was done jointly with T.G. Allen of ALPHATECH, Inc. David Krumme and Jon Polito of Tufts provided helpful comments and insights into this work. Jon Polito implemented the partitioning algorithms on the NCUBE and performed the performance experiments.

References

1. G. Cybenko and T.G. Allen, "Parallel algorithms for classification and clustering," *Proceedings of SPIE Conference on Advanced Architectures and Algorithms for Signal Processing,* San Diego, CA, 1987.
2. T.G. Allen, G. Cybenko, J. Polito, and C. Angelli, "Hypercube implementation of tracking algorithms," *Proceedings of JDL Workshop on Command and Control,* Washington, DC, 1987.
3. D.E. Keyes and W.D. Gropp, "A comparison of domain decomposition techniques for elliptic partial differential equations and their implementation," *SIAM Journal of Scientific and Statistical Computing,* vol. 8, pp. s166-s202, 1987.
4. M.J. Berger and S. Bokhari, "A partitioning strategy for non-uniform problems on multiprocessors," *IEEE Trans. on Computers,* vol. C-26, pp. 570-580, 1987.
5. M.R. Garey and D.S. Johnson, *Computers and Intractability: A Guide to the Theory of NP-Completeness,* p. W.H. Freeman, San Francisco, CA, 1979.
6. T.C.K. Chou and J.A. Abraham, "Load balancing in distributed systems," *IEEE Trans. on Software Engin.,* vol. 8, pp. 401-412, 1982.
7. F.C.H. Lin and R.M. Keller, "The gradient model load balancing method," *IEEE Trans. on Software Engin.,* vol. 13, pp. 32-37, 1987.
8. L.M. Ni, C.-W. Xu, and T.B. Gendreau, "A distributed drafting algorithm for load balancing," *IEEE Trans. on Software Engin.,* vol. 11, pp. 1153-1161, 1985.
9. S.B. Baden, "Dynamic load balancing of a vortex calculation running on multiprocessors," *Lawrence Berkeley Laboratory Research Report,* vol. 22584, 1986.
10. T.G. Allen and G. Cybenko, *Recursive binary partitions,* 1987. submitted to IEEE Trans. on Computers
11. G. Cybenko, *Dynamic load balancing for distributed memory multiprocessors,* 1987. Tufts University, Department of Computer Science, Technical Report
12. R.S. Varga, *Matrix Iterative Analysis,* Prentice-Hall, Englewood Cliffs, NJ, 1962.
13. F.P. Preperata and M.I. Shamos, *Computational Geometry,* Springer-Verlag, New York, 1985.
14. J.L. Bentley, "Multidimensional binary search trees used for associative searching," *Communications of the ACM,* vol. 18, pp. 509-516, 1975.
15. J.L. Bentley, "Multidimensional divide-and-conquer," *Communications of the ACM,* vol. 23, pp. 214-228, 1980.
16. J. Boris, "A vectorized "near neighbors" algorithm of order N using a Monotonic Logical Grid," *Journal of Computational Physics,* vol. 66, pp. 1-20, 1986.
17. P. Heidelberger, A. Norton, and J.T. Robinson, *Parallel quicksort using fetch-and-add,* 1987. IBM T.J. Watson Research Report RC 12576 (#56561)
18. L. Greengard, "The rapid evaluation of potential fields in particle systems," *Research Report-533,* Yale University, Department of Computer Science, 1987. (Ph.D. Thesis)

Systolic algorithms for digital signal processing and control

J.-P. Charlier, M. Vanbegin and P. Van Dooren

Philips Research Laboratory Brussels
Av. Van Becelaere 2 , B-1170 Brussels, Belgium

Abstract: In this paper we give an introduction to new developments in the interdisciplinary area of parallel algorithms for digital signal processing and related problems in control theory. It is not the purpose of the paper to be exhaustive in any sense, but rather to introduce the reader to the relevance of current developments in the area, and to guide him through the existing literature.

Keywords: Systolic algorithms, digital signal processing, control theory, linear algebra

1 Introduction

Parallel Algorithms (PA's) constitute presently a tentacular trend which extends its arms over a variety of mathematical fields. Indeed, parallelism seems to provide an adequate alternative means to speed up computation or data treatment, in consideration of the following facts: (i) the rate of growth of computing power available from a single processor has been slowing down more and more for one or two decades; (ii) at the same time, the hardware cost of executing an elementary operation has decreased dramatically; (iii) also, problems have grown in difficulty and complexity. Therefore, design of new algorithms, parallelization of existing ones, and study of their properties are in rapid development.

Digital Signal Processing (DSP) is an application area where early specific types of parallel implementation (namely *systolic arrays*) were proposed, partly because of the inherent resemblance between representations of digital filters and systolic arrays. Some of the techniques used for signal processing are also directly applicable to control theory. This is e.g. easily recognized in the duality between optimal control and optimal filtering. Both rely on recursive least squares techniques that are indeed very similar [1], [25].

The main reason why there is a need for highly performant parallel algorithms in both areas is that several applications now imply real time implementations with very high throughput rate of data and computational load. Typical examples of this are:

- *speech processing* where the sample rate can go up to 20,000/sec. and the number of operations may reach 1,500,000/sec. such as in speech recognition.

- *adaptive control* of industrial processes (such as a glass furnace) or of devices incorporated in commercial products (such as a TV set or a compact disc) which

NATO ASI Series, Vol. F47
Advanced Computing Concepts and Techniques
in Control Engineering
Edited by M. J. Denham and A. J. Laub

have to operate in real time.

- *medical scanners* which process up to 350,000 samples/sec. and perform up to 50,000,000 operations/sec.

- *radar processing* which seems to have the most ambitious processing requirements with 10,000,000 samples/sec. and 5,000,000,000 operations/sec.

Especially for the last two examples, the desired throughput rate and computational load can hardly be handled by conventional machines anymore. Even supercomputers – with a moderate level of parallelism but with very high speed processors – can hardly tackle these tasks. Moreover they would not exploit the special features of these problems which are:

- *repetitive nature of the computations* both on the level of the processing (e.g. every 10 msec. a portion of a scanner signal is transformed to the frequency domain via an FFT) as on the level of the algorithmic details (an FFT is intrinsically repetitive in its fine details).

- *use of elementary operations* such as fixed arithmetic, small wordlength (6 ... 16 bits for speech), cordic transformations, etc. is typical in this area.

- *dedicated hardware* allows to translate the special features of a specific application in (VLSI) hardware implementation.

In the next section we first give a brief outline of parallel architectures and algorithms. In section 3 we then discuss the interaction of this field with DSP, while in the last section we try to discern a number of relevant topics in the interdisciplinary research area of DSP/control and numerical linear algebra.

2 Parallel architectures and parallel processing

This section is an introduction to general aspects of parallelism: which architectures and how to measure the performance of parallel algorithms.

2.1 Parallel computers

Parallelism is neither a new idea nor a recently appeared reality. Human beings are highly parallelized. When they decided to build electronic computing machines, they already hoped to introduce some degree of parallelism in them. In most uniprocessor systems, a number of parallel processing mechanisms have been developed: examples are time sharing and the inclusion of multiple functional units in the CPU. Parallel processing is in contrast to sequential processing and can be defined as a cost-effective means to improve system performance through concurrent activities in the computer.

Parallel computers result from the general architectural trend to emphasize parallel processing. State-of-the-art parallel computers can be characterized into three structural classes: pipeline computers, array processors, and multiprocessor systems.

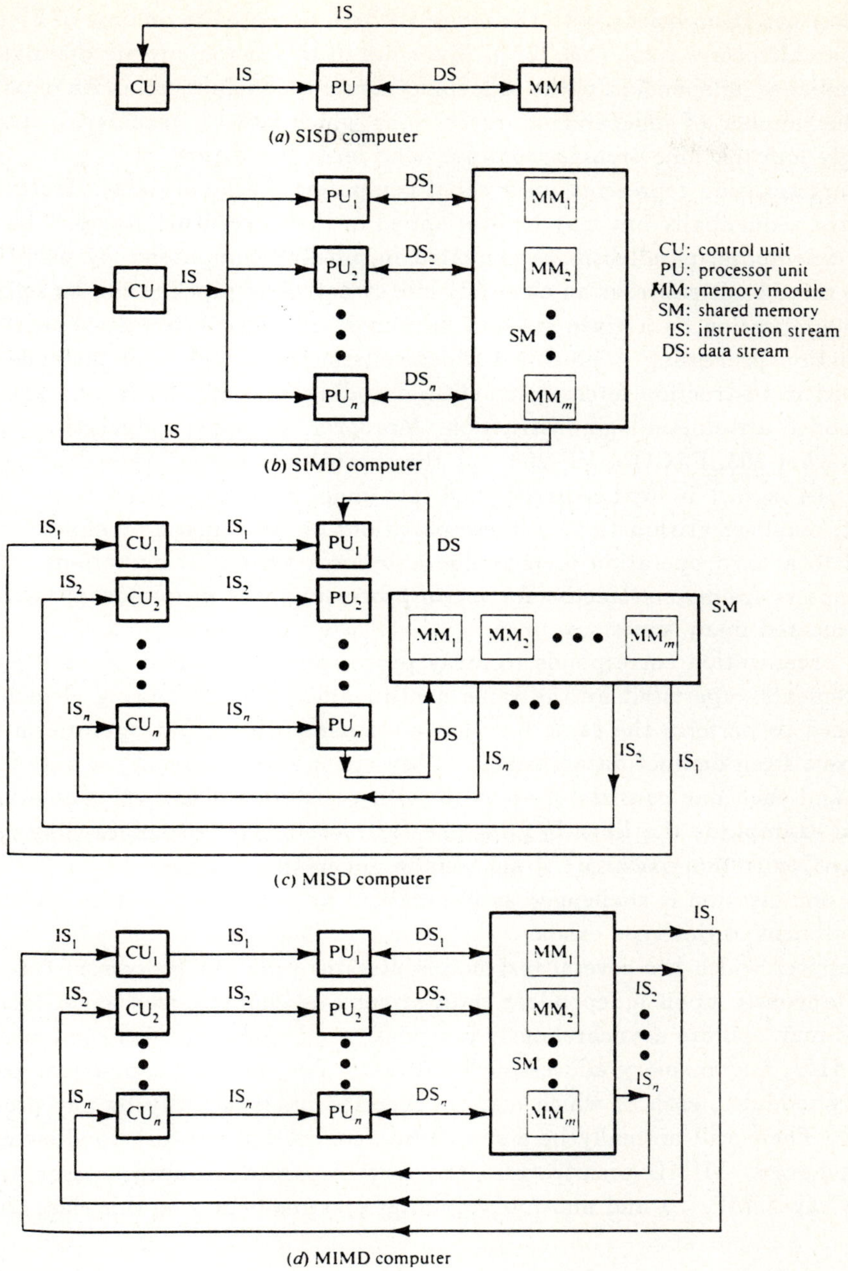

Fig. 1 : Flynn's classification of computer organizations (taken from [22])

We introduce them briefly with the aid of the well accepted taxonomy of Flynn for computer architectures ([24], [42], [22]). Machines in this taxonomy are distinguished by the number of independent paths to memory along which data can move in parallel, and by the number of independent instructions which can be executed in parallel. Accordingly four machine organizations are considered (see figure 1):

SISD organization represents most serial computers available today. Instructions are executed sequentially but may be overlapped in their execution stages. This overlapping, or temporal parallelism, is exploited in pipeline computers: by partitioning tasks into more basic subtasks an assembly line structure or pipeline can be set up for repetitive application on a given stream. In almost all high-performance computers, the instruction processing is pipelined ("instruction pipelining"), for instance being decomposed in instruction fetch, instruction decoding, operand fetch, and execution of the decoded arithmetic logic instruction. Moreover, in modern pipeline computers (Cray-1, Cyber-205, FACOM VP-200, ...) the execution phase of some floating-point operations ($+,\times$, ...) is further partitioned into simpler suboperations (exponent adjustement, mantissa arithmetic, ...); these partitions or "arithmetic pipelines" may be dedicated to a fixed operation or reconfigurable for a variety of operations. Clearly such computers are more attractive for vector processing, where component operations may be repeated many times.

SIMD organization corresponds to array processors, constituted by multiple processing elements supervised by the same control unit. The processing elements are synchronized to perform the same function at the same time, but operate on different data sets from distinct data streams. They are interconnected by a data-routing network, and each one consists of an ALU with registers and a local memory. The archetypal example is the Illiac-IV. Systolic (or systolic-like) machines (like systolic arrays, the Connection machine ...) are usually put in this category.

MISD organization is challenged as impractical by some computer architects. No real embodiment of this type exists.

A computer which has several instruction streams (one per processor) that operate simultaneously upon independent data streams is characterized as MIMD. The processors may operate asynchronously and must each contain an instruction decode unit, an ALU, and memory addressing hardware. The memory component consists of memory modules, each of which may be accessed independently by any processor in general. There will normally be a simple interconnection network to allow control signals exchange. MIMD architectures are very disparate. Multiprocessor systems (Cray-2, Cray-X MP, ...) and multiple computer systems belong to this class [20].

2.2 Systolic arrays and data flow machines

The rapid advent of VLSI technology has created new horizons in implementing parallel algorithms directly in hardware. Here we first describe features of systolic arrays and we justify them as our choice of architectural model in the sequel of this paper. Then we present the recent concept of data-driven computations. These are both to be classified under SIMD machines.

2.2.1 Systolic arrays

VLSI technology offers very fast and inexpensive computational elements which can be combined in highly parallel special-purpose structures for a wide range of potential applications. Nevertheless, general guidelines are necessary in order to achieve an efficient implementation of a given algorithm. The concept of *systolic* architecture has been developed at Carnegie-Mellon University as such a general methodology for mapping high-level computations into hardware structures. In a systolic system, "data flows from the computer memory in a rythmic fashion, passing through many processing elements before it returns to the memory, much as blood circulates to and from the heart" (figure 2) ([30], [29]). More precisely a systolic system is a computing network constituted by an array of processing elements or nodes (ideally implemented in a single chip), and possessing the following features [31]:

- *Synchrony.* The data are rhythmically computed (timed by a global clock) and passed through the array.
- *Regularity,* i.e. modularity and local interconnection. The array consists of modular processing units with regular and local interconnections. Moreover the computing network may be extended indefinitely.
- *Temporal locality.* There will be at least one unit-time delay allotted so that signal transactions from one node to the next can be completed.
- *Effective pipelinability.* Ideally, successive computations are to be pipelined, such that all processing elements are working at any time, which leads to a linear speed-up.

Regularity involves simplicity of design and control, thus allowing to cope with the large hardware resources offered by VLSI. Systolic arrays are suitable for implementing a diversity of parallel algorithms and can assume many different configurations (figure 3) for different algorithms.

The major limitation of systolic arrays, and of VLSI structures in general, is their restricted use to compute-bound algorithms. In a compute-bound algorithm, the number of computing operations is larger than the total number of input and output elements; otherwise the problem is I/O-bound. The I/O-bound problems are not appropriate for VLSI because VLSI packaging must be constrained with limited I/O pins. This limits the size of systolic arrays in practice. However, an alternative has been suggested to overcome that difficulty, by partitioning the initial problem into smaller parts and implementing them in a pipelined network [22]. Finally it is worthwhile to note that programmable systolic arrays are under study [11] in order to enlarge the set of algorithms that the same device (or collection of devices) can be used to solve. Also reconfigurable arrays have been proposed [22].

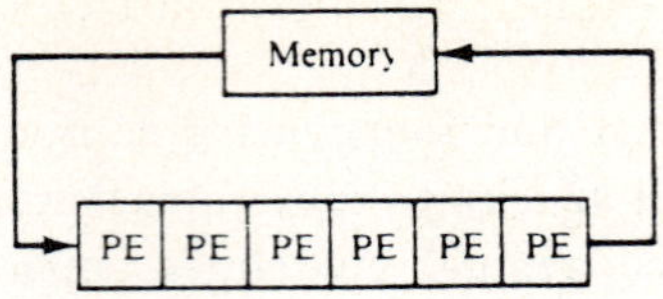

Fig. 2 : Basic principle of a systolic system (taken from [22])

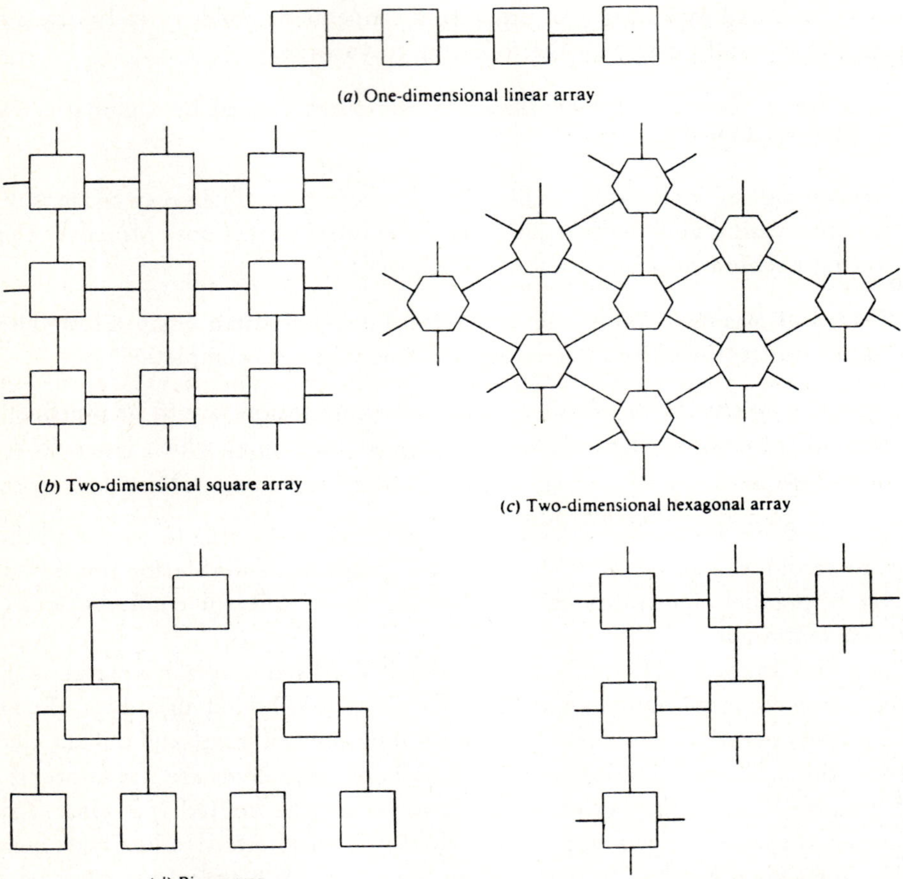

Fig. 3 : Some systolic array configurations (taken from [22])

In this paper we choose systolic arrays as an architectural reference, this founded on the following facts:

(i) They are in close relation with actual and probably future technology, and receive a great deal of attention from universities and industry.

(ii) By the inherently high and systematic degree of concurrency of their organization, and due to their formal definition, they constitute an abstract context in which existing algorithms can be efficiently parallelized and hopefully new algorithms could be designed.

(iii) Important applications of systolic arrays have been proposed in signal and image processing, matrix arithmetic, partial differential equations, pattern recognition, etc...

(iv) Procedures exist to express parallel algorithms in a form that can be easily implemented into systolic arrays (see for example [31], [32], [39], [38], ...).

2.2.2 Data-driven computations

The conventional von Neumann machines are called control flow computers because instructions are executed sequentially as controlled by a program counter. Sequential program execution is inherently slow. *Data flow* computers were suggested recently [8] to exploit maximal parallelism in a program. The basic concept is to enable the execution of an instruction whenever its required operands become available such that no program counters are needed in data-driven computations: the execution follows the data dependency constraints. Theoretically maximal concurrency can be exploited in such a data flow machine, constrained only by the hardware resource availability. Further advantage of data flow multiprocessors is a reduction of the use of centralized control and global memory. Programs for data-driven computations are represented by data flow graphs. A diversity of applications is illustrated or referenced in [31], [32], and [40].

An example of data flow multiprocessor is the *wavefront array* [31], which can be interpreted as a generalization of a systolic array by dropping synchrony and temporal locality features but keeping regularity and pipelinability. The name wavefront comes from the propagation of computations through the array in the same manner as electromagnetic wavefronts. The main difference between wavefront arrays and systolic arrays is the lack of synchrony: here the processing elements (PE's) each perform their computations as soon as all the required data are available. The PE's are thus not synchronized by a global clock, but the wave propagation of the data ensures by itself a kind of synchronization [31]. The concepts of *systolic array* and *wavefront array* are illustrated later on via specific examples.

In the rest of this paper, we put emphasis on algorithms for *systolic-like* arrays, by which we mean algorithms that have already been implemented or are likely to be implementable on systolic arrays, wavefront arrays, or closely related architectures.

2.3 Parallel algorithms

In studying PA's, we need some measure of their performance. Two related quantities are usually considered: *speed-up* measures the improvement in solution time using parallelism, while *efficiency* measures how well the processing power is being used. We also discuss here briefly issues of complexity, consistency and stability of parallel algorithms.

2.3.1 Speed-up and efficiency

The performance of a parallel algorithm strongly depends on the architectural features of the computer on which it is implemented. For instance, memory interference, interprocessor communication, and synchronization, lead to overhead and may affect seriously program execution times. In order to estimate precisely the importance of such factors, detailed computer models are necessary (see [22] for an introduction). In this text, those aspects will not really be taken into account in a quantitative way. Therefore performances will have to be interpreted as valid for an ideal implementation of the corresponding algorithms.

Let us examine measures of performance on array processors or multiprocessor systems. *Speed-up* (S) is defined by the ratio of the execution time T_s on a serial computer or uniprocessor to the execution time T_p on the parallel computer:

$$S = \frac{T_s}{T_p}. \tag{1}$$

In practice, T_p incorporates any overhead of architectural nature. In particular, if the processors can run asynchronously much of the time, a higher S can be expected than with a synchronized implementation. This has motivated the design of data flow machines (see above) as well as algorithmic research. In the definition (1), T_s and T_p often refer to the same algorithm. For a computer that can support P simultaneous processes, the ideal value of S is then P. In fact, to achieve a fair comparison, T_s would be the execution time of the fastest algorithm for the same problem on one processor (as done e.g. in [45], [19]). Since the two algorithms can be quite different, S could be greater than P in theory.

Related to S, *efficiency* E is defined by

$$E = \frac{S}{P} \tag{2}$$

for a system of P processors. E indicates the utilization rate of the available resources (the processors) and is given by the ratio of the actual speed-up S to the ideal speed-up P. Note that the maximum value 1 of E is always reached for $P = 1$, so the number of processors is not necessarily chosen in order to maximize E. Rather, the goal is to construct algorithms exhibiting linear speed-up in P and hence utilizing the processors efficiently. Speed-up of $kP/\log_2 P$ is also acceptable: the speed increase is almost two when the number of processors is doubled. More rapidly decreasing speed-up functions characterize algorithms that are poorly suited for parallelism.

2.3.2 Complexity, consistency and stability

As a rule, the evaluation of algorithms on serial computers is based upon computational complexity analysis, that is, arithmetic operation counts. For pipeline computers, this arithmetic complexity remains important, because every operation affects timing even if it is part of a vector operation. Therefore Lambiotte and Voigt [35] have introduced the notion of *consistency*: a parallel algorithm is said to be (asymptotically) *consistent* if its arithmetic complexity is of the same order of magnitude as that of the *best* serial algorithm. In general, consistent algorithms are expected to be faster on pipeline computers.

On array processors (or multiprocessor systems), consistency is less crucial. If one step is counted for each set of operations performed simultaneously, the most important consideration is the number of steps (T_p in (1)), not the total number of operations performed by all the processors. Indeed, non consistent algorithms are sometimes advantageous when the extra amount of computation does not lead to additional steps while communication needs between processors are reduced. E.g. with Csanky's method [7], the inversion of an $n \times n$ matrix can be obtained in $O(\log^2 n)$ time-units (the time required for 1 flop) but using $O(n^4)$ PE's.

The stability of parallel algorithms has not received much attention until now, but a growth of specific developments in this area will presumably take place in the near future. A recent study dealing with the evaluation of arithmetic expressions on a pipeline computer is given in [44]. Clearly, the behaviour of new algorithms is to be analyzed. Also the use of a non consistent algorithm may lead to a loss of accuracy, due to the greater number of operations that are required per result. Moreover, in the situation where the operations of a serial algorithm are purely rearranged in a new parallel order, stability properties may be affected. The fact that parallel processing of an algorithm may yield more accurate results than its serial version, is e.g. illustrated by the fan-in algorithm for sums and inner products [24], and by recursive least-squares updating and downdating algorithms ([14], [3]), but this property is certainly not true in general.

3 Parallel algorithms in digital signal processing

In this section we first point out the similarities between "flow graphs" often used in DSP and "computational networks" used for systolic arrays. Then we list various constraints that are relevant in the choice between different possible computational networks for a same DSP algorithm.

3.1 A simple example : linear convolution

In order to illustrate how PA's can efficiently be used in DSP problems we take one of the simplest possible examples. Consider the linear convolution of a signal $\{x_i;\ i = 0,\ldots,\infty\}$ with a finite impulse response (FIR) filter $w(z) = w_1 \cdot z^{-1} + \ldots + w_n \cdot z^{-n}$, (where z^{-1} stands for the delay operator) yielding the output signal $\{y_i;\ i = 1,\ldots,\infty\}$

defined by:

$$y_i = \sum_{j=1}^{n} w_j \cdot x_{i-j}. \tag{3}$$

A typical representation of the convolution (3) in DSP is the following "computational flow graph" (**D** represents a delay):

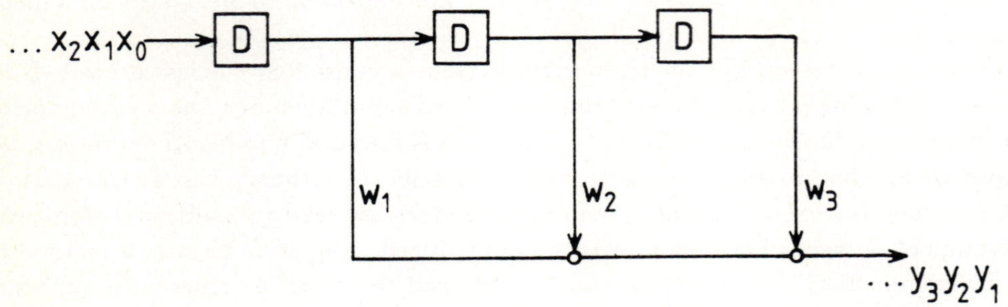

Fig. 4 : A simple example : FIR convolution for n=3.

where we have chosen $n = 3$ for illustrative simplicity. Several different systolic implementations of this formula are possible [28] and we mention here only a few. In each of them one clearly recognizes the classical "flow graph" representation of figure 4.

In figure 5 we have a first design where the x_i's are "broadcasted" simultaneously to all processors, each containing one value w_i, and the y_i's move systolically through the array of processors. In the second design the x_i's move in systolically and the y_i's are obtained by an addition of all current values of the processors. Both these designs need a *global* connection (a bus for broadcasting or a global adder for reconstructing y_i) which is a drawback for these two approaches.

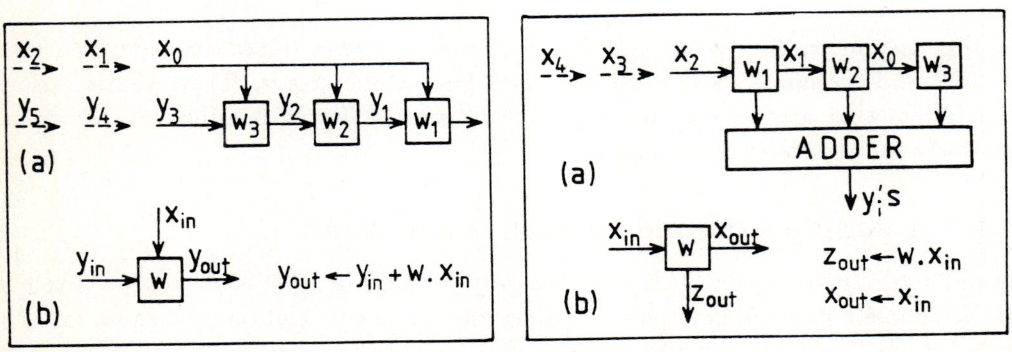

Fig. 5 : Linear processor arrays for convolution (modified from [29]).

In the following two designs (figure 6) this has been avoided by letting the w_i's move systolically as well. In the first one they move systolically in a direction opposite to that of the x_i's, while in the second one they move along with the x_i's, but at a different speed. As a result of this no global connections are needed anymore, but since the y_i stay in the array, a systolic output path (indicated by broken arrows) is needed to extract these values at the end of the convolution process.

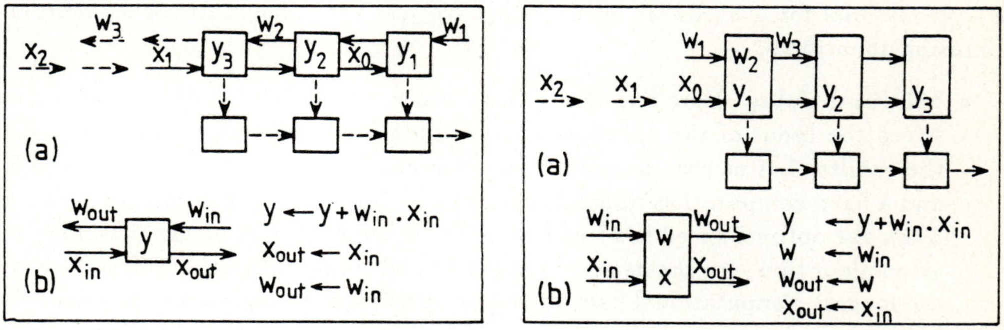

Fig. 6 : Two other designs for convolution (modified from [29]).

Finally, we consider in figure 7 two devices that avoid both these drawbacks. Here again the w_i are each connected to one cell and both the x_i and y_i signals move systolically: in the first device in a direction opposite to each other and in the second device in the same direction but at different speeds. Since the y_i already move out systolically here, no secondary path is needed to extract them.

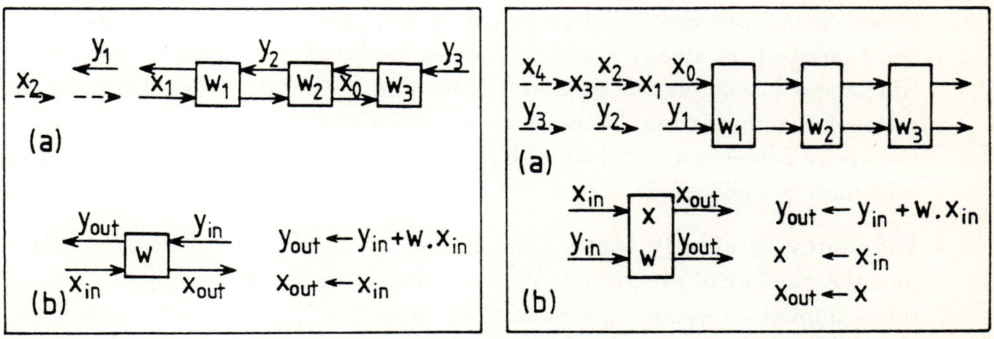

Fig. 7 : Two "optimal" designs for convolution (modified from [29]).

We use this example to illustrate that parallel algorithms in DSP are not "really" (one should perhaps say here "often") new algorithms but are rather clever implementations of existing DSP algorithms. In the next subsection we try to explain what exactly is meant by "clever".

3.2 Criteria for systolization

With *criteria for systolization* we mean those issues that have to be taken into consideration when trying to derive a "systolic" implementation which is *acceptable*, *good* or even *optimal* for a hardware realization. We just list some main issues here and comment upon them.

- **Minimal delay.** Here one wants to minimize the amount of time elapsing between the input of the data into the computational network and the output of the results. Let us assume a fixed transmission time between the different PE's and a fixed computation time for one flop (i.e. one scalar operation $y = a \cdot x + b$). Then the optimal speed is a function of the inherent degree of parallellism of the given algorithm – or the *depth* of the associated computational graph – since it is the longest computational path in that graph that determines the minimal delay between the input and the output of the scheme. The number of PE's to achieve this optimal speed (if it is known !) can be very large. In our example the longest path is easily shown to require three flops which could all be performed in one PE. Each PE then in fact computes one y_i and hence we require as many PE's as output data !

- **Speed-up in steady state.** Here one pursues roughly the same goal as above, but respecting the constraints of the input data flow. An example illustrates better this point. While the above "minimal delay" solution requires *each* x_i to be entered *at the same time* and at k PE's concurrently, this usually does not coincide with the *availability* of the $\{x_i\}$. Indeed, these are often the samples of some real-time process and thus arrive typically in a piecemeal fashion. If in (1), the times correspond to the execution of an "infinity" of identical computations either on one processor (T_s) or pipelined on a parallel computer (T_p), then S is the "speed-up in steady state". Since in this case one neglects the amount of time spent in loading the data and unloading the results, S is also given by the ratio of the throughput rates in sequential and parallel implementations. For the above problem a speed-up of 3 is then obtained in steady state by using the schedules of figures 5-7.

- **Efficiency in steady state.** This is the ratio of the speed-up in steady state and the number of processors. We note that optimal efficiency (in steady state) often implies to completely reorder or even rewrite the computation (see e.g. the FFT reformulation of a DFT), which thus makes it difficult to *prove* that a certain scheme is optimal. Techniques for deriving such schemes are borrowed from complexity theory or sometimes use graphical considerations. Results in this area, though, are far from complete.

- **Lay-out.** Here one wants to find a systolization that makes the inplementation in hardware as simple as possible. Typically one wants to avoid things as wire crossing (present in FFT, perfect shuffle, ...) or global connections (see second scheme of figure 5). One usually prefers *regular* 1-dimensional or 2-dimensional arrangements of the PE's with only few connections between them (nearest neighbors only) in order to yield compact hardware implementations. Typical examples are the linear and square arrays of figure 3.

- **Robustness.** With robustness one means that the implemented algorithm should indeed yield the desired answers or at least answers that are "reasonably close" to them. Therefore the design of the scheme should be such that overflows and underflows should not occur and that numerical errors should not propagate unboundedly (i.e. the scheme should be numerically stable in some sense). "Simple" correction schemes for recovering from (or at least detecting) possible loss of accuracy are needed when numerical stability can not be ensured. This is still a trouble spot of some of the popular systolic algorithms.

We illustrate these different issues by two short examples. The first one is the discrete Fourier transform, which is defined by

$$X_i = \sum_{k=0}^{n} w^{ik} \cdot x_k = \sum_{k=0}^{n} e^{-\frac{2ik\pi}{n+1} \cdot j} \cdot x_k, \tag{4}$$

where $j = \sqrt{-1}$. Performing a DFT of a signal $\{x_k;\ k = 0, \ldots, n\}$ is easily done with a linear array of processors, as shown in the figure below:

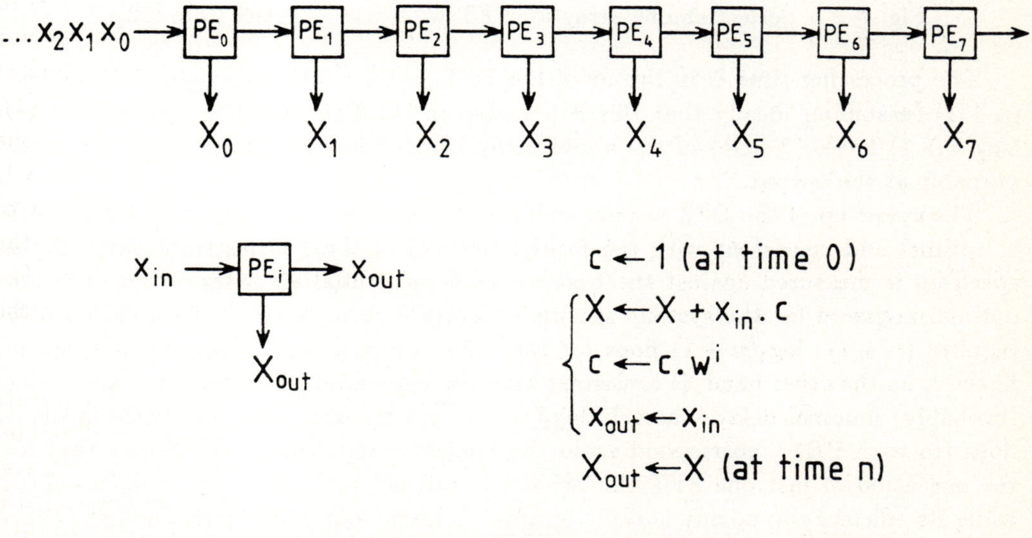

Fig. 8 : A linear array for DFT with $n = 7$ (modified from [32]).

where in each processor a time-varying weight factor $c = w^{ik}$ is updated, with k (the time step) starting from 0 as x_0 enters it. The Fourier transform $\{X_i;\ i = 0, \ldots, n\}$ stays in the node and is eventually pumped out from each processor when the last x_n has passed it.

Just as for sequential machines the FFT formulation allows for a faster implementation of formula (4) on a parallel machine. This is shown in figure 9, where c_i and d_i are appropriate powers of w stored in each processor and depending on the location of the processor in the graph. The drawback here, though, is that connections are not as regular anymore but require a *global* (Perfect Shuffle) communication network.

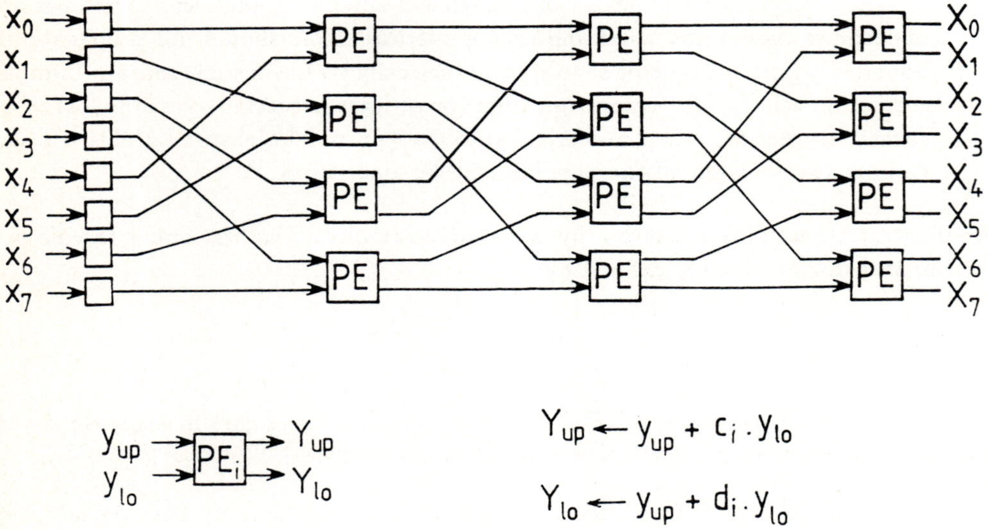

Fig. 9 : A perfect shuffle **array for FFT** with $n = 7$ (taken from [32]).

The processing time is in favour of the FFT, with a ratio of $\log_2(n+1)$ against $(n+1)$ (assuming ideally that this is a power of 2). The data have indeed to pass $\log_2(n+1)$ levels (3 here) of the above array before the signal $\{X_i;\ i = 0, \ldots, n\}$ is available at the output.

The speed-up of the DFT scheme in figure 8 is $O(n+1)$ using $(n+1)$ PE's which is optimal and consistent with the formulation (4) of the Fourier transform. If the speed-up is measured against the best available sequential algorithm, then it is not optimal anymore for the systolic scheme in figure 8 since the FFT formulation only requires $(n+1) \cdot \log_2(n+1)$ flops for the same computation. The scheme given in figure 9, on the other hand, is consistent with the sequential FFT algorithm and yields (probably) minimal delay – namely $\log_2(n+1) = 3$ transmissions and twice as many flops (in each PE) – corresponding to the longest computing path. Notice that for the execution of just one FFT this scheme is not efficient ($E = O(1/\log_2(n+1))$), while its efficiency in steady state is optimal. The lay-out of figure 9, though, is less appealing (interleaved connections) and the components $\{x_i\}$ of one data vector must be all available simultaneously, which is not the case in general.

If only a few FFT's have to be performed, a disadvantage of the scheme of figure 9 is its high number of processors, namely $\log_2(n+1)$ layers of $(n+1)/2$ processors. This can easily be replaced by *one* layer of $(n+1)/2$ processors through which the data circulate $\log_2(n+1)$ times, as shown in figure 10. This of course reduces the number of processors by a factor $\log_2(n+1)$ but also reduces its throughput rate by the same factor (each new data vector has to wait $\log_2(n+1)$ steps after the previous data vector before entering the array). The efficiency is now optimal both for one FFT *and* in steady state. On the other hand, the processors are now slightly more complicated since they must keep track of what layer the PE's are processing at each time step (the coefficients c_i and d_i now vary with time, while earlier they were fixed for each PE).

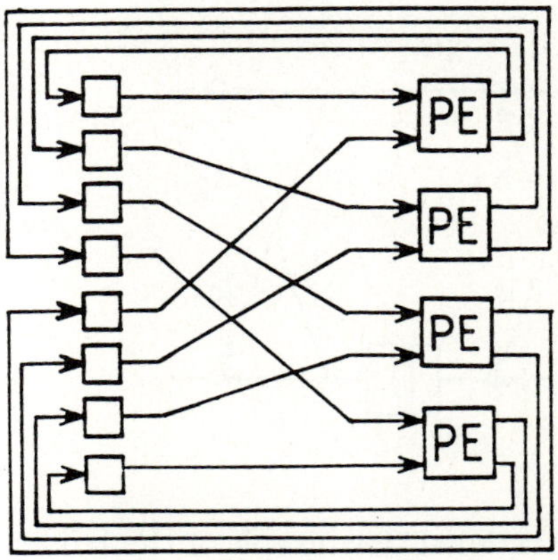

Fig. 10 : A one layer recursive array for FFT with $n = 7$ (taken from [32]).

The second example is the linear systolic array for implementing FIR digital filters. Consider a convolution with $a(z) = 1 + a_1 \cdot z^{-1} + \ldots + a_n \cdot z^{-n}$ applied to an input sequence $\{x_i;\ i = 0, \ldots, \infty\}$. A possible computational scheme for the output sequence $\{y_i;\ i = 0, \ldots, \infty\}$ is schematically represented by the signal flow graph of figure 11, which is often called a *transversal filter* representation of $a(z)$.

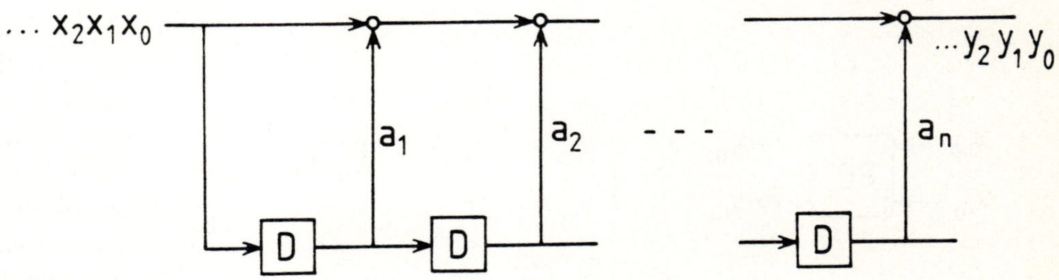

Fig. 11 : A computational flow graph of the transversal filter for $a(z)$.

Assume now $a(z)$ be a *stable* polynomial in z^{-1}, then it can as well be represented by its *reflection coefficients* $\{k_i;\ i = 1, \ldots, n\}$, linked with $a(z)$ via the following recursion:

$$\begin{aligned} & a_0(z) := 1, \\ & \text{for } i = 1 \text{ to } n, \qquad a_i(z) := a_{i-1}(z) + k_i \cdot z^{-i} \cdot a_{i-1}(z^{-1}), \\ & a(z) := a_n(z) \end{aligned} \tag{5}$$

It turns out that these coefficients k_i are then all bounded by 1 if and only if $a(z)$ is stable [36]. Moreover a computational graph for the output of the filter can be based on these coefficients as well. This is shown in figure 12 (see also [36], [34]).

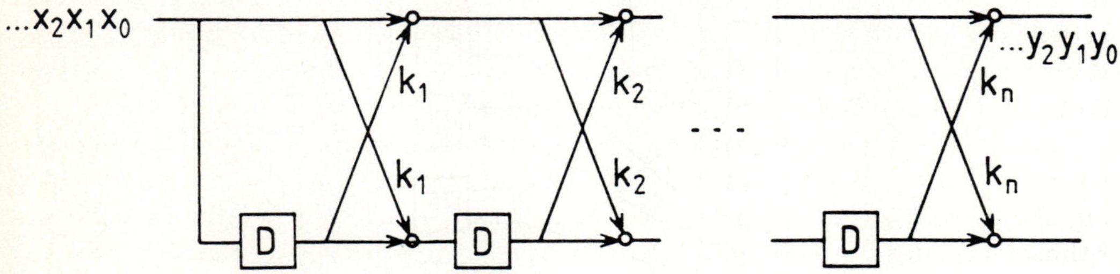

Fig. 12 : A computational flow graph of the ladder filter for $a(z)$.

Both flow graphs represent the same filter $a(z)$ but using a different parametrization. Yet there are clear similarities between both representations and even more so when one looks at their parallel implementation on a linear array of n processors. This is shown in figure 13, where the only difference between both schemes lies in the little programs stored in each processor.

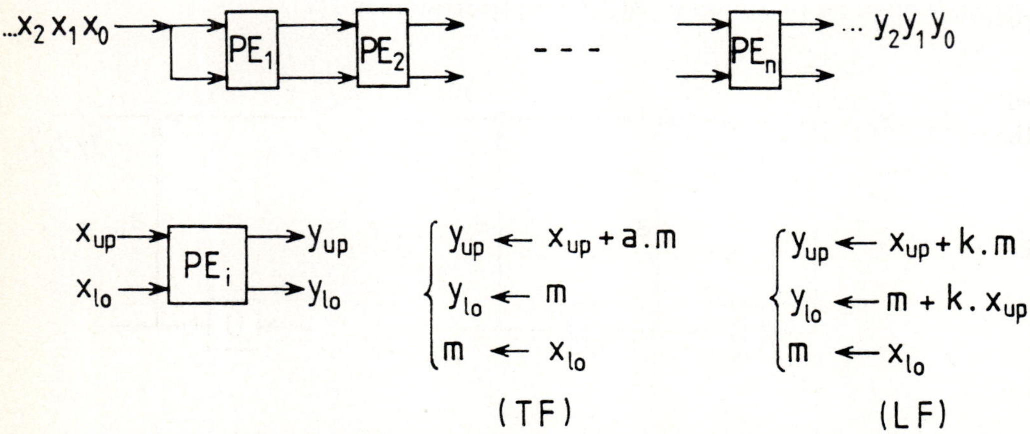

Fig. 13 : Systolic implementation for transversal filter (TF) and ladder filter (LF).

Notice that the throughput rate of both filters is the same although seemingly twice as much work (multiplications) has to be performed in the ladder filter. The fact is that the assignments in each of the processors could be performed in parallel as well (within each processor) and the execution time in each processor is then really determined by its slowest operation (one multiplication and one addition). In practice, preference is mostly given to the ladder filter for reasons of sensitivity (e.g. in speech synthesizers [36]).

4 Linear algebra and control

The purpose of this section is to show that some of the techniques that are typical to the interdisciplinary area of signal processing and parallel algorithms are also slowly getting more attention in solving linear algebra problems encountered in control theory. Problems being tackled now range from the solution of Lyapunov and Riccati equations [6] to the evaluation of the frequency response of a state space system [4].

One of the more classical examples, though, is the problem of updating and downdating Choleski and QR decompositions, and it appears in both digital signal processing and control theory. In control theory it is used e.g. in the fast square root algorithms for solving Riccati difference equations appearing in optimal regulator problems (both updating and downdating can be used for solving the Chandrasekhar equations [25]). In signal processing updating and downdating naturally appear in the recursive solution of least squares problems for filtering with sliding windows [21]. We do not aim here to give a complete derivation of these problems since they have already been treated extensively by several authors [15], [37], [3].

In updating a QR decomposition one starts from an $n \times n$ system of equations

$$R \cdot x = b \tag{6}$$

with R upper triangular, resulting from a previous QR decomposition. Adding a new equation

$$a^T \cdot x = c \tag{7}$$

with a^T of dimension $1 \times n$, one then wants to determine the least squares solution of the updated system [17]

$$\begin{bmatrix} a^T \\ R \end{bmatrix} \cdot x = \begin{bmatrix} c \\ b \end{bmatrix}, \tag{8}$$

i.e. to find the QR factorization of the "nearly triangular" matrix

$$\begin{bmatrix} a^T \\ R \end{bmatrix} = \begin{bmatrix} a_1 & a_2 & \dots & a_n \\ r_{11} & r_{12} & \dots & r_{1n} \\ & r_{22} & & \\ & & \ddots & \vdots \\ & & & r_{nn} \end{bmatrix} \tag{9}$$

It is shown in [16] that this only involves a sequence of n Givens rotations $G_{i,i+1}$ each operating on adjacent rows i and $i+1$:

$$G_{i,i+1} = \begin{bmatrix} 1 & .. & 0 & 0 & .. & 0 \\ : & & : & : & & : \\ 0 & .. & c_i & s_i & .. & 0 \\ 0 & .. & -s_i & c_i & .. & 0 \\ : & & : & : & & : \\ 0 & .. & 0 & 0 & .. & 1 \end{bmatrix}, \quad c_i^2 + s_i^2 = 1 \tag{10}$$

$$G_{n,n+1} \cdot \ldots \cdot G_{1,2} \cdot \begin{bmatrix} a^T \\ R \end{bmatrix} = \begin{bmatrix} R^+ \\ 0 \end{bmatrix}, \tag{11}$$

where R^+ is the updated triangular factor of the QR decomposition. In [15], [37], it is shown that these operations can nicely be implemented in parallel on a triangular array of processors, each containing just one element r_{ij} of the R matrix, to start with.

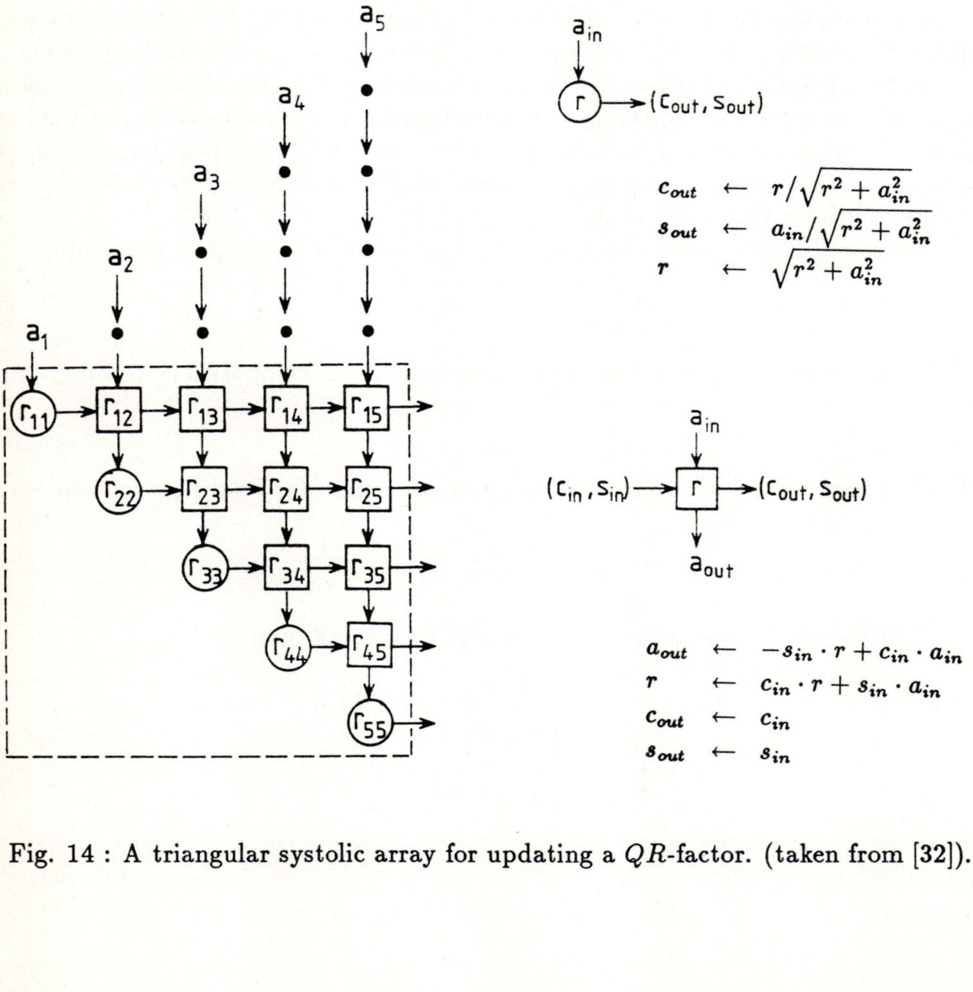

Fig. 14 : A triangular systolic array for updating a QR-factor. (taken from [32]).

Each round cell is a PE that generates an elementary orthogonal (Givens) rotation and each square cell is a PE that propagates these as row transformations. The transformation $G_{1,2}$ between the first row of R and a^T is entirely performed in the first row of the triangular array of processors displayed in figure 14. The diagonal processor constructs $G_{1,2}$ as soon as the element a_1 enters it, and it then propagates the rotation via the parameters $\{c_1, s_1\}$ to the processors on the right. Therefore, the new row a^T has to be fed into the array in a skewed fashion (as shown in figure 14) since the parameters $\{c_1, s_1\}$ have to reach the PE's simultaneously with a_j. All data required for performing the elementary rotation

$$\begin{bmatrix} r_{1i}^{+} \\ a_i^{(1)} \end{bmatrix} = \begin{bmatrix} c_1 & s_1 \\ -s_1 & c_1 \end{bmatrix} \cdot \begin{bmatrix} a_i \\ r_{1i} \end{bmatrix} \tag{12}$$

are then present in each processor at the right time. As a result of this a transformed (and shortened) row a^T is passed on to the second row of processors, where similarly $G_{2,3}$ is constructed and applied. This process is repeated in the whole array and one verifies that the new factor R_+ is gradually constructed as the row a^T "passes" through the array of figure 14 (see [15], [37] for the details).

An important advantage of the above architecture is that it can be applied as well for downdating as for updating a triangular factor. With downdating one wants to reverse the above process and derive the "old" triangular factor R from the new triangular factor R^+ and the row a^T. This in fact corresponds to *removing* or *deleting* an equation from a least squares problem. In [16], [3] it is shown how this involves the further reduction of the "nearly triangular" matrix

$$\begin{bmatrix} a^T \\ R^+ \end{bmatrix} = \begin{bmatrix} a_1 & a_2 & \dots & a_n \\ r_{11}^+ & r_{12}^+ & \dots & r_{1n}^+ \\ & r_{22}^+ & & \\ & & \ddots & \vdots \\ & & & r_{nn}^+ \end{bmatrix} \tag{13}$$

to a triangular form using a sequence of n skew-Givens rotations $S_{i,i+1}$ operating on adjacent rows i and $i+1$:

$$S_{i,i+1} = \begin{bmatrix} 1 & .. & 0 & 0 & .. & 0 \\ \vdots & & \vdots & \vdots & & \vdots \\ 0 & .. & \gamma_i & -\sigma_i & .. & 0 \\ 0 & .. & -\sigma_i & \gamma_i & .. & 0 \\ \vdots & & \vdots & \vdots & & \vdots \\ 0 & .. & 0 & 0 & .. & 1 \end{bmatrix}, \quad \gamma_i = 1/c_i, \quad \sigma_i = s_i/c_i, \quad c_i^2 + s_i^2 = 1 \tag{14}$$

$$S_{n,n+1} \cdot \ldots \cdot S_{1,2} \cdot \begin{bmatrix} a^T \\ R^+ \end{bmatrix} = \begin{bmatrix} R \\ 0 \end{bmatrix}. \tag{15}$$

Here R is the desired triangular matrix of the downdated QR decomposition. There is an apparent analogy between the updating equations (9-11) and the above downdating equations (13-15). Therefore it is clear that these can be implemented on the same architecture of figure 14 by a minor modification of the little programs in the two types of PE's. Moreover, it is shown in [3] that a carefully chosen implementation of the above skew rotations $S_{i,i+1}$ yields comparable stability properties with the rotations $G_{i,i+1}$ used for the updating problem.

A drawback of the parallel updating/downdating schemes, though, is their low efficiency. Indeed, as a^T passes through the triangular array – compare this to a wave traveling through a medium – only the processors on the wavefront are active simultaneously. It is therefore rather inefficient to use $2n$ time steps on $n(n+1)/2$ processors for something which requires only $2n^2$ operations on 1 processor ($E = O(1/n)$). But the nice feature of this architecture becomes apparent when performing *several* updates and/or downdates consecutively. These can indeed be *pipelined* in the triangular array one after the other as shown in figure 15. Each row to be updated/downdated is then fed into the array in a skewed fashion as indicated by the consecutive *data wavefronts* on the right, leading to an $O(1)$ steady state efficiency. Applying an update or a downdate on a single architecture can e.g. be realized by adjoining a flag to each data row a^T indicating which type of rotation to perform on it as it passes through the triangular array of processors. The triangular factor R of the final QR-factorization is then obtained in the triangular array after all rows have passed through the computational network. The speed-up is now roughly $n(n+1)/2$ (i.e. the number of processors) since all processors are always active at each time step.

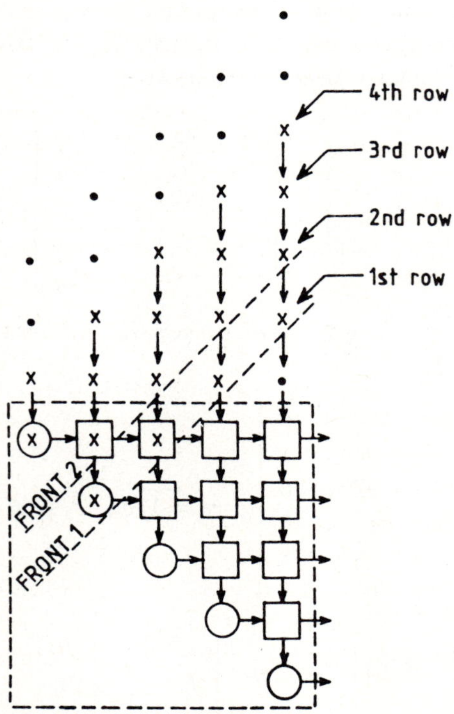

Fig. 15 : Wavefronts for consecutive updatings and downdatings. (taken from [32]).

From the above discussion it is clear that the parallel architecture of figures 14-15 nicely combines several desired properties: consistency, high performance, numerical stability, flexibility, appealing lay-out, etc. The algorithm is certainly one of the successful examples of the advantages of systolic implementations.

5 Concluding remarks

Digital signal processing is one of the application areas where the first systolic implementations were proposed, partly because the inherent resemblance between representations of digital circuits and systolic arrays (see e.g. [31]). Topics covered here are, among others, convolution [28], FFT [33], lattice or ladder filters [26], etc. For more specialized fields such as speech processing one also finds systolic designs at other levels, especially when a large amount of data has to be processed in a regular fashion. Typical examples are codebook processing in vector quantization, dynamic programming and time warping in word recognition ([2], [47], [12], [43], ...). Also in image processing there are typical techniques of a regular nature [48]: all pixels of an image are often treated in a similar "local" fashion, which only involves a few of the neighbouring pixels of the image. These techniques are therefore well suited for parallel implementation on systolic arrays. Examples are 2D convolution [28], statistical analysis of images [10], or real time processing ([9], [13]). From all this it is clear that Parallel Algorithms and DSP have had a serious impact on each other.

But systolic arrays are also being considered in related fields, such as systems and control theory. For some control processes parallel implementation becomes imperative either because of the size of the problem (distributed systems) or because of real-time constraints (adaptive control). When using state-space models one can then use linear algebra techniques that have been implemented on systolic arrays. Examples are parallel simulation of systems [46], Kalman filtering for adaptive identification and control [18], Riccati and Lyapunov equations [6], frequency domain evaluations [4], and many more.

Because of the interdisciplinary nature of PA's in DSP some new problems have also emerged. One is e.g. working on proofs of correctness of parallel algorithms [41], complexity theory of parallel algorithms, languages for systolic arrays, simulators of systolic arrays, automatic parallelization of algorithms, automatic systolization of parallel algorithms, hardware implementation of systolic arrays, and so on. In some of these areas significant advances have been made. In [31] the systematic derivation of systolic arrays from DSP-like signal flow graphs is proposed and some existence theorems in this context are derived. More systematic treatments are to be found in [38], [39], [23]. Geometric methods for deriving various systolic schemes from each other, have been proposed. Theories in this context are slowly emerging and are related to complexity theory and computational algebra [27]. Automatic translators are being considered as well.

Due to the constraints imposed by systolic-like arrays, a number of completely new methods have also emerged in DSP (altough they are not completely novel, they can certainly be called new to the DSP community). One of the most typical examples is certainly the Number Theoretic Transform (NTT) which is rather popular in discrete mathematics. It appears to be a good candidate for implementing in parallel a number of typical DSP problems. The derivation of new algorithms is also coming from areas such as differential equations or linear algebra [5]. The reason of this cross-fertilization is that PA's can be applied to these various fields and that common interests are being created.

Although PA's in DSP and control can hardly be called a new "discipline", it is difficult to deny that it has channeled a number of novel techniques and ideas into these fields.

References

[1] B. D. O. Anderson , J. B. Moore, *Optimal filtering*, Prentice Hall, Information and System Sciences Series, Englewood Cliffs, New Jersey, 1979.

[2] J.-P. Banatre, P. Frison, P. Quinton, "A systolic algorithm for connected word recognition", Int. Rept. No. 169, IRISA, Rennes, 1982 (also in Proceedings ICASSP 82).

[3] R. Brent, A. Bojanczyk, P. Van Dooren, F. de Hoog, "A note on downdating QR and Choleski decompositions", *SIAM Scisc*, Vol. 8, pp. 210-221, 1987.

[4] P. Capello, A. Laub, "Systolic computation of multivariable frequency response", Report Dept. Comp. Sc., Univ. California Santa Barbera, 1987.

[5] J.-P. Charlier, M. Vanbegin, P. Van Dooren, "An introductory survey of parallel algorithms for systolic-like arrays", Rept. R502, Philips Research Laboratory Brussels, Belgium, April 1986.

[6] J.-P. Charlier, P. Van Dooren, "A systolic algorithm for Riccati and Lyapunov equations", Report M194, Philips Research Laboratory Brussels, Belgium, June 1987 (also presented at ICIAM 87, Paris).

[7] L. Csanky, "Fast parallel matrix inversion algorithms", *SIAM J. Comput.*, Vol. 5, pp. 618-623, 1976.

[8] J. B. Dennis, "Data flow supercomputers", *Computer*, Vol. 13, pp. 48-56, 1980.

[9] M. Duff, *Computing structures for image processing*, Academic Press, New York, 1983.

[10] A. L. Fisher, "Systolic algorithms for running order statistics in signal and image processing", Int. Rept. CMU-CS-81-130, Carnegie Mellon University, 1981.

[11] A. L. Fisher, H. T. Kung, L. M. Monnier, and Y. Dohi, "The architecture of a programmable systolic chip", *J. VLSI and computer systems*, Vol. 1, pp. 153-169, 1984.

[12] P. Frison, P. Quinton, "An integrated systolic machine for speech recognition", Int. Rept., IRISA, Rennes, 1984.

[13] G. Gaillat, "Le calculateur CAPITAN : 600 MIPS pour l'imagerie temps réel", *Traitement du Signal*, Vol. 1, pp. 19-30, 1984.

[14] M. Gentleman, "Least squares computations by Givens transformations without square roots", *JIMA*, Vol. 12, pp. 329-336, 1973.

[15] W. Gentleman and H. Kung, "Matrix triangularization by systolic arrays", Proceedings SPIE Symp. 1981, Vol. 298, Real Time Signal Processing IV, pp. 19-26, 1981.

[16] G. H. Golub, "Matrix decompositions and statistical computations", pp. 365-397 in *Statistical Computation* (R. C. Milton and J. A. Nelder, eds.), Academic press, New York, 1969.

[17] G. H. Golub and C. F. Van Loan, *Matrix computations*, North Oxford Academic, Oxford, 1983.

[18] K. Hashimoto, H. Kimura, "A parallel architecture for recursive least squares identification", in Proceedings ICASSP86, Tokyo, 1986.

[19] D. Heller, "A survey of parallel algorithms in numerical linear algebra", *SIAM Review*, Vol. 20, pp. 740-777, 1978.

[20] R. W. Hockney, "MIMD computing in the USA - 1984", *Parallel computing*, Vol. 2, pp. 119-136, 1985.

[21] M. L. Honig and D. G. Messerschmitt, *Adaptive Filters*, Kluwer Academic, Hingham, 1984.

[22] K. Hwang and F. A. Briggs, *Computer architecture and parallel processing*, McGraw-Hill, 1984.

[23] H. V. Jagadish, S. K. Rao and T. Kailath, "Array architectures for iterative algorithms", *IEEE Proceedings*, Vol. 75, pp. 1304-1321, Sept. 1987.

[24] T. L. Jordan, "A guide to parallel computation and some Cray-1 experiences", in *Parallel computations* (G. Rodrigue ed.), Academic Press, New York, pp. 1-50, 1982.

[25] T. Kailath, *Linear Systems*, Prentice Hall, Information and System Sciences Series, Englewood Cliffs, New Jersey, 1980.

[26] T. Kailath, "Signal processing in the VLSI era", in *VLSI and modern signal processing* (S. Y. Kung, H. J. Whitehouse and T. Kailath, eds.), pp. 5-23, Prentice Hall, 1985.

[27] R. M. Karp, R. E. Miller and S. Winograd, "The organization of computations for uniform recurrence equations", *Journal ACM*, Vol. 14, pp. 563-590, 1967.

[28] H. T. Kung, "A two-level pipelined systolic array multi-dimensional convolution", Int. Rept. Dept. Comp. Sc., Carnegie-Mellon, 1982.

[29] H. T. Kung, "Why systolic architectures?", *Computer*, Vol. 15, pp. 37-46, 1982.

[30] H. T. Kung and C. E. Leiserson, "Algorithms for VLSI processor arrays", in *Introduction to VLSI systems* (C. Mead and L. Conway, eds.), Addison-Wesley, pp. 271-292, 1980.

[31] S. Y. Kung, "On supercomputing with systolic/wavefront array processors", *IEEE Proceedings*, Vol. 72, pp. 867-884, 1984.

[32] S. Y. Kung, "VLSI array processors", *IEEE ASSP Magazine*, Vol. 2, pp. 4-22, 1985.

[33] S. Y. Kung, K. S. Arun, R. J. Gal-Ezer, and D. V. Bhaskar Rao, "Wavefront array processor: language, architecture and applications", *IEEE Trans. Comp.*, Vol. CS- 31, pp. 1054-1066, 1982.

[34] S. Y. Kung, H. J. Whitehouse, T. Kailath, *VLSI and modern signal processing*, Prentice Hall, 1985.

[35] J. L. Lambiotte and R. G. Voigt, "The solution of tridiagonal linear systems on the CDC STAR-100 computer", *ACM Trans. Math. Software*, Vol. 1, pp. 308-329, 1975.

[36] J. D. Markel and A. H. Gray Jr., *Linear Prediction of Speech*, Springer Verlag, New York 1976.

[37] J. Mc Whirter, "Recursive least-squares minimization using a systolic array", Proceedings SPIE Symp. 1981, Vol. 298, Real Time Signal Processing IV, p. 105, 1981.

[38] D. I. Moldovan, "On the design of algorithms for VLSI systolic arrays", *IEEE Proceedings*, Vol. 71, pp. 113-120, 1983.

[39] D. I. Moldovan and J. A. B. Fortes, "Partitioning and mapping algorithms into fixed size systolic arrays", *IEEE Trans. Comput.*, Vol. C-35, pp. 1-12, 1986.

[40] D. P. O'Leary and G. W. Stewart, "Data-flow algorithms for parallel matrix computations", *Comm. ACM*, Vol. 28, pp. 840-853, 1985.

[41] M. Ossefort, "Correctness proof of communicating processes, three illustrative examples of the literature", *ACM Trans. Progr. Lang. & Syst.*, pp. 620-640, 1983.

[42] N. S. Ostlund, P. G. Hibbard, and R. A. Whiteside, "A case study in the application of a tightly coupled multiprocessor to scientific computations", in *Parallel computations* (G. Rodrigue ed.), Academic Press, New York, pp. 315-364, 1982.

[43] Y. Robert, M. Tchuente, "Réseaux systoliques pour des problèmes de mots", *RAIRO Th. Inf.*, Vol. 19, pp. 107-123, 1985.

[44] W. Rönsch, "Stability aspects in using parallel algorithms", *Parallel Computing*, Vol. 1, pp. 75-98, 1984.

[45] H. S. Stone, "Problems of parallel computation", in *Complexity of sequential and parallel numerical algorithms* (J. F. Traub ed.), Academic Press, New York, pp. 1-16, 1973.

[46] H. Tai, R. Saeks, "Parallel system simulation", *IEEE Trans. SMC*, Vol. SMC-14, pp. 177-183, 1984.

[47] N. Weste, D. Burr, B. Ackland, "Dynamic time warp pattern matching using an integrated multiprocessor array", *IEEE Trans. Comp.*, Vol. C-32, pp. 731-744, 1983.

[48] Special Issue "Computer architectures for image processing", *IEEE Computer*, Jan. 1983.

IV. CONTRIBUTED PAPERS

Microcomputer Based Expert Control and Adaptive PID Control[1]

Lawrence G. Lebow

Systems Research Center and Electrical Engineering Department
University of Maryland
College Park, Maryland 20742
USA

Abstract

The general concepts of expert control are briefly discussed. This is followed by a description of a specific implementation of an expert controller using available industrial controller and microcomputer technology. At this relatively early stage of development the expert controller is a reduced version of its intended form. It is essentially a rule based adaptive PID (proportional, integral and derivative) controller. As such, a large portion of this paper examines this methodology and its success. Results are outlined and the success of the implementation is examined.

[1]This research was supported in part by a grant of equipment and funds from the Industrial Systems Division of Texas Instruments, in part by the Engineering Research Center of the College of Engineering, University of Maryland, and in part by the Systems Research Center under NSF Grant CDR-85-00108.

NATO ASI Series, Vol. F47
Advanced Computing Concepts and Techniques
in Control Engineering
Edited by M. J. Denham and A. J. Laub

1 Introduction

Expert control as defined by Astrom, Anton and Arzen [7,6,1,2,3] involves the construction of a composite control structure for a complex process which includes supervisory functions, adaptive control algorithms and low level control laws all managed by an expert system which monitors process parameters and control system performance. In [1,2,3] a prototype expert controller was built using high level tools on a super-mini computer.[2] This work and the related work of Moore and colleagues [11] and others [10] has demonstrated the potential value of expert systems in management of the full range on-line control functions from alarms to single loop PID feedback elements. This paper reports on efforts to produce a practical implementation of an expert (industrial) controller on microcomputer based systems. As an initial step in the development of this implementation, an adaptive PID controller has been constructed.

Expert systems are computer programs designed to aid humans in complex tasks. By representing the knowledge about a given domain in the proper manner, and providing an "inference" control structure for access to the knowledge base, an expert system can "reason" to solve a problem or perform a difficult task. Such problems may have limited, conflicting or unreliable data and more than one solution. The expert system is usually designed to emulate human behavior by employing some heuristic "rules of thumb" to reduce many solution possibilities to a few "good" ones. Expert systems, in general, may handle large amounts of data and solve problems of high complexity. Most of the systems in current use are static and time invariant. This is not the case with expert controllers [7,2,3], which require the ability to interact with a dynamically changing environment.

2 Expert Control

An expert controller, as a decision making element in a feedback control loop requires much the same decision making ability needed in other expert systems, but there are significant differences. One crucial requirement is the need to produce expert behavior in "real time". Not only must the expert controller respond reasonably quickly, but its operations must interact with the process in a time varying environment. Also, the expert controller must be interfaced directly to a process and be equipped with the means for applying control to the process.

Many current industrial controllers include some heuristics for *safety net* procedures [7]. These heuristics may only handle extreme "alarm" type situations, and a prescribed solution may be a plant shutdown until human intervention can solve the problem. An expert controller should have the ability to *adapt* to changing situations and prevent most "alarms" from ever occurring.

An expert controller should have the capability of using several different control algorithms as well as the ability to *tune* the parameters of each algorithm to the process under control. Possible control algorithms might include Proportional, Integral and Derivative (PID), pole-placement, linear observers or algorithms designed for optimal control. The expert controller must provide control signals to the process (in "real time") in addition to "reasoning" about

[2] Specifically, the forward chaining production system YAPS and the object-oriented Flavors system running on a VAX 11/780.

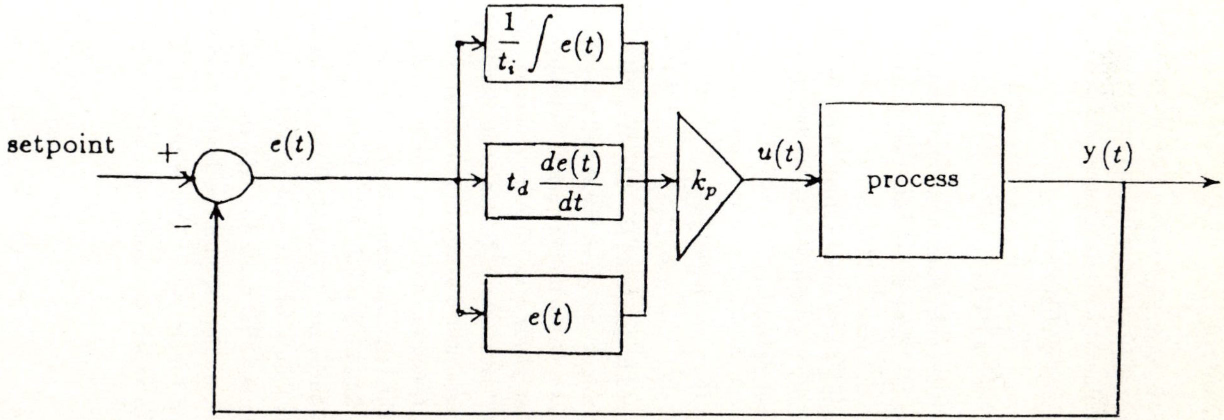

Figure 1: PID control loop.

3.1 Adaptive PID Control

The objective of PID control is to constrain a process response to follow input setpoints. In general, the output is compared to the setpoint and the error signal $e(t)$ is computed and fedback to a compensator that combines it with its integral and derivative to produce the control signal $u(t)$ as shown in equation (1) and Figure 1.

$$u(t) = k_p \left[e(t) + \frac{1}{t_i} \int e(t)\, dt + t_d \frac{de(t)}{dt} \right] \tag{1}$$

The key parameters in PID control are the proportional constant k_p, the integral constant t_i and the derivative constant t_d. These three "constants" can be manipulated to produce various response curves from a given process. The whole concept for adaptive PID control revolves around interactively changing these three values. The discrete time equivalent expression for PID control used in this study is shown in equation (2). In equation (2), t is the sample period for the microprocessor based controller.

$$u[k] = k_p \left[e[k] + \frac{t}{t_i} \sum_{i=1}^{n} e[i] + \frac{t_d}{t} \left(e[k] - e[k-1] \right) \right] \tag{2}$$

The overall approach used in designing the Adaptive PID Controller (APIDC) is based upon the methodology a human might use, given the same task. The situation or task he/she is faced with, can be stated as follows: Given no previous knowledge about the process to be controlled, find a PID control law that yields a satisfactory response. The first thing this person would ask is: How do you define "satisfactory"? This understanding of the type of process response deemed desirable must be established in order to proceed. The human operator would then need some initial values for k_p, t_i and t_d. Assuming this is accomplished by some means, the actual task of *manually* tuning the response would begin. This procedure would typically involve a repeated sequence of the following steps. (1) Choose k_p, t_i and t_d. (2) Moniter the process response to a "typical" disturbance. (3) Analyze the results and based on previous experience and knowledge, choose new values for k_p, t_i and t_d. This is in fact what must be down by a plant operator who manually tunes a PID controller. He/she must "turn the knobs", each associated with one of the three PID parameters, until the desired response is found. This entire sequence must be repeated for each tuning step. This is precisely the methodology employed by the APIDC. The details of the expert controller and specifically the APIDC are discussed in the following sections.

3.2 Configuration

The current configuration for the microcomputer based expert controller is depicted in Figure 2. Two major components are defined, the expert controller itself and the process. The expert controller block (dashed line in figure) is broken further into two parts. This division of tasks in the expert controller is one of the key aspects in this design. As first suggested by Arzen [1,2,3] a separation between control law implementation and expert system activities is desirable. The idea here is that the application of the numerical control algorithm, such as PID control, should be a continuous operation. The expert controller must go through an analysis and decision

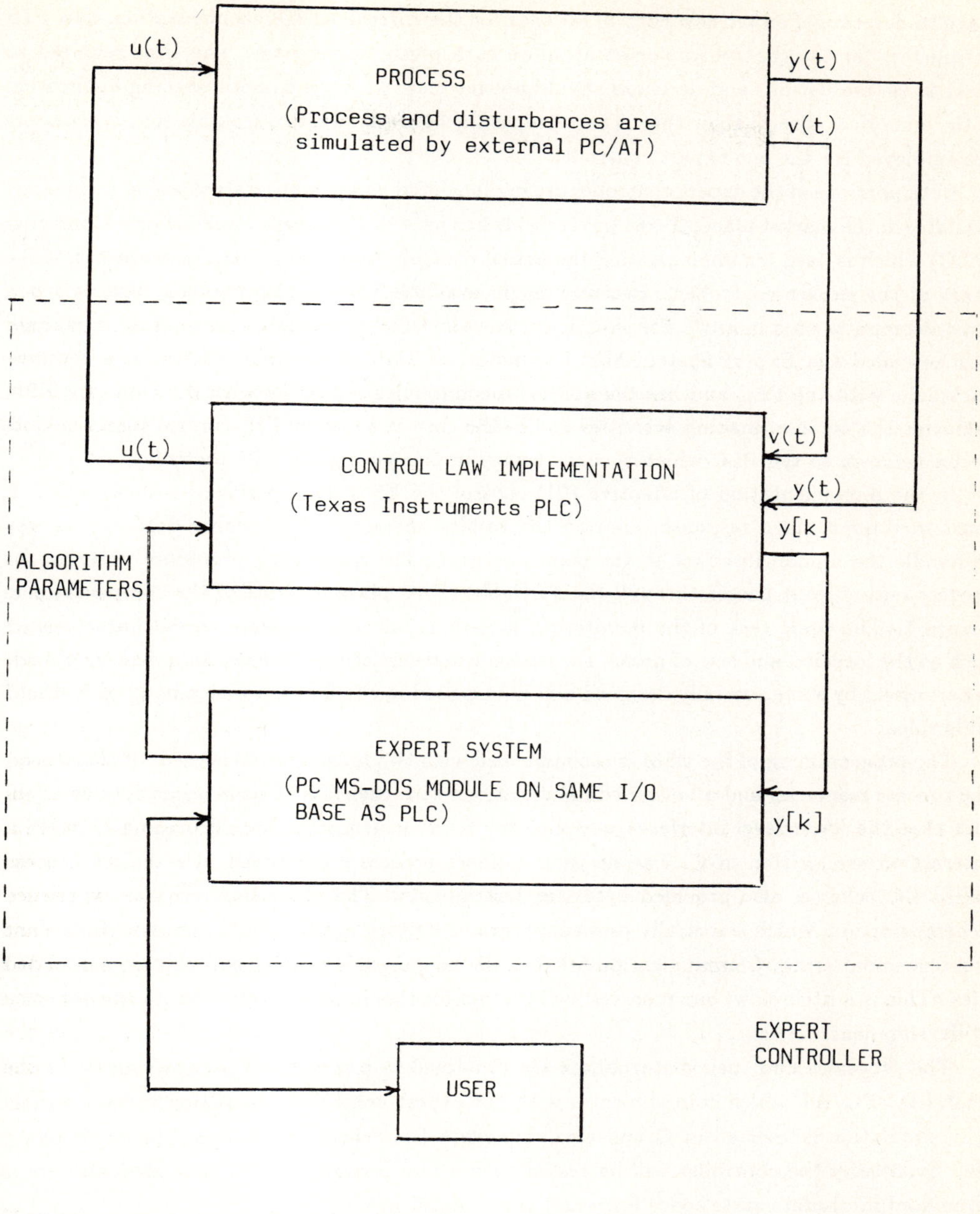

Figure 2: Current configuration of the microcomputer expert controller.

stage to determine, either new control laws, or for the current APIDC configuration, new PID parameters for the control implementation on each given tuning pass. The time required to make these evaluations and decisions should not interfere with the smooth ongoing application of the numerical control algorithm. Therefore, as the figure shows, two separate microprocessors are employed for the one expert controller.

Both portions of the expert controller are implemented using industrial controller technology available in the market place. Texas Instruments has provided a Programmable Logic Controller (PLC) which is used for implementing the actual control. Also, from Texas Instruments, is the heart of the expert controller. Recently made available in the market place, this unit is a "microcomputer on a board". For simplicity, this MS-DOS type personal computer on a board will be called the Expert System Microcomputer, or ESM. The ESM resides on a common backplane with the PLC and has the ability to communicate directly with it. Thus, the ESM performs the decision making activities and in the case of adaptive PID control then sends its decisions down to the PLC which is continuously implementing the PID control.

In the implementation of adaptive PID control the ESM itself performs several tasks. It must moniter a given response, analyze the results and make PID tuning decisions as well as handle the human interface to the plant operator. The monitering procedure observes the process output as it passes through the PLC. The PLC plays the role of the compensator in Figure 1. The basic task of the monitering stage is to identify response curve characteristics such as the location and size of peaks, the risetime and settletime. The decision making process is performed by some specially designed algorithms used in conjunction with a body of "if-then" style rules.

The programming of the ESM is accomplished with two languages. C is used for monitoring the process responses and all other communications with the PLC. This is primarily due to the fact that the lower level interfaces, supplied by Texas Instruments Inc., providing ESM/PLC interaction are written in C. The decision support processes are coded in a dialect of Lisp, called PC Scheme, also provided by Texas Instruments. The outer shell program is the PC Scheme portion, which essentially uses subprograms written in C for PLC communications and response monitering. Communication between the languages is handled through sequential disk files. This is a little slow, but it serves well enough for this implementation at its current stage of development.

The processes and their disturbances are simulated in a completely separate and independent IBM PC/AT which communicates with the expert controller via analog voltage signals. The simulation is written in C and also employs a PC Scheme shell to aid in result analysis. Eventually the controller will be tested with a real process, but for now processes are in simulated in discrete state space form and are modeled by

$$\begin{aligned} \vec{x}[k+1] &= \bar{A}\,\vec{x}[k] \;+\; \bar{B}\,(\,u[k] \;+\; v[k]\,) \\ \vec{y}[k] &= \bar{C}\,\vec{x}[k] \end{aligned} \tag{3}$$

In equation (3) $\vec{x}$ is a vector of state variables and $\vec{y}$ is the process output. $\bar{A}$, $\bar{B}$ and $\bar{C}$ are the system matrix, the input matrix and the output matrix of the process. Also, $u[k]$ is the current control signal and $v[k]$ is the current disturbance.

3.3 Pre-tuning the Adaptive PID Controller

Earlier, it was pointed out that the human employing manual PID tuning, as well as the APIDC, must have a starting point. That is, initial values for the PID parameters k_p, t_i and t_d must be defined by some means. One option is simply to make an educated guess. In many cases the plant operator will have some idea about good choices of parameters, although the final parameters chosen for a specific response might be unknown at start-up time. The second option, discussed below, requires that a deliberate disturbance be applied to the process while no control law is implemented. This natural response to a step disturbance (or setpoint change) is monitered and initial PID parameters can be computed based on analysis of this response. However, this may be a prohibited operation in a number of given plants. Therefore, in addition to this pre-tuning algorithm the means are provided for the plant operator to override its use and enter his/her own initial values for k_p, t_i and t_d.

The method for pre-tuning employed for the adaptive PID controller is called Zeigler-Nichols auto-tuning. The basic idea is to enter a known disturbance to the process; and, based on the system response, the three PID parameters can be estimated. The Zeigler-Nichols method requires only one test with a unit step disturbance applied and the open loop response monitored. As the response signal increases toward the setpoint, the maximum slope, s, is determined. The point on the time axis where the tangent to this slope intersects is called the deadtime, d. These two values yield PID parameters according to equation (4).

$$k_p = \frac{1.2}{s\,d} \qquad t_i = 2\,d \qquad t_d = \frac{d}{2} \tag{4}$$

The Zeigler-Nichols method is considered in this context to be an identification algorithm. PID control is preempted by zeroing the proportional constant k_p, and the required step disturbance to the natural (uncontrolled) process is generated by forcing the control input to the process itself to be a one. This natural response is then allowed to continue until it settles at the new setpoint of one. One advantage of working in a simulated environment lies in the allowance of this identification option that might be unrealistic in a given "real world" plant. Nonetheless, the ability to let this response complete its trajectory provides the APIDC with additional information about the process, yielding a simple identification scheme. At present the APIDC only uses the natural risetime and settletime to some advantage but more could potentially be achieved with this additional information.

3.4 Rule-based Tuning for Adaptive PID Control

Again, the objective of this tuning is to manipulate the PID control parameters such that the optimal response curve is achieved. It is assumed that some other means such as "pre-tuning" has been used to find the initial PID parameter settings. In this case, optimal is determined by the human operator and/or some generally accepted performance criteria. This specified criteria is based on measurable characteristics of the process response curve, such as the risetime and settletime.

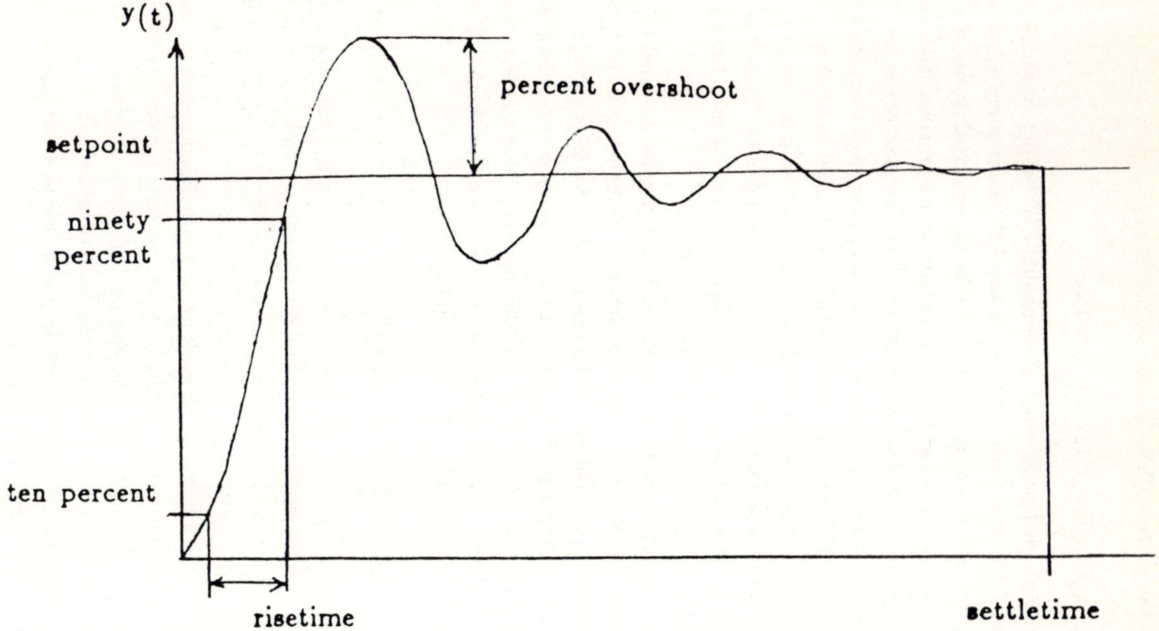

Figure 3: Example response trajectory.

response in this general form is what simple Zeigler-Nichols auto-tuning attempts to achieve.

The determination of whether a given response is "satisfactory" or not, requires a bit more sophistication. The approach used here is based on the use of compatibility or membership functions defined in fuzzy logic and set theory [12]. The curve in Figure 4(a) may be interpreted to signify the proximity of a given value for a response's settletime to some ideal settletime. Values below some point receive a rating of 1.0 on this continuous compatibility scale ranging from 0.0 to 1.0. Thus, settletimes less than a given value all get optimal ratings. Curve (b) is simply a generalization of curve (a), since each side is based upon the same curve shown in (a). This curve is useful to describe membership to properties that have an optimal value, and membership values must be available for both above and below this value. The other defining parameter of these curves can be thought of as an allowable bandwidth.

It is important to stop at this point and examine exactly what is being done here. Nothing *fuzzy* is actually being applied in any form. The curves defined here are used to evaluate process responses yielding something called membership ratings. These values are used to help tune the PID parameters in a very analytical specific way. A gaussian distribution, or for that matter many other probability distributions, could have served the same purpose. These membership functions from Zadeh were chosen for one simple reason. In generating there curves, Zadeh was attempting to model *human* definitions for words. Thus, these particular curves model the situation of how "good," in the human sense, a given response may be. Humans instinctively define a breaking point or bandwidth, at which point they decide a given value is too far from optimal. Above this point results are increasingly acceptable and below they fall off rapidly. Thus we are able to model human behavior in the tuning of PID loops fairly well with these particular curves. Nonetheless, we are only borrowing the curves themselves (and the terminology of membership ratings) *not* any of the actual application of fuzzy set theory.

At this time, six different criteria are used in the procedures that accomplish this curve evaluation using membership functions. Settletime needs no definition. Overshoot is clearly the percentage rise above the setpoint. (See Figure 3.) Risetime (as shown also in Figure 3) is the time required for the process response to go from ten percent to ninety percent of the setpoint. A value called peak height ratio is defined as the ratio of the second peaks overshoot to that of the first peak. Average peak spread is the mean value between positive and negative peaks. Finally, the peakcount itself is used as one of the defining criteria for a given process response curve.

Several properties are defined for each criteria. First, optimal values are chosen for each associated criteria. Second, bandwidths relative to the optimal value are designated. Third a weighting factor is associated with each criteria. And last, the curve type, or the defining function (one of the two in Figure 4) must be specified. All of these values can be chosen by the user or by internal defaults set up by the APIDC. (For examples of this information, see the portion of Appendix B labelled "Tuning Criteria at Start-up"). When a new process response has been monitered the actual current value for each of these criteria is recorded. These current values are then entered as input to the membership functions which return a value between zero and one. The curve type will apply to whether values falling off to one or both sides of optimal should be considered as having memberships less than one. The optimal

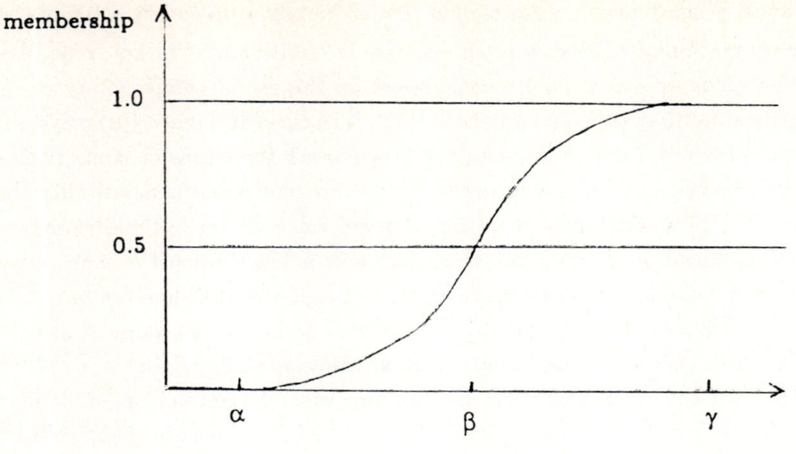

(a)

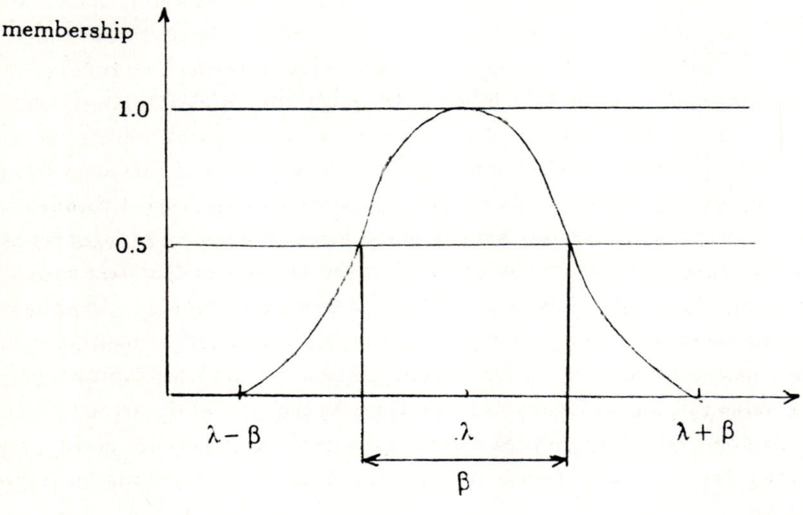

(b)

Figure 4: Compatibility and membership functions.

value represents the curve peak and is chosen by the user or the defaults as the best *reasonable* value for the criteria. The bandwidth is the breaking point in the curves from Figure 4. This is chosen to represent where the current response value is becoming no longer acceptable. One of the most powerful features of using these membership curves is the ease with which weighting factors can be applied. By applying exponential factors, greater or less than one, to a membership function the curve can be narrowed or widened respectively. Values greater than one make the membership rating more stringent. That is, membership ratings fall off from the same optimal value more rapidly. The inverse is true a weighting factor less than one which relaxes the evaluation of memberships. It is interesting to note that a weighting factor of zero results in a membership rating that is always a one.

Membership ratings, then, reflect how "good" a given response is, relative to the specification of "ideal" by the user. The ultimate measure of success in each tuning pass is indicated by a score which is quite simply the sum of the memberships for a given process response. With six criteria, the maximum achievable score is clearly six as maximum membership is unity. However, all this effort to evaluate a given curve goes far beyond computing a simple score. The information contained in the membership ratings tells the APIDC which criteria are far off the mark and which need not be tampered with if possible. A special algorithm for applying this information has been developed and successfully implemented. This special algorithm is the heart of the adaptive tuning process and is will be referred to as the Membership Based Tuning Algorithm (MBTA). The following paragraphs introduce and discuss this methodology.

3.4.1 Membership Based Tuning Algorithm

General observations may be made about the behavior of a given process to adjustments in the PID parameters k_p, t_i and t_d. Given enough experience in tuning PID controlled processes, a plant control engineer will often know which way to turn each of the "knobs" to accordingly influence the process output. For instance it may be possible to say, that if one wishes to decrease the percentage of overshoot, then k_p should be increased, t_i should be decreased and t_d should be increased. As it turns out, there are some general behaviors that can be noted and used in building a set of such rules for adaptive tuning.

Through simple, although extensive, trial and error analysis such overall trends have been observed. This trial and error procedure consisted of running tests with a significant cross-section of processes. The testing procedure was as follows. Starting values for the parameters k_p, t_i and t_d were chosen arbitrarily and the process response to a setpoint change (or unit step disturbance) was monitered and noted. Then each of the twenty-seven possible choices for turning the knobs was tried. The amount to turn knobs was varied between ten percent and twenty percent in different runs. By observing trends in the reactions of the six criteria of settletime, overshoot, risetime etc. a rule could be derived. Such a rule takes the general form given in the previous paragraph. As the testing became more complete, the two typical process response types, currently named A and A2, became somewhat clearly defined, and which direction to turn knobs for specific results became more evident. In addition, an indication as to the amount to turn each knob relative to the others could also be ascertained. As an example, the information taken from all the testing for a Type A process response is embodied in the

Crit.	k_p		t_i		t_d	
	inc.	dec.	inc.	dec.	inc.	dec.
Settle time	-.7	.7	-.3	.6	-.7	.7
% Overshoot	-.7	.7	.2	-.5	-.7	.7
Risetime	-.7	.7	.0	.0	.0	.0
Peak ratio	.0	.0	-.5	.5	-.7	.7
Peak spread	.3	-.3	.5	-.5	.0	.0
Peak count	.3	-.3	.0	.0	-.7	.7

Table 1: Type A Tuning Rules

structure labelled Type A in Table 1.

This strange looking table is used in the MBTA algorithm along with current membership ratings to generate a tuning decision consisting of the percentage and direction each PID parameter should be adjusted. The application of the information in Table 1 can be conceptually viewed as a body of twelve "if-then" type rules. However, the internal algorithm itself appears more numerical in nature with the "if-then" portions *buried* in the code. Such a rule is stated below, written in pseudo-code.

```
IF (settletime needs to be decreased) THEN
   {
     %kp_change = .7 * (1 - membership(current_settletime))
     %ti_change = .6 * (1 - membership(current_settletime))
     %td_change = .7 * (1 - membership(current_settletime))
   }
```

This rule uses two sources of information to compute the percentage changes for the PID parameters due to a need to decrease the settletime for a type A curve. (How a curve is determined to be type A or A2 is discussed later.) The .7, .6 and .7 are taken from Table 1, and are referenced by choosing the row for settletime and the subcolumn labelled "dec" (for decrease) under each of the respective columns for the three PID parameters. These values are the experience factors learned from the trial and error testing discussed above. The other crucial piece of information is the membership rating for the current value of the settletime. The rule for increasing settletime (although usually undesirable as a goal) has the same format except that .7, .6 and .7 are replaced respectively with −.7, −.3 and −.7. Two companion rules like these can be defined for each of the six criteria, resulting twelve in all. Of course twelve more such rules are implemented for A2 type processes.

For a typical tuning pass, six of the twelve possible rules will "fire" resulting in six indicated percentage adjustments for each of the three PID parameters. These values are added together to compute a total adjustment for each respective PID parameter. It is important to note that the experience factors in the table differ in sign. Thus some force a percentage increase in the parameter and some force a decrease. This corresponds to the coupled nature of different criteria. That is, action taken with the intention of decreasing settletime is likely increase the percentage overshoot. Hence, the six individual adjustments tend to counteract each other with the combined result representing a compromise between them. The real beauty and significance of the MBTA algorithm is that the membership rating of each criteria for a given pass directly influences the final indicated adjustments in the PID parameters. By including the arithmetic compliments for membership ratings in the computation, the resulting decision or adjustment is reacting directly to the current process response. Those criteria with the lowest ratings dominate the percentage changes, yet they are tempered by the adjustments indicated from all other criteria. The result is a suitable compromise for how far to *turn* each of the PID parameter knobs. The final summed up adjustment for each parameter is applied as a percentage change over the old value and the next pass will be implementing PID control with these new parameters. It is also interesting to note that this algorithm tends to converge. As memberships get higher, the resulting adjustments clearly get smaller and more refined.

As clever as the MBTA algorithm may or may not be, it is still not enough by itself. Not all processes fit so neatly into type A or type A2. In general, is it likely that a major adjustment in the parameters might be needed to get started on the right track. The adjustments in the MBTA algorithm, while they reflect the membership ratings, are often rather small. For that matter, having the adjustments too coarse would effect the ability for the algorithm to converge. Also situations such as instability may arise and must be dealt with directly. For these purposes a group of more general rules are used by the APIDC. It is also necessary to have a means to choose which type of MBTA tuning is to be be attempted. As it turns out the application of MBTA tuning is implemented as the result of one of these general rules firing. One rule exists for each of the two types.

3.4.2 General Rules for Adaptive PID Control

Rules are read into the APIDC memory from a floppy disk file that can be easily updated if necessary. Keeping the rules in a disk file facilitates the development of new rules in the design process and provides a future potential for users to add rules specific to their plant or process. Upon APIDC initialization, the rules are read from disk and arranged in a linked list style format. Examples of rules as they reside in the disk file are listed in Appendix A. There are actually four separate records associated with each rule. The first line or record is a symbol or acronym for the rule. This character string is used as an internal identifier to indicate when the rule has fired. The next "record," consisting in general of several lines, is the rule itself. As discussed earlier, the decision making portions of the APIDC are written in TI's PC Scheme, and hence so are the rules. The basic format is a conditional statement with one antecedent and one composite consequent that usually contains multiple executable statements. This is analogous to the basic "if-then" structure found in many other high-level languages. The last two records associated with each rule are pointers to other rules in the linked list. The first pointer is followed if the rule fires, and the second is taken if the it does not. Clearly, a rule fires when its anticedent, or conditional test, is found to be true. The following paragraphs describe several rules in detail and discuss an important feature of the APIDC called the history file.

The first rule in the list to be tried on every pass is the alarm rule. Instability is identified during the process monitering stage by detecting that peaks are growing in magnitude (top to bottom) rather than shrinking as time progresses. Instability is considered to be confirmed when this condition holds for three peaks in a row. At this point the monitering procedure will set k_p to zero in order to take the controller and its destructive effects on the process out of the feedback loop. The moniter procedure (written in C) then terminates and returns to the Scheme environment with the stability flag set to false (or in this case zero). The APIDC then enters a waiting phase to give the process a chance to settle naturally before any other action is taken. (For the present implementation is is assumed all processes to be controlled have stable natural responses.) Once the process has settled, the rules are tested starting with rule number one. This is of course the alarm rule and it fires accordingly. The pointer EOF (for End Of File) indicates that we skip to the end and test no further rules. Obviously, there is no need in trying any other rules if an alarm condition exists. The direct consequent of the alarm rule is to halve k_p , double t_d and leave t_i alone. This has been found by experience to be a good reaction to unstable responses, and has the general result of at least bringing the response back under control.

As examples, a few more rules will be discussed in some detail. However, a small digression is required first, in order to describe one of the most important features of the APIDC as well as the expert controller as a whole. This feature is a history file which maintains a body of information for each pass of the APIDC. At present the history file is used exclusively for information pertaining to adaptive PID control, but the concept and its usefulness remains the same for the expert controller when it extends beyond the adaptive PID controller configuration. The information saved on each pass contains the following: the acronym for each rule that fired, the k_p, t_i and t_d used as result of the rules that fired, the results from the process response for each of the six criteria, the membership ratings and direction off for each, and the score for this pass. (For examples of history file entries see the appropriate portion of Appendix B). This information plays a significant role in the conditions tested for by about half of the general rules in the system. (Sixteen rules are currently implemented). In addition to this extensive internal use of the history file, this feature provides a means for the user to review the events that have taken place up to a given point in time. The history file is stored internally in memory in a special matrix data structure. It is also periodically written to disk for safe keeping should the system crash.

Several rules pertain to the situation where it is time to stop actively trying to tune the process response. One of three conditions may apply. These are: (1) when the score has been steady for five passes, (2) when the score has been decreasing for five passes, and (3) when the current score is higher than 5.5 (a very high score) and has remained in this approximate region for at least two passes. The basic concept behind each of these rules is to have the controller recognize when it is no longer making useful progress (perhaps the MBTA algorithm has converged), or when it is actually headed in the wrong direction after already finding a satisfactory PID control. The action taken as a result of each of these rules firing is exactly the same. A data structure called "bestpass" which holds an integer (or index) pointer to the pass with the highest score on record is maintained by the APIDC. The APIDC accesses the values in the history file for k_p, t_i and t_d associated with this best score and assigns them as the values to be used for subsequent passes. A best score is also specifically saved and a dedicated monitering stage is entered by the assignment to one for the flag called "moniter score". Rule number two for monitering the score is then fired on every subsequent pass of the process, until some significant change occurs in the process output. Should the score for a given pass drop more than 0.5 below that with the best score, it is assumed the process has changed. As a result, dedicated monitering (via rule number two) is disabled and the entire tuning process, beginning with Zeigler-Nichols auto-tuning, is started again. Of course, if auto-tuning has been disallowed by the user, during controller initialization, then the user defined starting values are used for the three PID parameters.

4 Results and Conclusion

Results of testing the APIDC, and hence to some extent the expert controller, have been promising. A variety of processes have been satisfactorily controlled by the adaptive PID methodology. Tests to date have primarily been with second, third and fourth order linear processes. All processes tried can be modelled by simple transfer functions. (i.e. a constant over a polynomial in the frequency domain). Processes with system zeroes and/or non-linearities have not been tested to date. The greatest success has been achieved with processes containing only real poles. The controller can produce reasonable results with all such processes with results only deteriorating significantly with some fourth order processes. ("Reasonable" is as determined by the user, but scores of 3.5 or higher are generally accepted as reasonable). When processes involving complex poles are tested success is often achieved but with less consistency. Second order complex processes are usually no difficulty but as larger imaginary parts enter the process poles of third and fourth order processes, results fall off more significantly. Higher order processes have been briefly tested but no consistency has been found in the results. The following paragraphs describe some actual runs with the APIDC. It should be noted that a working APIDC indicates good expectations for the full-fledged expert controller using the current overall configuration and methodology.

Due to space limitations only one process tuning sequence will be reviewed. Appendix B contains the five curves discussed and an extract from the history file. The process for which the tuning passes are reviewed is a third order process with poles at -0.5, $-0.82 + 2.8i$ and $-0.82 - 2.8i$. The first portion of Appendix B refers to the values used to initialize the APIDC. The settletime and risetime are based on 66 percent and 33 percent of the natural response values respectively. As shown, optimal percent overshoot was chosen to be 40 percent or less, peak height ratio 0.25 etc. For complete information, please refer to Appendix B.

The first curve is associated with the natural response of the process. In this case, auto-tuning was allowed and the natural response is shown. The second iteration, or tuning pass, is a result of using the values 2.1429 for k_p, 1.4933 for t_i and .3733 for t_d. These were the values yielded by employing Zeigler-Nichols auto-tuning in iteration number one. In this case, we clearly got off to a very good start as this curve received a score of 4.3. Really the only criteria that had low membership ratings were peak height ratio and settletime.

Regular tuning actually begins on pass number three as this is the first time we actually access the body of sixteen general rules discussed above. Here rule five (which selects between Types A and A2) fires, and the resulting MBTA tuning for type A is applied. Unfortunately, the results have gotten worse, not better. However, on iteration four the APIDC has quickly ascertained the error in its ways and the "Getting UnStable" rule has fired. As it turns out, this was a very good decision resulting in an excellent score of 5.348. Here the only membership rating off much at all is average peak spread. This indicates the process is a little more overdamped than ideal. After a battery of small adjustments, with scores oscillating between 4.9 and 5.3 (not shown) the APIDC finally finds PID parameters that yield a very high score of 5.71 on iteration number fifteen. The PID parameters on this pass were 1.521 for k_p, 2.3745 for t_i and 0.4069 for t_d.

It is important to note that although it took a significant number of passes to settle upon the final PID parameters, the process was constantly under "good" PID control. That is, control was maintained while the controller tried small adjustments in an effort to find the optimal response curve. All coarse adjustments occurred in the first four or five passes.

In closing, there is one final point to be made about the topic of expert control. The ultimate key to success lies in identification. The more about a process that is ***known***, the more likely it becomes that optimal control can be achieved. Many techniques exist for process identification but most involve repeated "off line" testing of the process. Such testing takes time and often such repeated intentional disturbances to the plant may not be practical or even feasible. (Recall, disturbances occur at random with the APIDC "on line"; only the pre-tuning phase involves an intentional disturbance to the process system)."Off line" testing becomes completely unreasonable when the process is prone to change. The answer is to build an expert controller that can learn about the process "on line" while maintaining some level of control. From what it learns, an internal model or configuration of the current process can be built. Based on this model and its completeness and accuracy the best possible control can employed. The more the expert controller ***knows*** the better the control will be. Therefore, an expert controller requires the ability to ***learn*** about a process; and so, process identification is the crucial factor.

Appendix A

```
; Alarm rule - invoked when the response is unstable
**********
"1-ALM"
(COND ( (AND (EQUAL? stable 0)  (EQUAL? moniter_score 0))
        (SET! type "A2")
        (SET! kp (/ kp_old 2))
        (SET! ti ti_old)
        (SET! td (* 2 td_old)) ) )
"EOF"
2
**********

; Steady score rule - fires when score has been approximately
;                       the same for five passes
**********
"3-STDY_SCORE"
(COND ( (AND (>=? hist_count 5)
             (> (MATRIX-REF history bestpass 7) 3.5)
             (APPROX-EQUAL-5? .98
               score
               (MATRIX-REF history (- hist_count 1) 7)
               (MATRIX-REF history (- hist_count 2) 7)
               (MATRIX-REF history (- hist_count 3) 7)
               (MATRIX-REF history (- hist_count 4) 7) ) )
  (SET! kp (MATRIX-REF (MATRIX-REF history bestpass 3) 1 1))
  (SET! ti (MATRIX-REF (MATRIX-REF history bestpass 3) 2 1))
  (SET! td (MATRIX-REF (MATRIX-REF history bestpass 3) 3 1))
  (SET! bestscore (MATRIX-REF history bestpass 7))
  (SET! moniter_score 1) ) )
2
4
********

; Getting unstable rule - fires when response is headed for
;                               instability
*********
"6-GUS"
(COND ( (AND  (EQUAL?  type "A")
    (> hist_count 2)
    (EQUAL? (MATRIX-REF mem 1 2) 1)
    (< (MATRIX-REF mem 1 1) 0.15)
    (>= peak_ht_ratio
     (MATRIX-REF (MATRIX-REF history (- hist_count 1) 4) 4 1))
    (>= (MATRIX-REF (MATRIX-REF history (- hist_count 1) 4) 4 1)
     (MATRIX-REF (MATRIX-REF history (- hist_count 2) 4) 4 1))
    (>= settletime
     (MATRIX-REF (MATRIX-REF history (- hist_count 1) 4) 1 1))
    (>= (MATRIX-REF (MATRIX-REF history (- hist_count 1) 4) 1 1)
     (MATRIX-REF (MATRIX-REF history (- hist_count 2) 4) 1 1)) )
  (SET! kp (/ kp_old 2))
  (SET! ti ti_old)
  (SET! td td_old) ) )
"EOF"
7
**********
```

Appendix B

TUNING CRITERIA AT START-UP

Criteria	Optimal	Bandwidth	Weight	Curve-Type
Settletime	5.66	2.83	1.0	S
Overshoot	0.4	0.3	1.5	S
Risetime	1.213	1.213	0.5	S
Peak_ht_ratio	0.25	0.5	0.2	PI
Ave peak spread	1.075	0.2	0.5	PI
Peakcount	2	3	0.2	PI

EXTRACTS FROM THE HISTORY FILE

Iteration no. ==> 3

Rules that fired ==> 5-TYP 12-ATYPE

RESULTING PARAMETERS
kp = 2.3967 ti = 1.7235 td = 0.4436

RESULTS

Criteria	Value	Membership	Direction Off
Settletime	10.98	0.0	high
Overshoot	0.3525	1.0	low or equal
Risetime	0.7	1.0	low or equal
Peak_ht_ratio	0.7375	0.2619	high
Ave_peak_spread	1.0169	0.9116	low or equal
Peakcount	4	0.7402	high

SCORE = 3.9138

Iteration no. ==> 4

Rules that fired ==> 5-TYP, 6-GUS

RESULTING PARAMETERS
kp = 1.1983 ti = 1.7234 td= 0.4436

RESULTS

Criteria	Value	Membership	Direction Off
Settletime	5.86	0.9827	high
Overshoot	0.0525	1.0	low or equal
Risetime	0.98	1.0	low or equal
Peak_ht_ratio	0.0	0.8705	low or equal
Ave_peak_spread	0.945	0.495	low or equal
Peakcount	2	1.0	low or equal

SCORE = 5.3483

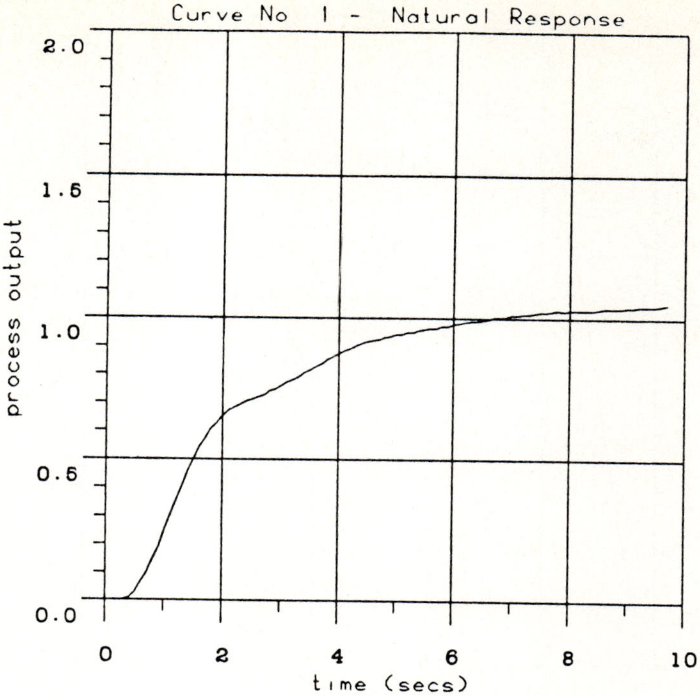

Curve No 1 - Natural Response
2.0
1.5
1.0
0.5
0.0
process output
0
2
4
6
8
10
time (secs)

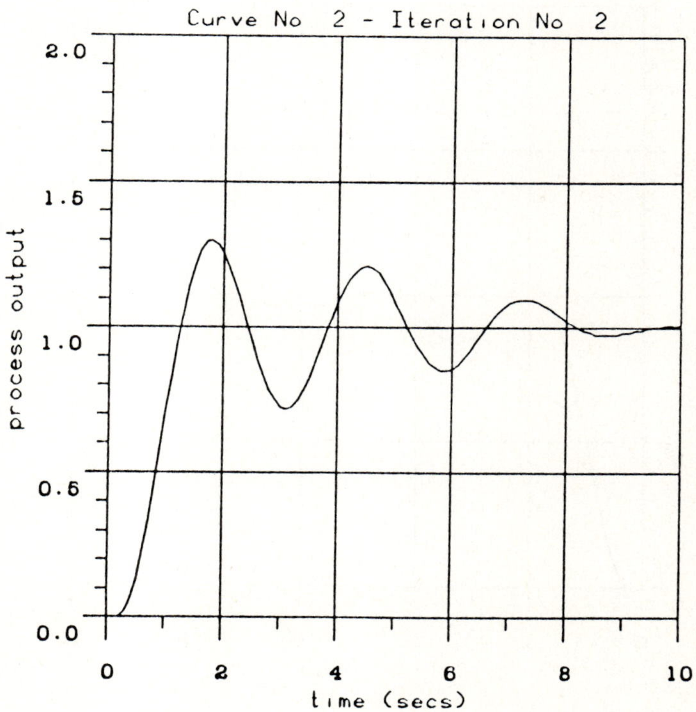

Curve No 2 - Iteration No 2
2.0
1.5
1.0
0.5
0.0
process output
0
2
4
6
8
10
time (secs)

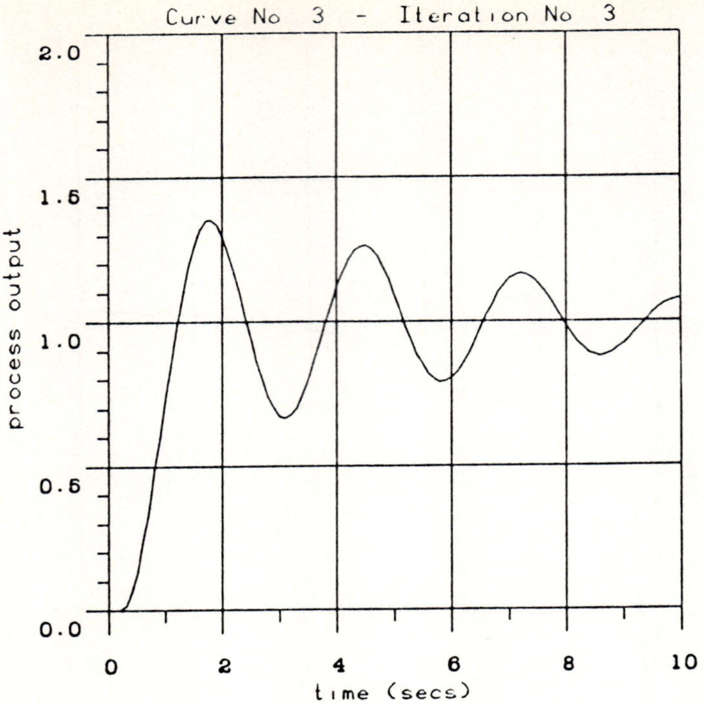
Curve No 3 - Iteration No 3
2.0
1.5
1.0
0.5
0.0
process output
0
2
4
6
8
10
time (secs)

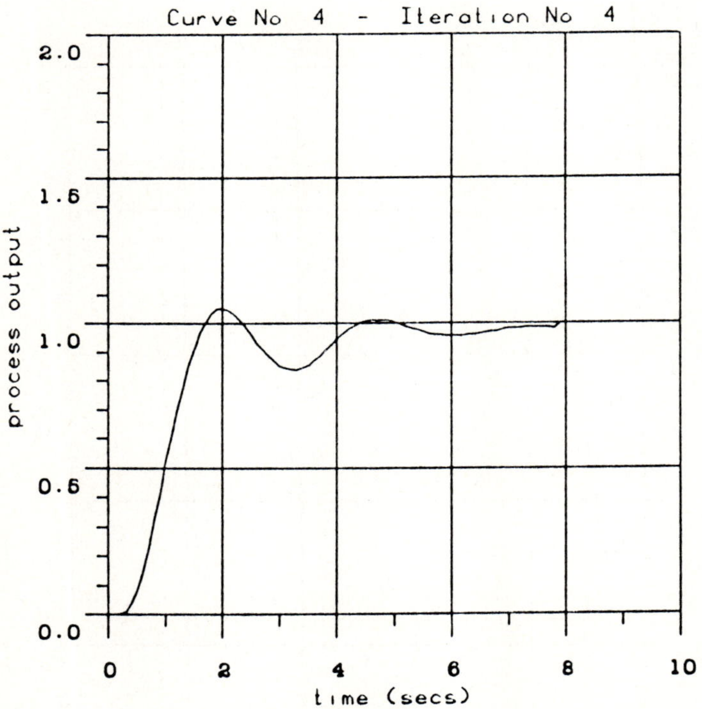
Curve No 4 - Iteration No 4
2.0
1.5
1.0
0.5
0.0
process output
0
2
4
6
8
10
time (secs)

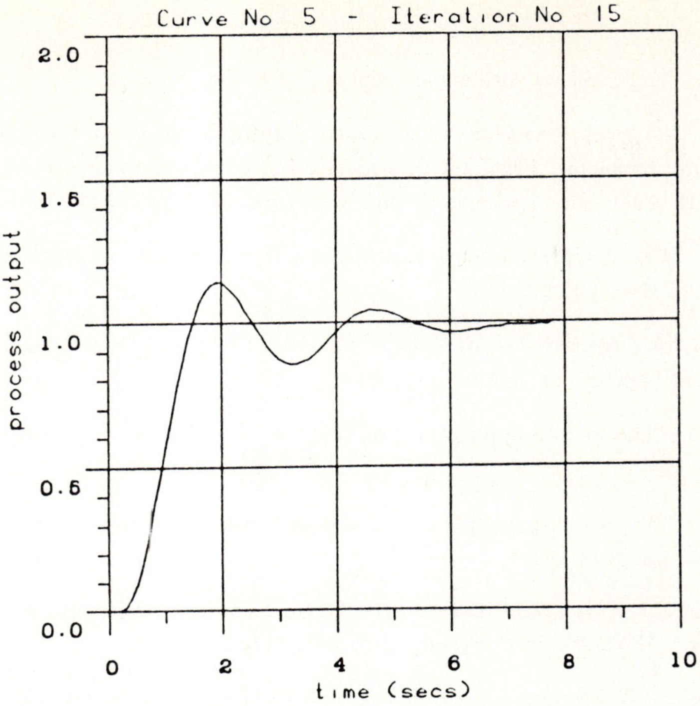
Curve No 5 - Iteration No 15
2.0
1.5
1.0
0.5
0.0
process output
0
2
4
6
8
10
time (secs)

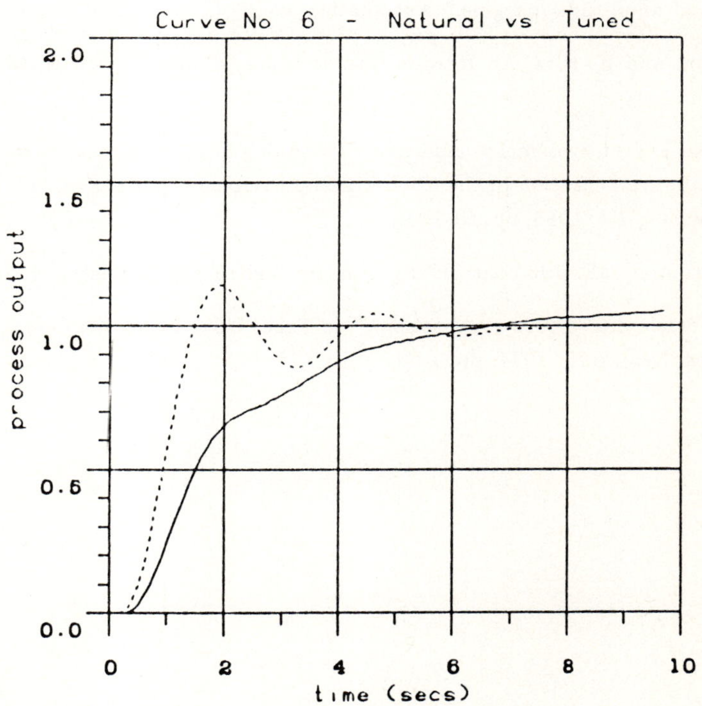
Curve No 6 - Natural vs Tuned
2.0
1.5
1.0
0.5
0.0
process output
0
2
4
6
8
10
time (secs)

References

[1] K.-E. Arzen, "Experiments with expert control," preprint, 1985.

[2] K.-E. Arzen, "Expert systems for process control," in Applications of Artificial Intelligence in Engineering Problems, Proc. 1st Internat. Conf., Southampton Univ., April 1986, D. Sriram and R. Adey, eds., Springer-Verlag, New York, 1986, pp. 1127-1138.

[3] K.-E. Arzen, "Use of expert systems in closed loop feedback control," Proc. Amer. Control Conf., Seattle, 1986, pp. 140-145.

[4] K.J. Astrom, "A Ziegler-Nicols auto-tuner," Report TFRT-3167, Department of Automatic Control, Lund Institute of Technology, 1982.

[5] K.J. Astrom, "Theory and applications of adaptive control," Automatica, vol. 19(1984), pp. 471-486.

[6] K.J. Astrom, "Auto-tuning, adaptation, and expert control," Proc. Amer. Control Conf., Boston, 1985, pp. 1514-1519.

[7] K.J. Astrom, J.J. Anton, and K.-E. Arzen, "Expert control," Automatica, vol. 22(1986). (See also Proc. IFAC World Congress, Budapest, 1984.)

[8] K.J. Astrom and T. Hagglund, "Automatic tuning of simple regulators with specifications on phase and amplitude margins," Automatica, vol. 20(1984), pp. 645-651.

[9] T. Fortmann and K. Hitz, An Introduction to Linear Control Systems, M. Dekker, New York, 1977.

[10] J. Litt, "An expert system for adaptive PID tuning based on pattern recognition techniques," in Instrumentation in the Chemical and Petroleum Industries, vol. 18, Proc. 1986 Conf., Secaucus, NJ, 1986, pp. 87-104.

[11] R.L. Moore, et al., "Expert control," Proc. Amer Control Conf., Boston, 1985, pp. 885-887.

[12] L. Zadeh, "Calculus of fuzzy restrictions," in Proc. US - Japan Seminar, Berkeley, Academic Press, New York, 1974, pp. 1-39.

SUPERVISORY CONTROL OF DISCRETE EVENT PROCESSES WITH ARBITRARY CONTROLS

C.H. Golaszewski and P.J. Ramadge *

Department of Electrical Engineering
Princeton University, Princeton, NJ 08544

Abstract

An extended model for a class of discrete event systems is proposed. Based on the model formulated by Ramadge and Wonham [1], we introduce the notion of a nondeterministic supervisor and extend the definition of a controllable language [1]. Some results on the existence of deterministic and nondeterministic supervisors to achieve a given closed loop language are presented. We briefly discuss closure properties of controllable languages in the extended model.

Keywords: Supervisory Control, Discrete Event Systems

1 Introduction

We will extend the framework for the control of DES (Discrete Event Systems) proposed by Ramadge and Wonham [1] to include more general control structures. In the original model two kinds of events are distinguished: uncontrollable events and controllable events. The latter are events which can be prevented from occurring by control actions. We will generalize this

* Research supported by the National Science Foundation through grant ECS-8504584

NATO ASI Series, Vol. F47
Advanced Computing Concepts and Techniques
in Control Engineering
Edited by M.J. Denham and A.J. Laub

framework by allowing a more general control structure. This is useful for certain forms of hierarchical control, or in situations where controlled events are logically coupled.

The results presented in this paper answer the basic problem in the supervisory control framework: when is it possible to find a supervisor (or controller) for the DES so that the generated strings of events lie within a set of acceptable behavior, i.e., only "legal" strings of events occur. We derive necessary and sufficient conditions for the solution of this problem in the extended framework.

The paper is organized as follows. In Sections 2,3 and 4 the extended model for discrete event processes is defined. In Section 5 we give a modified version of the definition for a controllable language. The main results on the existence of supervisors are contained in Sections 6 and 7. We conclude with a discussion of closure properties of controllable languages in Section 8.

2 Generators

Following [1] we define a *generator* to be a 5-tuple

$$\mathcal{G} = (Q, \Sigma, \delta, q_o, Q_m)$$

where Q is the set of *states* (not necessarily finite), Σ is the set of *output symbols*, $\delta: \Sigma \times Q \to Q$ is the *transition function* (partial function), $q_o \in Q$ is the *initial state* and $Q_m \subseteq Q$ is the set of *marker states.*

Denote by Σ^* the set of all finite strings s of elements of Σ, including the empty string $\mathbf{1}$, and extend the transition function δ to Σ^* in the usual fashion:

$$\delta(\mathbf{1}, q) = q, \quad q \in Q;$$

$$\delta(s\sigma, q) = \delta(\sigma, \delta(s, q)),$$

whenever $q\prime = \delta(s, q)$ and $q = \delta(\sigma, q\prime)$ are both defined.

Any subset of Σ^* will be called a *language* over Σ, and the strings of a language will often be referred to as *words.* The *closure* of $L \subseteq \Sigma^*$, denoted by $\overline{L}$, is the set of all strings which are prefixes of words of L, i.e.,

$$\overline{L} = \{s: s \in \Sigma^* \text{ and } \exists t \in \Sigma^*: st \in L\}.$$

If $L = \emptyset$, then $\overline{L} = \emptyset$, otherwise $\mathbf{1} \in \overline{L}$. A language is *closed* if $L = \overline{L}$.

The language *generated* by $\mathcal{G}$ is

$$L(\mathcal{G}) := \{w: w \in \Sigma^*, \delta(w, q_o) \text{ is defined}\}$$

and the language *marked* by $\mathcal{G}$ is

$$L_m(\mathcal{G}) := \{w: w \in L(\mathcal{G}) \text{ and } \delta(w, q_o) \in Q_m\}$$

Roughly, the language generated by $\mathcal{G}$ is the set of all strings of events that it can generate, and the language marked by $\mathcal{G}$ is that subset of the generated language which leaves the state of the generator in the set Q_m. These strings might model completed phases of operation, or identifiable subtasks.

3 Controlled Discrete Event Processes

3.1 Admissible controls

Consider a system whose behavior is modeled by a generator $\mathcal{G}$. In general there will be certain events over which we have control, e.g., the use of a resource, the activation of a machine, etc., whereas we may have no control over other events, e.g., the breakdown of a machine. To capture the different possibilities we allow the set of admissible controls to be a subset Γ of 2^{Σ} subject to the natural constraint:

$$\forall \sigma \in \Sigma \quad \exists \gamma \in \Gamma : \sigma \in \gamma.$$

We interpret the control $\gamma \in \Gamma$ as specifying the allowed "next" events in the evolution of the system. Any admissible set of controls Γ has the following three properties:

[i] Γ is a poset under $\subseteq$;
[ii] Γ has a set of maximal elements Γ_{max};
[iii] Γ has a set of minimal elements Γ_{min}.

The following three examples show the motivation for the above definition.

Example 3.1.

Ramadge and Wonham model In the original model of [1] Σ is divided into two disjoint subsets Σ_u and Σ_c: events in Σ_u are spontaneous (i.e., uncontrolled) while events in Σ_c can be disabled (i.e., prevented from occurring). The events in Σ_u are always enabled and a control consists of a specification of which events in Σ_c are also to be enabled.

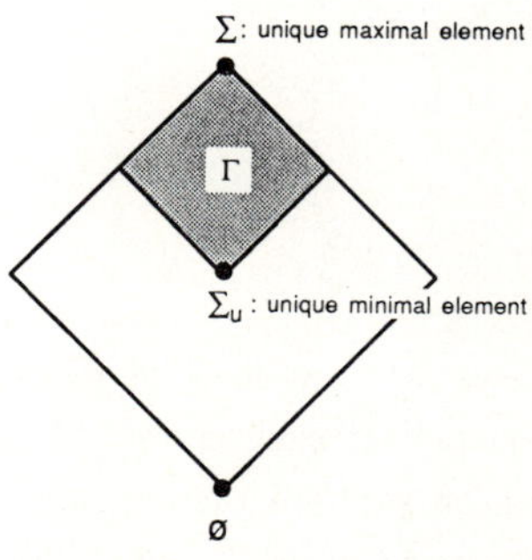

Thus the set of admissible controls consists of those subsets of 2^{Σ} that contain the set of uncontrolled events, i.e.,

$$\Gamma = \{\gamma : \gamma \in 2^{\Sigma} \text{ and } \Sigma_u \subseteq \gamma\}$$

Γ is closed under set intersection and set union and so forms a lattice under $\subseteq$. Σ_u is the unique minimal element of the lattice and Σ is the unique maximal element. This can be visualized in terms of the lattice diagram above.

Example 3.2.

Coupled Events Suppose Σ is divided into three disjoint subsets, Σ_u, Σ_c (with the same meaning as above) and Σ_k. Events in the latter set are controllable with the following restriction: Two events α, $\beta \in \Sigma_k$ cannot be enabled or disabled at the same time. α and β are coupled in the sense that enabling α corresponds to disabling β and vice versa. In this case the set of admissible controls is given by

$$\Gamma = \{\gamma : \gamma \in 2^{\Gamma} \text{ and } \Sigma_u \subseteq \gamma \text{ and } \gamma \cap \Sigma_k = \{\alpha\} \text{ some } \alpha \in \Sigma_k\}$$

The set Γ is not closed under union nor intersection. This example can be easily generalized to the case where not all of the events in Σ_k are coupled and some can be enabled simultaneously. The set of admissible controls can be visualized through the following lattice diagram:

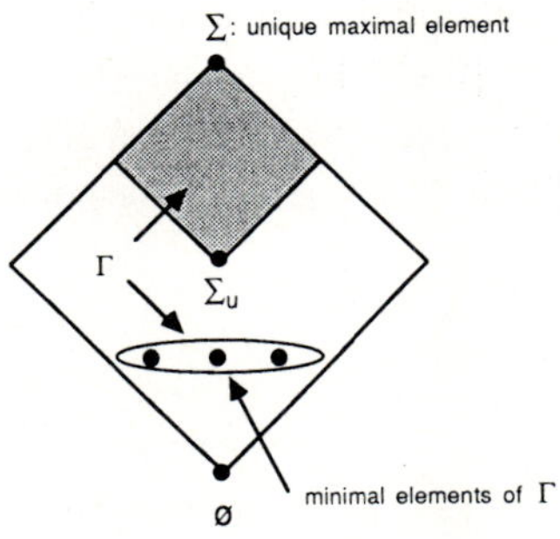

Example 3.3.

Hierarchical Control Let us consider a system consisting of several subsystems. Suppose $\mathcal{S}$ is a controller for a particular subsystem and as such is able to prevent certain events from occurring (c.f. Example 3.1). Furthermore suppose there is a controller $\mathcal{H}$ for the system, whose task it is to coordinate the operation of the subsystems. It is natural to assign a higher priority to control actions chosen by $\mathcal{H}$, than to the "local" control actions chosen by $\mathcal{S}$. For example, if $\mathcal{S}$ selects a control $\gamma_{\mathcal{S}}$ and $\mathcal{H}$ selects $\gamma_{\mathcal{H}}$ with higher priority, then the subsystem would be forced to execute an event from $\gamma_{\mathcal{H}}$, provided, of course, that both controls are admissible. A situation like this might correspond to partitioning Σ into three subsets Σ_u, Σ_c and Σ_f where Σ_u and Σ_c have the interpretation given in Example 3.1. Σ_f need not be necessarily disjoint from the other two sets. Events in Σ_f correspond to events which can be forced to occur by $\mathcal{H}$ although $\mathcal{S}$ has selected other events. In this example an event can be a forced event if selected by $\mathcal{H}$ or a regular event (in the sense of Example 3.1) if selected by $\mathcal{S}$ only. To illustrate this let us consider a special case where Σ_f is disjoint from Σ_u and Σ_c. Suppose $\mathcal{H}$ can only force a single event at any time. Thus the set of admissible controls is given by:

$$\Gamma = \{\gamma : \gamma \in 2^{\Sigma} \text{ and } \gamma = \{\sigma\} \text{ some } \sigma \in \Sigma_f \\ \text{or } \gamma \subseteq 2^{\Sigma_u \cup \Sigma_c} \text{ with } \Sigma_u \subseteq \gamma\}$$

Note that Γ is neither closed under union nor intersection.
The following lattice diagram illustrates this situation:

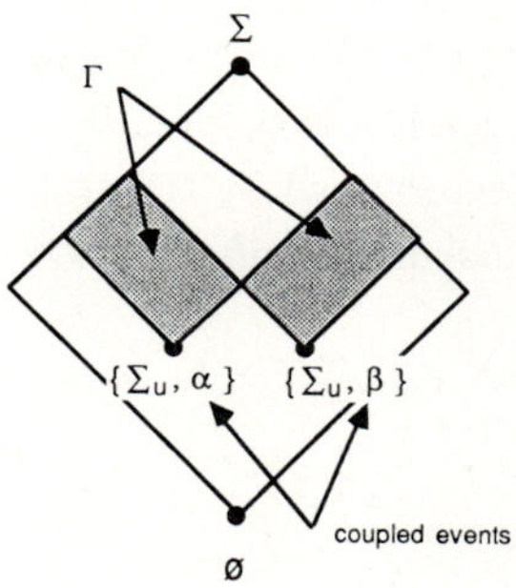

3.2 Controlled generators

Let Γ be a set of admissible controls for the generator $\mathcal{G}$. If $\delta : \Sigma \times Q \to Q$ is the transition function of $\mathcal{G}$, we define an augmented transition function $\delta_c : \Gamma \times \Sigma \times Q \to Q$, a partial function, according to

$$\delta_c(\gamma, \sigma, q) = \begin{cases} \delta(\sigma, q), & \text{if } \sigma \in \gamma, \\ \text{undefined}, & \text{otherwise.} \end{cases}$$

The generator

$$\mathcal{G}_c = (Q, \Gamma \times \Sigma, \delta_c, q_o, Q_m)$$

results from $\mathcal{G}$ by specifying Γ. For every fixed γ the resulting generator $\mathcal{G}(\gamma)$ is like the original generator $\mathcal{G}$ with certain events removed. $\mathcal{G}_c$ will be called a *controlled discrete event process* (CDEP).

4 Supervisors

A *supervisor* for $\mathcal{G}_c$ is a map $f: L(\mathcal{G}) \to \Gamma$ that specifies for each generated sequence the resultant control. As in [1] f can be realized by a pair

$$\mathcal{S} := (S, \phi)$$

where S is an automaton

$$S = (X, \Sigma, \xi, x_o)$$

and ϕ is a map $\phi: X \to 2^{\Sigma}$ such that for each $w \in L(\mathcal{G})$, $f(w) = \phi(\delta(w, x_o))$. S models the dynamic or memory element of the supervisor: the information about the past of a string relevant for the control is "stored" in the state of S.

In contrast to a generator, S does not run freely but is driven by strings $s \in \Sigma^*$. This allows us to couple S to a generator $\mathcal{G}$ in a feedback loop. The controlled discrete event process $\mathcal{G}_c$ is now constrained by the control determined by the states of S. This leads to the closed loop system $\mathcal{S}/\mathcal{G}$, called a *supervised discrete event process* (SDEP), defined by

$$\mathcal{S}/\mathcal{G} := AC(X \times Q, \Sigma, \xi \times \delta_c, (x_o, q_o), Q_m)$$

where

$$\xi \times \delta_c: \Sigma \times X \times Q \to X \times Q$$

(a partial function) is given by

$$(\sigma, x, q) \longmapsto (\xi(\sigma, x), \delta_c(\phi(x), \sigma, q)).$$

We are only interested in supervisors that have sufficient transitions defined to be capable of tracking events generated by $\mathcal{G}_c$. In [1] such supervisors are termed complete. Without loss of generality we assume that supervisors fulfill this technical condition (they can always be modified to do so).

The *controlled* language of $\mathcal{S}/\mathcal{G}$ is

$$L_c(\mathcal{S}/\mathcal{G}) := L(\mathcal{S}/\mathcal{G}) \cap L_m(\mathcal{G}).$$

This language corresponds to the strings of $L(\mathcal{G})$ that lead to completed "tasks" in the supervised discrete event processes.

The supervisors defined above are completely deterministic. It is also possible to consider a *nondeterministic supervisor* by allowing the map ϕ to be a set valued map $\phi: X \to 2^{\Gamma}$. In this case the supervisor can select any control from the subset $\phi(x) \subseteq \Gamma$ in a nondeterministic fashion. It is readily seen that if Γ is closed under set union, then to every nondeterministic supervisor there corresponds an equivalent deterministic supervisor. However, this is not the case if Γ is not closed under union, and in such situations nondeterministic supervisors can provide additional power.

The following examples illustrate the difference between the deterministic and nondeterministic supervisors.

Example 4.1.

Ramadge and Wonham model: since in this model Γ is closed under union nondeterministic supervisors are no more powerful than deterministic supervisors.

Example 4.2.

Consider the special case in Example 3.2. Suppose $\mathcal{H}$ has two controls σ, $\tau \in \Sigma_f$ to chose from. If $\mathcal{H}$ is deterministic then it could select one or the other but not both. In contrast a nondeterministic supervisor can select both events to occur.

5 Controllable Languages

In this section we extend the concept of a controllable language introduced in [1]. For each $K \subseteq \Sigma^*$ and each $s \in \bar{K}$ define the *active set of* K *after* s to be

$$\Sigma_K(s) = \{\sigma \in \Sigma : s\sigma \in \bar{K}\}$$

For example, in the case of the generator $\mathcal{G}$, $\Sigma_{L(\mathcal{G})}(s)$ is the set events possible at the state $\delta(s, q_o)$ of $\mathcal{G}$.

A *controllable language* is defined as follows:

Definition 5.1.

A language $K \subseteq L(\mathcal{G})$ *is* $(L(\mathcal{G}), \Gamma)$*-controllable if for all* $s \in \bar{K}$ *there exists a control* $\gamma \in \Gamma$ *such that*

1. $\Sigma_K(s) \in \gamma$; *and*
2. $s\gamma \cap L(\mathcal{G}) \subseteq \bar{K}$.

The second item will be recognized as the controllability condition from [1]. This condition ensures that there is a control to keep the evolution of the event string in $\bar{K}$. Item one ensures that this control can also be chosen to allow the full range of alternatives possible in K. The two conditions are equivalent to the following:

$$\forall s \in \bar{K}\ \exists \gamma \in \Gamma : \gamma \cap \Sigma_{L(G)}(s) = \Sigma_K(s)$$

Note that the empty set is always $(L(G),\Gamma)$-controllable, and that for $L(\mathcal{G})$ to be $(L(\mathcal{G}),\Gamma)$-controllable we require Γ to at least contain sets of events γ_q, $q \in Q$, such that $\Sigma_q = \{\sigma: \delta(\sigma,q)!\} \subseteq \gamma_q$

We illustrate the definition with the following examples:

Example 5.1.

For the Ramadge and Wonham model the controllability condition reduces to that given in [1]. To see this note that in this case item 2 in the definition is equivalent to $s\Sigma_u \cap L(\mathcal{G}) \subseteq \bar{K}$, since Σ_u is the unique minimal element of Γ. If Σ_u satisfies item 2, then $\gamma = \Sigma_K(s) \cup \Sigma_u$ will satisfy item 1 and item 2.

Example 5.2.

Consider the special case in Example 3.3: $K \subseteq L(\mathcal{G})$ is $(L(\mathcal{G}),\Gamma)$-controllable iff for each $s \in \bar{K}$ either

1. $\Sigma_K(s)$ contains no events that can be forced by $\mathcal{H}$ and $s\Sigma_u \cap L(\mathcal{G}) \subseteq \bar{K}$; or
2. $\Sigma_K(s)$ contains exactly one event that can be forced by $\mathcal{H}$.

6 Existence of Supervisors

The definition of the last section provides the basis for the proof of the main theorem. This is an extension of Theorem 5.1 in [1], and deals with the supervisor existence problem in a general setting. To state the theorem we need the following notation. For any set of controls Γ let $cl(\Gamma)$ denote the closure of Γ under set union.

Theorem 6.1.

Let $K \subseteq L(\mathcal{G})$. Then

1. there exists a deterministic supervisor $\mathcal{S}$ such that $L(\mathcal{S}/\mathcal{G}) = K$ if and only if K is closed and $(L(\mathcal{G}),\Gamma)$-controllable.
2. there exists a nondeterministic supervisor $\mathcal{S}$ such that $L(\mathcal{S}/\mathcal{G}) = K$ if and only if K is closed and $(L(\mathcal{G}), cl(\Gamma))$-controllable.

Proof.

1. (If) Clearly we need only define the required supervisor f on $\bar{K}$. By assumption for each $s \in \bar{K}$ there exists $\gamma_s \in \Gamma$ satisfying items 1 and 2 in the definition of controllability. Define $f: \bar{K} \to \Gamma$ by $f(s) = \gamma_s$. Then it is readily shown by an induction argument similar to that given in [1] that this supervisor ensures that the closed loop behavior is K.
(Only if) Assume that there exists a supervisor f such that the controlled system has closed loop behavior K. Then for each $s \in \bar{K}$ we must have $\Sigma_K(s) \subseteq f(s)$ in order to ensure that all

possible continuations of s in K are possible in the closed loop system. On the other hand we must have $sf(s) \cap L(\mathcal{G}) \subseteq \bar{K}$ in order to ensure that only those events permitted by K occur.

2. Similar. ■

7 Nonblocking Supervisors

The previous result on the existence of supervisors can be extended along the lines of [1]. Using Theorem 6.1 we prove the following proposition:

Proposition 7.1.

Let $K_1 \subseteq L_m(\mathcal{G})$ and $\emptyset \neq K_2 \subseteq L(\mathcal{G})$. There exists a supervisor $\mathcal{S}$ such that, for the closed loop system $\mathcal{S}/\mathcal{G}$,

$$L_c(\mathcal{S}/\mathcal{G}) = K_1;$$
$$L(\mathcal{S}/\mathcal{G}) = K_2,$$

if and only if

(1) $K_1 = K_2 \cap L_m(\mathcal{G})$;

(2) K_2 is closed and controllable.

Proof.

Follows directly from Theorem 7.1, and the definition of $L_c(\mathcal{G})$. ■

Definition 7.1.

A supervisor will be called nonblocking if

$$\overline{L_c(\mathcal{S}/\mathcal{G})} = L(\mathcal{S}/\mathcal{G}).$$

If a supervisor fails to be nonblocking then there exists a string generated by $\mathcal{S}/\mathcal{G}$ that can never be completed to a word $st \in L_c(\mathcal{S}/\mathcal{G})$. Thus the supervisor might block the CDEP from ever completing a "task". Before we conclude our discussion on the existence of supervisors with a theorem on nonblocking supervisors we need the following definition:

Definition 7.2.

Let K and L be two languages. We say that K is L-closed iff

$$K = \overline{K} \cap L.$$

Theorem 7.1.

Let $K \subseteq L_m(\mathcal{G})$, $K \neq \emptyset$. There exists a nonblocking supervisor $\mathcal{S}$ such that

$$L_c(\mathcal{S}/\mathcal{G}) = K$$

if and only if K is $L_m(\mathcal{G})$-closed and controllable.

Proof.

K is controllable and $L_m(\mathcal{G})$-closed $\Longleftrightarrow$ the pair

$$(K_1, K_2) := (K, \overline{K})$$

satisfies conditions (1) and (2) of Proposition 8.1 $\Longleftrightarrow$ there exists a supervisor $\mathcal{S}$ such that

$$(L_c(\mathcal{S}/\mathcal{G}), L(\mathcal{S}/\mathcal{G})) = (K, \overline{K})$$

and this means that $\mathcal{S}$ is nonblocking. ∎

8 Closure Properties of Controllable Languages

For a fixed CDEP $\mathcal{G}_c$ let $\mathcal{C}_\mathcal{G}(L)$ denote the family of $(L(\mathcal{G}), \Gamma)$-controllable sublanguages of $L \subseteq L(\mathcal{G})$, i.e.,

$$\mathcal{C}_\mathcal{G}(L) := \{K\colon K \subseteq L \text{ and } K \text{ is } (L(\mathcal{G}), \Gamma)-\text{controllable}\}$$

It was shown in [1] that the structure of this family of languages often played a useful role in the solution of basic control synthesis problems. Here we investigate the closure properties of this class under set union.

Proposition 8.1.

If Γ is closed under set union, then $\mathcal{C}_\mathcal{G}$ is closed under set union.

Note that $\mathcal{C}_\mathcal{G}(L)$ is nonempty since the empty set is a sublanguage of L and is controllable. Thus if Γ is closed under set union, then the previous proposition immediately implies the existence of a supremal controllable sublanguage, sup $\mathcal{C}_\mathcal{G}(L)$, contained inside L.

If Γ is not closed under union, then the above result makes nondeterministic supervisors attractive since we are then guaranteed the existence of a unique largest achievable closed loop behavior contained within any given language.

9 Conclusions

An extended model for discrete event systems has been presented. It preserves the features of the original model introduced in [1] and increases the range of potential applications since the new model admits a more general class of controls. The main change to the existing results was an extension of the definition of a controllable language.

If we admit nondeterministic supervisors or if the set of controls Γ is closed under set union, then there exists a unique supremal achievable closed loop behavior contained in any prescribed language. However, this property may be lost if the set of controls is not closed under set union and we restrict attention to deterministic supervisors.

10 References

[1] Ramadge, P.J., and Wonham, W.M., "Supervisory Control of a class of Discrete-Event Processes", *SIAM J. Control and Optimization*, Vol. 25, No. 1, January 1987.

[2] Wonham, W.M., and Ramadge, P.J., "On the Supremal Controllable Sublanguage of a Given Language", *SIAM J. Control and Optimization*, Vol. 25, No. 3, May 1987.

[3] Golaszewski, C.H., and Ramadge, P.J., "Control of Discrete Event Processes with Forced Events", to appear: *Proceedings of the 26th Conference on Decision and Control*, Los Angeles, December 1987.

[4] Cieslak, R.,Desclaux, C., Fawaz, A., and Varaiya, P., "Supervisory Control of Dicsrete-Event Processes with Partial Observations", Memorandum No. UCB/ERL M86/63, August 1986, University of California, Berkley, CA 94720.

[5] Lin, F., and Wonham, W.M., "Decentralized Supervisory Control of Discrete-Event Processes", Systems Control Group Report No. 8612, July 1986, Dept. of Electrical Engineering, University of Toronto, Canada.

[6] Lafortune, S., and Wong, E., "A State Model for the Concurrency Control Problem in Data Base Management Systems", Memorandum No. UCB/ERL M85/27, April 1985, University of California, Berkley, CA 94720.

[7] Maimon, O., and Tadmor, G., "Efficient Low Level Control of FMS", Draft Technical Report LIDS-P-1571, June 1986, Laboratory for Information and Decision Systems, MIT, Cambridge, MA 02139.

[8] Ramadge, P.J., and Wonham, W.M., "Modular Feedback Logic for Discrete Event Systems", *SIAM J. Control and Optimization*, Vol. 25, No. 5, September 1987.

[9] Wonham, W.M., and Ramadge, P.J., "Modular Supervisory Control for Discrete Event Systems", *Mathematics of Control, Signals and Systems*,Vol. 1, No. 1, 1988.

High Speed RLS Parameter Estimation by Systolic-like Arrays

Luigi Chisci
Dipartimento di Sistemi e Informatica
Universita' di Firenze
Via di Santa Marta, 3
50139 Firenze
Italy

Abstract - The parallel implementation of recursive least-squares parameter estimation on systolic-like arrays is considered and the efficiency in terms of estimate updating rate is discussed. New implementations are proposed, which allow higher throughput rates (O(1) estimate updates per time unit), not depending on the model order. Since in one of them, a distortion with respect to exact RLS is introduced, its performance is investigated, both analytically and experimentally. Tradeoffs between complexity and performance are discussed.

Keywords: adaptive control/ parameter estimation/ recursive least squares/ parallel processing/ systolic arrays.

1. Introduction

Recursive parameter estimation plays a key role in adaptive signal processing and control. Among the wide variety of available methods [1], RLS is no doubt the most commonly used. Traditional square root algorithms result in $O(n^2)$ operations per time update, n being the model order. Recently, a great deal of fast algorithms (see [2] for a comprehensive review) have also been proposed. They exploit the "shift invariance" property, fulfilled by many practical models, in order to reduce complexity to O(n). The same O(n) complexity but lower computational requirements are exhibited by the so called Least Mean Square (LMS) algorithm [3]. However, LMS presents, compared to RLS, a worse transient behaviour and slower convergence, which may become unacceptable for highly correlated models [2].

The breakthrough of VLSI technology has provided powerful tools for parallel processing of adaptive algorithms. In this respect, the most significant figure of merit is the throughput rate (more briefly throughput), generally defined as the number of data output per time unit. Throughout the paper we shall refer to throughput as the estimate updating

NATO ASI Series, Vol. F47
Advanced Computing Concepts and Techniques
in Control Engineering
Edited by M. J. Denham and A. J. Laub

rate and, being interested mainly in its dependence on n, we shall denote its value by O(f(n)) analogously to computational complexity.

New computing architectures, such as systolic arrays [4] and wavefront arrays [5], hereafter referred to with the unique term of "systolic-like arrays", have been successfully exploited in the implementation of square root or other type RLS algorithms. However the problem of their efficiency has not been fully addressed, since in planar arrays (with $O(n^2)$ Processing Elements (PE's)) only $O(n^{-1})$ throughputs have been obtained [6,7]. This is probably due to the fact that, in applications so far considered (i.e. adaptive beamforming and noise cancellation), the estimate of the whole parameter vector is not necessary. However there exist other cases, like parameter estimation in adaptive control, where all the model parameters are required.

In this paper the difficulties met in high throughput RLS parameter estimation by systolic-like arrays are pointed out and new implementations, providing a constant (O(1), not depending on n) throughput, are proposed.

The paper is organized as follows. After a brief overview of RLS in Sect.2, Sect. 3 describes traditional as well as new parallel implementations. In Sect. 4 the performance of an "approximate" implementation is analyzed. Finally, in Sect.5, they are compared with standard RLS and LMS algorithms through computer simulations.

2. RLS overview

Given a discrete-time linear regression model of the form

$$y(t) = \varphi'(t)\theta + e(t), \quad t=1,2,....... \tag{1}$$

where $y(t)$ is the observation, $\varphi(t)\in\mathbb{R}^n$ is the regressor, $\theta\in\mathbb{R}^n$ is the unknown parameter vector and $e(t)$ the observation error, the exponentially windowed RLS algorithm provides in a recursive way the LS estimate $\hat{\theta}(t)$ minimizing, w.r.t. (with respect to) θ, the cost functional

$$J_\lambda(t,\theta) = \sum_{k=1}^{t} \lambda^{t-k}\epsilon(k/\theta), \qquad \epsilon(t/\theta) = y(t) - \varphi'(t)\theta \tag{2}$$

Numerically robust RLS algorithms [8] are essentially based on Cholesky or UD factorization of either the so called "covariance matrix",

$$P(t) = (\lambda^t P^{-1}(0) + \sum_{k=1}^{t} \lambda^{t-k}\varphi(k)\varphi'(k))^{-1}, \quad P(t) = P'(t) > 0 \tag{3}$$

or of its inverse $P^{-1}(t)$, namely the "information matrix". Accordingly, they are referred to as "square root covariance filter" (SRCF) or "square root information filter" (SRIF). In this paper we shall consider algorithms based on UD factorization, viz $P^{-1}(t) := U'(t)D(t)U(t)$, with $U(t)$ unit upper triangular (u.u.t.) and $D(t)$ diagonal. With this respect, in the SRIF, the LS estimate $\hat{\theta}(t)$ is provided, at each time instant t, by the following

SRIF updating procedure

Step 1a): Triangularization update - Given $U(t-1)$, $D(t-1)$ and $z(t-1)$, perform the triangularization

$$Q(t)\begin{bmatrix}\lambda D(t-1) & 0\\ 0 & 1\end{bmatrix}^{1/2}\begin{bmatrix}U(t-1) & z(t-1)\\ \varphi'(t) & y(t)\end{bmatrix} = \begin{bmatrix}D(t) & 0\\ 0 & \delta(t)\end{bmatrix}^{1/2}\begin{bmatrix}U(t) & z(t)\\ 0 & w(t)\end{bmatrix}$$

with the orthogonal matrix $Q(t)$

Step 2a): Matrix vector product evaluation - Compute $\hat{\theta}(t)$ according to:

$$\hat{\theta}(t) = U^{-1}(t)z(t). \blacksquare$$

Let us define :

$$\bar{V} := \begin{bmatrix} U^{-1} & \hat{\theta} \\ 0 & 1 \end{bmatrix}, \quad \bar{D}^{-1} := \text{block-diag}\,\{D^{-1}, 0\}, \quad \bar{\varphi}' := [\varphi', -y]$$

With these notations, the LS estimate via SRCF is, therefore, provided by:

SRCF updating procedure

Step 1b): Matrix vector product evaluation - Compute

$$f(t) := \bar{V}'(t-1)\bar{\varphi}(t)$$

Step 2b): Triangularization update - Perform the triangularization

$$\begin{bmatrix}\bar{V}(t-1) & 0\\ f'(t) & 1\end{bmatrix}\begin{bmatrix}(\lambda\bar{D}(t-1))^{-1} & 0\\ 0 & 1\end{bmatrix}^{1/2}Q(t) = \begin{bmatrix}\bar{V}(t) & k(t)\\ 0 & 1\end{bmatrix}\begin{bmatrix}\bar{D}^{-1}(t) & 0\\ 0 & r(t)\end{bmatrix}^{1/2}$$

with the orthogonal matrix $Q(t)$. $\blacksquare$

Notice that the SRCF provides, besides the new estimate $\hat{\theta}(t)$ contained in the last column of $\bar{V}(t)$, also the Kalman gain vector $k(t)$ and the normalized prediction variance $r(t)$. From a numerical point of view, the triangularization update can be efficiently performed by the Gentleman algorithm [9], which has the twofold advantage to avoid square root computations as well as to involve less operations than Cholesky-type algorithms.

Let us now make some comparison between the two algorithms.

1) - Since U is u.u.t., $\hat{\theta}$ in step 2a) can be simply obtained by backsubstitution, without computing U^{-1} explicitly. Hence 2a) and 1b) have the same complexity.

2) - Apart from dimensionality, the triangularization steps 1a) and

2b) differ only in the way in which the orthogonal trasformation Q is applied: in 1a) the bottom row vector r' is annihilated by elementary Givens row rotations whereas in 2b) f' is annihilated by column rotations.

3) - Despite these similarities, the overall SRIF differs from SRCF in the dependence between the first and the second step at different time instants (see the graph in fig.1). Indeed step 1a) can be recursively performed independently of 2a). Conversely, this is not possible for either 1b) or 2b).

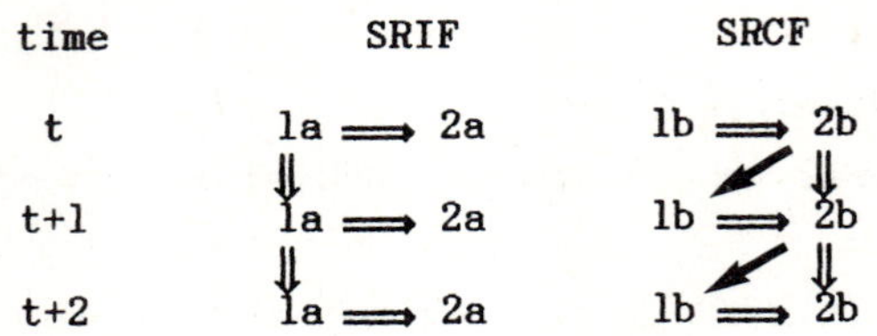

Fig. 1 - Dependence graph for SRIF and SRCF steps

3. Systolic RLS processing

Parallel implementations of RLS on systolic-like arrays have already been devised in the last few years, using either SRIF [6] or SRCF [7] algorithms. As far as their efficiency is concerned the current state of the art shows that, whether $O(n^2)$ or $O(n)$ processors are employed, the best throughput so far obtained is $O(n^{-1})$. Only in applications in which the parameter estimates are not of direct interest, O(1) throughput has been exploited.

The main goal of this paper is to analyze the reasons of such a bottleneck affecting RLS parallel implementation and to show how, with a modest increase of hardware complexity, a constant throughput can be obtained even for parameter estimation.

Two standard systolic modules (see fig.2) are employed for RLS parallel processing: the basic triangular module (BTM) and the basic rectangular module (BRM). Being supplied with proper temporally skewed input data matrices, they can perform (see Fig.3 for details):

- matrix triangularization;
- matrix multiplication;
- unit triangular linear system solution.

This set of matrix operations represents all that is required for both SRIF and SRCF implementations. With regard to SRIF, the basic architecture consists of a BTM(n), storing the UD factored information matrix, plus a BRM(n,1) storing z(t) (see Fig. 4). In [6] it has been shown how this architecture can provide

- O(1) updating rate for the stored data U(t),D(t),z(t) (step 1a);
- O(1) residual extraction rate.

With the same architecture, the step 2a) (parameter extraction) can be performed, essentially, in three ways:

- parameter "flushing" [10];
- matrix inversion [11];
- backsubstitution.

Parameter "flushing" - This method follows from the residual extraction technique. Since the output residual $\epsilon(t) = y(t) - \varphi'(t)\hat{\theta}(t)$, the generic parameter $\hat{\theta}_i(t)$ can be extracted by simply introducing $\varphi(t) = e_i$ and $y(t) = 0$. This way, the entire vector $\hat{\theta}(t)$ can be extracted from the OP_0 port by inputting an identity matrix (IP_1-IP_n ports) and a zero vector (IP_{n+1} port). Notice that the state of the array must be frozen during the parameter extraction in order to prevent any undesired adaptation.

Matrix inversion - The BTM(n) provides $U^{-T}(t)$, according to the scheme of fig. 3.c.I ($Z=I$, $X=U^{-T}(t)$), and the BRM(n,1) computes the matrix-vector product 2a) according to fig. 3.b ($A=U^{-T}(t)$, $B=0_{nn}$, $C=1_{nn}$, $Y=0_n$, $Z=\bar{Z}=z(t)$, $\bar{Y}=-\hat{\theta}(t)$). Hence the parameter vector can be extracted sequentially from the OP_1 port.

Backsubstitution - The parameter vector can also be obtained by backsolving the linear system $U(t)\hat{\theta}(t) = z(t)$ according to the scheme of fig. 3.c.2 ($Z=z(t)$, $X=\hat{\theta}(t)$). The parameters can, therefore, be extracted in parallel from the array top. This method can be implemented on the array of fig.4 by providing suitable means for extracting the z(t) vector together with additional right-to-left data links for backsubstitution. ■

This discussion clearly indicates why, for the SRIF algorithm, O(1) throughputs cannot be obtained by the array shown in Fig. 4: even if steps 1a) or 2a) can be performed separately with O(1) throughput, they cannot be processed together at the same rate, since either they need different input data (cfr. triangularization with parameter "flushing" or matrix inversion), or their processing wavefronts propagate in different directions (cfr. triangularization with backsubstitution).

Similar considerations hold also for the SRCF implementation: in this case a BTM(n+1) is required. It can be seen [12] that, once again, the processing wavefronts of the steps 1b) and 2b) move in different directions.

However (for SRIF), slight modifications to the basic architecture of fig.4 and a different timing of input data allow to improve throughput. Four different implementations will be presented and discussed hereafter. They will be referred to as

- SIRLS: Systolic Interleaved RLS
- SSRLS: Systolic Simultaneous RLS;
- SERLS: Systolic Exact RLS
- SARLS: Systolic Approximate RLS.

1) *SIRLS* - The key idea consists of interleaving step 2a) (via "flushing" or matrix inversion) during the normal information processing. This can be accomplished by pipelining the following input matrices:

$$\begin{array}{c} \begin{array}{ccc} IP_0 & IP_1 - IP_n & IP_{n+1} \end{array} \\ \left[\begin{array}{ccc} 0 & I & 0 \\ \hline 1 & \varphi'(t+1:t+p) & y(t+1:t+p) \end{array}\right] \begin{array}{c} n \\ p \end{array} \end{array} \qquad t = k(n+p) \quad \text{for } k = 0,1,2,\ldots..$$

where $x(i;j) := [x(j), x(j-1), \ldots, x(i+1), x(i)]'$ $(i \leq j)$.
The corresponding output matrices are then extracted (*'s denote don't care data) :

$$\begin{array}{c} \begin{array}{cc} OP_0 & OP_1 \end{array} \\ \left[\begin{array}{cc} -\hat{\theta}(t+p) & * \\ \hline \epsilon(t+1:t+p) & w(t+1:t+p) \end{array}\right] \begin{array}{c} n \\ p \end{array} \end{array}$$
, if the "flushing" technique is being used

$$\left[\begin{array}{cc} * & -\hat{\theta}(t+p) \\ \hline \epsilon(t+1:t+p) & w(t+1:t+p) \end{array}\right]$$
, if the matrix inversion technique is being used

Assuming that input data acquisition rate is O(1), this interleaved processing mode induces a loss of information. In fact, n data out of n+p are not processed by the SIRLS(p) estimator. A suitable choice of p allows a tradeoff between percentage of processed data (100p/(n+p) %) and estimate updating rate (1/n+p). ■

2) *SSRLS* - This technique differs from SIRLS in that parameter extraction and information processing are carried out simultaneously on separate input

data. In this way all available data are processed. On the other side, this is paid with an increased architectural complexity. In fact SSRLS requires, w.r.t. the basic architecture, the following additional features:

- separate inputs for identity matrices and regressors;
- double PE complexity;
- double number of ports and data links;
- control data-flow for periodically freezing the state of the array.

For the sake of brevity, we omit the explanation of how SSRLS may be practically realized. Nevertheless, the technique is quite straightforward and may be implemented in a very direct manner. ■

Parameter estimates produced by the above implementations are not updated at each time instant since they are extracted sequentially, parameter by parameter, from a single output port. Conversely, when the backsubstitution procedure is employed for a parallel extraction, a computational interference between wavefronts propagating in opposite directions occurs. This problem is addressed by the following two implementations.

3) *SERLS* - The basic idea underlying SERLS consists of storing past values of the recursively evolving matrix U(t) inside the corresponding PE's and of properly using them instead of actual values in the backsubstitution process. This can be accomplished by providing internal PE's with FIFO buffers of proper size. Insofar, the SERLS estimator differs from the ideal one (ERLS) only for an O(n) response delay. Practical details are found in [13]. ■

4) *SARLS* - The SARLS technique consists of carrying out the backsubstitution process with the actual values of the matrix U instead of the correct ones.The rationale for this approach is that the elements of the U matrix undergo only slow variations. Obviously, this introduces an approximation in the parameter estimate computation. Denoting by $\bar{\theta}(t)$ the approximate estimate obtained by this approach, namely the SARLS estimate, it can be easily verified that $\bar{\theta}(t)$ satisfies

$$\bar{U}(t)\bar{\theta}(t) = z(t) \tag{4}$$

where the coefficient matrix $\bar{U}(t)$ is such that

$$\bar{u}_{ij}(t) = u_{ij}(t+4n+1-2i-2j) \quad , 1\leq i\leq n,\ i+1\leq j\leq n \tag{5}$$

Hence SARLS is equivalent to introduce a perturbation over the coefficient matrix in the linear system of step 2a). The amount of such an approximation will be investigated in the next section. ■

Let us finally consider the SRCF. Due to its different dependence graph, SRCF is not amenable to O(1) throughput systolic implementation. In fact, since the "exact" f(t) vector is not immediately available at each time instant, the triangularization update procedure cannot occur at an O(1) rate. Therefore the SRCF counterparts of SIRLS, SSRLS and SERLS do not exist. As far as the SRCF counterpart of SARLS is concerned, since approximations would necessarily be introduced in both the steps 1b) and 2b), a satisfactory performance (i.e. with bounded estimation errors) for the overall algorithm cannot be guaranteed.

4. SARLS performance evaluation

In the following analysis the time index will be omitted for simplicity, except when explicitly required. According to (4), the SARLS estimate is given by $\bar{\theta}=(U+\tilde{U})^{-1}z$ where $\tilde{U}=\bar{U}-U$ is the "perturbation" matrix. The corresponding parameter estimation error $\tilde{\theta}:=\theta-\bar{\theta}$ becomes

$$\tilde{\theta}:=\theta-\bar{U}^{-1}D^{-1}U^{-\prime}\left(\sum_{i=1}^{t}\lambda^{i}\varphi(t-i)(\varphi'(t-i)\theta+e(t-i)\right) \tag{5}$$

Exploiting (3) and assuming that, in an appropriate matrix norm, $\|\tilde{U}\| \ll \|U\|$

$$\tilde{\theta}\cong U^{-1}\tilde{U}\theta+\bar{U}^{-1}D^{-1}U^{-\prime}\sum_{i=0}^{t}\lambda^{i}\varphi(t-i)e(t-i):=\tilde{\theta}_{SA}+\tilde{\theta}_{N} \tag{6}$$

The second term in the RHS of (6), denoted as $\tilde{\theta}_N$, can be considered as a zero-mean perturbation produced by the observation noise e(t). Its variance, when $\lambda=1$ (φ,U and D are assumed as deterministic quantities) is

$$E[\tilde{\theta}_N\tilde{\theta}_N'] = \sigma_e^2(\bar{U}'D\bar{U})^{-1} \tag{7}$$

and does not substantially differ from the corresponding variance of ERLS estimation error, viz $\sigma_e^2(U'DU)^{-1}$. Therefore $\tilde{\theta}_{SA}:=U^{-1}\tilde{U}\theta$ represents the part of the estimation error due to the systolic approximation. A precise statistic characterization of this contribution is difficult to obtain. However some remarks concerning its properties can be easily made:

1) *Asymptotic behaviour*. - Under stationarity of the input signal, stability of the plant and weak ergodicity assumptions, when $\lambda = 1$, the matrix U(t) converges, in probability, to the corresponding factor U_∞ of the asymptotic regressor covariance matrix ϕ, i. e.

$$\lim_{t\to\infty} U(t) = U_\infty \; ; \quad U_\infty'D_\infty U_\infty=E[\varphi(t)\varphi'(t)]:=\phi \tag{8}$$

Hence $\tilde{U}(t)$ and, by (6), $\tilde{\theta}_{SA}(t)$ asymptotically vanish. Conversely, for $\lambda<1$, the limit in (8) is a random matrix with non-zero variance. Hence $\tilde{U}$ and $\tilde{\theta}_{SA}$

cannot converge exactly to zero. From this point of view, the behaviour of $\tilde{\theta}_{SA}$ and $\tilde{\theta}_N$ is similar.

2) *S/N ratio dependence* - From (7) it follows that the variance of $\tilde{\theta}_N$ is proportional to the inverse of input and output variances. Conversely, $\tilde{\theta}_{SA}$ is insensitive to any regressor rescaling. Hence it should be expected that SARLS will behave satisfactorily, w.r.t. ERLS, if the S/N ratio is low enough.

3) *Input-output processes correlation dependence* - $\tilde{\theta}_{SA}$ depends on the condition number of U. In fact, using the spectral norm, we get

$$\frac{\|\tilde{\theta}_{SA}\|}{\|\theta\|} \leq K(U) \frac{\|\tilde{U}\|}{\|U\|}$$

where the condition number $K(U) := \|U\| \|U^{-1}\|$ is, by definition, $\lambda_{max}(U'U)/\lambda_{min}(U'U)$. Notice that, in general, the condition number of U is comparable or inferior to that of $D^{1/2}U$ which, in turn, tends to the maximum-to-minimum eigenvalue ratio of ϕ. Hence one should expect a better behaviour of SARLS when the eigenvalue spread of ϕ is not large, i.e. for weakly correlated input and output processes.

5. Simulation results

Simulation experiments have been carried out in order to assess the previous results and to compare SARLS with the other RLS parallel implementations. Moreover also the LMS algorithm [3] has been considered as a benchmark. Indeed, due to its simplicity, the LMS performance can be considered as an upper bound for any RLS implementation of some potential utility. The ERLS performance has been, conversely, assumed as a lower bound reference. All the simulations deal with the identification of an ARX(ℓ,1) model, i.e.

$$A(q)y(t) = b_1u(t-1) + e(t)$$

where e(t) is a zero-mean white noise with unit variance, and A(q) is a (stable) polynomial in the backward shift operator q,

$$A(q) := 1+a_1q+\ldots+a_\ell q^\ell$$

With reference to (1), the regressor and the corresponding parameter vector become

$$\varphi(t) := [u(t-1), y(t-\ell), \ldots, y(t-1)]' \quad ; \qquad \theta := [b_1, -a_\ell, \ldots, -a_1]'$$

In all simulations, u(t) is a zero-mean white noise; the coefficient b_1 has been adjusted to those of the A polynomial so that $\sigma_y^2 = \sigma_u^2$; this variance denotes also the S/N ratio. Comparisons are carried out by plotting (in dB on the vertical axis) for each iteration step, either $\|\tilde{\theta}(t)\|^2$, the squared norm of the parameter error vector, or the Excess Mean Square Error (EMSE) [3], i.e. $E[(y(t) - \varphi'(t)\hat{\theta}(t))^2] - \sigma_e^2 = \tilde{\theta}(t)'\Phi\tilde{\theta}(t)$

The first simulation (Fig.5) concerns the convergence capability of the algorithms, since no exponential windowing is present (λ=1): SARLS estimates, even if more noisy, behave satisfactorily, when compared to ERLS and LMS.

In the subsequent simulations (Figs.6-8) λ=0.98; the covariance matrix P(t) is initialized with its asymptotic mean value (which depends on λ, the input signal and the model parameters); moreover also the LMS gain has been tuned in order to get the same steady state error of the ERLS. Therefore these simulations deal with the tracking capability of the algorithms (with a steady state gain) to a sudden parameter change. From the simulations it turns out that SARLS transient behaviour almost coincides with that of ERLS. Conversely the steady-state error level of SARLS is higher (in particular, it increases with the S/N ratio and the eigenvalue spread of Φ). On the other side SARLS is faster and more accurate than LMS, especially with low S/N ratio or large eigenvalue spread. SARLS superiority is enhanced if EMSE is replaced by $\|\tilde{\theta}(t)\|^2$ (cfr. Fig.9).

Finally all parallel implementations are compared in Fig.10. SIRLS (due to the limited number of processed data), is the slowest. SSRLS is the best, since its performance differs from the optimal one only for a delay of O(n) sampling periods in the estimate updating. SARLS is in between: for low S/N ratio and eigenvalue spread it almost coincides with SSRLS.

6. Conclusions

RLS algorithm parallel implementations on systolic-like architectures have been revisited in order to optimize their throughput.

Differences concerning high throughput implementability of SRIF and SRCF algorithms have been first highlighted. Subsequently, several SRIF based parallel implementations have been proposed, each tackling in a different way the "parameter extraction" problem. Tradeoffs between complexity and performance have been discussed. From analytical considerations as well as simulation experiments, the following conclusions can be drawn:

- The solution with the highest throughput (SERLS) is too complex (it requires FIFO buffers of different lenght on each PE);
- the simplest SIRLS implementation (not requiring any additional hardware) requires a difficult compromise between alertness (low p) and converging speed; an on-line adaptive choice of this parameter seems necessary;
- the "approximate" SARLS implementation has low complexity and shows good tracking capability. However its steady-state performance is too-heavily dependent on the statistics of the incoming data.

As a conclusion, the SSRLS solution, with moderate hardware complexity and satisfactory overall performance, yields the best compromise.

References

[1] Ljung L., Soderstrom T.(1983). Theory and practice of recursive identification. The MIT Press, Cambridge, USA.
[2] Cioffi J.M., Kailath T.(1984). Fast recursive least-squares transversal filters for adaptive filtering. IEEE Trans.on ASSP, vol.31, no.6, pp.1394-402.
[3] Widrow B., Stearns S.D.(1985). Adaptive signal processing. Prentice Hall, Englewood Cliff, N.J., USA.
[4] Kung H.T.(1982). Why systolic architectures? IEEE Computer, pp.65-82.
[5] Kung S.Y., Arun K.S., Gal-Ezer R.J., Bhaskar Rao D.V.(1982). Wavefront array processor: language, architecture and applications. IEEE Trans. on Computers, vol.31, no.11, pp.1054-65.
[6] Mc.Whirter J.G.(1983). Recursive least squares minimization using a systolic array, Proc.SPIE Symp., RTSP VI, pp.105-13.
[7] Jover J.M., Kailath T.(1986). A parallel architecture for Kalman filter measurement update and parameter estimation. Automatica, vol.22, no.1, pp.43-57.
[8] Bierman G.J.(1977). Factorization methods for discrete sequential estimation. Academic Press, New York, USA.
[9] Gentleman W.M.(1973). Least-squares computations by Givens transformations without square roots. J.Inst.Maths Applics., no.12, pp.329-36.
[10] Chen M.J., Yao K.(1986). On realizations of least-squares estimation and Kalman filtering by systolic arrays. Proc. 1-st Int.Workshop on Systolic Arrays, Oxford, U.K.
[11} Ward C.R., Hargrave P.J., Mc.Whirter J.G.(1987). A novel algorithm and architecture for adaptive beamforming. IEEE Trans.on Antennas and Propagation, vol.24, no.3, pp.338-46.
[12] Chisci L., Mosca E.(1987). Parallel architectures for RLS with directional forgetting. Int. Journal of Adaptive Control and Signal Processing, vol.1, no.1, pp.69-88.
[13] Chisci L., Zappa G. (1987). Parallel architectures for the recursive least squares problem. Proc.10-th Int.Conf. on Digital Signal Processing, Firenze, Italy.

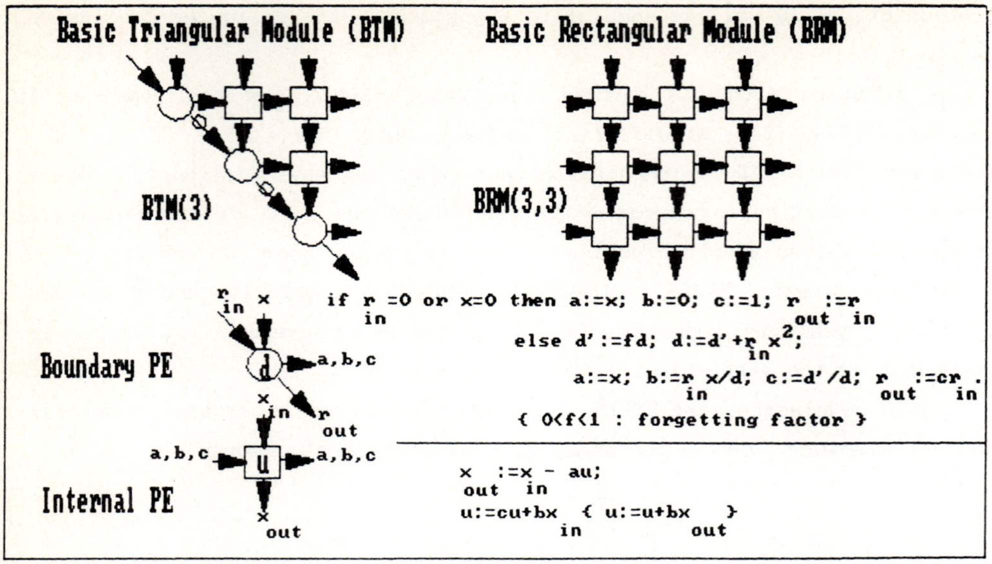

Fig.2 - Basic systolic-like arrays for RLS.

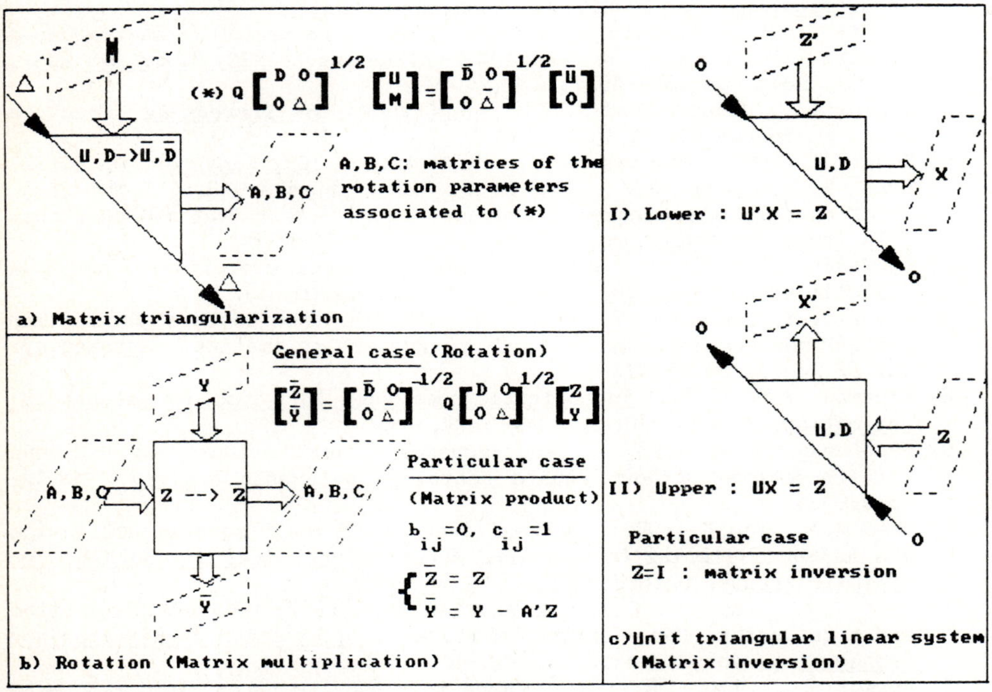

Fig.3 - Mapping some basic matrix operations onto the BTM and BRM.

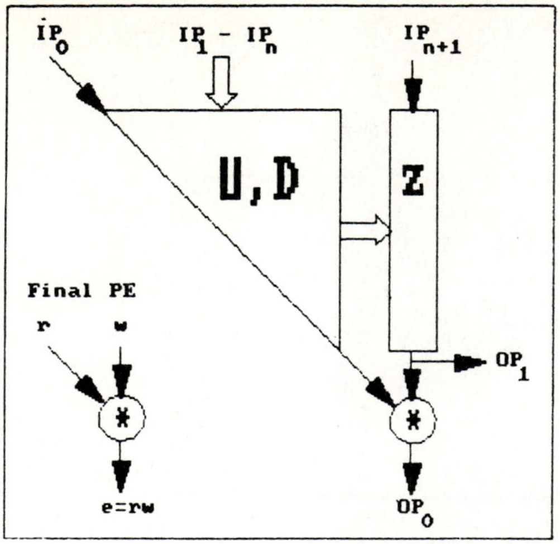

Fig.4 - Architecture for SRIF RLS.

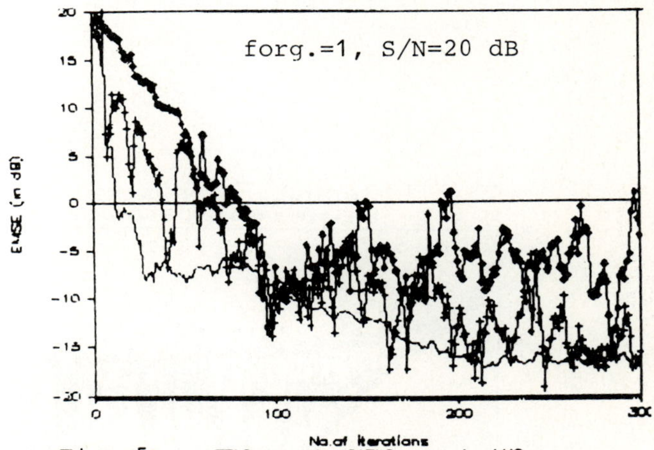

Fig. 5 —— ERLS + SARLS ◊ LMS

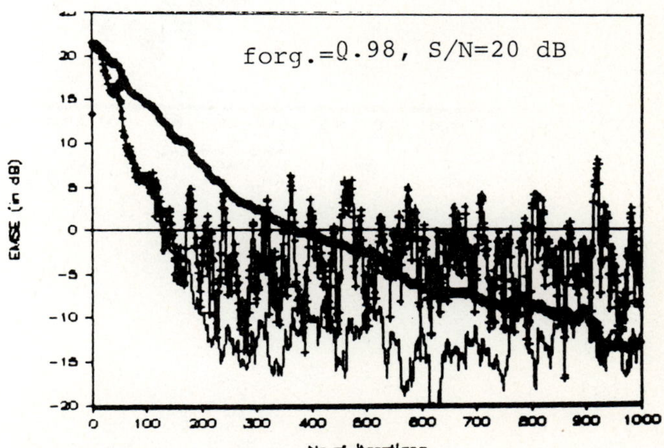

Fig. 6 —— ERLS + SARLS ◊ LMS

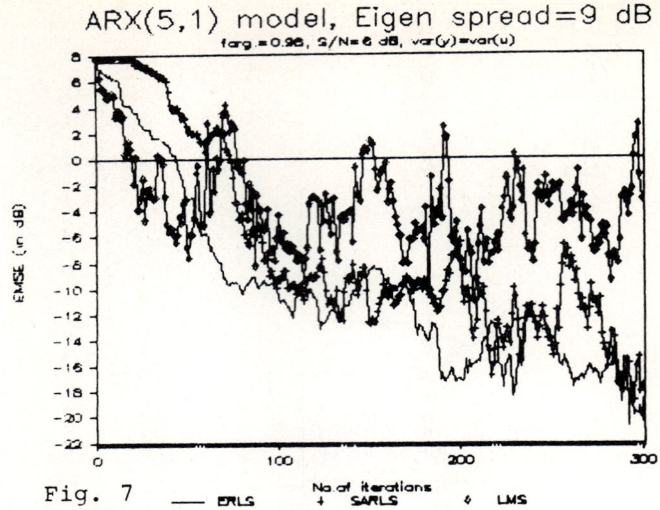
ARX(5,1) model, Eigen spread=9 dB
forg.=0.98, S/N=6 dB, var(y)=var(u)
EMSE (in dB)
No.of iterations
—— ERLS
+ SARLS
◊ LMS

Fig. 7

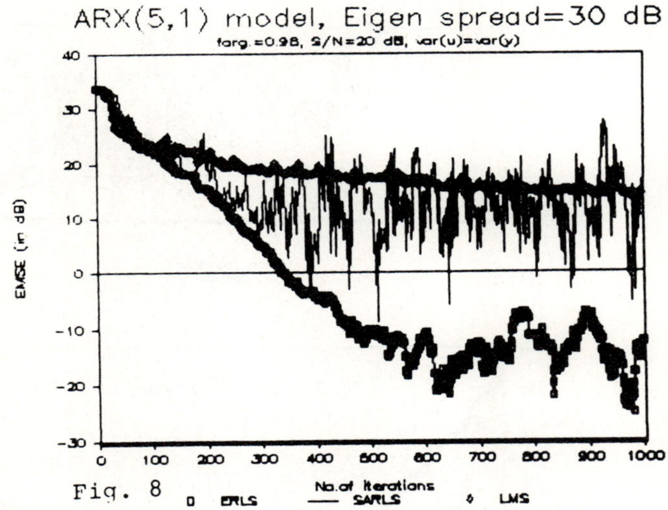
ARX(5,1) model, Eigen spread=30 dB
forg.=0.98, S/N=20 dB, var(u)=var(y)
EMSE (in dB)
No.of iterations
□ ERLS
—— SARLS
◊ LMS

Fig. 8

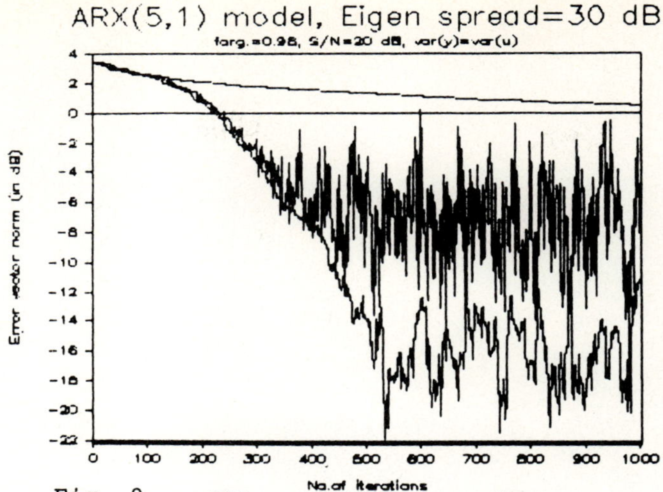

Fig. 9 —— ERLS —— SARLS —— LMS

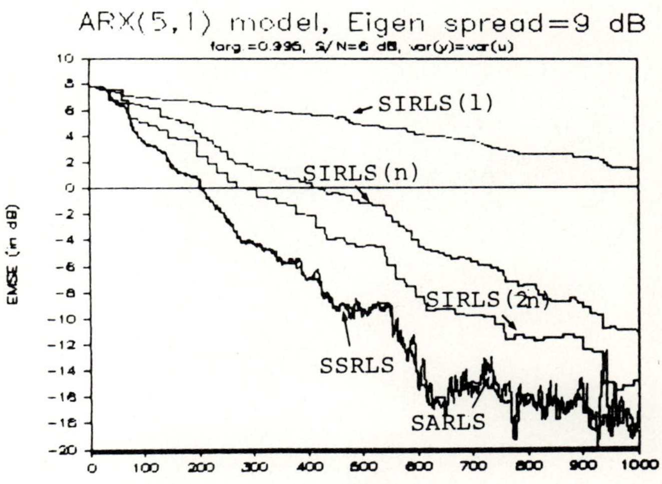

Fig. 10

Intelligent Control Specification Language Compilers, the Q-parameterization, and Convex Programming: Concepts for an Advanced Computer-Aided Control System Design Method

David G. Meyer

Robotics and Control Laboratory

Department of Electrical Engineering

University of Virginia

Charlottesville, VA 22901

V. Balakrishnan, C. H. Barratt, S. P. Boyd

N. M. Kraishi, X. M. Li, S. A. Norman

Information Systems Laboratory

Stanford University

Keywords: computer-aided design / Q-parameterization / convex programming control specification language / compiler / optimization

Abstract

A new computer-aided method for the design of linear controllers is outlined. Rather than designing the controller directly, the method designs the Q-parameter that parameterizes all stabilizing controllers. Because closed-loop maps are *affine* in Q, specifications on the closed-loop system translate to *convex* constraints on Q. The translation is done with an intelligent compiler that reads specifications written in a control specification language and generates a standard convex program for Q that is readily solved.

The constraint and objective constructions in the control specification language are natural for the control engineer and allow a wide variety of specifications including asymptotic tracking, limits on peak excursions of variables, decoupling, step response settling and overshoot, frequency domain inequalities, and others.

NATO ASI Series, Vol. F47
Advanced Computing Concepts and Techniques in Control Engineering
Edited by M.J. Denham and A.J. Laub

1 Introduction

Especially now, with the explosion in VLSI technology making implementation much easier, the *design* phase for a digital linear controller represents the major fraction of work, time, and resources in the construction of a complete control system. As anyone with experience knows, designing a linear controller to stabilize a given plant and meet a set of engineering specifications can be quite a challenge; as computers became more powerful and cheap, it was natural to develop computer-aided methods to help meet this challenge. Hence a variety of packages and techniques came to be at the system designer's diposal.

These available computer-aided design (CAD) methods generally fell into two broad catagories: "analytic methods" and "parameter optimization methods." The analytic methods currently known today include the venerable LQG with its newer accompanyment, LTR([1]), the recent H_∞ techniques ([5]), and the even more recent ℓ_1 techniques ([3]). We refer to these methods as analytic because in each case, a "closed-form" solution to the problem has been found.[1] Such methods are nice because they are *guaranteed* to find a stabilizing controller; on the other hand, they suffer from a sort of myopia because they are, of necessity, designed with only one (optimal) objective in mind. Although these optimal objectives have some free "design variables" (such as weight matrices in LQG) it is often very difficult (and more art than science) to use such freedom to reflect other specifications of interest. Moreover, one has little control over the structure of the resulting controller, and cannot enforce, say, the requirement that it have an integrator for asymptotic tracking of a step input, or that it be decentralized (diagonal).

Parameter optimization methods ([7,8,9]) sit at almost the opposite end of the spectrum. Indeed, with these methods one characteristically has *complete* control over the structure of the compensator, which is chosen beforehand. Parameters in this given structure are then adjusted to minimize some cost functional (hence the name) usually by non-linear programming methods. The use of non-linear programming techniques allows, in sharp contrast with available analytic methods, a cost functional that can be completely general, incorporating a wide spectrum of concerns in the design such as overshoot in a given channel, cost of manufacture, response time, etc. The use of an a-priori fixed structure and non-linear programming means, however, that parameter optimization methods are *not* guaranteed to find a stabilizing controller, and are quite heuristic — with a strong dependence on initialization values for parameters and local minima. Though they often work spectactularly they just as often dramatically fail.

The method we describe here ([2]) lies, in principle, somewhere in the middle. Like analytic methods, it is guaranteed to find a stabilizing controller that meets specifications.[2] Yet like parameter optimization methods it allows the specification, to some degree, of a controller structure and admits a very general cost functional.

[1] Of course, actually *computing* these solutions is a job for a computer. Many control system "CAD packages" are simply routines to do just this.

[2] Or it will *tell* you that your specifications are "too tight" and that no such controller exists.

One aspect of the new method not shared with any other current CAD methods for control systems design is the use of a *compiler* to translate engineering specifications on the closed-loop system into numerical constraints on the controller.

2 Q-Parameterization

Consider the basic feedback system shown in Figure 1 The vector signal w is the *exogenous inputs* which includes all reference commands, noises, disturbances, and anything else from the outside world, including fictitious signals injected anywhere in the plant. The vector signal u is the *actuator inputs* and includes every signal that the controller can use to control the plant, P. At the output, the vector signal y is the *measured variables* and includes all the signals which the controller can have access to —like physical sensor outputs and external reference commands. Finally, the vector signal z is the *regulated variables* which includes every signal we are interested in controlling, again including possibly "fictitious" variables like combinations of states or even filtered signals. Note that z and y may share signals (e.g. a measured output we would like to control), and likewise z and u may share signals. Finally, as mentioned, w and y can share a signal if there is an external reference command that the controller has access to.[3]

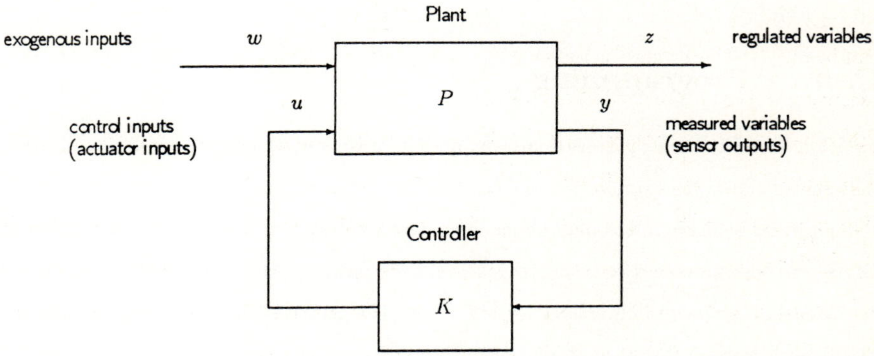

Figure 1: Basic Setup

From these conventions of input and output signals, it is by definition that the controller can use y only as input and produce u only as output, and the cloosed-loop transfer function from w to z (H_{ZW}) contains every map about which we would like to express a specification.

An amazing recent result about the basic feedback system shown in Figure 1 is that if we can find **one** controller, K_{nom}, which stabilizes the closed-loop, then we can find them **all** in terms of a single parameter, Q, which is allowed to take on any value in the set of stable transfer matrices. This result is called the "Q-parameterization" [12,4,11].

[3] In principle, y and u could also share a signal but this is rare.

An important consequence of the Q-parameterization is a characterization of the set of achievable closed-loop maps. If H_{zw} denotes a transfer matrix achievable with a stabilizing compensator, then

$$H_{\mathrm{zw}} = H_0 + H_1 Q H_2 \tag{1}$$

where H_0, H_1, H_2 are stable transfer matrices that can be found easily once a single stabilizing controller, K_{nom}, is given.

Equation (1) is the basis for new the method. Notice that H_{zw} is *affine* in the parameter Q. From this affinity stems a crucial observation: Most[4] constraints on the closed-loop map H_{zw} result in *convex* constraints on Q. To make this more precise, suppose we specify Q by its impulse response coefficients — $(Q_0, Q_1, Q_2, \ldots)$. This sequence represents a point in a (matrix) Hilbert space, $\mathcal{H}$, of all real (matrix) sequences. If $\mathcal{S}$ is the set of such points that meet a collection of closed-loop specifications on H_{zw} then $\mathcal{S}$ will be convex in $\mathcal{H}$.

It is important to realize that constraints on H_{zw} *do not* lead to convex constraints on the controller, K. Indeed, it is quite easy to see that H_{zw} varies in a linear fractional way with K, and this complicated dependence on the parameters of K is what necessitates the non-linear/non-convex optimization approaches currently used. By designing in the Q parameter and exploiting the simple dependence (1) of H_{zw} on Q we can use *effective* optimization techniques.

3 Convex Programming

In this section we talk about optimization techniques on a convex set. As we noted above, designing in the Q parameter allows the exploitation of convexity.

Although so far we have talked only about constraints, noting that the Q-parameterization allows us to specify most constraints on closed-loop maps as convex constraints on Q's impulse response sequence, a similar discussion applies to objectives. If $\tilde{\Phi}: \mathcal{D} \mapsto R^+$ is a positive-real valued convex functional on a suitable domain,[5] then on the set of H_{zw}'s achievable with a stabilizing compensator, equation (1) shows that $\tilde{\Phi}[H_{\mathrm{zw}}]$ is equivalent to $\Phi\left[(Q_0, Q_1, \ldots)\right]$ where Φ is also convex.

The preceding discussion implies that the problem of minimizing a convex objective functional[6] subject to specifications on the closed-loop is the problem:

$$\begin{array}{ll} \text{minimize} & \Phi(x) \\ \text{subject to:} & \\ & G(x) \leq \Theta \end{array} \tag{2}$$

[4] "Most" is not an exageration; the list is quite exhaustive— see [2].

[5] We are being purposely vague here. $\tilde{\Phi}$ can be a function of time or frequency quantities for any closed-loop map from any exogenous input to any regulated output, as well as a function of Q.

[6] Again, most common engineering objectives, including H_∞, LQG, ℓ_1, overshoot, etc. are convex. For a complete list, see [2].

Where Φ is a convex functional and G is a convex mapping on the space $\mathcal{H}$ introduced in the previous section. We have nothing new to add to the theory of problems such as (2), which is a general *convex program.* (For an excellent reference source about such problems, see [6].) However, there are a few points we wish to note here. First, (2) has a *global* optimum solution; second, since the Lagrange functional generates a supporting hyperplane for the primal, its representation in $\mathcal{H}$ (Riesz Theorem, [10]) has an interpretation as a vector whose ith component is a "partial derivative of the objective with respect to the ith constraint," and thus tends to identify costly constraints and constraints that could be tightened without much additional cost. This is especially useful when (2) is *infeasible* since offending constraints (too tight) can generally be identified by a large associated Lagrange multiplier. Third, and lastly, current methods for solving (2) make the problem finite dimensional by choosing an integer M and truncating Q's impulse response at M. Intuitively, seeing that Q must be stable and so has a decaying impulse response, this approximation should be good for large M. However, both in theory and in practice, the consequences of this approximation are currently unknown.

4 Compilers

We have described, in the previous two sections, how specifications on and objectives for the closed-loop map H_{ZW} give rise to convex constraints and functionals on the Q-parameter's impulse response sequence, $(Q_0, Q_1, \ldots)$. While the translation of specifications on H_{ZW} to constraints on Q is mathematically straightforward, it is tedious to perform such a translation by hand. We propose, therefore, a *compiler* which would accept as input a description (say in state-space form) of the plant, P, a single stabilizing controller, K_{nom} and a set of specifications on H_{ZW} written in a *control specification language* and would return a standard convex program — i.e. the functional Φ, the map $G(\cdot)$, and the vector Θ as in equation (2).

We now give a brief discussion of possible forms for a control specification language. The overall structure might be:

```
minimize     {
                 description of objective;
}
subject_to   {
                 first constraint;
                 ⋮
                 last constraint;
}
```

In the language we might have primitive constructions allowing us to refer to the impulse response, step response, or magnitude frequency response from the jth exogenous input to the ith regulated variable at a time, t:

h$[i][j](t)$

u_step$[i][j](t)$

Mag_H$[i][j](t)$

and between these allow inequalities of the form

$$e_1 \leq f \leq e_2$$

where e_1 and e_2 represent *scalar expressions and f is a construction such as one of the ones listed above.*

More complicated constructions such as

undershoot$[i][j]$

overshoot$[i][j]$

Max_Mag_H$[i][j]$

Peak_Gain_H$[i][j]$

Peak_Slew_H$[i][j]$

which are the step-response undershoot, step-response overshoot, maximum magnitude of the frequency response, peak gain, and peak slew-rate of the transfer function from jth exogenous input to ith regulated variable could be easily implemented. Even higher level constructions are possible at the expense only of computation time and implementation burden.

Similarly, objective terms might include everything described above and in addition, terms involving the impulse response squared, frequency domain magnitude squared, impulse response absolute value, and analgous quantities for the Q-parameter:

h_sqr$[i][j](t)$

Mag_H_sqr$[i][j](t)$

Abs_h$[i][j](t)$

q_sqr$[i][j](t)$

Abs_q$[i][j](t)$

Looping, comments and standard pre-processor features such as **#define** might be allowed. A complete specification file might thus look like:

```
                                /* exogenous inputs */
#define  REF                    1            /* reference input */
#define  NOISE                  2            /* noise input */
#define  U_LOOP_IN              3            /* loop test input */

                                /* regulated variables */
#define  POSN                   1            /* position of large mass */
#define  ACTUATOR               2            /* plant input u */

                                /* Q FIR input numbers */
#define REF_OUT                 1            /* command */
#define OBS_ERR                 2            /* observer error */

declarations {
        n_exog = 3;
        n_reg = 2;
        n_sens = 2;
        n_act = 1;
        n_sample = 80;
        n_tap = 5;

        h0_coeffs = {
\#include "h0"
        };

        h1_coeffs = {
\#include "h1"
        };

        h2_coeffs = {
\#include "h2"
        };
}
minimize {
        from t=0 to n_sample-1 sum {
                h_sqr[POSN][NOISE](t);
        }
}
constraints {
```

```
        /* bounds on step response from REF to POSN */
        for t=0 to 30 do
                -0.2 <= u_step[POSN][REF](t) <= 1.05;
        1.00 <= u_step[POSN][REF](15) <= 1.05;
        for t=21 to n_sample-1 do
                1 - 0.90^t <= u_step[POSN][REF](t) <= 1 + (0.90^t);

        /* asymptotic tracking */
        Re_H[POSN][REF](1,0) == 1;

        /* M circle radius requirement */
        for w = 0.0 to %pi step 0.02 do {
                mag_H[ACTUATOR][U_LOOP_IN](1, w) <= 2.0;
        }

        /* bounds on the control effort */
        for t=0 to n_sample-1 do
                -18.5 <= u_step[ACTUATOR][REF](t) <= 18.5;

        for t=0 to n_tap-1 do {
                abs(q[1][OBS_ERR](t)) <= 1000;
                abs(q[1][REF_OUT](t)) <= 1000;
        }
}
```

Notice that the control specification language and accompanying compiler allow the control engineer to express his specifications on closed-loop behavior in very natural fashion. Hopefully, this type of "natural language interface" will allow much more effective use of a designer's time and skill.

5 Conclusions

We have outlined a new computer-aided control system design method. This method, which is based on three key ideas — the Q-parameterization, convex programming, and intelligent compilers — combines some of the good features of analytic and parameter optimization methods that are currently used. The compiler accepts a state-space description of the plant, specifications on the closed-loop system in a "natural language" *control specification language*, a single nominal stabilizing controller, K_{nom}, and then generates a convex program for the impulse response coefficients, $(Q_0, Q_1, \ldots)$ of the Q-parameter. This convex program is solved by *convex programming* techniques, which guarentees that the method is effective— i.e. guaranteed to find a stabilizing controller if one exists. Once the Q-parameter is known, a simple manipulation yields the desired controller.

In practice and in theory, the effects of truncation of the convex program for $(Q_0, Q_1, \ldots)$ at some finite integer M, and dependence on the selection of the nominal controller, K_{nom} are unknown.

References

[1] G. Stein and M. Athans, "The LQG/LTR Procedure for Multivariable Feedback Control Design," *IEEE Trans. Automat. Contr.*, vol AC-32, pp. 105-114, 1987.

[2] S. P. Boyd, V. Balakrishnan, C. H. Barratt, N. M. Kraishi, X. M. Li, D. G. Meyer, and S. A. Norman, "A new CAD method and associated architectures for linear controllers,"*IEEE Trans. Automat. Contr.*, vol AC-33, no. 3, 1988.

[3] M. A. Dahleh and J. B. Pearson, "ℓ_1-optimal controller for MIMO discrete-time systems," *IEEE Trans. Automat. Contr.*, vol. AC-32, no. 4, pp. 314-323.

[4] C. A. Desoer, R.-W. Liu, J. Murray, and R. Saeks,"Feedback system design: The fractional representation approach to analysis and synthesis," *IEEE Trans. Automat. Contr.*, vol AC-29, no. 3, pp. 399-412, 1984.

[5] B. A. Francis and G. Zames, "On H^∞-optimal sensitivity theory for SISO feedback systems," IEEE *Trans. Automat. Control,* AC-29, pp. 9-17, 1984.

[6] D. G. Luenberger, "Optimization by Vector Space Methods," John Wiley & Sons, Inc., New York, 1969.

[7] U.-L. Ly, A. E. Bryson, and R. H. Cannon, "Design of low-order compensators using parameter optimization, In *Applications of Nonlinear Programming to Optimization and Control,* IFAC, 1983.

[8] D. Q. Mayne, E. Polak, and A. Sangiovanni - Vincentelli, "Computer aided design via optimization," *Control Applications of Non-linear Programming,* IFAC, 1979.

[9] E. Polak, , P. Siegel, T. Wuu, W. T. Nye, and D. Q. Mayne, "DELIGHT.MIMO: An interactive optimization-based multivariable control system design package," *IEEE Cont. Sys. Magazine,* No. 4, Dec., 1982.

[10] W. Rudin, *Real and Complex Analysis,* McGraw-Hill, New York, 1974.

[11] M. Vidyasagar, *"Control System Synthesis: The Factorization Approach,"* Cambridge, MA., MIT Press, 1985.

[12] D. C. Youla, H. A. Jabr, and J. J. Bongiorno, "Modern Wiener-Hopf design of optimal controllers-Part II: The multivariable case," *IEEE Trans. Aut. Contr.*, vol. AC-21, no. 6, pp. 319-338, 1976.

A DESIGN ENVIRONMENT FOR COMPUTER-AIDED CONTROL SYSTEM DESIGN VIA MULTI-OBJECTIVE OPTIMISATION

A.C.W.Grace, P.J.Fleming
University College of North Wales, Bangor, Gwynedd, UK

Keywords

Multi-objective optimisation/Computer-Aided Control System Design/MATLAB/Design by Evolution/Integral Quadratic Measures.

Abstract

Control system design problems are characterised by competing multiple objectives. A Computer-Aided Control System Design environment is described for tackling such problems via multi-objective optimisation. The package consists of an upgraded version of MATLAB linked to an optimisation suite, ADS. Control design is performed through selection of control structure and performance measures in an evolutionary design process. Integral quadratic measures of performance are used in the initial stages of the design. An example is given which illustrates this procedure.

1. Introduction: The Control Problem In Perspective

The problem of designing feedback controllers is one that can be tackled by a wide variety of methods and techniques. The problem is typically characterised by multiple and competing objectives and the controller may take on a number of structures. Examples of typical control objectives are given in Fig 1., which also shows a general form of controller and plant which may include state feedback.

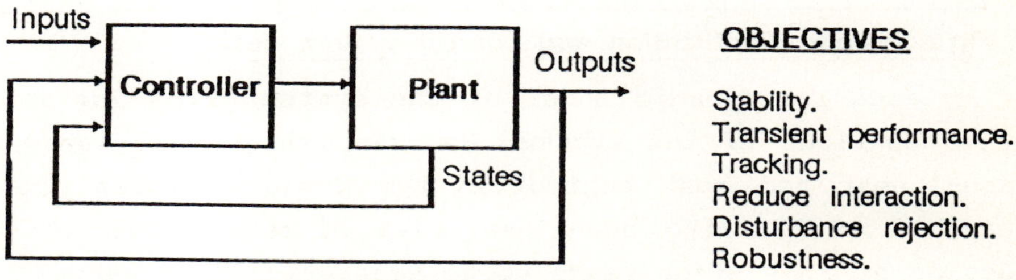

Fig 1. The Control Problem.

NATO ASI Series, Vol. F47
Advanced Computing Concepts and Techniques
in Control Engineering
Edited by M.J. Denham and A.J. Laub

The problem is coupled tightly to the identification of a non-linear model of the plant and subsequent linearisation and reduction of this model. There is a diversity of methods for control system design at various stages of the modelling process. Fig. 2 shows examples of control design techniques and where they are performed with respect to system identification and model reduction.

The control design may be performed without modelling of the plant by on-line techniques such as Ziegler-Nichols tuning or adaptive methods. These techniques often fail to give adequate performance and stability, due to lack of system information, and are inappropriate for many classes of system, where integrity is an important issue or on-line tuning is impractical or hazardous.

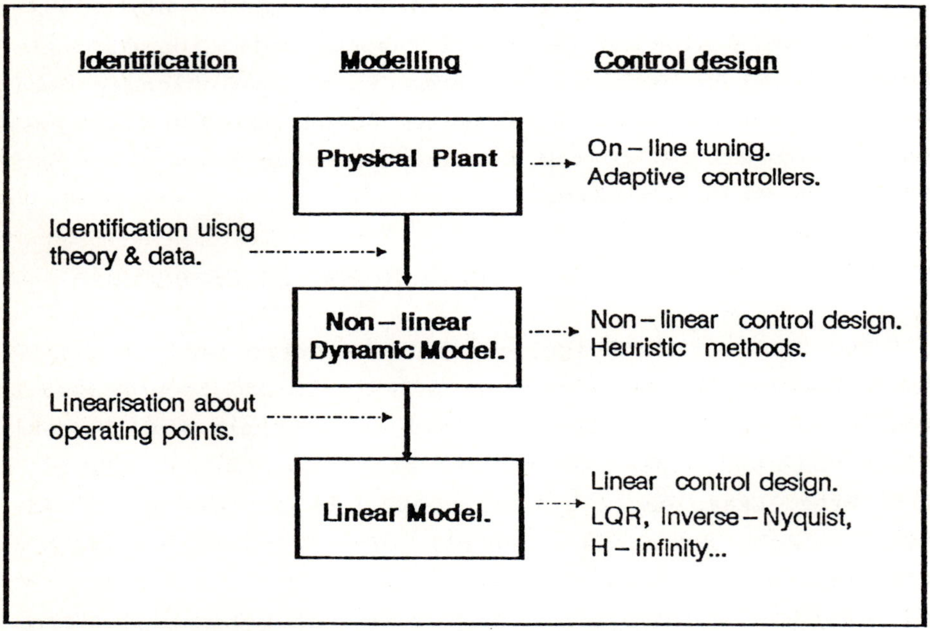

Fig 2. System identification and control system design.

An accurate dynamic model of the system gives insight into the behaviour of the system. However, this model tends to be non-linear and most control system design methods are not capable of handling non-linearities directly. Control system design at this stage tends to be heuristic (i.e. manual tuning of the control structure and control parameters). This proves to be difficult where the system is complex and interactive.

Linearising the non-linear model at various operating points allows linear methods to be used. Here, one problem is tackling non-linearity and this may be achieved by sensitivity reduction to plant parameter variations or through gain-scheduling and adaptive techniques. A large class of non-linearities may also be tackled through the use of linear converters in the form of algorithms or look-up tables (e.g.polar conversion). This essentially converts the problem to a linear one. One non-linearity characteristic to all plants, which cannot be tackled in this manner, is the restriction on the control gains arising from actuator limits. Controller complexity is also an important issue and is limited for reasons of practicality and realizability.

It is apparent that the control design, irrespective of the modelling process stage at which it is performed,is characterised by multiple and competing objectives, as well as constraints on control gains and controller complexity. A good control design method must be able to tackle a design specification characterised by multiple objectives and be able to produce controllers which are both realizable and economically viable.

It is not surprising that there has been a large gap between the heuristic methods which have tended to be used by industry and the theoretical methods developed by academia, which often fail to take into account all aspects of a design. Very often, true objectives are obscured by fuzzy ill-defined criteria, such as describing the handling qualities of an aircraft. Specifying the control problem is a difficult task in itself and one-pass synthesis design methods are not the solution to complicated design problems. It is better to work within a design environment which allows the problem to be built up as a series of multiple objectives and which allows design trade-offs to be examined in an interactive manner. It is such an environment which this paper describes.

2. Design Environment

A computer package has been developed which offers the user a flexible tool for tackling control problems and is particularly suited to problems involving the application of optimisation methods. With reference to Fig. 3, the kernel of the package is a Fortran version of MATLAB[1], which has been upgraded to form a high-level procedural language, MATLANG. MATLAB is a matrix-manipulation package which provides a very user-friendly command-driven environment and is increasingly becoming the de facto standard for linear control system design. MATLANG parallels to a large extent the development of PRO-MATLAB[2], which provides the user with powerful programming tools and an increased range of functions. MATLANG, however, differs from PRO-MATLAB in that programs and subroutines may be tightly coupled into the package. This provides a fast link which proves to be essential for the iterative nature of optimisation problems.

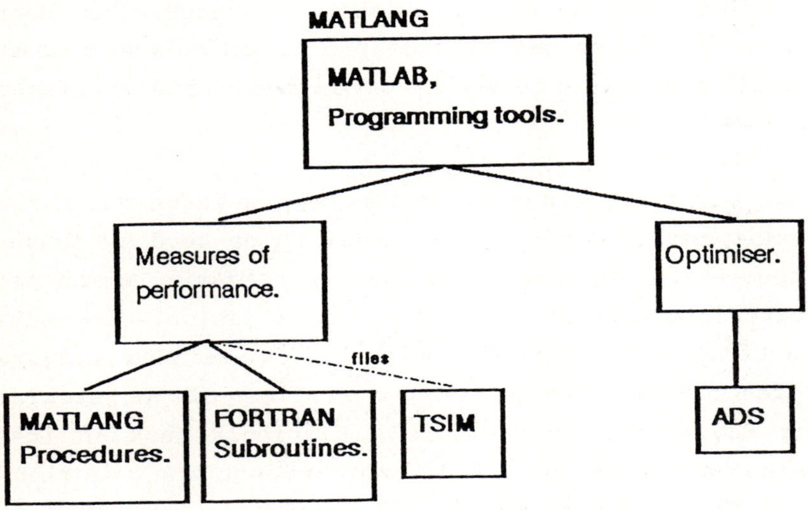

Fig 3. The Design Environment.

MATLANG has been linked to a sophisticated set of optimisation routines contained in the ADS package[3] and a number of FORTRAN subroutines for fast evaluation of performance criteria. The design method is formulated as a set of MATLANG procedures which form a toolbox for handling a range of performance measures and controller structures. The package has also been linked to a non-linear dynamic simulator, TSIM[4], for access to non-linear simulations and direct down-loading of linearised models.

2.1 Design Cycle

The design cycle for the application of multi-objective optimisation is illustrated in Fig. 4. Having entered the model, goals and objectives, a multi-objective problem is formulated (see Section 3). Parameters are initialised which dictate the controller structure and optimisation strategy.

During the optimisation process, design goals and constraints are evaluated at each iteration of the design. These are calculated in FORTRAN subroutines, where computing execution speed is important or MATLANG procedures, where the ease of programming outweighs the execution speed advantage. Design objectives arising from non-linear simulations performed by the package, TSIM, may also be accessed via file, though this option is left to the later stages of the design due to excessive computing demands. At each iteration the optimiser is accessed by passing down control parameters and objective information in the form of non-linear constraints, a cost function and any gradient information available. This cycle is then repeated until satisfactory convergence to a minimum has been achieved. At this stage the design environment permits the inspection of the design and re-formulation or changes to the design objectives. New controller structures may also be introduced at this stage. Trade-offs may be inspected in this way.

3. Multi-objective Problem Formulation[5]

Multiple objectives and constraints typically specify an acceptable design. The usual optimization approach is to condense these objectives into a one-dimensional objective function which is then optimised with respect to selected controller parameters, subject to the set of constraints. Since the design objectives are likely to be competing with one

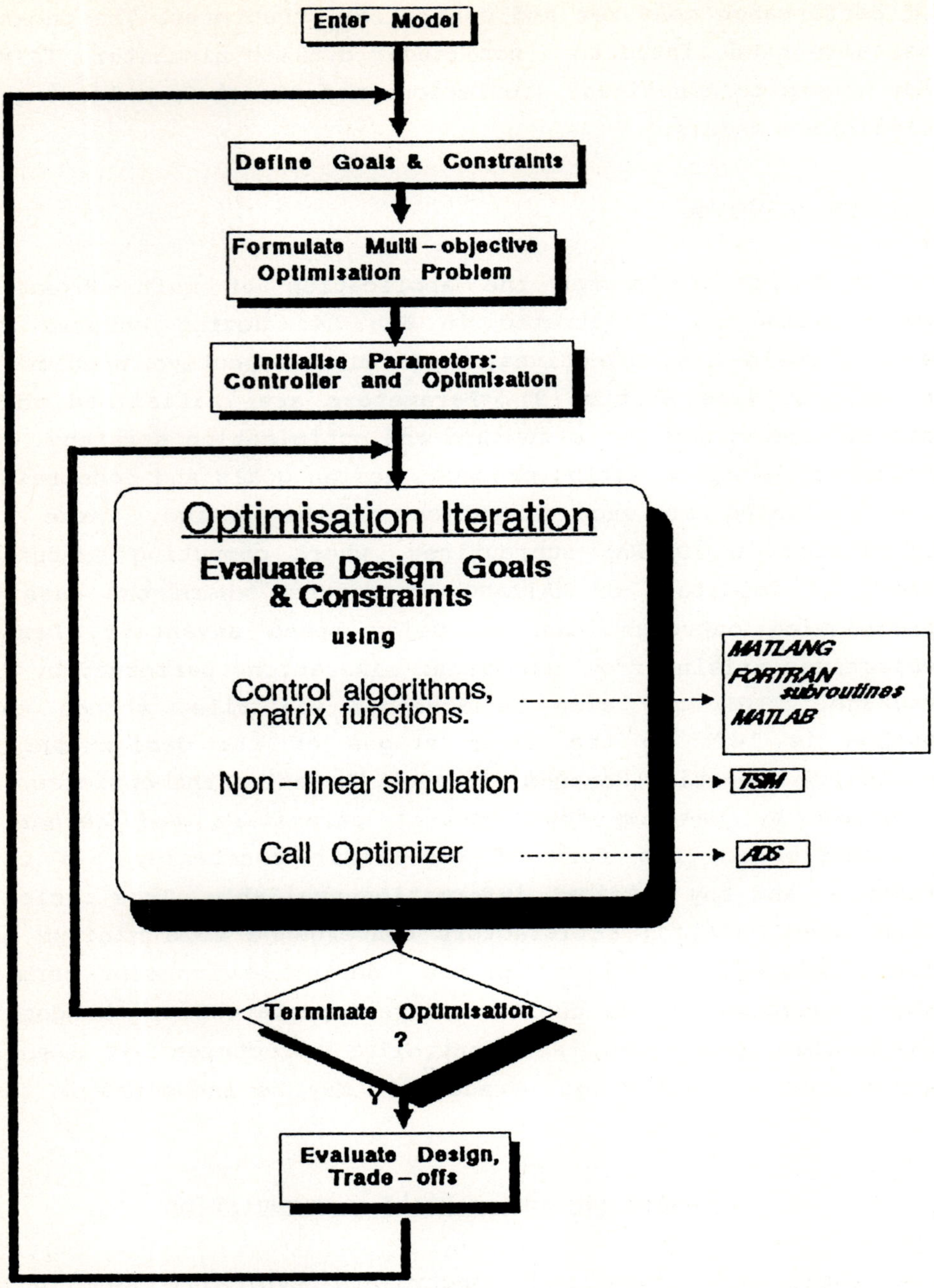

Fig 4. Multi-objective Optimisation Design Cycle.

another, one cannot aggregate these objectives into a single function until the relative importance of the objectives and their trade-offs is well understood. Such information tends to be obscured by the single objective function approach.

Multi-objective optimisation seeks to simultaneously optimise a number of objectives with respect to the free controller parameters. Here, there is no unique "optimal" solution, but rather a set of "non- inferior" solutions, i.e. solutions for which improvement in one of the objectives leads to a degradation in at least one of the remaining objectives. Identification of members of this set of noninferior solutions informs the designer of the nature of the problem and the character of the necessary trade-offs.

3.1 Problem Statement

Let $\underline{k}$ = col $(k_1,k_2, \ldots k_n)$ denote the set of n minimising controller parameters and let $\underline{F}$ = col $(F_1,F_2, \ldots, F_m)$ represent the m design objectives. Also, assume that the following constraints must be satisfied:- $\underline{g}(\underline{k}) < 0$, $\underline{h}(\underline{k}) = 0$. The feasible region, Ω, in the parameter space is the set of all minimising parameters which satisfy these constraints.

The multi-objective optimisation problem is thus stated as:

$$\min_{\underline{k}\varepsilon\Omega} \underline{F} ,$$

where individual components of $\underline{F}$ are simultaneously minimized, subject to the given constraints. Since there is no unique solution to this problem, noninferior solutions are sought.

3.2 Strategies For Multi-objective Optimisation.

3.2.1. Weighted Sum

This familiar technique (cf. LQR design) converts the multi-objective problem into a scalar one by forming the function F, where:

$$\bar{F} = \sum_{i=1}^{m} w_i F_i ,$$

and $w_i > 0$ are the weighting coefficients representing the relative importance of the objectives. The problem now becomes that of minimizing F, subject to the existing constraints:

$$\min_{\underline{k}} \bar{F}$$

A major design drawback to this approach is the difficulty in selecting suitable weighting coefficients. Besides, if the noninferior solution boundary is non-convex, certain sets of solutions will not be accessible via this approach.

Design methods of this type (eg. LQR) are characterised by many iterations of the design process for different values of weighting coefficients in order to meet design requirements. To ask the designer to specify precisely trade-off information for each objective over feasible regions of values would be an unreasonable request and so some compromise is sought.

3.2.2 Goal Attainment Strategy[6]

This method requires the specification of a set of desired goals, $\underline{F}^*$, for the objective function, $\underline{F}$ and, as such, is consistent with the usual approach to control system design. The nonlinear programming problem to be solved is:

$$\min_{\lambda, \underline{k}} \lambda$$

such that $F_i(\underline{k}) - w_i < F_i^*$, $i = 1, 2, \ldots, m$, where $w_i > 0$ are weighting coefficients and λ is an unrestricted scalar variable. The variable, λ, introduces a "slackness" element into the problem. The quantity, w_i, may be interpreted as the degree of under- or over- attainment of the goal F_i^*.

Points in the noninferior solution set are found through selection of new values of w_i. In this way, a trade-off surface is derived. This strategy offers the advantage of providing active constraint information at the solution, together with the ability to access solutions on a nonconvex boundary. Through its ability to under- or over-attain goals, problem specification is straightforward.

4.Integral Quadratic Measures Of Control.

Whilst multi-objective optimisation allows the designer more freedom in the choice of objectives, the design procedure requires that these objectives be reliably quantified. Further, for an interactive design environment the optimisation phase is required to execute within a reasonable time cycle. For this reason, in the initial stages of an evolutionary design process, integral quadratic measures are used, exploiting the existence of fast and reliable algorithms. Integral quadratic measures such as $\int x_1^2\,dt$, $\int u_2^2\,dt$, $\int \underline{y}^T Q \underline{y}\,dt$ are used.

Minimisation of quadratic performance objectives through correct problem formulation produces an inherently stable design and one which can be tailored to tackle a range of problems and performance criteria[7]. Problems, such as regulator and tracking (servo) design, may be undertaken with consideration to criteria such as noise and disturbance rejection, reduction of interaction in multivariable systems, actuator limits and sensitivity reduction. The multi-objective use of quadratic measures at this intermediate design stage is typically used to identify and address specific design requirements such as disturbance rejection or interaction in a broad sense.

5.Design By Evolution

The overall design is performed by a method of evolution. Starting with very simple objectives and a simple controller the problem is systematically evolved to accommodate a wider set of objectives and increasingly complex controllers. In a typical design, the problem is posed initially using a low-order model, a simple constant gain output feedback controller and simple objectives such as stability requirements. Fig 5 shows how following phases incorporate more complex controller structures and objectives.

Fig 5. Design by Evolution.

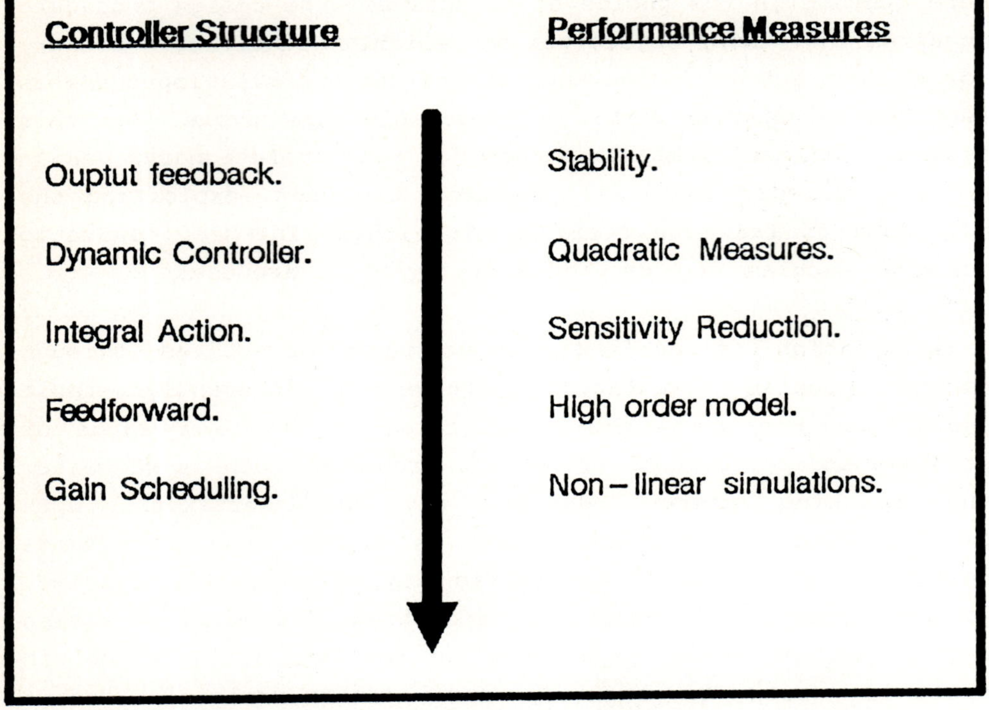

Results from previous design phases are used as starting values for the next phase. At present, to accommodate an increase in controller order and to exploit the solution point arising from the evolved design, a method of pole/zero cancellation is used to restart the design cycle.

Through design by evolution a growing appreciation of the trade-offs associated with control order and competing objectives is gained. Further, since computationally expensive goals are added later in the design cycle, this approach serves to reduce the computational burden and also reduces the likelihood of encountering local minima. The technique is illustrated in the following Section.

6. Design Example.

We consider here an unstable third order 2-input 2-output plant with eigenvalues at (0,1.0310,-5.254). The objectives of the design are to find a stabilising controller that gives a fast response, tracks step reference input changes and results in low interaction between the inputs. Controller energy is also minimised to prevent actuator saturation.

Fig. 6 shows the plant and controller configuration used in this example. The feedback controller is allowed to be of any order and is systematically increased throughout the design. A constant gain precompensator is used to ensure tracking of the input. Although integral action would normally be the preferred form of control it was found that this had an unacceptable de-stabilising effect on the plant.

Fig 6. Controller Configuration.

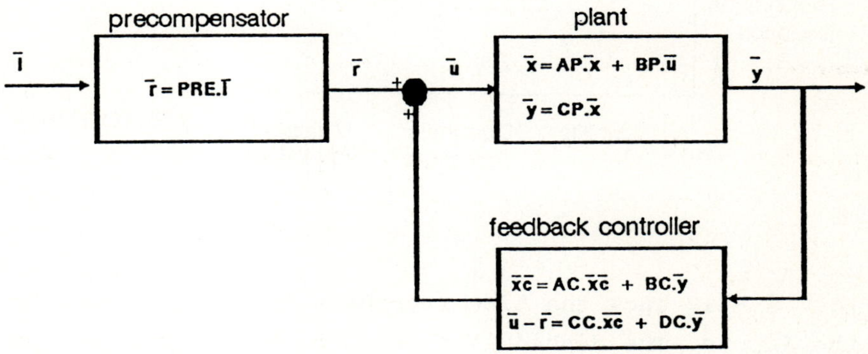

A five phase design approach is demonstrated which might form the basis of a larger design cycle. Fig. 7 shows how both the controller and performance requirements are to be evolved using the design environment. The first design requires only that a stabilising controller is found for a proportional output feedback controller. The following design uses integral quadratic measures which are then used on a first order dynamic controller. Multiple objectives are used here to address each of the outputs and control inputs separately in order to achieve the desired time responses and these are used on both a first and second order controller.

Fig 7. Evolutionary Design Approach.

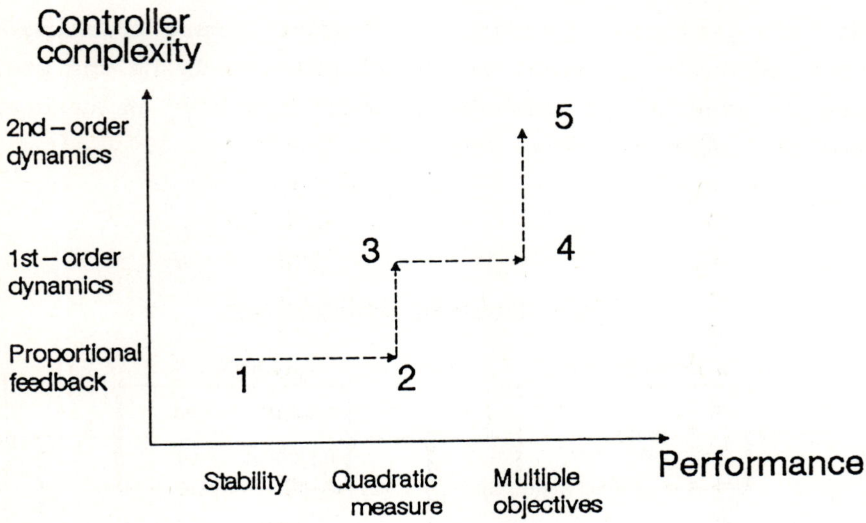

Figs. 8 - 12 show the time responses for step inputs to each of the inputs. The dashed line on each of the graphs represent the interaction which it is required to minimise as far as possible. Primary design objectives are listed together with the designer reaction to the responses.

PLANT DESCRIPTION

$$\dot{X} = \begin{vmatrix} 0. & 1.0000 & 0. \\ 0. & -2.0680 & 10.0240 \\ 0. & 0.9850 & -2.155 \end{vmatrix} X + \begin{vmatrix} 0 & 0 \\ 1 & 0 \\ 0 & 1 \end{vmatrix} U$$

$$Y = \begin{vmatrix} 1 & 0 & 0 \\ 0 & 0 & 1 \end{vmatrix} X$$

CONTROLLER DESCRIPTION

$$\dot{XC} = AC.XC + BC.Y$$
$$U-R = CC.XC + DC.Y$$
$$R = PRE.I$$

DESIGN 1: STABILISATION OF PLANT

OBJECTIVE : Stabilise the system

OPTIMISED CONTROLLER: DC = -0.0089 -9.8387
0.0000 -3.1703

DESIGNER REACTION: Response to input 1 too slow.
Actuator saturation in input 2.

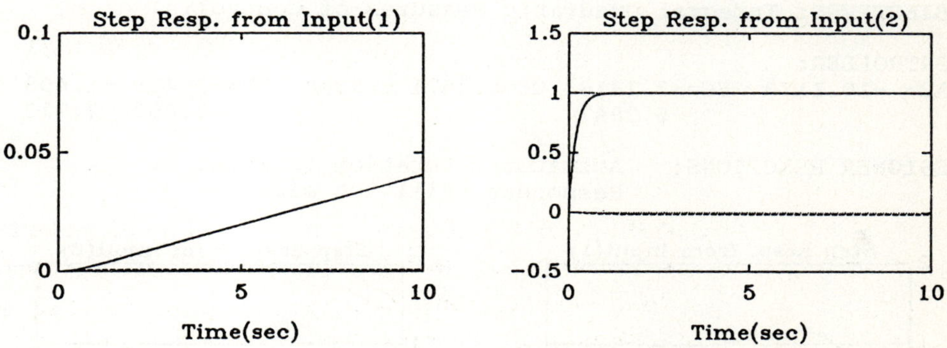

Fig 8. Design 1: Step Responses for Stabilising Controller

DESIGN 2: SCALAR INTEGRAL QUADRATIC MEASURES OF CONTROL

OBJECTIVES: Minimise cost functional for step input to i_1 and i_2. $\int_0^\infty (\underline{y}^T.Q.\underline{y} + \underline{u}^T.R.\underline{u})dt$

where $Q=\langle \begin{matrix} 1 & 0 \\ 0 & 1 \end{matrix} \rangle$ $R=\langle \begin{matrix} 1 & 0 \\ 0 & 1 \end{matrix} \rangle$

CONTROLLER: DC = $\begin{matrix} -0.0093 & -0.4643 \\ -0.5805 & -6.3699 \end{matrix}$

DESIGNER REACTION:
Interaction too high.
Responses still too slow.
Try increasing order of controller.

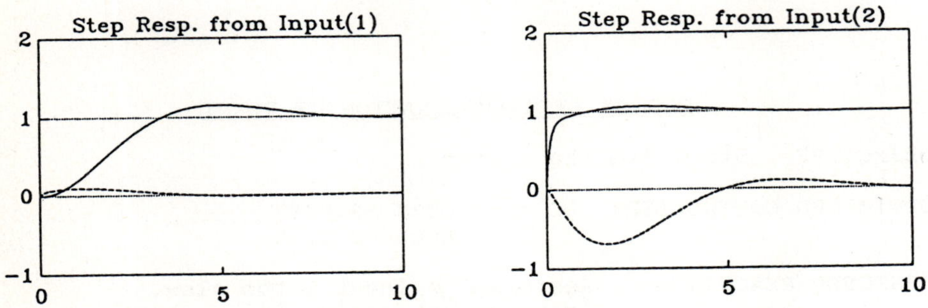

Fig. 9. Step Responses for Design 2

DESIGN 3: FIRST ORDER DYNAMIC CONTROLLER

OBJECTIVES: Integral Quadratic Measures of Control.

CONTROLLER:
AC= -10.1179 BC=$\begin{matrix} -3.1818 \\ -6.2847 \end{matrix}$ CC=0.8671 5.9348 DC=$\begin{matrix} -3.439 & -1.458 \\ -5.007 & -2.913 \end{matrix}$

DESIGNER REACTIONS:
Actuator saturation in input 1.
Responses still too slow.

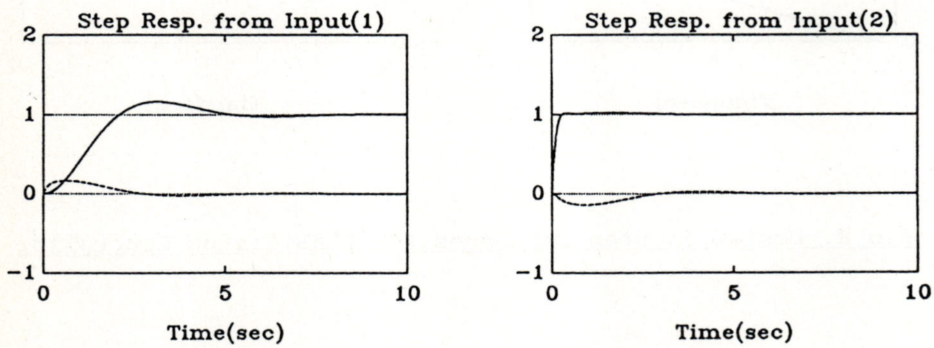

Fig. 10. Step Responses for Design 3

DESIGN 4: MULTIPLE INTEGRAL QUADRATIC MEASURES

OBJECTIVES: Set design goals $\int_0^\infty y_1{}^2dt \quad \int_0^\infty y^2{}_2dt \quad \int_0^\infty u_1{}^2dt \quad \int_0^\infty u_2{}^2dt$

CONTROLLER:
AC=-4.9089 BC=-3.9250 CC=0.8087 4.5172 DC=-1.694 -2.440
-3.8775 -1.711 -2.4364

DESIGNER REACTIONS: Responses now adequate.
Compare trade-offs with a higher order controller.

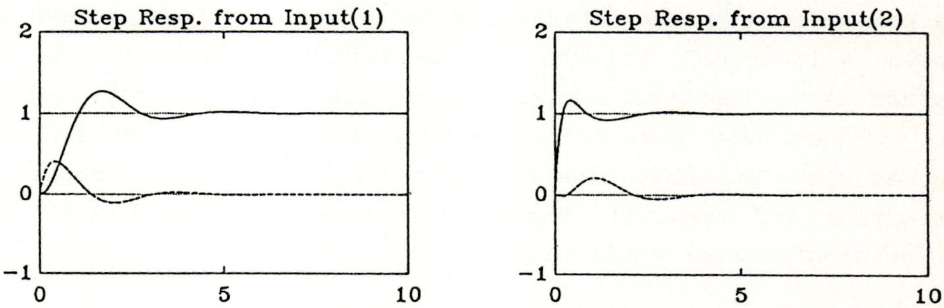

Fig. 11. Step Responses for Design 4

DESIGN 5: MULTI-OBJECTIVE SECOND ORDER CONTROLLER

OBJECTIVES: Set design goals $\int_0^\infty y_1{}^2dt \; i_1 \quad \int_0^\infty y_2{}^2dt \quad \int_0^\infty u_1{}^2dt \quad \int_0^\infty u_2{}^2dt$

CONTROLLER: AC=-0.8570 2.2436 BC=-2.8053 1.6057
-7.7934 -9.9972 -4.1416 0.5834

CC=0.4605 2.4038 DC=-1.6481 -2.1731
-3.4523 -4.8713 -7.7934 -9.9972

DESIGNER REACTION: Performance requirements have been achieved.

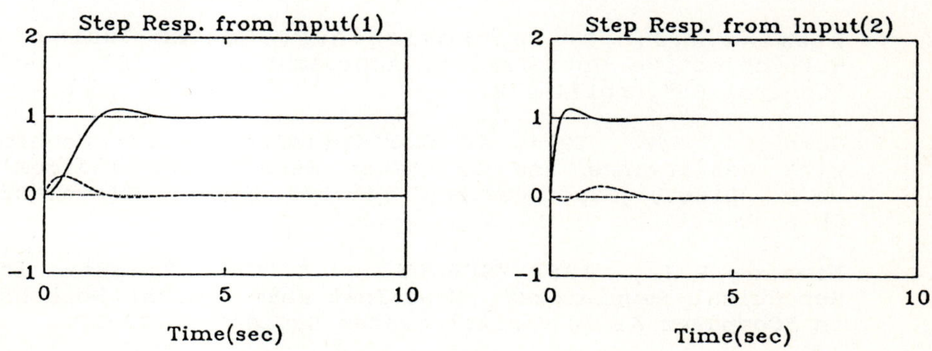

Fig 12. Step Responses for Design 5

7.Concluding Remarks

This paper has described an interactive design environment for multi-objective optimisation applications. Work is currently in progress on the development of a generalised toolbox which will enable the application of multi-objective optimisation to a wide range of problems. Efforts are also being made at links to PRO-MATLAB via inter-process communication which is seen as the gateway to wider use of the package. For a truly interactive environment, a fast execution speed is important. Efforts will be made to address this design aspect using improved hardware facilities and the optimisation of software via compilation and hard coding of the MATLANG procedures. The objective of this programme is to create an effective and versatile design environment, accessible by the practicing control engineer.

REFERENCES

1. Moler,C., 1980, 'MATLAB Users Guide', Dept. Computer Science, University of New Mexico, Alberquerque.

2. Moler, C., Little,J., Bangert,S., Kleinman,S., 1987, 'PRO-MATLAB Users Guide', The Mathworks, Inc.

3. Vanderplaats, G.N., 1983, 'ADS - A Fortran Program for Automated Design Synthesis', Naval Postgraduate School, Monterey, California, U.S.A.

4. Winter, J.S., Corbin, M.J. and Murphy, L.M., 1983, 'TSIM2 - A software package for Computer Aided Design of Flight Control Systems', RAE Tech. Report 83007.

5. Fleming,P.J. & Pashkevich.A.P.,"CACSD using a Multiobjective Optimization Approach",Proc. IEE. Conf. "Control 85", pp174-179.

6. Gembicki, F.W., 1974, 'Vector Optimization for Control with Performance and Parameter Sensitivity Indices', Ph.D. Diss., Case Western Reserve Univ., Cleveland, Ohio, U.S.A.

7. Fleming,P.J., 1984,"SUBOPT - A CAD Program for Suboptimal Regulators", Proc.Inst.Meas.Control Workshop on "Computer Aided Control System Design",pp.13-20.

MAÑANA: a Real-time Language for Control

J.R.B. Cockett
and
The Back Row at Il Ciocco

Mañana is a powerful new real-time language suitable for implementing sophisticated control strategies. Some of its important features highlighted in this paper are: a **very** abstract fuzzy specification language, the use of **deep** reasoning in deriving its control actions, the employment of possibilistic inference to give optional control, and the application of mutual delusion protocols for casual communication.

Keywords: real-time programming / communication protocols / deep reasoning / soft constraints / fuzzy guards.

1 Introduction

Designers of real-time and other programming languages are increasingly tending to simplify their constructs. This tendency has meant that those seldom-used baroque features with unclear semantics[1], and those exciting new research ideas that never quite worked[2], are invariably omitted in favor of more conservative constructs. This has led to the proliferation of rather dull programming languages.

While there are obvious reasons for having a language which uses standard constructs (for example it works) there are also strong arguments for languages which embody those *other* ideas: namely the baroque, the almost possible, the unreasonable, and the apparently downright impossible. Such languages have great pedagogical value and are at the forefront of research.

Mañana addresses these fundamental issues. It embodies exciting new ideas with unclear, or simply unspecified, semantics and shows how these extend naturally into real-time programming. A Mañana programmer is never bored: he finds it a major and exciting challenge to accomplish even the simplest of tasks. Mañana, thus, has revitalized the art of programming, which seemed to be degenerating so dangerously into a science.

While the designers of Mañana only modestly intended it to be a research language, its reception has been such that it must be seriously considered as a real contender for adoption by all major on-going multiple-year projects[3]. It has been observed that, once Mañana has engaged the attention and competitive spirit of programmers, the rate of detectable software errors decreases dramatically.

Mañana is a bold new real-time programming language which fills a void. It is an extremely high-level language: in particular it supports deep reasoning[4] which leads to **very** succinct

[1] Whatever happened to the **come from** statement?

[2] Why did the future conditional [1] (**if** a **will be** 5 **then** ..) never catch on?

[3] As yet no major project employing Mañana has ever been completed. However, those projects negotiated on a "cost plus" basis are doing **extremely well**.

[4] That is, not only reasoning but also meta-reasoning, meta-meta-reasoning *etc.* (as in [2]: but now we can do it much faster).

NATO ASI Series, Vol. F47
Advanced Computing Concepts and Techniques
in Control Engineering
Edited by M.J. Denham and A.J. Laub

and powerful programs[5]. Mañana also supports some important software engineering principles which facilitate the production of error-free code. In particular it has its own powerful specification language which is based on the Mañana operator, fuzzy guards, and soft constraints. In the area of real-time programming it supports many different levels of communication introducing a degree of flexibility absent in comparable languages.

2 Mañana and Software Engineering

Mañana exploits a number of software engineering principles. The support of the first principle given below needs no further comment.

Software Engineering Principle I:

A programmer always rises above his tools.

In fact the more impossible the programming language the better the programs. This principle explains, for example, why the best programs are invariably written in the worst languages (**FORTRAN** or **COBOL**) and why personal computer manufacturers still support **BASIC**.

Software Engineering Principle II:

Most "bugs" are a result of a poor specification and thus are not software bugs.

This has often led to the employment of specification methodologies which totally cramp the programmer's style. This is clearly disastrous: it produces frustrated programmers and thence a flood of (deliberate?) bugs. What is needed is a specification language which will not tie down the implementation. Mañana provides just this: a *very* abstract *fuzzy* specification language, in which it is impossible to say anything definite. This has many advantages: the programmer can get on with what he wants to do and the designer can happily specify designs without having to worry that they may be incorrectly implemented.

In the unlikely event that either is called to task it will be difficult to show that either party is to blame for any error. The programmer has two lines of defense: as the semantics of Mañana programs are not completely specified there is some leeway over how a program should be interpreted. As the compiler was developed in Mañana and has a deep reasoning switch[6] minor coding errors will not affect the correctness of the eventual code. If this argument is not sufficient to satisfy his management he can always fall back on the specification itself.

In this way Mañana supports defensive programming at both the specification and the coding level. This is a major advance over most current languages, which only support the programmer at the latter level (and have superficial compilers).

[5]If the advantage of neural computing is that no programming is necessary as a neural system learns any task, the advantage of deep programming is that neither programming nor learning is necessary. Once the program knows some basic facts such as "I think" it can quickly infer consequences such as "I am", etc. until it determines that it must apply a particular real-time control action such as a torque of $2.0791 gmrads/sec^2$ to drive shaft DR0072.

[6]This switch has never actually been employed as it is computationally **very** expensive. However, it would certainly be used to obtain production code for a project.

Perhaps the best known software engineering principle is that the *more* the programmer has to write the *more* errors he will make:

Software Engineering Principle III
The number of errors is directly proportional to the number of lines of code.

The designers of Mañana, in order to reduce the lines of code (and thus reduce the frequency of errors), decided not to support comments [7] [8]. Mañana programs are, therefore, self-documenting.

3 The Mañana Specification Language

A Mañana specification consists of a number of statements which must be true of the production program. The statements can be in any order. All statements involve the Mañana enabling operator, $\Rightarrow$. This operator makes the well-known diamond, $\diamond$, of temporal logic, pale into insignificance. Approximately translated it means "then eventually enable". The time of enabling thus may thus be *very* much later than the time at which the values of the guards are obtained.

A typical specification statement may take the following form:

$$\mathbf{dog}(x_1) \wedge \mathbf{man}(x_2) \wedge \mathbf{bite}(x_2, x_1, z) \Rightarrow [1yr]\mathbf{news}(z).$$

A rough English translation of this is:

> If man bites dog then within a year of its eventually becoming known, it will become news (possibly).

All the predicates $\mathbf{dog}(x_1), \mathbf{man}(x_2)$, and $\mathbf{bite}(x_2, x_1, z)$ are fuzzy guards on the transition of the event z from unknown and ignored to known and newsworthy. The time stamp of an enabled guard is simply the time of enabling and $[1yr]$ is the constraint that the event has to, within a year of that time, take place.

In fact, the situation is even more complex: each guard evaluates to a fuzzy value, that is a number, v_i, in the unit interval. The minimum of these v_i is the fuzzy truth of the guard and gives the *possibilistic* value of the transition eventually happening. Of course *possibility* should not be confused with *probability*: they are completely different concepts[9]. Thus, if the possibility of an event is 1 then this simply means that it is *possible* that it will eventually happen. On the other hand, if the possibility of an event is 0 then this simply means that it is *possible* that it will *not* happen (even eventually).

[7] One may suppose that comments cannot cause errors. This is not so. Firstly, obviously, a programmer can forget to delimit his comments correctly. Secondly, more seriously, as programmers cannot spell they run their program through spelling checkers (so that the comments will be correct): in this process usually the names of some of their variables are also "corrected". Thirdly, a programmer may actually state what he *intended* the program to do: this would completely obviate the defensive programming mechanisms and probably throw the deep reasoning compiler completely off track.

[8] This decision was actually a compromise. The initial suggestion was to disallow both comments *and* blanks (the latter are calculated to be responsible for increasing the length of programs by almost a factor of five).

[9] The authors are grateful to P.J. Ramadge for bringing this confusion to their attention.

4 Communication Protocols in Mañana

Perhaps the most interesting feature of Mañana is its extensive support for real-time programming and specification. In particular Mañana has a wide range of powerful communication primitives some of which are listed below:

- the (super) secure handshake,
- the secret handshake,
- the forced handshake,
- the casual send and receive.

These primitives are supported at a low level by the *mutual delusion* protocols which give an innovative and very efficient implementation for message handling.

4.1 The (Super) Secure Handshake

The secure communication primitives are based on the fundamental principle of secure system communication:

Principle of Secure Communication:
Every message sent must be acknowledged.

This mode of communication is supported by two primitives: secure-send and secure-receive. Consider the following two tasks:

Task (A)	AND	Task (B)
...		...
secure-send(M_1,B);		secure-receive(M_1,A);
...		...

Notice that for secure communication each task must know the name of the task with which it is communicating and the message to be exchanged. The latter is very important to ensure that the wrong message cannot be received.

A secure communication is achieved by a secure handshake. Task A sends M_1 to B, B must now acknowledge the receipt by sending an acknowledgment to A. Next A must acknowledge the acknowledgment which B sent (in accordance with the principle of secure communication), *etc.*

4.2 The Secret Handshake

Mañana was designed to support projects in which secrecy plays a significant role. In this respect Mañana is a unique language[10]. Its secret communication protocols are unmatched. They are based on the following principle:

[10] Indeed the documentation is so highly secret that nobody has sufficient clearance to read it.

Principle of Secret Communication
If you never send a message it will remain secret.

This mode of communication is supported by two primitives: dont-send and dont-receive:

Task (A)	AND	Task (B)
...		...
dont-send(M_1,B);		dont-receive(M_1,A);
...		...

Mañana has orthogonal language primitives, thus features such as secure-dont-send and secure-dont-receive are fully supported. These latter primitives are extremely popular in secret applications.

4.3 Casual Communication

For casual communication Mañana implements not only broadcasting but also the dual of broadcasting. The latter allows one to receive a message from the mass of active tasks without specifying any specific sender or location from which the message is to be obtained. This a powerful facility in its own right, but the Mañana designers have gone much further:

Message Clairvoyance
A casual message can be received before it is sent (provided it will be sent).

It is clear that this sort of casual communication can be very efficient provided a task is not thrown into a deep reasoning cycle in order to discover whether it knows if a message will be sent. Fortunately this does not happen.

As Mañana programs are fair (as everything happens eventually) a task knows that any message which it sends will eventually be received provided there is a receiver. However, if there were not a receiver then the send statement would have been redundant. The deep compiler would have readily spotted this trivial fact. Furthermore, it would have worked out what was intended by the code. This would mean that either the send would have been removed (end of problem) or a receiver which the programmer had intended would have been created (end of problem). Thus every casual send will be received.

By a dual argument, a task when it wishes to receive a given message, knows that a task which should send the message will eventually get round to sending it. These observations give:

Mutual Delusion Principle
If a task expects a message then it will be sent; if a task sends a message then it will be received.

This has led to a very efficient implementation for casual communication[11]. The communication primitives for casual communication are: casual-receive and casual-send. They both have

[11] In fact the implementation does not even require the tasks to be running on physically connected machines.

only one argument, the message, as no communicating task need be specified.

5 Conclusions

The idea that real-time programs and languages should have a precise semantics has long been a millstone around the programming community's neck. The use of very abstract fuzzy specifications, approximate semantics, and a deeeep reasoning compiler in Mañana directly addresses these problems. This language is perhaps the first which does not inhibit the inventive genius of the programmer. Furthermore it does so without adversely affecting the incorrectness of the programs[12].

The language has been highly successful. It is the front-running contender for adoption in the Strategic Detergent Initiative. The language has already gone through a series of specification levels each of which called for a more fuzzy and approximate version than the last:

- Spaghettiman
- Cannelloniman
- Rigatoniman
- Pizzaman.

Currently we are at the rigatoniman level and preliminary systems have been shipped to β-sites. The pizzaman call is expected anytime and adoption is expected mañana.

References

[1] Nylin W. C. and Harvill J. B., *Multiple tense computer programming.* SIGPLAN Not. 11(12), 74-93 (1976)

[2] Carroll L., *What the tortoise said to Achilles.* Mind N.S.IV, No. 14, 278-280 (1895)

[12] Deep compile that!

NATO ASI Series F

Vol. 1: Issues in Acoustic Signal – Image Processing and Recognition. Edited by C. H. Chen. VIII, 333 pages. 1983.

Vol. 2: Image Sequence Processing and Dynamic Scene Analysis. Edited by T. S. Huang. IX, 749 pages. 1983.

Vol. 3: Electronic Systems Effectiveness and Life Cycle Costing. Edited by J. K. Skwirzynski. XVII, 732 pages. 1983.

Vol. 4: Pictorial Data Analysis. Edited by R. M. Haralick. VIII, 468 pages. 1983.

Vol. 5: International Calibration Study of Traffic Conflict Techniques. Edited by E. Asmussen. VII, 229 pages. 1984.

Vol. 6: Information Technology and the Computer Network. Edited by K. G. Beauchamp. VIII, 271 pages. 1984.

Vol. 7: High-Speed Computation. Edited by J. S. Kowalik. IX, 441 pages. 1984.

Vol. 8: Program Transformation and Programming Environments. Report on an Workshop directed by F. L. Bauer and H. Remus. Edited by P. Pepper. XIV, 378 pages. 1984.

Vol. 9: Computer Aided Analysis and Optimization of Mechanical System Dynamics. Edited by E. J. Haug. XXII, 700 pages. 1984.

Vol. 10: Simulation and Model-Based Methodologies: An Integrative View. Edited by T. I. Ören, B. P. Zeigler, M. S. Elzas. XIII, 651 pages. 1984.

Vol. 11: Robotics and Artificial Intelligence. Edited by M. Brady, L. A. Gerhardt, H. F. Davidson. XVII, 693 pages. 1984.

Vol. 12: Combinatorial Algorithms on Words. Edited by A. Apostolico, Z. Galil. VIII, 361 pages. 1985.

Vol. 13: Logics and Models of Concurrent Systems. Edited by K. R. Apt. VIII, 498 pages. 1985.

Vol. 14: Control Flow and Data Flow: Concepts of Distributed Programming. Edited by M. Broy. VIII, 525 pages. 1985.

Vol. 15: Computational Mathematical Programming. Edited by K. Schittkowski. VIII, 451 pages. 1985.

Vol. 16: New Systems and Architectures for Automatic Speech Recognition and Synthesis. Edited by R. De Mori, C.Y. Suen. XIII, 630 pages. 1985.

Vol. 17: Fundamental Algorithms for Computer Graphics. Edited by R.A. Earnshaw. XVI, 1042 pages. 1985.

Vol. 18: Computer Architectures for Spatially Distributed Data. Edited by H. Freeman and G. G. Pieroni. VIII, 391 pages. 1985.

Vol. 19: Pictorial Information Systems in Medicine. Edited by K. H. Höhne. XII, 525 pages. 1986.

Vol. 20: Disordered Systems and Biological Organization. Edited by E. Bienenstock, F. Fogelman Soulié, G. Weisbuch. XXI, 405 pages. 1986.

Vol. 21: Intelligent Decision Support in Process Environments. Edited by E. Hollnagel, G. Mancini, D.D. Woods. XV, 524 pages. 1986.

Vol. 22: Software System Design Methods. The Challenge of Advanced Computing Technology. Edited by J.K. Skwirzynski. XIII, 747 pages. 1986.

NATO ASI Series F

Vol. 23: Designing Computer-Based Learning Materials. Edited by H. Weinstock and A. Bork. IX, 285 pages. 1986.

Vol. 24: Database Machines. Modern Trends and Applications. Edited by A. K. Sood and A. H. Qureshi. VIII, 570 pages. 1986.

Vol. 25: Pyramidal Systems for Computer Vision. Edited by V. Cantoni and S. Levialdi. VIII, 392 pages. 1986.

Vol. 26: Modelling and Analysis in Arms Control. Edited by R. Avenhaus, R. K. Huber and J. D. Kettelle. VIII, 488 pages. 1986.

Vol. 27: Computer Aided Optimal Design: Structural and Mechanical Systems. Edited by C. A. Mota Soares. XIII, 1029 pages. 1987.

Vol. 28: Distributed Operating Systems. Theory und Practice. Edited by Y. Paker, J.-P. Banatre and M. Bozyiğit. X, 379 pages. 1987.

Vol. 29: Languages for Sensor-Based Control in Robotics. Edited by U. Rembold and K. Hörmann. IX, 625 pages. 1987.

Vol. 30: Pattern Recognition Theory and Applications. Edited by P. A. Devijver and J. Kittler. XI, 543 pages. 1987.

Vol. 31: Decision Support Systems: Theory and Application. Edited by C. W. Holsapple and A. B. Whinston. X, 500 pages. 1987.

Vol. 32: Information Systems: Failure Analysis. Edited by J. A. Wise and A. Debons. XV, 338 pages. 1987.

Vol. 33: Machine Intelligence and Knowledge Engineering for Robotic Applications. Edited by A. K. C. Wong and A. Pugh. XIV, 486 pages. 1987.

Vol. 34: Modelling, Robustness and Sensitivity Reduction in Control Systems. Edited by R. F. Curtain. IX, 492 pages. 1987.

Vol. 35: Expert Judgment and Expert Systems. Edited by J. L. Mumpower, L. D. Phillips, O. Renn and V. R. R. Uppuluri. VIII, 361 pages. 1987.

Vol. 36: Logic of Programming and Calculi of Discrete Design. Edited by M. Broy. VII, 415 pages. 1987.

Vol. 37: Dynamics of Infinite Dimensional Systems. Edited by S.-N. Chow and J. K. Hale. IX, 514 pages. 1987.

Vol. 38: Flow Control of Congested Networks. Edited by A. R. Odoni, L. Bianco and G. Szegö. XII, 355 pages. 1987.

Vol. 39: Mathematics and Computer Science in Medical Imaging. Edited by M. A. Viergever and A. Todd-Pokropek. VIII, 546 pages. 1988.

Vol. 40: Theoretical Foundations of Computer Graphics and CAD. Edited by R. A. Earnshaw. XX, 1246 pages. 1988.

Vol. 41: Neural Computers. Edited by R. Eckmiller and Ch. v. d. Malsburg. XIII, 566 pages. 1988.

Vol. 42: Real-Time Object Measurement and Classification. Edited by A. K. Jain. VIII, 407 pages 1988.

NATO ASI Series F

Vol. 43: Sensors and Sensory Systems for Advanced Robots. Edited by P. Dario. XI, 597 pages. 1988.

Vol. 44: Signal Processing and Pattern Recognition in Nondestructive Evaluation of Materials. Edited by C.H. Chen. VIII, 344 pages. 1988.

Vol. 45: Syntactic and Structural Pattern Recognition. Edited by G. Ferraté, T. Pavlidis, A. Sanfeliu, H. Bunke. XVI, 467 pages. 1988.

Vol. 46: Recent Advances in Speech Understanding and Dialog Systems. Edited by H. Niemann, M. Lang, G. Sagerer. X, 521 pages. 1988.

Vol. 47: Advanced Computing Concepts and Techniques in Control Engineering. Edited by M.J. Denham, A.J. Laub. XI, 518 pages. 1988.